Methods in Enzymology

Volume 161

BIOMASS

Part B

Lignin, Pectin, and Chitin

METHODS IN ENZYMOLOGY

EDITORS-IN-CHIEF

John N. Abelson   Melvin I. Simon

Methods in Enzymology

Volume 161

Biomass

Part B
Lignin, Pectin, and Chitin

EDITED BY

Willis A. Wood

THE SALK INSTITUTE BIOTECHNOLOGY/INDUSTRIAL ASSOCIATES, INC.
SAN DIEGO, CALIFORNIA

Scott T. Kellogg

THE SALK INSTITUTE BIOTECHNOLOGY/INDUSTRIAL ASSOCIATES, INC.
SAN DIEGO, CALIFORNIA

ACADEMIC PRESS, INC.
Harcourt Brace Jovanovich, Publishers
San Diego New York Berkeley Boston
London Sydney Tokyo Toronto

ACADEMIC PRESS, INC.
San Diego, California 92101

United Kingdom Edition published by
ACADEMIC PRESS, INC. (LONDON) LTD.
24-28 Oval Road, London NW1 7DX

LIBRARY OF CONGRESS CATALOG CARD NUMBER: 54-9110

ISBN 0-12-182062-9 (alk. paper)

PRINTED IN THE UNITED STATES OF AMERICA
88 89 90 91 9 8 7 6 5 4 3 2 1

Table of Contents

Section I. Lignin

A. Preparation of Substrates for Ligninases

B. Assays for Ligninases

C. Chemical Methods for Characterization of Lignin

D. Chromatographic Methods for Lignin and Related Compounds

E. Nucleic Acid Preparations Related to Lignin Degradation

F. Purification of Lignin-Degrading Enzymes

Section II. Pectin

A. Assays for Pectin-Degrading Enzymes

B. Purification of Pectin-Degrading Enzymes

Section III. Chitin

A. Preparation of Substrates for Chitin-Degrading Enzymes

B. Assay for Chitin-Degrading Enzymes

C. Analytical Methods for Chitin

D. Purification of Chitin-Degrading Enzymes

Contributors to Volume 161

Article numbers are in parentheses following the names of contributors.
Affiliations listed are current.

MARGARET ALIC (9), *Department of Chemical and Biological Sciences, Oregon Graduate Center, Beaverton, Oregon 97006*

YOSHIO ARAKI (66, 67), *Department of Chemistry, Faculty of Science, Hokkaido University, Sapporo, Hokkaido 060, Japan*

ANGEL ARROYO-BEGOVICH (58), *Departamento de Microbiología, Instituto de Fisiología Celular, Universidad Nacional Autónoma de México, México, D.F., 04510 México*

PAUL R. AUSTIN (44), *College of Marine Studies, University of Delaware, Newark, Delaware 19716*

JUN-ICHI AZUMA (2), *Department of Wood Science & Technology, Faculty of Agriculture, Kyoto University, Kitashirakawa, Oiwake-cho, Sakyo-ku, Kyoto 606, Japan*

BÄRBEL BAUCH (68), *Department of Biology, University of Kaiserslautern, D-6750 Kaiserslautern, Federal Republic of Germany*

FRITHJOF-HANS BERNHARDT (29), *Abteilung Physiologische Chemie, Rheinisch-Westfälische Technische Hochschule Aachen, D-5100 Aachen, Federal Republic of Germany*

ECKHARD BILL (29), *Institut für Physik, Medizinische Universität zu Lübeck, D-2400 Lübeck 1, Federal Republic of Germany*

JOHN BLACKWELL (51), *Department of Macromolecular Science, Case Western Reserve University, Cleveland, Ohio 44106*

THOMAS BOLLER (50, 60), *Abteilung Pflanzenphysiologie, Botanisches Institut der Universität Basel, CH-4056 Basel, Switzerland*

PAUL BRODA (18, 20, 21), *Department of Biochemistry and Applied Molecular Biology, University of Manchester Institute of Science and Technology, Manchester M60 1QD, England*

GÖSTA BRUNOW (8), *Department of Organic Chemistry, University of Helsinki, SF-00100 Helsinki 10, Finland*

JOHN A. BUSWELL (27, 28, 30), *Department of Biology, Paisley College of Technology, Paisley, Renfrewshire PA1 2BE, Scotland*

ENRICO CABIB (48, 55, 56, 63), *National Institute of Diabetes and Digestive and Kidney Disease, National Institutes of Health, Bethesda, Maryland 20892*

JOHN E. CASTLE (62), *College of Marine Studies, University of Delaware, Lewes, Delaware 19958*

CHEN-LOUNG CHEN (14, 15), *Department of Wood and Paper Science, North Carolina State University, College of Forest Resources, Raleigh, North Carolina 27695*

LISA C. CHILDERS (69), *Department of Biology and Molecular Biology Institute, San Diego State University, San Diego, California 92182*

ALAN COLLMER (35), *Department of Botany and Center for Agricultural Biotechnology of the Maryland Biotechnology Institute, University of Maryland, College Park, Maryland 20742*

DON L. CRAWFORD (3, 5, 16, 17, 24), *Department of Bacteriology and Biochemistry, University of Idaho, Moscow, Idaho 83843*

RONALD L. CRAWFORD (3, 11, 26), *Department of Bacteriology and Biochemistry, University of Idaho, Moscow, Idaho 83843*

A. STEPHEN DAHMS (69), *Department of Chemistry and Molecular Biology Institute, San Diego State University, San Diego, California 92182*

DONALD H. DAVIES (52), *Department of Chemistry, Saint Mary's University, Halifax, Nova Scotia, Canada B3H 3C3*

KARL-ERIK ERIKSSON (27, 28, 33, 34), *Department of Chemistry, Swedish Pulp and Paper Research Institute, S-114 86 Stockholm, Sweden*

LUDGER ERNST (6), *NMR-Laboratorium der Chemischen Institute, Technische Universität Braunschweig, D-3300 Braunschweig, Federal Republic of Germany*

HELGA FÖRSTER (38), *Department of Plant Pathology, University of California, Riverside, Riverside, California 92521*

ANNETTE GEHRI (60), *Abteilung Pflanzenphysiologie, Botanisches Institut der Universität Basel, CH-4056 Basel, Switzerland*

JEFFREY K. GLENN (9, 25), *Department of Neurobiology and Anatomy, The University of Rochester Medical Center, Rochester, New York 14642*

MICHAEL H. GOLD (9, 25), *Department of Chemical and Biological Sciences, Oregon Graduate Center, Beaverton, Oregon 97006*

KONRAD HAIDER (6), *Institut für Pflanzenernährung und Bodenkunde, Bundesforschungsanstalt für Landwirtschaft (FAL), D-3300 Braunschweig, Federal Republic of Germany*

ERNEST R. HAYES (52), *Department of Chemistry, Acadia University, Wolfville, Nova Scotia, Canada B0P 1X0*

RICHARD HAYLOCK (21), *Department of Biology, University of Ulster at Coleraine, Coleraine County Londonderry, BT52 1SA, Northern Ireland*

FARAH HEDJRAN (69), *Eukaryotic Regulatory Biology Program, University of California, San Diego, La Jolla, California 92093*

KEVIN B. HICKS (46), *Eastern Regional Research Center, Agricultural Research Service, United States Department of Agriculture, Philadelphia, Pennsylvania 19118*

TAKAYOSHI HIGUCHI (13, 19), *Research Section of Lignin Chemistry, Wood Research Institute, Kyoto University, Gokasho, Uji, Kyoto 611, Japan*

SHIGEHIRO HIRANO (45), *Department of Agricultural Biochemistry, Tottori University, Tottori 680, Japan*

WILLIAM H. HUISMAN (69), *Department of Chemistry and Molecular Biology Institute, San Diego State University, San Diego, California 92182*

VAN-BA HUYNH (11, 26), *Chemical Abstracts Service, Columbus, Ohio 43216*

EIJI ITO (66, 67), *Department of Chemistry, Faculty of Science, Hokkaido University, Sapporo, Hokkaido 060, Japan*

E. JAEGER (31), *Hygiene Institut Eschweiler, D-5180 Eschweiler, Federal Republic of Germany*

MEHRDAD JANNATIPOUR (69), *Agouron Institute, La Jolla, California 92037*

AKIRA KAJI (39), *Kagawa University, Takamatsu City 760, Japan*

HEINRICH KAUSS (68), *Department of Biology, University of Kaiserslautern, D-6750 Kaiserslautern, Federal Republic of Germany*

ROBERT L. KELLEY (10, 32), *Biotechnology, Institute of Gas Technology, Chicago, Illinois 60616*

HARTMUT KERN (6), *Institut für Biotechnologie, Kernforschungsanlage Jülich, D-5170 Jülich 1, Federal Republic of Germany*

T. KENT KIRK (1, 8, 12, 23), *Forest Products Laboratory, Forest Service, United States Department of Agriculture, Madison, Wisconsin 53705*

HARUYOSHI KONNO (41, 42), *Institute for Agricultural and Biological Sciences, Okayama University, Kurashiki-shi, Okayama 710, Japan*

LIONEL MARCUS (40), *Department of Botany, George S. Wise Faculty of Life Sciences, Tel-Aviv University, Ramat-Aviv, 69978 Tel-Aviv, Israel*

FELIX MAUCH (50, 60), *Abteilung Pflanzenphysiologie, Botanisches Institut der Universität Basel, CH-4056 Basel, Switzerland*

OTAKAR MIKEŠ (43), *Institute of Organic Chemistry and Biochemistry, Czechoslo-*

vak Academy of Science, 166 10 Prague 6, Czechoslovakia

MASARU MITSUTOMI (54), *Department of Agricultural Chemistry, Faculty of Agriculture, Saga University, Saga 840, Japan*

BERNARD MONTIES (4), *Laboratoire de Chimie Biologique, Institut Nationale de la Recherche Agronomique, Centre de Biotechnologie Agro-Industrielle, Institut National Agronomique Paris-Grignon, Centre de Grignon, F-78850 Thiverval-Grignon, France*

MARK S. MOUNT (35), *Department of Plant Pathology, University of Massachusetts, Amherst, Massachusetts 01003*

FUMIAKI NAKATSUBO (7), *Department of Wood Science & Technology, Faculty of Agriculture, Kyoto University, Kitashirakawa, Oiwake-cho, Sakyo-ku, Kyoto 606, Japan*

ATSUMI NISHIDA (34), *Forest Products Chemistry Division, Forestry and Forest Products Research Institute, Kenkyu, Danchi-Nai, Ibaraki 305, Japan*

JOHN R. OBST (1, 12), *Forest Products Laboratory, Forest Service, United States Department of Agriculture, Madison, Wisconsin 53705*

AKIRA OHTAKARA (49, 54, 57, 64, 65), *Department of Agricultural Chemistry, Faculty of Agriculture, Saga University, Saga 840, Japan*

ANDRZEJ PASZCZYŃSKI (11, 26), *Department of Biochemistry, University of M. Curie-Skłodowska, 20-031 Lublin, Poland*

ALISTAIR PATERSON (18), *Department of Bioscience and Biotechnology, Food Science Division, University of Strathclyde, Glasgow G1 1SD, Scotland*

G. F. PEGG (59, 61), *Department of Horticulture, University of Reading, Reading RG6 2AU, England*

D. PITT (37), *Department of Biological Sciences, Washington Singer Laboratories, University of Exeter, Exeter EX4 4QG, England*

ANTHONY L. POMETTO III (5, 16, 17, 24), *Department of Bacteriology and Biochemistry, University of Idaho, Moscow, Idaho 83843*

UTE RAEDER (20), *Department of Biochemistry and Applied Molecular Biology, University of Manchester Institute of Science and Technology, Manchester M60 1QD, England*

C. ADINARAYANA REDDY (22, 32), *Department of Microbiology and Public Health, Michigan State University, East Lansing, Michigan 48824*

LUBOMÍRA REXOVÁ-BENKOVÁ (43), *Institute of Chemistry, Slovak Academy of Science, 842 38 Bratislava, Czechoslovakia*

DOUGLAS W. RIBBONS (30), *Centre for Biotechnology, Imperial College of Science and Technology, London SW7 2AZ, England*

JEFFREY L. RIED (35), *Department of Botany, University of Maryland, College Park, Maryland 20742*

DANIELLE ROBERT (15), *Département de Recherche Fondamentale (DRF), Laboratoires de Chimie, Centre d'Etudes Nucléaires de Grenoble, F-38041 Grenoble, France*

TAKUO SAKAI (36), *College of Agriculture, University of Osaka Prefecture, Sakai, Osaka 591, Japan*

ADRIANA SBURLATI (55), *National Institute of Diabetes and Digestive and Kidney Disease, National Institutes of Health, Bethesda, Maryland 20892*

ABEL SCHEJTER (40), *Sackler Institute of Molecular Medicine, Sackler Faculty of Medicine, Tel-Aviv University, Ramat-Aviv, 69978 Tel-Aviv, Israel*

KENZO SHIMAHARA (47), *Department of Industrial Chemistry, Faculty of Engineering, Seikei University, Musashino-shi, Tokyo 180, Japan*

RAFAEL W. SOTO-GIL (69), *Department of Biology and Molecular Biology Institute, San Diego State University, San Diego, California 92182*

KIYOSHI TAGAWA (39), *Department of Bioresource Science, Faculty of Agriculture,*

Kagawa University, Miki-cho, Kagawa 761-07, Japan

YASUYUKI TAKIGUCHI (47), *Department of Industrial Chemistry, Faculty of Engineering, Seikei University, Musashino-shi, Tokyo 180, Japan*

MITSUHIKO TANAHASHI (13), *Research Section of Lignin Chemistry, Wood Research Institute, Kyoto University, Gokasho, Uji, Kyoto 611, Japan*

KOSHIJIMA TETSUO (2), *Research Section of Wood Chemistry, Wood Research Institute, Kyoto University, Gokasho, Uji, Kyoto 611, Japan*

MING TIEN (23), *Department of Molecular and Cell Biology, Pennsylvania State University, University Park, Pennsylvania 16802*

ALFRED XAVER TRAUTWEIN (29), *Institut für Physik, Medizinische Universität zu Lübeck, D-2400 Lübeck 1, Federal Republic of Germany*

HANS TWILFER (29), *Institut für Anatomie, Medizinische Universität zu Lübeck, D-2400 Lübeck 1, Federal Republic of Germany*

YASUSHI UCHIDA (64), *Department of Agricultural Chemistry, Faculty of Agriculture, Saga University, Saga 840, Japan*

TOSHIAKI UMEZAWA (19), *Research Section of Lignin Chemistry, Wood Research Institute, Kyoto University, Gokasho, Uji, Kyoto 611, Japan*

URS VÖGELI (60), *Abteilung Pflanzenphysiologie, Botanisches Institut der Universität Basel, Ch-4056 Basel, Switzerland*

J. VOLC (33), *Department of Experimental Mycology, Institute of Microbiology, Czechoslovak Academy of Science, 142 20 Prague, Czechoslovakia*

ARNOLD C. M. WU (53), *Research and Development Department, Fishery Products, Inc., Danvers, Massachusetts 01923*

YI-ZHENG ZHANG (22), *Biotechnology Department, Sichuan University, Chengdu, Sichuan, People's Republic of China*

JOHN P. ZIKAKIS (62), *Department of Animal Science and Agricultural Biochemistry, Delaware Agricultural Experiment Station, and College of Marine Studies, University of Delaware, Newark, Delaware 19717*

WOLFGANG ZIMMERMANN (18), *Institut für Biotechnologie, Eidgenössische Technische Hochschule (ETH), CH-8093 Zürich, Switzerland*

JUDITH W. ZYSKIND (69), *Department of Biology and Molecular Biology Institute, San Diego State University, San Diego, California 92182*

Preface

Volumes 160 and 161 of *Methods in Enzymology* collate for the first time an array of procedures related to the enzymatic conversion of plant structural biomass polymers into their constituent monomeric units. This collection of methods for the hydrolysis of cellulose and hemicellulose (Volume 160) and of lignin, as well as related methods for pectin and chitin (Volume 161), is timely because of the increasing tempo of investigation in this area. This is in response to an immediate interest in the conversion of biomass monosaccharides into fuel ethanol and the longer term concern for maintaining supplies of liquid fuels and chemicals with eventual petroleum depletion.

Enzymatic treatment of plant biomass involves special methods due to the insolubility of the lignocellulosic complex and other similar polymers. These methods include substrate preparation, measurement of chemical changes, and culturing of organisms that produce the enzymes. Many of the methods are published in applied and special purpose journals not routinely seen by investigators and hence are not highly visible.

The ability to clone genes, transform cells, and express and secrete heterologous proteins in industrially important microorganisms presents opportunities to produce biomass enzymes in large quantity and at low prices. When this capacity is developed, enzymes will not be selected because of better production in a wild-type organism. Instead, the enzymes will be chosen for their superior catalytic capability and compatibility with the conditions of an industrial process. Since genes from various and often obscure organisms may produce enzymes better suited to such purposes, we have attempted to include methods for the preparation of enzymes in each class, for instance endocellulases, from a wide variety of sources so that investigators seeking to develop useful processes may make use of the options available.

We wish to acknowledge the expert secretarial assistance of Ms. Karen Payne in preparation of these volumes.

WILLIS A. WOOD
SCOTT T. KELLOGG

METHODS IN ENZYMOLOGY

EDITED BY

Sidney P. Colowick and Nathan O. Kaplan

VANDERBILT UNIVERSITY
SCHOOL OF MEDICINE
NASHVILLE, TENNESSEE

DEPARTMENT OF CHEMISTRY
UNIVERSITY OF CALIFORNIA
AT SAN DIEGO
LA JOLLA, CALIFORNIA

I. Preparation and Assay of Enzymes
II. Preparation and Assay of Enzymes
III. Preparation and Assay of Substrates
IV. Special Techniques for the Enzymologist
V. Preparation and Assay of Enzymes
VI. Preparation and Assay of Enzymes (*Continued*)
Preparation and Assay of Substrates
Special Techniques
VII. Cumulative Subject Index

METHODS IN ENZYMOLOGY

EDITORS-IN-CHIEF

Sidney P. Colowick and Nathan O. Kaplan

VOLUME VIII. Complex Carbohydrates
Edited by ELIZABETH F. NEUFELD AND VICTOR GINSBURG

VOLUME IX. Carbohydrate Metabolism
Edited by WILLIS A. WOOD

VOLUME X. Oxidation and Phosphorylation
Edited by RONALD W. ESTABROOK AND MAYNARD E. PULLMAN

VOLUME XI. Enzyme Structure
Edited by C. H. W. HIRS

VOLUME XII. Nucleic Acids (Parts A and B)
Edited by LAWRENCE GROSSMAN AND KIVIE MOLDAVE

VOLUME XIII. Citric Acid Cycle
Edited by J. M. LOWENSTEIN

VOLUME XIV. Lipids
Edited by J. M. LOWENSTEIN

VOLUME XV. Steroids and Terpenoids
Edited by RAYMOND B. CLAYTON

VOLUME XVI. Fast Reactions
Edited by KENNETH KUSTIN

VOLUME XVII. Metabolism of Amino Acids and Amines (Parts A and B)
Edited by HERBERT TABOR AND CELIA WHITE TABOR

VOLUME XVIII. Vitamins and Coenzymes (Parts A, B, and C)
Edited by DONALD B. MCCORMICK AND LEMUEL D. WRIGHT

VOLUME XIX. Proteolytic Enzymes
Edited by GERTRUDE E. PERLMANN AND LASZLO LORAND

VOLUME XX. Nucleic Acids and Protein Synthesis (Part C)
Edited by KIVIE MOLDAVE AND LAWRENCE GROSSMAN

VOLUME XXI. Nucleic Acids (Part D)
Edited by LAWRENCE GROSSMAN AND KIVIE MOLDAVE

VOLUME XXII. Enzyme Purification and Related Techniques
Edited by WILLIAM B. JAKOBY

VOLUME XXIII. Photosynthesis (Part A)
Edited by ANTHONY SAN PIETRO

VOLUME XXIV. Photosynthesis and Nitrogen Fixation (Part B)
Edited by ANTHONY SAN PIETRO

VOLUME XXV. Enzyme Structure (Part B)
Edited by C. H. W. HIRS AND SERGE N. TIMASHEFF

VOLUME XXVI. Enzyme Structure (Part C)
Edited by C. H. W. HIRS AND SERGE N. TIMASHEFF

VOLUME XXVII. Enzyme Structure (Part D)
Edited by C. H. W. HIRS AND SERGE N. TIMASHEFF

VOLUME XXVIII. Complex Carbohydrates (Part B)
Edited by VICTOR GINSBURG

VOLUME XXIX. Nucleic Acids and Protein Synthesis (Part E)
Edited by LAWRENCE GROSSMAN AND KIVIE MOLDAVE

VOLUME XXX. Nucleic Acids and Protein Synthesis (Part F)
Edited by KIVIE MOLDAVE AND LAWRENCE GROSSMAN

VOLUME XXXI. Biomembranes (Part A)
Edited by SIDNEY FLEISCHER AND LESTER PACKER

VOLUME 73. Immunochemical Techniques (Part B)
Edited by JOHN J. LANGONE AND HELEN VAN VUNAKIS

VOLUME 74. Immunochemical Techniques (Part C)
Edited by JOHN J. LANGONE AND HELEN VAN VUNAKIS

VOLUME 75. Cumulative Subject Index Volumes XXXI, XXXII, XXXIV–LX
Edited by EDWARD A. DENNIS AND MARTHA G. DENNIS

VOLUME 76. Hemoglobins
Edited by ERALDO ANTONINI, LUIGI ROSSI-BERNARDI, AND EMILIA CHIANCONE

VOLUME 77. Detoxication and Drug Metabolism
Edited by WILLIAM B. JAKOBY

VOLUME 78. Interferons (Part A)
Edited by SIDNEY PESTKA

VOLUME 79. Interferons (Part B)
Edited by SIDNEY PESTKA

VOLUME 80. Proteolytic Enzymes (Part C)
Edited by LASZLO LORAND

VOLUME 81. Biomembranes (Part H: Visual Pigments and Purple Membranes, I)
Edited by LESTER PACKER

VOLUME 82. Structural and Contractile Proteins (Part A: Extracellular Matrix)
Edited by LEON W. CUNNINGHAM AND DIXIE W. FREDERIKSEN

VOLUME 83. Complex Carbohydrates (Part D)
Edited by VICTOR GINSBURG

VOLUME 84. Immunochemical Techniques (Part D: Selected Immunoassays)
Edited by JOHN J. LANGONE AND HELEN VAN VUNAKIS

VOLUME 97. Biomembranes [Part K: Membrane Biogenesis: Assembly and Targeting (Prokaryotes, Mitochondria, and Chloroplasts)]
Edited by SIDNEY FLEISCHER AND BECCA FLEISCHER

VOLUME 98. Biomembranes (Part L: Membrane Biogenesis: Processing and Recycling)
Edited by SIDNEY FLEISCHER AND BECCA FLEISCHER

VOLUME 99. Hormone Action (Part F: Protein Kinases)
Edited by JACKIE D. CORBIN AND JOEL G. HARDMAN

VOLUME 100. Recombinant DNA (Part B)
Edited by RAY WU, LAWRENCE GROSSMAN, AND KIVIE MOLDAVE

VOLUME 101. Recombinant DNA (Part C)
Edited by RAY WU, LAWRENCE GROSSMAN, AND KIVIE MOLDAVE

VOLUME 102. Hormone Action (Part G: Calmodulin and Calcium-Binding Proteins)
Edited by ANTHONY R. MEANS AND BERT W. O'MALLEY

VOLUME 103. Hormone Action (Part H: Neuroendocrine Peptides)
Edited by P. MICHAEL CONN

VOLUME 104. Enzyme Purification and Related Techniques (Part C)
Edited by WILLIAM B. JAKOBY

VOLUME 105. Oxygen Radicals in Biological Systems
Edited by LESTER PACKER

VOLUME 106. Posttranslational Modifications (Part A)
Edited by FINN WOLD AND KIVIE MOLDAVE

VOLUME 107. Posttranslational Modifications (Part B)
Edited by FINN WOLD AND KIVIE MOLDAVE

VOLUME 108. Immunochemical Techniques (Part G: Separation and Characterization of Lymphoid Cells)
Edited by GIOVANNI DI SABATO, JOHN J. LANGONE, AND HELEN VAN VUNAKIS

VOLUME 109. Hormone Action (Part I: Peptide Hormones)
Edited by LUTZ BIRNBAUMER AND BERT W. O'MALLEY

VOLUME 110. Steroids and Isoprenoids (Part A)
Edited by JOHN H. LAW AND HANS C. RILLING

VOLUME 111. Steroids and Isoprenoids (Part B)
Edited by JOHN H. LAW AND HANS C. RILLING

VOLUME 112. Drug and Enzyme Targeting (Part A)
Edited by KENNETH J. WIDDER AND RALPH GREEN

VOLUME 113. Glutamate, Glutamine, Glutathione, and Related Compounds
Edited by ALTON MEISTER

VOLUME 114. Diffraction Methods for Biological Macromolecules (Part A)
Edited by HAROLD W. WYCKOFF, C. H. W. HIRS, AND SERGE N. TIMASHEFF

VOLUME 115. Diffraction Methods for Biological Macromolecules (Part B)
Edited by HAROLD W. WYCKOFF, C. H. W. HIRS, AND SERGE N. TIMASHEFF

VOLUME 116. Immunochemical Techniques (Part H: Effectors and Mediators of Lymphoid Cell Functions)
Edited by GIOVANNI DI SABATO, JOHN J. LANGONE, AND HELEN VAN VUNAKIS

VOLUME 117. Enzyme Structure (Part J)
Edited by C. H. W. HIRS AND SERGE N. TIMASHEFF

VOLUME 118. Plant Molecular Biology
Edited by ARTHUR WEISSBACH AND HERBERT WEISSBACH

VOLUME 119. Interferons (Part C)
Edited by SIDNEY PESTKA

VOLUME 120. Cumulative Subject Index Volumes 81–94, 96–101

VOLUME 121. Immunochemical Techniques (Part I: Hybridoma Technology and Monoclonal Antibodies)
Edited by JOHN J. LANGONE AND HELEN VAN VUNAKIS

VOLUME 122. Vitamins and Coenzymes (Part G)
Edited by FRANK CHYTIL AND DONALD B. MCCORMICK

VOLUME 123. Vitamins and Coenzymes (Part H)
Edited by FRANK CHYTIL AND DONALD B. MCCORMICK

VOLUME 124. Hormone Action (Part J: Neuroendocrine Peptides)
Edited by P. MICHAEL CONN

VOLUME 125. Biomembranes (Part M: Transport in Bacteria, Mitochondria, and Chloroplasts: General Approaches and Transport Systems)
Edited by SIDNEY FLEISCHER AND BECCA FLEISCHER

VOLUME 126. Biomembranes (Part N: Transport in Bacteria, Mitochondria, and Chloroplasts: Protonmotive Force)
Edited by SIDNEY FLEISCHER AND BECCA FLEISCHER

VOLUME 127. Biomembranes (Part O: Protons and Water: Structure and Translocation)
Edited by LESTER PACKER

VOLUME 128. Plasma Lipoproteins (Part A: Preparation, Structure, and Molecular Biology)
Edited by JERE P. SEGREST AND JOHN J. ALBERS

VOLUME 129. Plasma Lipoproteins (Part B: Characterization, Cell Biology, and Metabolism)
Edited by JOHN J. ALBERS AND JERE P. SEGREST

VOLUME 130. Enzyme Structure (Part K)
Edited by C. H. W. HIRS AND SERGE N. TIMASHEFF

VOLUME 131. Enzyme Structure (Part L)
Edited by C. H. W. HIRS AND SERGE N. TIMASHEFF

VOLUME 132. Immunochemical Techniques (Part J: Phagocytosis and Cell-Mediated Cytotoxicity)
Edited by GIOVANNI DI SABATO AND JOHANNES EVERSE

VOLUME 133. Bioluminescence and Chemiluminescence (Part B)
Edited by MARLENE DELUCA AND WILLIAM D. MCELROY

VOLUME 134. Structural and Contractile Proteins (Part C: The Contractile Apparatus and the Cytoskeleton)
Edited by RICHARD B. VALLEE

VOLUME 148. Plant Cell Membranes
Edited by LESTER PACKER AND ROLAND DOUCE

VOLUME 149. Drug and Enzyme Targeting (Part B)
Edited by RALPH GREEN AND KENNETH J. WIDDER

VOLUME 150. Immunochemical Techniques (Part K: *In Vitro* Models of B and T Cell Functions and Lymphoid Cell Receptors)
Edited by GIOVANNI DI SABATO

VOLUME 151. Molecular Genetics of Mammalian Cells
Edited by MICHAEL M. GOTTESMAN

VOLUME 152. Guide to Molecular Cloning Techniques
Edited by SHELBY L. BERGER AND ALAN R. KIMMEL

VOLUME 153. Recombinant DNA (Part D)
Edited by RAY WU AND LAWRENCE GROSSMAN

VOLUME 154. Recombinant DNA (Part E)
Edited by RAY WU AND LAWRENCE GROSSMAN

VOLUME 155. Recombinant DNA (Part F)
Edited by RAY WU

VOLUME 156. Biomembranes (Part P: ATP-Driven Pumps and Related Transport: The Na,K-Pump)
Edited by SIDNEY FLEISCHER AND BECCA FLEISCHER

VOLUME 157. Biomembranes (Part Q: ATP-Driven Pumps and Related Transport: Calcium, Proton, and Potassium Pumps)
Edited by SIDNEY FLEISCHER AND BECCA FLEISCHER

VOLUME 158. Metalloproteins (Part A)
Edited by JAMES F. RIORDAN AND BERT L. VALLEE

VOLUME 159. Initiation and Termination of Cyclic Nucleotide Action
Edited by JACKIE D. CORBIN AND ROGER A. JOHNSON

VOLUME 160. Biomass (Part A: Cellulose and Hemicellulose)
Edited by WILLIS A. WOOD AND SCOTT T. KELLOGG

VOLUME 161. Biomass (Part B: Lignin, Pectin, and Chitin)
Edited by WILLIS A. WOOD AND SCOTT T. KELLOGG

VOLUME 162. Immunochemical Techniques (Part L: Chemotaxis and Inflammation)
Edited by GIOVANNI DI SABATO

VOLUME 163. Immunochemical Techniques (Part M: Chemotaxis and Inflammation)
Edited by GIOVANNI DI SABATO

VOLUME 164. Ribosomes (in preparation)
Edited by HARRY F. NOLLER, JR. AND KIVIE MOLDAVE

VOLUME 165. Microbial Toxins: Tools for Enzymology
Edited by SIDNEY HARSHMAN

VOLUME 166. Branched-Chain Amino Acids (in preparation)
Edited by ROBERT HARRIS AND JOHN R. SOKATCH

VOLUME 167. Cyanobacteria (in preparation)
Edited by Lester Packer and Alexander N. Glazer

VOLUME 168. Hormone Action (Part K: Neuroendocrine Peptides) (in preparation)
Edited by P. MICHAEL CONN

VOLUME 169. Platelets: Receptors, Adhesion, Secretion (Part A) (in preparation)
Edited by JACEK HAWIGER

VOLUME 170. Nucleosomes (in preparation)
Edited by PAUL M. WASSARMAN AND ROGER D. KORNBERG

VOLUME 171. Biomembranes (Part R: Transport Theory: Cells and Model Membranes) (in preparation)
Edited by SIDNEY FLEISCHER AND BECCA FLEISCHER

VOLUME 172. Biomembranes (Part S: Transport: Membrane Isolation and Characterization) (in preparation)
Edited by SIDNEY FLEISCHER AND BECCA FLEISCHER

Section I

Lignin

[1] Isolation of Lignin

By JOHN R. OBST and T. KENT KIRK

General Considerations

Introduction

Significant gains have recently been made in understanding the biochemistry of the microbial degradation of lignin.[1] Further advances will be facilitated through studies using isolated lignins. This chapter presents some of the most useful methods for lignin isolation.

Critique of Lignin Preparations. Probably the best known isolated lignin is Klason lignin, which is obtained by treating wood with sulfuric acid. The polysaccharides are hydrolyzed to water-soluble sugars, and the lignin is recovered as an insoluble residue. Although this method for lignin isolation has great utility as an analytical means of determining lignin content (see chapter [12], this volume), the highly condensed and altered Klason lignin is generally unsuited for either chemical characterization or studies of biological modification and degradation. For such studies, what is needed is an isolated lignin that is representative of the lignin in the original lignocellulose (which is sometimes referred to as protolignin).

Because the methods for isolating lignin have been devised by wood scientists, the procedures discussed here are those used to isolate lignin from wood (see Table I). Often these methods will be suitable for other lignocellulosics, but sometimes modifications will be required. In particular, it may be desirable to treat certain plant materials, such as forages and immature woody tissues, to remove protein prior to lignin isolation. This can be accomplished by treatment with proteases or by extraction with hot, neutral detergent (see chapter [12], this volume).

The most useful lignin preparation is Björkman lignin, also known as Björkman milled wood lignin or simply milled wood lignin (MWL). Milled wood lignin is purified from the aqueous *p*-dioxane extract of finely milled wood, which has been first extracted with organic solvents to remove extraneous components. Although it has not been rigorously proved that MWL is representative of protolignin, it is considered to be appropriate for most chemical and biological studies. Milled wood lignin can be obtained

[1] T. Higuchi (ed.), "Biosynthesis and Biodegradation of Wood Components," Chap. 19–21. Academic Press, San Diego, California, 1985.

TABLE I
LIGNIN ISOLATION METHODS

Preparation	Methodology	Remarks
Milled wood lignin (MWL)	Aqueous dioxane extraction of finely milled wood	Obtained in about 20% yield; considered to be representative of the original lignin
Milled wood enzyme lignin (MWEL)	Residue left after polysaccharidase hydrolysis of the carbohydrates in finely milled wood	Ninety five plus percentage yield, but contains 10–12% carbohydrate; not completely soluble in common lignin solvents
Cellulase enzyme lignin (CEL)	Solvent-soluble fraction of MWEL	Similar to MWL
Brauns' native lignin	Ethanol extract of ground wood (fine sawdust-size particles)	Lower yield and lower molecular weight than MWL
Brown rot lignin	Ethanol or aqueous dioxane extract of brown-rotted wood	Probably not severely altered, but some demethylation of methoxyls and oxidation of side chains has occurred
Chemical lignins (kraft and sulfite)	Dissolution of lignin at high temperature and pressure with chemicals	Not representative of the original lignin; major by-products in pulp production to make paper
Klason lignin	Insoluble, condensed residue left after hydrolysis of polysaccharides with sulfuric acid	Not representative of the original lignin; often used as a measure of lignin content (see chapter [12], this volume)

in 20–30% yields, based on the total lignin. This method requires either a vibratory or rotary ball mill.

One lignin preparation which does not require any ball-milling equipment is Brauns' lignin (sometimes referred to as Brauns' native lignin or native lignin, which should not be confused with protolignin). The wood is first extracted with cold water and then with ether to remove extraneous components. Subsequent extraction of some of the lignin with ethanol followed by purification steps gives Brauns' native lignin. This lignin preparation has fallen into disfavor among wood chemists, who consider its low yield (about 8% based on the total lignin) and its low molecular weight to be disadvantageous for investigations of structure and reactivity. However, the low-molecular-weight distribution of Brauns' lignin may be beneficial in some lignin biodegradation studies by providing greater accessibility of the substrate and increased degradation rates. Basically, the structure of Brauns' native lignin is similar to that of MWL except for its molecular weight and associated properties. Brauns' native lignin may be used as the first lignin substrate, and successful investigations can then move on to use other more representative isolated lignins.

Ball-milled wood, prepared in the same manner as that used for MWL extractions, may be treated with polysaccharidase enzymes to solubilize the carbohydrate components. In this way, a lignin residue is produced which contains nearly all of the lignin in the wood. This lignin, termed milled wood enzyme lignin (MWEL), has not been severely modified by any chemical treatment. Milled wood enzyme lignin is the most representative of all the isolated lignins. Unfortunately, it contains a relatively high residual carbohydrate content of 10–12%, a result of covalent linkages between lignin and polysaccharide fragments. Also, due to its high molecular weight, it is not completely soluble in common lignin solvents, such as aqueous dioxane, acetic acid, dimethylformamide, and dimethyl sulfoxide. This insolubility presents experimental difficulties in handling, purifying, and analyzing MWELs.

Fractionation of MWEL, based on solubility in dioxane–water, is a means of preparing a soluble lignin which can then be purified in the same manner as MWL. This lignin was originally termed cellulase enzyme lignin (CEL). It is thought to be more representative of protolignin than MWL but has a lower yield than MWEL. Milled wood lignin is probably adequate for most studies, and the additional steps in preparing CEL are usually not justifiable.

There are numerous other lignin preparations, including hydrochloric acid lignin, periodate lignin, cuoxam lignin, enzymatically liberated lignin, alcohol–HCl lignin, thioglycolic acid lignin, acetic acid lignin, dioxane–HCl lignin, phenol lignin, and hydrogenolysis lignin.[2] These preparations usually are not adequate as substrates for biochemical studies of protolignin. However, "enzymatically liberated lignin," or brown rot lignin, may be useful in certain circumstances. When wood is rotted by brown rot fungi, the lignin is not substantially degraded while the carbohydrates are removed. The rotted wood, largely lignin, is extracted with lignin solvents and the lignin is purified; yields of over 20% of the original lignin are obtained. Although brown rot fungi demethylate aromatic methoxyl groups and cause a limited amount of oxidation, the lignin is not otherwise severely damaged. The demethylation may even be considered advanta-

[2] D. Fengel and G. Wegener, "Wood, Chemistry, Ultrastructure, Reactions," p. 50. de Gruyter, Berlin, Federal Republic of Germany, 1983.

geous: subsequent methylation of phenolic hydroxyls using a ^{14}C label would provide a substrate for monitoring lignin degradation by measuring the evolution of labeled carbon dioxide or for studying demeth(ox)ylation.

Another class of lignins is produced by chemical pulping processes. Most of the chemical pulp produced in the United States is by the kraft process. Kraft lignin is highly modified; it is lower in molecular weight, has a higher phenolic content and a lower methoxyl content, and has undergone extensive side-chain reactions.

Lignosulfonate, as implied by its name, is the sulfonated lignin removed from wood by sulfite pulping. Lignosulfonates have higher molecular weight than kraft lignin but are not representative of protolignin. Hydrolysis reactions have occurred, and sulfonation (to give a water-soluble product) can be extensive.

Whereas kraft lignin and lignosulfonates are not suitable for studies modeling the behavior of protolignin, they are important in their own right as industrial by-products, and research is warranted on their biodegradation and bioconversion.

Finally, after a lignin has been isolated and purified, it is essential to analyze it to be certain that it is not grossly contaminated. Often, determinations, such as carbohydrate content, methoxyl content, and ultraviolet absorption, are sufficient, but sometimes more detailed analyses are required. An overview of quantitative lignin determinations is given in chapter [12] in this volume.

Isolation of Lignin

Milled Wood Lignin (Björkman Milled Wood Lignin, Björkman Lignin)

The wood used for lignin isolation should be sapwood; heartwood often contains polyphenolics which are difficult to remove and may even have condensed with the lignin. Air-dried wood is milled in a Wiley mill to pass 40 mesh and extracted first with acetone : water (9 : 1, v : v) by percolation at room temperature and then with ethanol : benzene (2 : 1, v : v). Lignin is heat-sensitive, so extractions should not be at the boiling point of the solvents. For this reason also, the extracted wood should never be oven-dried, but rather dried in a vacuum desiccator over phosphorus pentoxide or other efficient desiccant.

The dry, extracted wood is then milled either in a vibratory ball mill[3] or

[3] Siebtechnik vibrating ball mills may be obtained from Tema, Inc., 11584 Goldcoast Drive, Cincinnati, Ohio 45249. We have successfully used mill type USM 12 to grind 200 g of wood at one time. For smaller amounts of wood (1 – 10 g), we have used the smaller

in a conventional rotating jar ball mill. The time of milling can vary from 1 hr in a very efficient vibratory ball mill, such as the NBS-type mill first used by Björkman,[4] to 12 hr in a large Siebtechnik vibratory mill, to 3 weeks using a rotating jar mill. Thus, milling time and the amount of material milled must be optimized for each type of mill and grinding medium. Because the temperature of the milling jar will increase, it is desirable to cool the mill to avoid damaging the lignin.[3]

Often the ball milling is done in a nonswelling solvent such as toluene. The solvent excludes oxygen, and the milled wood may be recovered by centrifugation. However, the wood may be milled dry, preferably under carbon dioxide or nitrogen. In this case, the milled wood may conveniently be removed from the balls after conditioning in a high humidity environment by shaking the balls on screens in a Ro-Tap sieve shaker.[5]

The milled wood is dispersed in dioxane : water (96 : 4, v : v) and mechanically stirred; the ratio of wood to solvent is chosen to be convenient, for example, 10 g of milled wood and 250 ml of dioxane : water. After 1 day, the suspension is centrifuged, and the residue is redispersed in fresh dioxane : water and stirred for an additional day. Although lignin would continue to be extracted for many subsequent extractions, the bulk of the MWL is removed in the first two. The extracts are combined and then freeze-dried (or simply dried in a rotary vacuum evaporator) to give a crude MWL in about 20–30% yield; this lignin contains up to about 10% residual carbohydrate. (The milled wood may also be extracted with 9 : 1 dioxane : water, giving a higher yield of MWL but with more carbohydrate.) As is, this crude MWL is useful for many experiments.

In most cases, it is desirable to purify the crude MWL. This is accomplished by dissolving the lignin (the dioxane : water is removed by vacuum evaporation) in 90% acetic acid, using 20 ml of solvent for each gram of lignin. The acetic acid solution is then added dropwise, with stirring, to water (about 220 ml of water per gram of lignin). The precipitated lignin is centrifuged and then freeze-dried or air-dried, followed by drying in a vacuum oven. It is then dissolved with stirring in a mixture of 1,2-dichloro-

Siebtechnik mill or a custom-made mill patterned after the NBS mill.[4] The temperature increase of the mill jars on the large mill is minimized by milling for no longer than 1 hr at a time, followed by 1 hr of cooling. A large fan is used to cool the jars during the entire process. The smaller mill may be placed in a cold room.

[4] A. Björkman, *Sven. Papperstidn.* **59,** 477 (1956).

[5] For the preparation of large amounts of milled wood, we have found that mechanical shaking on a sieve is a convenient way to remove the wood from the balls. We use a No. 7 (7-mesh) stainless-steel screen with a custom-made stainless inner collar to minimize ash contamination. Sieves and the Ro-tap may be obtained from W. S. Tyler, Inc., Mentor, Ohio.

ethane:ethanol (2:1, v:v) and centrifuged to remove solids. The lignin solution is added dropwise to *anhydrous* ethyl ether to precipitate the lignin. About 20 ml of solvent and 230 ml of ether are used for 0.5–1 g of lignin. After centrifugation, the insoluble MWL is washed three times with fresh ether. The yield of the purified MWL may be half that of the crude preparation, but its residual carbohydrate content is about 4%.[4] When prepared from light-colored woods, MWL is cream colored.

There are several ways significantly to reduce the carbohydrate content of MWL. However, yield losses may be substantial. Two such methods are those of Freudenberg and Neish[6] and Lundquist and Simonson.[7]

Milled Wood Enzyme Lignin

The wood is extracted and then ball milled as for milled wood lignin (see above).

For digestion of the carbohydrate in 100 g of milled wood, 3 g of Cellulysin (Calbiochem-Behring Corp., La Jolla, California 92307), which is a mixture of polysaccharidase enzymes, or a comparable preparation,[8] is dissolved in 40 ml of 0.5 *M* acetate buffer (pH 4.6) and 200 ml of distilled water. The enzyme solution is centrifuged to remove undissolved materials. Water is added to the suspension of milled wood in the enzyme solution to bring the total volume to about 1.5 liters. Several drops of toluene are added as a preservative. The digestion is carried out with stirring in a suitable glass container for a week to 10 days at 48°. The suspension is then centrifuged; the residue is washed with water and redigested in the same way two more times. The final residue is thoroughly washed and then freeze-dried. Yields based on the lignin in the wood are

[6] K. Freudenberg and A. C. Neish, "Constitution and Biosynthesis of Lignin," p. 52. Springer-Verlag, Berlin and New York, 1968.

[7] K. Lundquist and R. Simonson, *Sven. Papperstidn.* **78,** 390 (1975).

[8] Solubilization of the polysaccharides in milled wood and other lignocellulosic materials requires the concerted action of the cellulase system (endo- and exo-1,4-glucanases) plus hemicellulose-depolymerizing enzymes. The latter include enzymes that hydrolyze substituted 1,4-xylans and substituted glucomannans, also 1,4-linked. Such mixtures of enzymes are produced commercially with the fungus *Trichoderma reesei (Trichoderma viride)* grown on delignified lignocellulosic substrates such as newsprint or on finely milled lignocellulosics, which have the necessary constituents to induce the enzymes. By suitable experimentation, the researcher should be able to produce the enzyme mixture without undue difficulty. General references to the production of cellulases and hemicellulases by *T. reesei* are as follows: K.-E. Eriksson and T. M. Wood, "Biosynthesis and Biodegradation of Wood Components" (T. Higuchi, ed.), pp. 469–503. Academic Press, San Diego, California, 1985; and R. F. H. Dekker, *in* "Biosynthesis and Biodegradation of Wood Components" (T. Higuchi, ed.), pp. 505–533. Academic Press, San Diego, California, 1985.

over 95%. The MWEL will have a carbohydrate content of approximately 10–12%.[9]

The MWEL may be fractionated to remove some of the residual carbohydrate and to give a solvent-soluble lignin with lower molecular weight, termed CEL.[10] To prepare CEL, MWEL is twice extracted with 96% (or 90%) dioxane. This extract may be purified in the same manner as milled wood lignin as described above. The residual carbohydrate content of the CEL-96 is about 4%, whereas that of the CEL-90 is higher. The extracted residue may be further extracted with 50% dioxane. However, this extract is not completely soluble in dichloroethane: ethanol and cannot be purified by the MWL procedure.[10]

Brauns' Native Lignin (Brauns' Lignin, Native Lignin)

The wood is ground in a Wiley mill to pass a 100- to 150-mesh screen and extracted first with cold water and then with ethyl ether for 48 hr to remove extraneous components. The wood is then extracted by percolation with 95% ethanol at room temperature for 8–10 days or until the extract is colorless. A small amount of calcium carbonate is added to the extract to neutralize wood acids, and the alcohol is removed under reduced pressure. Water is added to the residue, and the evaporation continues to remove traces of the alcohol. The lignin residue is triturated alternatively with water and ether until it becomes solid. The solid is filtered and dried over an efficient desiccant. The dry lignin is extracted with anhydrous ether in a Soxhlet apparatus. The residue is dissolved in dioxane to give a 10% solution, and it is precipitated by dripping into stirred distilled water (about 15 times the volume of the dioxane). If a colloidal solution forms instead of a precipitate, a little sodium sulfate is added and the solution vigorously stirred to coagulate the lignin. The precipitate is filtered, washed with water, and dried in a desiccator. It is then dissolved in dioxane to give a 10% solution, centrifuged, and filtered. The solution is slowly dripped, with stirring, into anhydrous ethyl ether. The Brauns' native lignin separates as a fine tan-colored powder. It is washed sequentially with ether, high-boiling petroleum ether, and low-boiling petroleum ether and then dried in a desiccator over sulfuric acid and paraffin shavings. Precipitation into ether may be repeated until the methoxyl content of the Brauns' native lignin is constant. The yield is about 8% based on the lignin in the wood.[11]

Similar lignin preparations can be obtained from ground samples with

[9] J. R. Obst, *Tappi* **65,** 109 (1982).

[10] H.-M. Chang, E. B. Cowling, and W. Brown, *Holzforschung* **29,** 153 (1975).

[11] F. E. Brauns, "The Chemistry of Lignin," p. 51. Academic Press, New York, 1952.

other lignin solvents, including acetone : water, 9 : 1 (v : v),[12] and aqueous dioxane.

Brown Rot Lignin [Enzymatically Liberated Lignin (ELL)]

The wood is decayed in soil block chambers (ASTM D 2017-81) by a brown rot fungus, such as *Gleophyllum trabeum, Lentinus lepideus,* or *Poria vaillantii,* to weight losses of about 60–70%.[13] The dried, decayed wood is ground in a Wiley mill to pass a 60-mesh screen. Brown rot lignin is then obtained, employing the purification methods used for milled wood lignin or Brauns' native lignin. Alternatively, the Wiley-milled decayed wood may be extracted with 50% aqueous *p*-dioxane and the lignin purified by gel permeation chromatography on Sephadex G-25.[13]

Chemical Lignins (Kraft Lignin and Lignosulfonate)

Two types of commercially available lignin are kraft lignin and lignosulfonate.[14] Although the lignins from commercial sources may be adequate for many studies, it is recommended that this type of lignin be isolated from laboratory pulping experiments whenever possible. In this way, the entire history of the lignin is known and controlled.

Kraft Lignin (Thiolignin, Sulfate Lignin)

Kraft pulping is accomplished by degrading and dissolving the lignin in hot alkaline sodium sulfide solution ("white liquor"). Kraft white liquor is prepared by dissolving 16 g of sodium sulfide per liter of 1 *N* sodium hydroxide. The extracted wood, either in chip form or Wiley-milled form, plus white liquor at a 4 : 1 (w : w) liquor-to-wood ratio are sealed in a stainless-steel bomb. Cooking temperature for most hardwoods (angiosperm woods) is about 155°, whereas 170–180° is required for softwoods (gymnosperm woods). The bomb is usually heated in an oil bath and rotated, end over end, to ensure mixing. If industrial conditions are to be mimicked, the time to raise the bath from room temperature to the maximum cooking temperature should be about 90 min. Time at temperature will depend on the pulp yield desired; 1–2 hr is typical.

[12] T. K. Kirk and H.-M. Chang, *Holzforschung* **28,** 217 (1974).

[13] T. K. Kirk, *Holzforschung* **29,** 99 (1975).

[14] Kraft lignin, and modified kraft lignins, may be obtained from Westvaco, Chemical Division, Box 70848, Charleston Heights, South Carolina 29415. Lignin sulfonates may be obtained from Reed Lignin, Inc., 100 Highway 51 South, Rothschild, Wisconsin 54474, and from Crown Zellerbach Corp., P.O. Box 4266, Vancouver, Washington 98662.

The bomb is then cooled with water, and the "black liquor" containing the lignin, some hemicellulose, carbohydrate degradation products, and inorganic chemicals is filtered from the pulp. The lignin may be precipitated by the addition of acid. Generally it is better to use acetic acid rather than mineral acids. Traces of sulfuric acid or hydrochloric acid are hard to remove, and they may cause the lignin to undergo condensation reactions when it is dried. Carbon dioxide may also be used to precipitate the lignin, but yields are usually lower than with acetic acid. The precipitated lignin should be washed thoroughly with distilled water and freeze-dried.

The kraft lignin may be purified through solvent (pyridine : acetic acid : water) fractionation.[15] Alternatively, kraft lignin may be purified by repetitive dissolution in 0.1 *N* sodium hydroxide and precipitated with acetic acid. Finally, it is washed with distilled water and freeze-dried.

Lignosulfonate (Lignin Sulfonate, Sulfite Lignin)

The sulfite pulping of wood is accomplished by treating wood at high temperatures with aqueous sodium sulfite. The cook may be acid, neutral, or alkaline. An example of neutral sulfite cooking conditions is as follows. The wood chips are heated from ambient temperature to 175° over 90 min in sulfite liquor at a 3 : 1 (w : w) liquor to wood ratio. The liquor contains 15% sodium sulfite and 1.5% sodium carbonate, based on the dry wood. Time at temperature is 1 hr or more.

The sulfonated, water-soluble lignin (lignosulfonate) cannot be isolated by precipitation from the spent liquor with acid as are kraft lignins. However, this lignin may be purified by complexing with amines. For example, spent sulfite liquor, 408 ml containing 175 g of solids, is heated to 70–85° with mild stirring. *N,N*-Dimethylhexadecylamine (Armak, Chicago, Illinois) is added, and then the solution is adjusted to pH 3.5 with 10 *N* sulfuric acid. Four hundred grams of 1-octanol is added, and the solution is stirred for 5 min more, then left to stand. One hour is required for the layers to separate. The bottom, aqueous acidic layer is discarded.

To the alcohol layer is added 122 g of water, with stirring, and 28 g of 50% sodium hydroxide (the solution should be about pH 9.5). The solution is heated at 60–70° with stirring. The layers are allowed to separate over 30 min. The bottom, aqueous layer contains about 100 g of sodium lignin sulfonate and a trace of octanol. The alcohol is removed by vacuum evaporation.[16]

[15] K. Lundquist and T. K. Kirk, *Tappi* **63,** 80 (1980).
[16] S. Y. Lin (Reed Lignin, Inc., Rothschild, Wisconsin), personal communication.

Acknowledgment

The use of trade, firm, or corporation names in this publication is for the information and convenience of the reader. Such use does not constitute an official endorsement or approval by the U.S. Department of Agriculture of any product or service to the exclusion of others which may be suitable.

[2] Lignin–Carbohydrate Complexes from Various Sources

By JUN-ICHI AZUMA and KOSHIJIMA TETSUO

Some polysaccharides in the cell walls of lignified plants are linked to lignin to form lignin–carbohydrate complexes.[1–4] Three types of evolutionally different plants, softwoods (angiosperm), hardwoods (gymnosperm), and graminaceous plants (grass), contain structurally different molecular species of hemicellulose and lignin.[5–8] This implies the existence of variations in the sugars linking to lignin in these different types of plants. The major obstacle in the characterization of lignin–carbohydrate complexes is the difficulty in isolating these complexes in a homogeneous state. A simple procedure described below has been developed for the isolation and fractionation of water-soluble lignin–carbohydrate complexes from various types of lignified plants.[9–13]

[1] A. Björkman, *Sven. Papperstidn.* **60,** 243 (1957).
[2] O. P. Grushnikov and N. N. Shorygina, *Russ. Chem. Rev. (Engl. Transl.)* **39,** 684 (1970).
[3] Y. Z. Lai and K. V. Sarkanen, *in* "Lignins" (K. V. Sarkanen and C. H. Ludwig, eds.), pp. 165–240. Wiley (Interscience), New York, 1971.
[4] E. Adler, *Wood Sci. Technol.* **11,** 169 (1977).
[5] K. V. Sarkanen and H. L. Hergert, *in* "Lignins" (K. V. Sarkanen and C. H. Ludwig, eds.), pp. 43–94. Wiley (Interscience), New York, 1971.
[6] K. C. B. Wilkie, *Adv. Carbohydr. Chem.* **36,** 215 (1979).
[7] G. O. Aspinall, *in* "The Biochemistry of Plants" (J. Preiss, ed.), Vol. 3, pp. 473–500. Academic Press, New York, 1980.
[8] T. Higuchi, *in* "Plant Carbohydrates II" (W. Tanner and F. A. Loewus, eds.), pp. 194–224. Springer-Verlag, Berlin and New York, 1981.
[9] J. Azuma, N. Takahashi, and T. Koshijima, *Carbohydr. Res.* **93,** 91 (1981).
[10] S. Mukoyoshi, J. Azuma, and T. Koshijima, *Holzforschung* **35,** 233 (1981).
[11] J. Azuma, N. Takahashi, and T. Koshijima, *Mokuzai Gakkaishi* **31,** 587 (1985).
[12] J. Azuma, T. Nomura, and T. Koshijima, *Agric. Biol. Chem.* **49,** 2661 (1985).
[13] A. Kato, J. Azuma, and T. Koshijima, *Holzforschung* **38,** 141 (1984).

Isolation Procedure

The following procedure employs about 1 kg of plant meal. The purification step by solvent extraction, however, is usually carried out on 10 g of the crude lignin–carbohydrate complex.

Step 1: Preparation of Extractive-Free and Depectinated Plant Meal. One kilogram of air-dried plant is milled to 24–80 mesh, extracted with ethanol–benzene (1:2, v/v) for 48 hr, and depectinated with 10 liters of 0.25% (w/v) aqueous potassium acetate at 60° for 24 hr.[14] The plant meal is then vibromilled for 48 hr under nitrogen with external cooling using tap water. Exhaustive drying of the plant meal is necessary to avoid coagulation during milling. More than 80% of the particles accumulated are in the 12- to 36-μm size range.[14] Unless otherwise specified, all subsequent operations are performed at room temperature.

Step 2: Extraction of Lignin–Carbohydrate Complexes. The finely divided plant meal is extracted with 10 liters of 80% aqueous 1,4-dioxane for 48 hr. The residue is washed and reextracted with the same solvent. The combined filtrates and washings are concentrated by evaporation at 40° to about 2 liters and dialyzed against distilled water. The precipitate formed during dialysis (milled wood lignin) is removed by centrifugation at 8,000 *g* for 15 min and washed three times with distilled water. The combined aqueous solution is concentrated to about 200 ml by evaporation below 40° and lyophilized.

Step 3: Purification by Solvent Extraction. Ten grams of the crude lignin–carbohydrate complex is dissolved in 200 ml of 50% aqueous 1,4-dioxane and is extracted with 200 ml of chloroform to remove any lignin not complexed to carbohydrate. The chloroform and the intermediate turbid layers are washed five times with 100 ml of distilled water. The combined aqueous solution is dialyzed against distilled water, concentrated to a small volume by evaporation, and lyophilized. Ten grams of this lignin–carbohydrate complex is solubilized in 56 ml of pyridine–acetic acid–water (9:1:4, v/v/v)[15] and is extracted with 6 vol of chloroform. After addition of 400 ml of distilled water, the aqueous layer is recovered by centrifugation (8,000 *g* for 15 min). The chloroform and intermediate jellylike layers are washed four times with 400 ml of distilled water. The combined aqueous layer is dialyzed against distilled water, concentrated to a small volume by evaporation, and lyophilized. This final step yields a lignin–carbohydrate complex (LCC-W) which is completely soluble in water.

[14] T. Koshijima, T. Taniguchi, and R. Tanaka, *Holzforschung* **26,** 211 (1972).
[15] K. Lundquist and R. Simonson, *Sven. Papperstidn.* **78,** 390 (1975).

Fractionation of the Lignin–Carbohydrate Complexes

Gel Filtration. One gram of the lignin–carbohydrate complex is solubilized in 15 ml of 25 m*M* sodium phosphate buffer, pH 6.8, and is applied to a 95.0 × 5.2-cm column of Sepharose 4B equilibrated with the same buffer. Elution with the same buffer is carried out at 60 ml/hr, and fractions of 20 ml are collected. The position of the lignin–carbohydrate complex peaks is monitored by the phenol–sulfuric acid carbohydrate assay:[16] lignin is measured by absorbance at 280 nm, and phenolic acid is measured by absorbance at 315 nm. All lignin–carbohydrate complexes (LCC-W) are separated into three distinct subfractions (W-1, W-2, and W-3) (Fig. 1), which are pooled, thoroughly dialyzed against distilled water, and lyophilized. The purity of the separated lignin–carbohydrate subfractions is determined by glass-fiber electrophoresis and analytical ultracentrifugation, and each gives a single component in both tests.

Hydrophobic Chromatography. Hydrophobic chromatography is carried out on gel columns (15 × 1.8 cm) of phenyl- and octyl-Sepharose CL-4B (ligand concentration of 40 μm/ml, Pharmacia) at 25°. The columns are equilibrated with 25 m*M* sodium phosphate buffer, pH 6.8, containing 50% (w/v) ammonium sulfate which enhances the adsorbing ability of softwood lignin–carbohydrate complexes. Ammonium sulfate is not used for hardwood and grass lignin–carbohydrate complexes, because it prevents complete solubilization.[11,12] The lignin–carbohydrate complex (500 mg) is solubilized in 10 ml of the equilibration buffer and is applied to the column. The column is eluted with this buffer (1 ml/min) until the absorbance at 280 nm returns to the background level and then is eluted with 50% (v/v) 2-ethoxyethanol in 25 m*M* sodium phosphate buffer, pH 6.8. Elution with the 2-ethoxyethanol solution reveals one large lignin peak which superimposes on the carbohydrate distribution. Both unadsorbed and adsorbed fractions are dialyzed against distilled water, concentrated to a small volume, and lyophilized.

Properties of Lignin–Carbohydrate Complexes

Chemical and Physical Properties. Table I summarizes the chemical properties of the lignin–carbohydrate complexes (LCC-W) and their subfractions (W-1, W-2, and W-3) from bald cypress *(Taxodium distichum),* birch *(Betula platyphylla),* and rice straw *(Oryza sativa),* selected as representative of softwood, hardwood, and grass, respectively. The other lignin–carbohydrate complexes from softwoods (normal[9] and compres-

[16] M. Dubois, K. A. Gilles, J. K. Hamilton, P. A. Rebers, and F. Smith, *Anal. Chem.* **28,** 350 (1956).

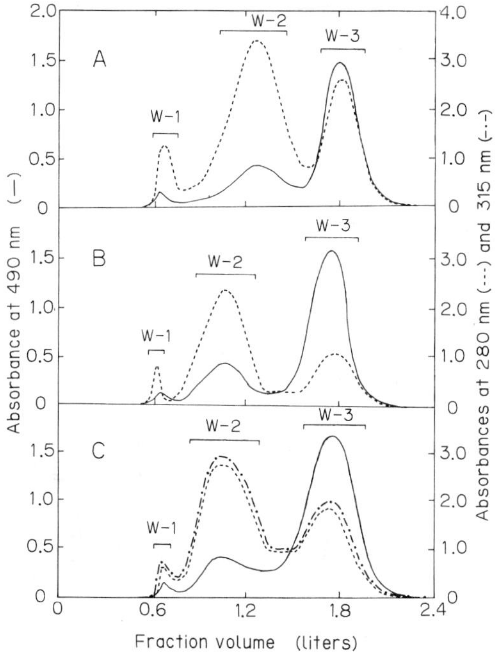

FIG. 1. Gel filtration profiles of lignin–carbohydrate complexes from (A) bald cypress, (B) birch, and (C) rice straw on Sepharose 4B. See text for details. The separated lignin–carbohydrate subfractions (W-1, W-2, and W-3) are pooled as indicated.

sion wood[10] of *Pinus densiflora, Cryptomeria japonica, Chamaecyparis obtusa,* and *Metasequioa glyptostroboides),* hardwoods *(Fagus crenata,*[11] *Populus euramericana, Acacia auriculaeformis,* and *Ochroma lagopus),* and grasses *(Phyllostachys pubescens,*[12] *Saccharum officinarum,*[13] and *Cocos nucifera)* have chemical properties similar to those listed in Table I, except lignin–carbohydrate complexes from poplar which contain 1.0–6.0% of *p*-hydroxybenzoic acid. The yields of LCC-W are about 0.1–1.4% (softwoods), 0.3–1.5% (hardwoods), and 3.0–5.8% (grasses), based on the depectinated plant meals.

All the unfractioned lignin–carbohydrate complexes (LCC-W) are composed of 50–65% neutral sugar, 2–10% uronic acid, and 24–45% lignin. The W-2 fractions contain more lignin than the W-3 fractions, while the W-3 fractions are rich in carbohydrates. The molecular weights of the W-2 and W-3 fractions are in the range of $1.5-8.5 \times 10^5$ and $1.5-10.0 \times 10^3$ and show no species-specific characteristics. However, comparative chemical, methylation, Smith degradation, and ^{13}C NMR spectroscopic analyses and 2,3-dichloro-5,6-dicyanobenzoquinone treatment

TABLE I

CHEMICAL AND PHYSICAL PROPERTIES OF THE LIGNIN–CARBOHYDRATE COMPLEXES

Components	Bald cypress				Birch				Rice straw			
	LCC-W	W-1	W-2	W-3	LCC-W	W-1	W-2	W-3	LCC-W	W-1	W-2	W-3
Relative distribution ratio (%)	100	8.2	34.5	45.9	100.0	8.2	22.8	52.8	100.0	7.0	26.7	49.3
Carbohydrate content (%)[a]	51.4	25.4	28.1	63.8	69.1	32.5	48.9	87.1	63.9	26.2	51.4	80.1
Neutral sugar composition (%)												
L-Arabinose	32.3	19.8	19.3	26.0	1.5	2.1	2.0	0.9	13.0	11.0	11.9	15.3
D-Xylose	21.4	23.1	27.3	24.0	95.2	86.2	90.3	97.4	80.1	77.3	80.7	78.4
D-Mannose	20.8	22.9	23.1	21.9	0.5	2.2	1.0	0.2	0.4	1.6	0.5	0.4
D-Galactose	16.8	24.1	21.3	20.6	1.2	2.9	2.7	0.6	2.3	2.9	1.9	2.8
D-Glucose	8.7	10.0	8.9	7.4	1.6	6.6	3.8	0.8	4.3	7.2	5.0	3.1
Uronic acid content (%)[b]	4.8	2.5	1.6	8.7	9.5	5.5	6.5	10.6	2.8	1.6	2.5	3.5
Lignin content (%)[c]												
Klason	42.0	70.9	61.6	26.9	20.0	51.0	40.2	7.6	27.7	46.3	36.8	11.7
Acid-soluble	2.5	3.8	3.6	4.5	2.4	1.7	2.2	2.3	5.6	3.0	3.3	6.4
Acetyl content (%)[d]	1.5	0.9	0.8	2.4	5.9	0.8	1.7	7.9	4.2	2.9	3.6	4.6
Phenolic acid content (%)[e]												
trans-p-Coumaric acid	—	—	—	—	—	—	—	—	4.0	7.0	5.8	1.3
trans-Ferulic acid	—	—	—	—	—	—	—	—	0.8	0.2	0.6	1.6
$[\alpha]_D^{25}$ (*c* 0.5, 50% dioxane, degree)	11.2	4.3	7.6	21.1	−35.0	−10.0	−23.3	−40.1	−42.8	−19.1	−26.0	−58.5
$\overline{M}_w$ ($\times 10^{-3}$)[f]	ND[g]	ND[g]	180.0	2.8	ND[g]	ND[g]	580.0	9.0	ND[g]	ND[g]	705.0	4.0
s_{20}°(S)	ND[g]	ND[g]	8.5	0.3	ND[g]	ND[g]	23.5	1.0	ND[g]	ND[g]	21.3	0.5
Relative extent of binding (%)[h]												
Phenyl-Sepharose CL-4B	84.8	ND[g]	86.0	75.0	38.0	ND[g]	92.5	30.5	34.1	ND[g]	82.5	27.8
Octyl-Sepharose CL-4B	82.5	ND[g]	84.8	60.2	28.5	ND[g]	90.5	28.0	30.9	ND[g]	80.3	26.0

[a] Determined by the phenol–sulfuric acid method.[16]

[b] Determined by the method of J. T. Galambos [*Anal. Biochem.* **19,** 119 (1967)].

[c] Determined by the standard methods of Tappi T 222os − 74 and UM250.

[d] Determined according to M. Tomoda, S. Kaneko, and S. Nakatsuka [*Chem. Pharm. Bull.* **23,** 430 (1975)].

[e] Determined according to R. D. Hartley and E. C. Jones [*J. Chromatogr.* **107,** 213 (1975)].

[f] Estimated by gel filtration on Sepharose 4B using dextrans of known molecular weights.

[g] ND, Not determined.

[h] Amount of the adsorbed material was determined as described in the text.

indicate that the softwood lignin–carbohydrate complexes are distinctively different from the others in that their carbohydrate portions consist of galactoglucomannan, arabino-4-*O*-methylglucuronoxylan, and arabinogalactan which are linked to lignin at benzyl positions.[9,10,17] In contrast, carbohydrate portions of hardwood and grass lignin-carbohydrate complexes are composed exclusively of 4-*O*-methylglucuronoxylan and arabino-4-*O*-methylglucuronoxylan, respectively.[11,12] It has been suggested that the neutral sugars present in the lignin–carbohydrate complexes are associated with the linkages with lignin[17,18] and that uronic acid residues are esterified to lignin at the benzyl position of the lignin.[18] Saponified *trans-p*-coumaric acid is esterified at C_α and C_γ of the side chain of bamboo lignin.[19] *p*-Hydroxybenzoic acid is also esterified to popular lignin.[20] In contrast, *trans*-ferulic acid is ether linked to lignin.[21] Both *trans-p*-coumaric and *trans*-ferulic acids are also esterified to O-5 of the L-arabinofuranosyl residues of arabinoxylan of the bagasse lignin–carbohydrate complex,[22] *Zea mays* shoot,[23] barley aleurone layers and straw,[24,25] and wheat bran.[26] There is no evidence concerning the linkages between *p*-hydroxybenzoic acid and carbohydrate in poplar. These results indicate that *p*-coumarate and ferulates might form cross-links between lignin and hemicellulose in the native cell walls of graminaceous plants.

Hydrophobic Properties. Lignin–carbohydrate complexes show amphiphatic or surface-active properties because of the presence of hydrophilic carbohydrate and hydrophobic lignin in the same molecule. The amount of lignin–carbohydrate complex bound to phenyl-Sepharose is consistently higher than for octyl-Sepharose (Table I). Since the hydrophobicity of a phenyl group is equivalent to a linear chain hydrocarbon with four or five carbon atoms, the hydrophobic interactions operate as the exclusive force for adsorption in the case of octyl-Sepharose, and aromatic–aromatic interactions participate as an additional force for adsorption in the case of phenyl-Sepharose. Some extended applications of the hydrophobic chro-

[17] T. Koshijima, T. Watanabe, and J. Azuma, *Chem. Lett.* p. 1737 (1984).
[18] Ö. Eriksson and D. A. I. Goring, *Wood Sci. Technol.* **14,** 267 (1980).
[19] M. Shimada, T. Fukuzuka, and T. Higuchi, *Tappi* **54,** 72 (1971).
[20] D. C. C. Smith, *J. Chem. Soc.* p. 2347 (1955).
[21] A. Scalbert, B. Monties, J.-Y. Lallemand, E. Guittet, and C. Rolando, *Phytochemistry* **24,** 1359 (1985).
[22] A. Kato, J. Azuma, and T. Koshijima, *Chem. Lett.* p. 137 (1983).
[23] Y. Kato and D. J. Nevins, *Carbohydr. Res.* **137,** 139 (1985).
[24] M. M. Smith and R. D. Hartley, *Carbohydr. Res.* **118,** 65 (1983).
[25] I. Mueller-Harvey, R. D. Hartley, P. J. Harris, and E. H. Curzon, *Carbohydr. Res.* **148,** 71 (1986).
[26] F. Gubler, A. E. Ashford, A. Bacic, A. B. Blakeney, and B. A. Stone, *Aust. J. Plant Physiol.* **12,** 307 (1985).

matography for fractionation of lignin–carbohydrate complexes on the basis of lignin content have already been published.[27,28]

Comments

Our knowledge regarding the linkages between lignin, phenolic acid, and carbohydrate is still very fragmentary. Further investigation is necessary to characterize the whole structure and function of lignin–carbohydrate complexes.

[27] N. Takahashi, J. Azuma, and T. Koshijima, *Carbohydr. Res.* **107,** 161 (1982).
[28] J. Azuma and T. Koshijima, *Mokuzai Gakkaishi* **31,** 383 (1985).

[3] [^{14}C]Lignin-Labeled Lignocelluloses and ^{14}C-Labeled Milled Wood Lignins: Preparation, Characterization, and Uses[1]

By RONALD L. CRAWFORD and DON L. CRAWFORD

Preparation

Lignin is synthesized in vascular plants by way of a branching sequence of reactions during which CO_2 is converted first to shikimic acid.[2] Shikimic acid is converted in several steps to prephenic acid, where the pathway branches to yield (in several more steps) L-tyrosine and L-phenylalanine. These two amino acids ultimately are converted to three phenylpropanoid compounds that comprise the primary building blocks of lignin: sinapyl alcohol, coniferyl alcohol, and *p*-coumaryl alcohol. Intermediates between the amino acids and the lignin alcohols include cinnaminc acid, *p*-hydroxycinnamic acid, caffeic acid, ferulic acid, 5-hydroxyferulic acid, and sinapic acid. A convenient method for labeling the lignins of a variety of plants is to feed lignifying tissues ^{14}C-labeled lignin precursors,[3] such as

[1] This work was performed at the University of Minnesota and at the University of Idaho under support from the United States National Science Foundation.
[2] T. Higuchi, M. Shimada, F. Nakatsubo, and M. Tanahashi, *Wood Sci. Technol.* **11,** 153 (1977).
[3] R. L. Crawford, "Lignin Biodegradation and Transformation." Wiley (Interscience), New York, 1981.

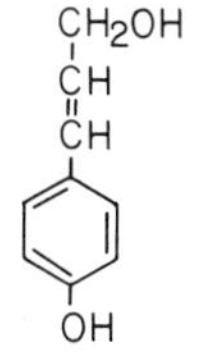

p-COUMARYL ALCOHOL

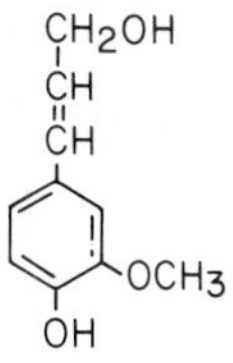

CONIFERYL ALCOHOL

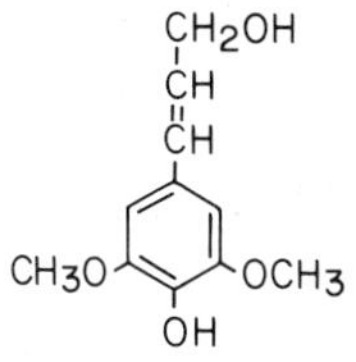

SINAPYL ALCOHOL

L-phenylalanine, cinnamic acid, or ferulic acid. These precursors usually are incorporated preferentially into a plant's lignin, and with the proper workup procedures to remove low-molecular-weight ^{14}C-labeled contaminants, one may prepare [^{14}C]lignin-labeled lignocelluloses.[4] Additional effort will yield ^{14}C-labeled milled wood lignins. The following summarizes methods used to prepare, characterize, and use these two types of specifically ^{14}C-labeled lignins.

Sources of ^{14}C-Labeled Lignin Precursors

The most readily available precursor is L-[U-^{14}C]phenylalanine, which can be purchased from numerous suppliers of radiolabeled compounds. This precursor will work well for many plants, especially rapidly lignifying hardwood species such as poplar (*Populus* sp.). However, some plants will incorporate too much L-phenylalanine into protein components of their tissues, and in these instances, precursors further along the lignin biosynthetic sequence are required. Alternate precursors that are prepared with relatively little difficulty include cinnamic acid and ferulic acid. *trans*-[^{14}C]Cinnamic acid may be prepared by enzymatic deamination of L-[^{14}C]phenylalanine, as described by Pometto and Crawford[5] and also recently summarized in Commanday and Macy.[6] ^{14}C-Side chain-labeled ferulic acid may be prepared in a single step by condensation of vanillin with [^{14}C]malonic acid.[7] If one desires to label lignins specifically in only certain parts of the polymer, then specifically labeled lignin precursors must be administered to plants. Preparation of specifically labeled ferulic acids is a recommended approach. For example, *trans*-[^{14}C-*methoxyl*]ferulic acid, ^{14}C-ring-labeled *trans*-ferulic acid, and ^{14}C-side chain-labeled *trans*-ferulic acid were prepared by Haider and colleagues and used to label

[4] D. L. Crawford and R. L. Crawford, *Appl. Environ. Microbiol.* **31,** 714 (1976).
[5] A. L. Pometto and D. L. Crawford, *Enzyme Microb. Technol.* **3,** 73 (1981).
[6] F. Commanday and J. M. Macy, *Arch. Microbiol.* **142,** 61 (1985).
[7] K. Kratzl and K. Buchtela, *Monatsh. Chem.* **90,** 1 (1959).

lignins of corn *(Zea mays)*.[8-10] Unless specific labeling is required and someone familiar with the techniques of radiosynthesis is available to the project, it generally is recommended that L-[U-^{14}C]phenylalanine, ^{14}C-side chain-labeled ferulic acid, or [U-^{14}C]cinnamic acid be employed as lignin precursors for labeling of natural lignins.

Administration of Labeled Compounds to Plants

A typical procedure for administering labeled compounds to plants involves the use of plant cuttings. The procedure has been summarized by Crawford.[3] Briefly, it involves the following steps: (1) remove a small limb or stalk from the plant of interest and cut the surface while submerged under water; (2) immerse the cut surface in a small volume of sterile, dilute buffer (pH 7) containing a dissolved ^{14}C-labeled lignin precursor of high specific radioactivity (10–50 μCi of precursor is a commonly employed amount); (3) allow the plant to absorb the solution, but do not let the container become dry; (4) allow the plant to metabolize, alternating 10–12 hr of light with 10–12 hr of darkness, until it wilts (usually 2–7 days) (this should be done in a hood or a greenhouse); (5) remove the sapwood (bark removed, if possible) from tree specimens or use the intact stalk of grasses, dry the material at 60°, and store in a desiccator. It is assumed that all of the above procedures and all procedures that follow are undertaken with standard precautions to prevent contamination of workers or work areas with radioactive solutions or plant particles. Some variations of administration of radiolabeled compounds to plant cuttings recently have been presented by Benner *et al.*[11] Other possible routes of administration of labeled compounds to plants include slow injection through hypodermic needles,[9] absorption through the severed ends of adventitious roots,[6] and simple absorption through leaf surfaces.[12]

Removing Nonlignin ^{14}C-Labeled Contaminants

Labeled plant material, prepared as described above, is ground to pass a 40-mesh sieve and then subjected to the following series of solvent extractions: water at 60–80° for 4 hr (repeated once), benzene–ethanol (1 : 1) in a Soxhlet apparatus for 8 hr (removes numerous extractives and Brauns' native lignin fraction), absolute ethanol (Soxhlet) for 4 hr (repeated until the ethanol extract is visually colorless), water at 60–80° until removal of

[8] K. Haider and J. Trojanowski, *Arch. Microbiol.* **105,** 33 (1975).

[9] K. Haider, J. P. Martin, and E. Rietz, *Soil Sci. Soc. Am. J.* **41,** 556 (1977).

[10] K. Haider and J. P. Martin, *in* "Lignin Biodegradation: Microbiology, Chemistry, and Applications" (T. K. Kirk and T. Higuchi, eds.). CRC Press, Boca Raton, Florida, 1980.

[11] R. Benner, A. E. Maccubbin, and R. E. Hodson, *Appl. Environ. Microbiol.* **47,** 381 (1984).

[12] J. Trojanowski, K. Haider, and V. Sundman, *Arch. Microbiol.* **114,** 149 (1977).

additional radioactivity becomes insignificant. The extracted material is then dried at 60° and stored in tight containers in a desiccator. An additional extraction recommended for most [^{14}C]lignin-labeled lignocellulose involves treatment of the material with proteolytic enzymes. This should always be done if the plant material employed has been labeled by feeding plants L-phenylalanine, since some of the label administered will be incorporated into plant proteins. Labeling of plant proteins usually is not a problem when lignin precursors, such as cinnamic acid or ferulic acid, are employed.[11] Protease treatment as outlined by Reid and Seifert[13] or Benner *et al.*[11] should suffice to remove most ^{14}C-labeled protein from the lignocelluloses.

Extraction procedures for removing nonlignin contaminants may vary somewhat from the procedures described above. For example, Haider *et al.*[9] used the following techniques to remove extractive from their ^{14}C-labeled corn lignocellulose: (1) plant tissues were fed labeled lignin precursors, harvested, ground in liquid nitrogen, and freeze-dried; (2) the freeze-dried powder was extracted three times with 80% boiling ethanol and dried prior to use. Trojanowski *et al.*[12] followed about the same procedures, but added an extraction with cold diluted NaOH. These procedures seemed to be reasonably effective, but the omission of a benzene extraction might be expected to leave in some types of plant tissues some nonpolar extractive removed by the more extensive solvent series. Treatments with NaOH (though under mild conditions) probably will slightly delignify lignocellulosic materials and swell the substrates' fibers, making it somewhat more susceptible to microbial attack than similar material not treated with NaOH.[3] It should be noted, however, that even extraction of plant tissues with hot water will remove some lignin and esterified phenolic compounds.[3] It is necessary to accept these small lignin loses and the loss of the Brauns' lignin fraction to ensure that the final product contains minimal nonlignin ^{14}C-labeled contaminants.

After completing the above labeling and extraction procedures, one should have a specifically lignin-labeled product containing 500–50,000 dpm of radiolabel per milligram of material.[3,14,15] The amount of ^{14}C in the substrate is determined by oxidizing several small samples of product to $^{14}CO_2$ which is trapped for counting by standard liquid scintillation counting procedures.[3,11,13–15]

[^{14}C]Lignin-labeled lignocelluloses must be characterized very carefully before use as experimental substrates. It must be confirmed that labeling of the lignin components of such materials has been specific. Techniques employed to characterize labeled lignocelluloses are discussed below.

[13] I. D. Reid and K. A. Seifert, *Can. J. Bot.* **60,** 252 (1982).

[14] J. Trojanowski, K. Haider, and A. Hüttermann, *Arch. Microbiol.* **139,** 202 (1984).

[15] I. D. Reid, *Appl. Environ. Microbiol.* **45,** 830 (1983).

^{14}C-Labeled Milled Wood Lignin: Preparation from [^{14}C]Lignin-Labeled Lignocellulose

Milled wood lignin (Björkman lignin) is a largely unmodified lignin extracted from powdered plant tissues using neutral solvents at low temperature.[16,17] Dioxane–water (9 : 1) usually is used to extract lignin from extractive-free plant tissues that have been reduced to very small particles by ball milling. The extracted lignin must be purified by a series of precipitations. Minimally, the lignin is precipitated from dichloroethane : ethanol (1 : 1) into diethyl ether to remove low-molecular-weight phenolic components. The precipitate is collected, redissolved in 90% acetic acid, and precipitated into water to remove carbohydrates and other water-soluble contaminants.[18] The average molecular weight of a milled wood lignin prepared from a wood such as spruce is 15,000–16,000.[19] Milled wood lignins usually contain a few percentage of contaminating carbohydrate, but some preparations may contain as little as 0.05% carbohydrate.[20] These lignins are thought to be representative of the bulk of lignin in a plant from which they are derived,[3] except that they may contain increased amounts of hydroxyl substituents as a result of the ball-milling process.[19]

Milled wood lignins have not been prepared from many of the plants commonly labeled with radioactive lignin precursors; therefore, care should be taken to follow very closely the original procedures of Björkman[16] and Björkman and Person[17] when preparing milled wood lignins from plants that have not been subject to this procedure before. Each new plant may have its own peculiarities as regards milled wood lignin preparation. For example, one might expect that grasses may be more difficult to ball mill than hardwoods or softwoods, since grass cell walls often contain appreciable quantities of silica. Any milled wood lignin preparation should be well characterized prior to use in microbiological or other experiments (see below).

^{14}C-Polysaccharide-Labeled Lignocelluloses

It also is possible to prepare lignocelluloses labeled specifically in the polysaccharidic components.[3–5,11] This is accomplished by feeding plant cuttings D-[U-^{14}C]glucose instead of a radiolabeled lignin precursor. The procedures described above are followed without substantial modification,

[16] A. Björkman, *Sven. Papperstidn.* **59,** 477 (1956).

[17] A. Björkman and B. Person, *Sven. Papperstidn.* **60,** 158 (1957).

[18] D. L. Crawford, *Appl. Environ. Microbiol.* **35,** 1041 (1978).

[19] H.-M. Chang, E. Cowling, W. Brown, E. Adler, and G. Mikschi, *Holzforschung* **29,** 153 (1975).

[20] K. Lundquist and R. Simonson, *Sven. Papperstidn.* **78,** 390 (1975).

other than the use of radiolabeled glucose, which is commercially available from numerous sources.

Characterization

Once the specifically labeled lignocelluloses have been prepared, they must be characterized thoroughly to confirm that incorporated ^{14}C is present only or predominantly in the substrates' lignin components. If milled wood lignins are prepared from [^{14}C]lignin-labeled lignocelluloses, these lignins also must be characterized prior to using them in microbiological or other experiments. The following summarizes recommended procedures for characterization of radiolabeled natural lignins.

Klason Hydrolysis

The distribution of label between the lignin and polysaccharide components of lignocelluloses may be estimated roughly by hydrolyzing the polysaccaridic components with acid at moderately high temperature, the so-called Klason procedure.[21,22] Amounts of ^{14}C solubilized (primarily hydrolyzed polysaccharides) are compared to amounts left in the insoluble residue (lignin). In a typical procedure,[11] 100 mg of extractive-free [^{14}C]lignocellulose of known specific activity and total disintegrations per minute (three replicates) are hydrolyzed with 72% H_2SO_4 (2 ml) at 30° for 1 hr with continuous shaking. Water (56 ml) is added (final acid, 3%) and samples are autoclaved at 15 lb/in^2 steam pressure (121°) for 1 hr. A detailed procedure is described in reference 22. After hydrolysis, samples are cooled, filtered (Whatman No. 1), and washed thoroughly with hot water. The amount of ^{14}C remaining in the acid-insoluble fraction (lignin) is determined by combustion of the entire dried residue and filter paper to $^{14}CO_2$ which is trapped in an ethanolamine-supplemented liquid scintillation fluid for counting.[4] The total number of disintegrations per minutes recovered is divided by the initial disintegrations per minute per 100 mg of [^{14}C]lignocellulose, and the decimal fraction obtained multiplied by 100 to obtain the percentage recovery of ^{14}C in the lignin fraction. The amount of ^{14}C in the soluble fraction (polysaccharides and some proteins) is determined by assaying neutralized 1-ml samples of the filtrate in an appropriate liquid scintillation fluid (e.g., 10 ml of Scintiverse II from Fisher Scientific Company or its equivalent). The total volume (hydrolyzate plus washings) must be known in order to calculate the total disintegrations per minute of ^{14}C present in the soluble fraction. The percentage recovery of

[21] B. L. Browning, "Methods of Wood Chemistry," Vol. 2. Wiley (Interscience), New York, 1967.

[22] M. J. Effland, *Tappi* **60,** 143 (1977).

^{14}C in the soluble fraction is calculated by multiplying disintegrations per minute per milliliter times the total volume. The value obtained is divided by the initial disintegrations per minute per 100 mg of [^{14}C]lignocellulose, and the resulting decimal fraction is multiplied by 100 to obtain the percentage recovery of ^{14}C in the soluble fraction. All values are reported as averages of three replicates plus/minus the standard deviation.

It should be recognized that the Klason procedure works well with some plant species (e.g., with softwoods like spruce the procedure separates lignin and other major wood components cleanly), but only moderately well with other species (e.g., hardwood lignins often are partially soluble in acid.)[3] Also, plant proteins and/or nonlignin polyphenols sometimes may condense with the lignin fraction during acid hydrolysis. The behavior of many species to acid hydrolysis has not been determined. The lignin precursor used to label the plant tissue may affect the results of the Klason procedure with some plants.[11] The use of labeled precursors farthest along the lignin biosynthetic pathway to label plants often minimizes cross-labeling of nonlignin tissues.[11] In summary, the Klason analysis should be performed to give a general idea of the distribution of label between lignin and nonlignin plant components; however, results of the procedure should be interpreted carefully and backed up by additional analyses (see below).

Chromatography of Sugars

The distribution of radioactivity in sugars released by acid hydrolysis of [^{14}C]lignocelluloses should be examined. The filtrate from the Klason hydrolysis (above) is a convenient source of hydrolyzed sugars for analysis. Cellulose thin-layer chromatography[23] is recommended. Known volumes of the neutralized lignocellulose hydrolyzate and known sugar standards are spotted on the chromatography sheets, and the sheets are developed with an appropriate solvent [e.g., butanol/pyridine/water (10/3/3)[24]]. The hydrolyzate solutions may have to be concentrated prior to chromatography.[24] After drying the chromatograms, sugar spots are located by spraying with silver nitrate reagent[24] (composed of equal volumes of 0.1 *N* $AgNO_3$ and 5 *N* NH_4OH). The chromatogram is heated at 105° for 5–10 min; sugars appear as brown spots on the chromatogram. The spots are scraped from the sheets for counting by liquid scintillation counting techniques. Radioactivity in spots may be determined most accurately by combustion and trapping of $^{14}CO_2$. Sugars examined should always include the primary sugars found in cellulose and hemicellulose: glucose, xylose, and mannose. If a plant being studied contains other sugars in significant

[23] D. W. Vomhof and T. C. Tucker, *J. Chromatogr.* **17,** 300 (1965).

[24] R. L. Crawford and D. L. Crawford, *Dev. Ind. Microbiol.* **19,** 35 (1978).

amounts, these also should be examined for radioactivity. From these analyses, one should obtain a value for the percentage of total radioactivity in the [^{14}C]lignin-labeled lignocellulose contained in sugars (normally <2–4%)[3,11] and the amount of radioactivity associated with particular sugars.

Protease Treatment

[^{14}C]Lignin-labeled lignocelluloses may have significant amounts of radioactivity associated with their protein components, particularly if the plant under study contains appreciable quantities of protein and it was labeled by feeding the plant L-[^{14}C]phenylalanine[3] (precursors like cinnamic acid or ferulic acid rarely label protein significantly[11]). In all cases, however, the possible labeling of proteins should be examined. This may be accomplished by digesting powdered [^{14}C]lignocellulose with one or more proteases and by analyzing solubilized material for radioactivity. For example, Benner *et al.*[11] treated their lignocelluloses (100 mg) with a pepsin solution (4 ml, 1% pepsin in 0.1 *N* HCl at 40° for 20 hr). After treatment, the insoluble residue was washed thoroughly and the washings combined with filtrate for determination of released radioactivity. Reid and Seifert also have described an acceptable protease treatment.[13] If significant amounts of protein are observed to have been labeled, then the whole batch of [^{14}C]lignocellulose should be subjected to protease digestion, or the original labeled precursor fed to the plant should be changed to one that yields less cross-labeling of proteins.

Examination of Esterified ^{14}C-Labeled Components

Some lignins, particularly grass lignins, contain aromatic acids like *p*-coumaric acid and ferulic acid esterified to the periphery of the lignin macromolecule. Though these esterified acids might be considered to be part of the lignin molecule, they probably are more readily degradable by microorganisms than the bulk of the polymer. Thus, [^{14}C]lignin-labeled lignocelluloses should be characterized as to the extent of esterification of aromatic acids to the lignin backbone. To release esterified acids for determination of released radioactivity, a mild base hydrolysis is used.[25] Typically, about 40 mg of ^{14}C-labeled material is hydrolyzed by stirring in 10 ml of 1 *N* NaOH at 20° for 24–48 hr (see reference 3). The suspension then is filtered, the residues are washed thoroughly with water, and the washings and filtrate are combined. An aliquot of this solution (e.g., 1 ml) is added to liquid scintillation fluid and its radioactivity is determined in a liquid scintillation counter. This gives an estimate of the maximal amount

[25] T. Higuchi, Y. Ito, M. Shimada, and I. Kawamura, *Phytochemistry* **6,** 1551 (1967).

(some lignin also will dissolve) of radioactivity in the original lignocellulose that might be contained as esterified components. The remaining solution is acidified to pH 2 (HCl, 6 *N*) and is extracted with diethyl ether (equal volumes, three times). The combined ether extracts are evaporated to dryness, and the residue obtained is redissolved in a small, known volume of ethanol or acetone. A few milligrams of *p*-coumaric acid and ferulic acid are added as carriers to trap radioactive acids freed from the lignocellulose by base hydrolysis. Known aliquots of this solution are spotted on thin-layer chromatography sheets, which are developed in an appropriate solvent (*n*-butanol/acetic acid/water, 62/15/23, works well when cellulose thin-layer sheets are employed[11,26]; silica gel sheets also perform well, using toluene/ethyl acetate/formic acid, 94/5/1). Spots of *p*-coumaric acid and ferulic acid are scraped from the chromatograms and trapped radioactivity is counted by liquid scintillation procedures. This gives an estimate of the amount of radioactivity in the original lignocellulose that was esterified as *p*-coumaric acid and ferulic acid. The best [^{14}C]lignin-labeled lignocellulose preparations will contain very little in the way of radioactive esterified components (<2–3% of total radioactivity in the material). If amounts of esterified ^{14}C-labeled acids are high (e.g., >3–5%), the whole batch of lignocellulose should be treated with dilute base to remove these compounds, though some core lignin may be lost to this procedure.

Radioactivity in Lignin Aldehydes

It is desireable to examine each [^{14}C]lignin-labeled lignocellulose to confirm that incorporated ^{14}C is located predominately within basic lignin substructures. This may be accomplished by degrading the lignin component of the substrate under alkaline oxidizing conditions to the level of lignin aldehydes.[27,28] Oxidations are performed in stainless-steel bombs. About 50 mg of [^{14}C]lignocellulose is added to a small bomb along with 5 ml of 2 *N* NaOH and 0.5 ml of nitrobenzene. The bomb then is heated to 180° for 2 hr. The oxidized residue is cooled, filtered, and the filtrate extracted three times with equal volumes of dichloromethane (to remove nitrobenzene). The solution then is adjusted to pH 2.0–2.5 (6 *N* HCl), saturated with NaCl, and extracted three times with equal volumes of dichloromethane. Combined dichloromethane extracts are evaporated to dryness and the residue obtained is redissolved in a small, known volume

[26] R. D. Hartby, *Phytochemistry* **12,** 661 (1973).

[27] W. S. Gardner and D. W. Menzel, *Geochim. Cosmochim. Acta* **38,** 813 (1974).

[28] K. V. Sarkanen and C. H. Ludwig (eds.), "Lignins: Occurrence, Formation, Structure, and Reactions." Wiley (Interscience), New York, 1971.

of ethanol. Aliquots of the ethanol solution are subjected to thin-layer chromatography according to the procedures of Brand[29] (silica gel sheets; solvent, *n*-hexane/isoamyl alcohol/acetic acid, 100/16/0.25) to separate the primary lignin alcohols (vanillin, *p*-hydroxybenzaldehyde, and syringaldehyde) which have been freed by nitrobenzene oxidation. Spots are scrapped from the chromatograms and the radioactivity in each aldehyde, as a percentage of total radioactivity in the lignocellulose, is determined by liquid scintillation counting procedures. The best [^{14}C]lignin-labeled lignocelluloses will contain much of their radioactivity in these aldehydes. If label is distributed within all three aldehydes, then one may have confidence that radioactivity has been distributed through the lignin molecule.[11]

An Ideal [^{14}C]Lignin-Labeled Lignocellulose

Under ideal labeling conditions, most often obtained using rapidly lignifying, low-protein plant species such as certain trees,[9] the [^{14}C]lignin-labeled lignocellulose produced should contain approximately the following characteristics, based on the analyses above: (1) specific radioactivity, about 10,000–50,000 dpm/mg; (2) radioactivity released by Klason hydrolysis, <5% with softwoods but variable with other plants; (3) radioactivity released by protease treatment, <2–3%; (4) radioactivity released by mild saponification of esters, <2–3%; (5) distribution of radioactivity in lignin aldehydes, 50–75%, depending on labeling pattern of precursors fed to plants, and distributed within all three aldehydes; and (6) radioactivity within hydrolyzed wood sugars, <2–3%. If substantial amounts of radioactivity are observed in proteins or esterified acids, these conditions can be remedied to a large extent by treating the whole preparation with protease and/or dilute base. If substantial amounts of radioactivity are observed in wood sugars and little radioactivity is found in lignin aldehydes, such preparations should not be used. It is of utmost importance to remember that every plant behaves differently during labeling procedures and labeling may vary greatly depending on which lignin precursor is employed in plant feedings. Every single preparation must be characterized as completely as possible.

Characterization of ^{14}C-Labeled Milled Wood Lignins

If milled wood lignins are prepared carefully according to the procedures of Björkman,[16,17] they should be quite pure and representative of the bulk of lignin in the plants from which they were isolated. However, milled wood lignins prepared from [^{14}C]lignin-labeled lignocellulose should be

[29] J. M. Brand, *J. Chromatogr.* **21**, 424 (1966).

characterized at least by the following criteria: (1) disintegrations per minute ^{14}C per milligram, (2) percentage of ^{14}C associated with lignin aldehydes, (3) carbohydrate content, (4) radioactivity associated with any contaminating sugars, (5) molecular-weight distribution and its correspondence with distribution of radioactivity, and (6) radioactivity associated with esterified aromatic acids. These determinations can be made, respectively, as follows: (1) combustion of a few milligrams of lignin and trapping and counting of $^{14}CO_2$, (2) isolation of lignin aldehydes and determination of radioactivity associated with each following nitrobenzene oxidation of a small lignin sample (see above), (3) analysis of reducing sugars released on acid hydrolysis against a standard curve prepared using glucose,[21,30] (4) thin-layer chromatography of sugars released on acid hydrolysis (see above, *Chromatography of Sugars*), (5) gel-exclusion chromatography[31] on Sephadex LH-60 using dimethylformamide as solvent with determination of radioactivity present in each column fraction, and (6) dissolution of a known amount of milled wood lignin in 1 *N* NaOH followed by thin-layer chromatography of solvent extracts of the solution (see above, *Examination of Esterified* ^{14}*C-Labeled Components*).

An excellent ^{14}C-labeled milled wood lignin should have a molecular weight of 15,000–16,000 and contains at most a few percentage carbohydrate.[3] Radioactivity should be associated with the lignin polymer (as evidenced by gel chromatography and analyses of lignin aldehydes), and hydrolyzed sugars and esterified aromatic acids should contain no significant ^{14}C.

Characterization of ^{14}*C-Polysaccharide-Labeled Lignocelluloses*

As discussed previously, specifically polysaccharide-labeled lignocelluloses can be prepared by feeding plants D-[U-^{14}C]glucose instead of a labeled lignin precursor.[3,4,11] These preparations must be characterized with the same rigor as [^{14}C]lignin-labeled lignocelluloses. The procedures described above are appropriate, except that they are used to demonstrate that ^{14}C is localized in the plants' polysaccharides, not lignin.

Uses

As discussed above, lignocelluloses may be prepared such that they contain ^{14}C specifically in either their lignin or carbohydrate components. In addition, ^{14}C-labeled milled wood lignins (largely carbohydrate-free and

[30] W. E. Moore and D. B. Johnson, "Procedures for the Chemical Analysis of Wood and Wood Products. Forest Products Laboratory Protocols." U.S. Department of Agriculture, Forest Products Laboratory, Madison, Wisconsin, 1967.

[31] W. J. Connors, S. Sarkanen, and J. L. McCarthy, *Holzforschung* **34,** 80 (1980).

unmodified lignins extracted from powdered plant tissues) may be prepared from [^{14}C]lignin-labeled lignocelluloses. These radiolabeled preparations are very useful in studies of microbiological lignin degradation and transformation. Here are summarized some of the experimental questions that have been approached using radiolabeled natural lignins and lignocelluloses.

Mineralization of Radiolabeled Lignins by Microorganisms

Procedures employed to study mineralization of ^{14}C-labeled lignins and lignocelluloses to $^{14}CO_2$ are similar to procedures long used by microbial ecologists to study biodegradation of organic compounds by the microfloras of soil and water.[24] Samples of soil or water, or pure microbial cultures in a defined or complex medium, are incubated in the presence of the ^{14}C-labeled substrate under experimental conditions appropriate to the questions being asked. Incubation vessels are equipped with a means to trap any evolved $^{14}CO_2$ or $^{14}CH_4$ for counting by liquid scintillation counting techniques.[32] A common experimental apparatus would be a glass vessel (usually a flask or test tube) equipped with gassing ports which allow either continuous or intermittent flushing of the vessel contents with air, nitrogen, or other gas phase. This may be accomplished with or without maintenance of aseptic conditions. Evolved $^{14}CO_2$ is trapped either in a NaOH solution (usually 1 *N*) or directly in a scintillation cocktail containing an organic base, such as ethanolamine[2-4] or hyamine hydroxide.[33] Evolved $^{14}CH_4$ may be trapped directly in most scintillation fluids, as it is soluble in organic solvents.[6] If a NaOH solution is used to trap $^{14}CO_2$, an aliquot may be counted directly [e.g., the combination of 1 ml NaOH, 4 ml H_2O, and 10 ml of Aquasol-2 (New England Nuclear) to form a gel works very nicely]. Alternatively, the NaOH can be acidified to release the trapped $^{14}CO_2$ which is retrapped by purging through an ethanolamine-supplemented scintillation fluid. Two traps in the line and scintillation fluid without added base, followed by scintillation fluid containing ethanolamine (or a NaOH trap), permit trapping and quantification of both radiolabeled carbon dioxide and methane from a single experimental vessel.[34] Evolved radiolabeled gases also may be quantitated by gas chromatography/gas-flow proportional counting techniques.[35] Special biometer flasks have been designed for studying evolution of $^{14}CO_2$ from soils

[32] T. K. Kirk, W. J. Connors, R. D. Bleam, W. F. Hackett, and J. G. Zeikus, *Proc. Natl. Acad. Sci. U.S.A.* **72,** 2515 (1975).

[33] J. S. Hubbard, *Bull Ecol. Res. Comm., NFR* **17,** 199 (1973).

[34] J. Ferry and R. S. Wolf, *Arch. Microbiol.* **107,** 33 (1976).

[35] W. F. Hackett, W. J. Connors, T. K. Kirk, and J. G. Zeikus, Appl. Environm. Microbiol. **33,** 43 (1977).

amended with ^{14}C-labeled substrates, and several of these should work well for many applications.[36,37] Examples of vessels used to study biodegradation of radiolabeled substrates have been summarized by Atlas and Bartha.[38]

Transformation of Radiolabeled Lignocelluloses by Microorganisms: Biotransformation Products

Microorganisms may transform lignin during decay of lignocellulose without oxidizing it completely to carbon dioxide. In such cases, biotransformation products may accumulate. When one employs [^{14}C]lignin-labeled lignocelluloses in biotransformation experiments, ^{14}C-labeled products accumulate, and these may be isolated and studied. ^{14}C-labeled low-molecular-weight phenols and other water-soluble lignin biotransformation products may be isolated from incubation media and examined by standard techniques of gas chromatography, thin-layer chromatography, and gel-exclusion chromatography.[39,40] Some microorganisms, particularly the actinomycetes, solubilize lignin as they degrade lignocellulose, producing a high-molecular-weight modified lignin known as acid-precipitable polymeric lignin (APPL).[41] This polymer material may be isolated in radiolabeled form and used for additional types of biotransformation experiments.[40]

General Uses for Radiolabeled Lignins and Lignocelluloses

Radiolabeled lignins and lignocelluloses have been applied successfully to a large variety of research questions. These include determination of the range of microbial taxa that are able to mineralize or transform lignin,[3,42] kinetics of lignin biodegradation in natural environments,[11] enumeration of lignin-degrading microorganisms in natural environments,[43] examination of the physiology of lignin decay by white-rot fungi,[15] examination of

[36] R. Barth and D. Pramer, *Soil Sci.* **100,** 68 (1965).

[37] J. T. Marvel, B. B. Brightwell, J. M. Malik, M. L. Sutherland, and M. L. Rueppl, *J. Agric. Food Chem.* **26,** 1116 (1978).

[38] R. M. Atlas and R. Bartha, "Microbial Ecology: Fundamentals and Applications." Addison-Wesley, Reading, Massachusetts, 1981.

[39] I. O. Reid, G. D. Abrams, and J. M. Pepper, *Can. J. Bot.* **60,** 2357 (1982).

[40] A. L. Pometto III and D. L. Crawford, *Appl. Environ. Microbiol.* **51,** 171 (1986).

[41] D. L. Crawford, A. L. Pometto III, and R. L. Crawford, *Appl. Environ. Microbiol.* **45,** 898 (1983).

[42] K. Haider, J. Trojanowski, and V. Sundman, *Arch. Microbiol.* **119,** 103 (1978).

[43] R. L. Crawford, D. L. Crawford, C. Olofsson, J. Wikstrom, and J. W. Wood, *J. Agric. Food Chem.* **25,** 704 (1977).

the biodegradability of industrial lignins (prepared from [^{14}C]lignin-labeled lignocelluloses),[44] comparisons of the biodegradabilities of different species of wood lignins,[3] and biodegradation of lignin within the guts of invertebrates.[45]

[44] D. L. Crawford, S. Floyd, A. L. Pometto III, and R. L. Crawford, *Can. J. Microbiol.* **23,** 434 (1977).

[45] J. H. A. Butler and J. C. Buckerfield, *Soil Biol. Biochem.* **11,** 507 (1979).

[4] Preparation of Dioxane Lignin Fractions by Acidolysis

By BERNARD MONTIES

General Principles

A convenient method for the preparation of lignin fractions from biomass is provided by the combination of mild solvolytic conditions with a good lignin solvent.

Arylalkyl ether bonds, the most frequent intermonomeric linkage occurring in lignins, are cleaved under acid conditions[1] with the formation of low-molecular-weight phenols and depolymerization of the lignin network and/or the lignin–carbohydrate complexes. The term acidolysis has been coined[1] for designating the solvolysis of lignin by refluxing with 0.2 *M* hydrogen chloride in a dioxane–water mixture (9 : 1, v : v). Dioxane is a particularly interesting isolating agent of lignin fractions because of its solubility parameter, which is very close to the optimum value for isolated lignin preparations.[2] Systematic studies have been thus carried out using a 9 : 1 mixture of dioxane–water and dilute hydrochloric acid for isolation of lignin both at ambient[3] and high temperature.[4] Preparation of dioxane acidolysis lignins has the main advantage of providing, in the shortest time, lignin fractions in high yields and in a form suitable for further chemical and biotechnological investigations. These lignins generally contain only a very low percentage of associated polysaccharides. However, the procedure suffers from the disadvantage that condensation reactions occur inside the

[1] E. Adler, J. M. Pepper, and E. Erikson, *Ind. Eng. Chem.* **49,** 1391 (1957).

[2] C. Schuerch, *J. Am. Chem. Soc.* **74,** 5061 (1952).

[3] K. Freudenberg, *in* "Moderne Methoden der Pflanzenanalyse," Vol. 3, p. 509. Springer-Verlag, Berlin, Federal Republic of Germany, 1955.

[4] J. M. Pepper and M. Siddiqueullah, *Can. J. Chem.* **39,** 390 (1961); and J. M. Pepper, P.E.T. Baylis, and E. Adler, *Can. J. Chem.* **37,** 1241 (1959).

lignin network during its solvolysis and increase to a significant extent with increasing temperature and acidity. Another disadvantage may be the addition of chlorine to dioxane lignin fraction[3] reported approximately 1.5–2%.

Two procedures for preparation of dioxane lignins are reported with their quantitative and qualitative comparisons.

Method

Principle

Two procedures for isolation have been adapted: the first one is carried out at ambient temperature according to Freudenberg,[3] and the second is performed at reflux temperature, approximating the procedure of Pepper *et al.*[4] Acidolysis conditions are, however, similar to those chosen by Lapierre *et al.*[5] for characterization of lignins by analysis of acidolysis products.

Reagent and Materials

1,4-Dioxane. This is a commercial product (RP Normapur grade, PROLABO) stabilized against peroxidation by addition of about 0.03% of di-*tert*-butyl-2,6-*p*-cresol and used without further purification.[6]

Hydrochloric Acid. This is a commercial product (PROLABO M Normadose) titrated to 2 *N* in water.

Biomass (Wood or Straw) Samples. Samples of air-dried xylem (wood without bark) or culm (stems from straw without leaves) were ground with a rotating knife mill until the entire sample was able to pass through a 0.5-mm screen. The powder was extracted exhaustively with toluene–ethanol (2 : 1, v : v), then with ethanol, and finally with water until the extracts were colorless. The solvent-free sample, parietal residue (PR), was then freeze-dried.

Procedure

The procedure followed in this laboratory for isolation of dioxane lignin preparation on a gram scale uses 0.2 *N* HCl in dioxane, freshly prepared before use.

[5] C. Lapierre, C. Rolando, and B. Monties, *Holzforschung* **37,** 189 (1983).

[6] Comparative acidolysis experiments have been done with RP grade (PROLABO) dioxane and with dioxane purified according to A. I. Vogel, *in* "Practical Organic Chemistry," 3rd Ed., p. 177. Longmans, London, 1964. Under these conditions, no qualitative changes were observed among monomeric acidolysis products. When using purified dioxane, the total yield in acidolysis monomers was, however, increased by about 10%. Lignin preparations were performed with commercial dioxane only.

Acidolysis Procedure for Reflux Dioxane Lignin (RDL). Twenty grams of PR is stirred with 250 ml of reagent in a 1-liter, two-necked flask fitted with a reflux condenser and a nitrogen bubbler. After permeation, an additional 250 ml of reagent is added and a slow stream of nitrogen is bubbled through the suspension for 10 min before heating. Reflux is maintained for 30 min under nitrogen without bubbling. The reaction mixture is allowed to cool under nitrogen and is then rapidly filtered through a large 13-cm diameter, sintered glass funnel (coarse number 3 porosity). The residual pulp is washed twice with 100 ml of reagent.

The filtrate is neutralized with an excess of sodium bicarbonate;[7] the excess is removed by filtration. The dioxane–water filtrate is concentrated under reduced pressure at 40° until precipitation commences. Because the gummy mass of lignin tends to precipitate, just enough dioxane–water mixture (9 : 1, v : v) is added to obtain a clear solution. The final volume obtained ranges between 20 and 25 ml.

Acidolysis Procedure for Ambient Temperature Dioxane Lignin (ADL). A 50-g sample of PR is suspended in 1 liter of reagent in a 2-liter glass-stoppered flask, under nitrogen with mild magnetic stirring. Acidolysis is allowed to proceed for 20 days at 20°,[3] or at room temperature in the dark. Isolation and concentration of acidolysis lignin are performed as for RDL.

Purification of Dioxane Lignins. Solutions of dioxane lignin, RDL or ADL, are precipitated by injection, through an hypodermic syringe as a fine stream, into 2 liters of water very vigorously stirred with a magnetic stirrer (1000 rpm) in a large beaker (diameter, 15 cm). When the solution is injected in the direction of flow along the beaker wall, a very fine haze (precipitate) is obtained which sediments quickly, leaving a clear solution.[8] Precipitated dioxane lignin is recovered by centrifugation (1 hr, 2500 *g*, ambient temperature). Pellets are dissolved in dioxane (50 ml) and lignin is again precipitated by injection into diethyl ether (2 liters). Precipitated lignin is recovered, after decantation of the clear mother solution, by centrifugation in closed tubes at low temperature between −10 and −15°.[9]

The resulting lignin preparations are freeze-dried and stored in the dark at a temperature of about −18°.

Discussion

The preparation procedure for dioxane lignin by acidolysis under reflux allows for the isolation of RDL from gymnosperm (*Abies* sp.) wood, from

[7] About 20 g of $NaHCO_3$ (RP, Normapur, PROLABO) is sufficient for neutralization which can be controlled by pH measurement after dilution of the filtrate with water (1 : 4, v : v).

[8] A rapid coagulation of lignin flocs is obtained regularly only if the volume of injected dioxane is significantly lower than 5% of the volume of precipitating water.

[9] For safety reasons, purification in diethyl ether may be replaced by thorough washing with water and repeated centrifugations.

angiosperm (*Populus* sp.) wood, and from gramineae (*Triticum* sp.) straw; the total yields are about 15, 30, and 65% Klason lignin, respectively. These data are in complete agreement with the results of Pepper and Wood.[10]

Yields for ADL isolated from the same sample as RDL were found to be significantly lower: the ADL yield was about 15% of Klason lignin for poplar and only 3% in the case of fir. The same differences have been reported for another angiosperm, *Betula* sp., and gymnosperm, *Picea* sp., by Desmet.[11]

Qualitative differences previously reported by UV[12] and IR[13] spectroscopy between RDL lignins of these three types of plants were also observed.

Comparison of poplar ADL and RDL has been done in this laboratory. IR spectra analysis failed to show very clear differences between 1800 and 1000 cm^{-1}, particularly for the monomeric composition at 1330, 1270, 1120, and 1030 cm^{-1} and total carbonyl groups at 1710 and 1660 cm^{-1}. However, significant differences were recorded between UV spectra of ADL and RDL, when lignins were reduced with $NaBH_4$. The occurrence of nonreducible carbonyls in RDL was suggested by the presence of a shoulder between 320 and 400 nm, this shoulder being absent for ADL.

These latter data are indicative of the occurrence of discrete differences, at least in the case of poplar, between the two types of preparations; they confirm the "heterogeneity represented in the group of dioxane acidolysis lignins."[14]

Recent data from gel permeation over μBondagel E 125 and E 500 columns, using pure tetrahydrofuran as solvent, have given very clear evidence of the same type of heterogeneity inside these corresponding RDL and ADL poplar lignin preparations. The use of a high-speed diode array ultraviolet detector, according to Nicholson *et al.*,[15] allows one to assess the variations in the UV spectra (220–400 nm) of the eluted dioxane lignin fractions in relation to their elution volume. Very significant differences were noted between the shape of the ultraviolet spectra of the fractions eluted before and after the maximum of the elution curve, even when a standard unimodal distribution was obtained by size-exclusion chromatography of RDL and ADL preparations. Fractions eluting after the maximum showed significantly different absorbancy shoulders between 300 and

[10] J. M. Pepper and P. D. S. Wood, *Can. J. Chem.* **40,** 1026 (1962).

[11] J. Desmet, Ph.D. thesis. Université Scientifique et Médicale, Grenoble, France, 1971.

[12] D. F. Arseneau and J. M. Pepper, *Pulp Pap. Mag. Can.* **66,** T-415 (1965).

[13] H. L. Hergert, *J. Org. Chem.* **25,** 405 (1960).

[14] K. V. Sarkanen, *in* "Lignins: Occurrence, Formation, Structure, and Reactions" (K. V. Sarkanen and C. H. Ludwig, eds.), p. 187. Wiley (Interscience), New York, 1971.

[15] J. C. Nicholson, J. J. Meister, D. R. Patil, and L. R. Field, *Anal. Chem.* **56,** 2447 (1984).

360 nm; such changes were not found for fractions eluted before the peak maximum (B. Monties, unpublished results). These results emphasize the necessity to consider carefully the isolation procedure of dioxane lignin fractions in relation to qualitative differences in their molecular and macromolecular properties.

[5] Acid-Precipitable Polymeric Lignin: Production and Analysis

By DON L. CRAWFORD and ANTHONY L. POMETTO III

The initial product of lignin degradation by *Streptomyces viridosporus* is a water-soluble polymeric intermediate, acid-precipitable polymeric lignin (APPL).[1] *Streptomyces viridosporus* T7A (ATCC No. 39115) degrades lignin by an oxidative depolymerization mechanism involving the cleavage of intermonomeric β-ether linkages and the introduction of phenolic hydroxyl, α-carbonyl, and carboxylic acid groups into the lignin polymer.[2,3] APPLs also contain some nonlignin components including organic nitrogen (2–3%), carbohydrate (2–3%), and ash (4–5%).

The net result of enzymatic attack on lignin by *S. viridosporus,* when the culture is growing in solid-state fermentations on lignocelluloses such as those derived from grasses, is the generation of water-soluble polymeric lignin fragments (APPLs) of high molecular weight (>80,000) which can accumulate to levels approaching 30% or more of the initial lignin present in the lignocellulosic substrate.[1,4] It appears that *S. viridosporus* produces APPL primarily as a mechanism for gaining access to polysaccharides in the lignocellulose.[4,5] These polysaccharides are degraded with active polysaccharidases.[6] While the APPLs are further metabolized after their release, they are degraded very slowly by the *Streptomyces* that produce them.[5]

[1] D. L. Crawford, A. L. Pometto III, and R. L. Crawford, *Appl. Environ. Microbiol.* **45,** 898 (1983).

[2] D. L. Crawford, M. J. Barder, A. L. Pometto III, and R. L. Crawford, *Arch. Microbiol.* **131,** 140 (1982).

[3] D. L. Crawford, A. L. Pometto III, and L. A. Deobald, "Recent Advances in Lignin Biodegradation Research" (T. Higuchi, H.-M. Chang, and T. K. Kirk, eds.), pp. 78–95. Uni, Tokyo, 1983.

[4] J. R. Borgmeyer and D. L. Crawford, *Appl. Environ. Microbiol.* **49,** 273 (1985).

[5] A. L. Pometto III and D. L. Crawford, *Appl. Environ. Microbiol.* **51,** 171 (1986).

[6] D. L. Crawford, T. M. Pettey, B. M. Thede, and L. A. Deobald, *Biotechnol. Bioeng. Symp.* **14,** 214 (1984).

APPLs can readily be recovered from aqueous culture supernatants or from hot water extracts of solid-state fermentation residues by acidifying them to below pH 5.[1]

In general, APPLs are more oxidized and more phenolic than native lignin;[7] however, the specific chemistry of APPLs produced by different *Streptomyces* may differ markedly. For example, APPLs produced by *S. viridosporus* are still relatively ligninlike in structure, whereas those produced by *Streptomyces badius* (ATCC No. 39117) are so modified that they bear little resemblance to the lignin from which they were derived.[4] Therefore, it is important that APPLs or APPL-like products be thoroughly characterized chemically when one is examining lignin degradation in previously unstudied microorganisms.

This chapter describes the procedures for growing two distinctly different lignin-degrading *Streptomyces* under conditions optimal for lignocellulose degradation and APPL production by each. One species, *S. viridosporus,* is grown in solid-state fermentations, and the second, *S. badius,* is grown in submerged cultures after preincubation for a period of time as a solid-state fermentation.[4] Direct mycelial contact with the substrate is required for growth of both of these *Streptomyces* on lignocellulose. Also, the procedures for harvesting and chemically characterizing residual partially degraded lignocelluloses and the APPLs produced as a result of lignin depolymerization are described. After their harvest by aqueous extraction from partially decomposed lignocellulosic residues, the APPLs and the residual insoluble lignocelluloses are analyzed for their Klason lignin, carbohydrate, protein, and ash contents. The APPLs are further characterized chemically by degradative procedures including acidolysis, permanganate oxidation, and alkaline ester hydrolysis. These procedures break the APPL down to the level of single-ring aromatics which can be analyzed by gas–liquid chromatography (GLC).[1,2,4] When native lignin (Björkman lignin purified from an undegraded lignocellulose) is used as a comparative control, these analyses allow one to elucidate some of the key chemical changes occurring in lignin as a result of its degradation by ligninolytic microorganisms.[1,2]

Preparation of Inoculum and Lignocellulose Substrates

Lignocellulose Substrate Preparation. The lignocellulosic substrates most readily degraded by actinomycetes are those prepared from grasses, although hardwood and softwood lignocelluloses are also degraded by these microorganisms.[3] The plant material to be used is first air-dried at

[7] A. Björkman, *Sven. Papperstidn.* **59,** 477 (1956).

80–100°, and then it is ground in a Wiley mill to pass a 40-mesh screen. The ground material is next extracted with hot water (using a Büchner funnel and Whatman No. 54 filter paper), air-dried, and then extracted with benzene–ethanol (1 : 1) followed by ethanol only (using a Soxhlet extractor). Following a final cold water extraction to remove the ethanol and an additional hot water extraction, the extractive-free lignocellulose is air-dried as above prior to use. At each step discussed above, the lignocellulose is extracted until the filtrate is clear, and deionized water is used for the water extractions.

For solid-state fermentations using corn stover *(Zea mays)* lignocellulose, 5 g of dry, extracted lignocellulose is placed into a 1-liter reagent bottle and autoclaved for 1 hr uncovered and then for 15 min more after addition of a cotton plug. For submerged cultures, 5 g of extracted corn lignocellulose is added to a 2-liter flask and autoclaved as above. As the amount of lignocellulse to be autoclaved increases, autoclaving time must also be increased.

Inoculum Preparation. Actinomycetes such as *S. viridosporus* and *S. badius* are maintained at 4° as sporulated cultures on agar slants of a rich organic medium, such as yeast extract–malt extract agar or other suitably rich medium.[8] The spores from a single stock slant are used to inoculate 1 liter of sterile liquid medium containing a mineral salts solution (5.3 g of Na_2HPO_4, 1.98 g of KH_2PO_4, 0.2 g of $MgSO_4 \cdot 7H_2O$, 0.2 g of NaCl, 0.05 g of $CaCl_2 \cdot 2H_2O$, plus 1.0 ml of trace element solution[8] per liter of deionized H_2O; pH 7.1–7.2) and 0.6% (w : v) yeast extract (Difco Laboratories, Detroit, Michigan). The culture is then incubated aerobically by shaking at 100 rpm at 37° for 48–72 hr, at which time the cells will have entered late logarithmic growth phase. This late log-phase culture will consist of spherical hyphal balls, which can be concentrated by simply allowing them to settle to the bottom of the flask.

Cultivation of Actinomycetes on Lignocellulose

Solid-State Fermentation. The solid-state fermentation described here is composed of a lignocellulosic substrate dampened to its water holding capacity with an inoculum of log-phase *Streptomyces* cells suspended in mineral salts, yeast extract growth medium. It is important that the incubating lignocellulose not be water saturated and that the thickness of dampened lignocellulose in the culture vessel not be too great so that good oxygen transfer rates can be maintained during culture incubation. The desired lignocellulose thickness is about 0.1 cm, which can be achieved by

[8] T. G. Pridham and G. Gottlieb, *J. Bacteriol.* **56,** 107 (1948).

spreading the inoculated lignocellulose over the inside walls of the reagent bottle or across the bottom of a flask. Establishing the fermentations at the water holding capacity of the substrate is more complex. For different lignocellulose types, water holding capacity is determined by measuring the volume of 1 g of dried extracted (40-mesh) lignocellulose. This density value (milliliters per gram) will vary for different ground, extracted lignocelluloses depending on the plant source used. For example, the density of a typical corn stover lignocellulose is 6.6 ml/g. For spruce lignocellulose *(Picea pungens),* the value is 5.0 ml/g, and for peanut hull lignocellulose, it is 3.0 ml/g.[9] Therefore, the volume of liquid inoculum added per gram of initial lignocellulose is specific for each type of lignocellulose due to substrate density differences, and this volume must, therefore, be calculated. The following formula is used.

$$V = 1.5\,(D)$$

where V is the volume (milliliters) of the inoculum and D is the density (milliliters per gram) of the dried ground and extracted lignocellulose.[9]

For the present discussion, a 1-liter reagent bottle (Corning No. 1460 or Nalgene No. 2006-0032) is considered the vessel of choice. When the lignocellulose is sterilized by the procedure above, five to six marbles can be added to the vessel along with the lignocellulose. These marbles will help later after inoculation in the dispersal of lignocellulose over the inside surface of the vessel. The sterile lignocellulose is inoculated with a specific volume (determined as above) of log-phase cell suspension, and then it is rolled on its side to disperse the lignocellulose evenly over the inside walls of the vessel. One can also calculate the amount of lignocellulose to add to the reagent bottle in order to achieve a final average 0.1-cm thickness of substrate by using the formula.

$$W = \frac{2dh\,(0.1\ \text{cm})}{2.5/D}$$

where W is the weight (grams) of lignocellulose to be used, d is the diameter (centimeters) of the cylindrical vessel, h is the height (centimeters) of the cylindrical vessel, 0.1 cm is the desired final thickness of the dampened lignocellulose on the internal surface, and D is the density (milliliters per gram) of the dried extract lignocellulose.[9] Overall, for a specific lignocellulose, the weight of lignocellulose per culture vessel, the volume of inoculum per gram of lignocellulose, and the size and shape of the culture vessel should all be considered before one begins producing APPL by growing *Streptomyces* on lignocellulose.

[9] D. L. Crawford, A. L. Pometto III, and R. L. Crawford, *Biotechnol. Adv.* **2,** 217 (1984).

For inoculation of corn stover lignocellulose with *S. viridosporus* cells, the following protocol is recommended. A 50-ml volume of gravity-concentrated cell suspension is aseptically pipetted from the bottom of the inoculum growth flask into a 1-liter reagent bottle containing 5.0 g of sterile ground lignocellulose, also containing six marbles. The vessel is then rolled on its side to spread the lignocellulose, and then it is incubated for 6 weeks at 37° in a humid incubator.[1,4] Uninoculated sterile controls are incubated alongside inoculated cultures. Controls are carried through the same treatment as inoculated cultures, but receive 50 ml of sterile 0.6% (w : v) yeast extract–mineral salts medium.

Submerged Culture System. With *Streptomyces* such as *S. badius*, preincubation as a solid-state fermentation is required to ensure that, during its initial growth, this filamentous organism becomes firmly attached to the lignocellulose substrate.[4] Therefore, the same considerations as to inoculation volumes, substrate densities, etc., employed with solid-state fermentations also apply here. For example, 5 g of sterile corn stover lignocellulose is appropriate for a 2-liter flask. The sterile lignocellulose is inoculated with 50 ml of a gravity-concentrated, log-phase cell suspension of *S. badius* cells. Then the inoculated lignocellulose is spread over the bottom surface of the flask and incubated as a solid-state fermentation for 1–2 weeks at 37° in a humid incubator. Then, 1 liter of sterile 0.6% (w : v) yeast extract–mineral salts medium is added, and the flask is incubated aerobically at 37° with shaking at 100 rpm for a total incubation time of up to 6 weeks. For submerged cultures of *S. badius* growing on corn lignocellulose, a final concentration of 5 g of lignocellulose per liter of medium in a 2-liter flask gives excellent results. Sterile uninoculated controls are also incubated. They are carried through the same procedure, but initially receive 50 ml of sterile medium instead of the live cell inoculum.

APPL Harvest

Solid-State Fermentations. After 6 weeks of incubation, 500 ml of water is added to each solid-state fermentation flask, and the mixture is placed into a boiling water bath or steamer for 1 hr. The suspension is then filtered (Whatman No. 54) to recover the APPL. Alternatively, 100 ml of 1 *N* NH_4OH or 1 *N* NaOH per 5 g of initial lignocellulose can be added to the residues, and after thorough mixing, the soluble APPL can be recovered by filtration. The advantages of using the base extraction as opposed to water alone are that less volume of extraction liquid is required, and there is no need to steam the mixture to maximize recovery of APPL. APPLs are quite soluble in basic solutions, and essentially all of the APPL is recovered by this procedure. However, in studies with new lignin-degrading microorganisms, the hot water extraction is recommended in order to avoid possible base-mediated modifications of the APPL or residual lignocellulose. Each filtrate is acidified to below pH 5 with concentrated HCl, at which time, the APPLs precipitate as their free acids.[1,4] The use of H_2SO_4 should be avoided because the precipitated APPLs will contain sufficient inorganic salts to increase their ash contents to 4–5%, whereas APPLs precipitated with HCl normally contain only 1–2% ash. If base extraction is used, it is possible to remove NH_4OH prior to the acidification step by aerating the basic solutions, thus reducing the amount of acid required to precipitate the APPL. After precipitation, APPLs are collected by centrifugation (16,000 *g*), washed several times with acidic water, and then air-dried at 50°. The dry APPL is then ground to a fine powder for chemical analysis.

Submerged Culture System. After shaking incubation for up to 6 weeks, each flask is heated to 100° in a boiling water bath or steamer for 1 hr. The residual lignocellulose is removed by filtration (Whatman No. 54). The filtrate is then acidified to below pH 5, and the precipitated APPLs are collected and dried as described above.

Recovery of APPL without Acidification. There are some acid-mediated condensation reactions that occur in APPLs when they are subjected to the acid-precipitation procedure.[1,4] These condensations result in an increase in the average molecular weight of the APPLs. If such changes are undesirable, it is possible to collect the APPLs using a dialysis procedure. In this case, the APPL-containing solutions are dialyzed for several days against a large volume of water to remove low-molecular-weight materials and salts. The resulting solution is frozen and then freeze-dried to a golden crystalline-looking powder. APPLs collected in this way are of lower average molecular weight than acid-precipitated APPLs, but their ash and carbohydrate contents tend to be higher.[4]

Compositional Characterizations of Residual Lignocelluloses and APPLs

At the end of each incubation, it is important to determine the overall composition of each partially degraded lignocellulosic residue and APPL and to determine the amounts of lignin and cellulose degradation that have occurred as a result of microbial metabolism. Data should be compared for inoculated cultures versus uninoculated controls. One must first determine the total weight loss of lignocellulose resulting from microbial degradation. Lignocellulose weight losses are measured by collecting and air-drying the partially degraded residual or control lignocelluloses onto preweighed filter paper disks, which are then reweighed. To specifically calculate lignin and carbohydrate losses, the lignin and carbohydrate contents of the degraded residues are determined and compared to those of the starting, undegraded lignocellulose. Then, by utilizing the weight loss, carbohydrate content, and lignin content values, one can calculate the total amount of lignin and carbohydrate removed from the lignocelluloses as a result of microbial degradation (or nonbiological leaching in the case of controls). It is particularly important that the amount of lignin degradation be determined, so that lignin loss can be checked for its predicted correlation with APPL production.

The lignin and carbohydrate contents of the lignocelluloses and APPLs are determined by Klason lignin and Somogyi carbohydrate assays as summarized below. If desired, one can also determine the crude protein contents of the residues as compared to the starting lignocellulose in order to estimate the amount of microbial biomass present in each residue. In addition, by determining the crude protein contents of the APPLs, one can assess the degree to which they are contaminated by protein that may have coprecipitated with the APPLs when they were harvested. Organic nitrogen determinations are performed by the micro-Kjeldahl procedure as discussed below. Finally, in order to determine the amount of inorganic contamination of the APPLs, their ash contents should be determined as described below.

Modified Klason Lignin Assay. Lignin content is assayed by a modified Klason procedure,[10,11] which measures lignin as the acid-insoluble fraction of lignocellulose samples subjected to hydrolysis by strong acid. In this assay, 50 mg of lignocellulose residue or APPL is placed in a 50-ml flask. One milliliter of concentrated H_2SO_4 is added, and the mixture is allowed to stand for 1 hr with occasional mixing. Then, 28 ml of distilled water is added, and the suspension is autoclaved (121°) for 1 hr. After cooling, acid-insoluble material (Klason lignin) is collected by filtration onto small preweighed filter paper disks (Whatman No. 1). Then, after several washes with small volumes of distilled water, the lignin and filter paper are dried at 70–80° for 48–72 hr. The filtrate is saved for later use in the Somogyi assay. After drying and equilibration to room temperature, the lignin-containing filter paper is reweighed to determine the amount of acid-insoluble Klason lignin present. The lignin content of the residue or APPL is calculated as a percentage of the initial weight of the sample assayed, and values should be based on averages of at least three replicates. Total lignin loss

[10] P. Klason, *Sven. Papperstidn.* **26,** 319 (1923).

[11] W. E. Moore and D. B. Johnson, "Procedures for the Analysis of Wood and Wood Products." U.S. Department of Agriculture, Forest Products Laboratory, Madison, Wisconsin, 1967.

from the lignocellulose residues is based on the change in the total lignin content of the recovered residue as compared to the total lignin present in the starting lignocellulose.

Somogyi Carbohydrate Assay. The carbohydrate contents of the lignocellulose residues or APPLs are determined by measuring the reducing sugar content of the Klason filtrates, using a modified Somogyi–Nelson procedure.[12] The Klason filtrates retained from the assay above contain the monosaccharide sugar components produced by acid hydrolysis of the polysaccharides originally present in the residues or APPLs. To neutralize each solution, 18 ml of 2 *N* NaOH is first added. The volume of each solution is then brought up to 50 ml in order to simplify later calculations of total reducing sugars present. One milliliter of neutralized filtrate (water for the blank) is placed into a 6 in. screw cap tube, and 3 ml of assay solution C is added. Solution C is prepared fresh by combining four parts of solution A and one part of solution B. (Solution A contains 180 g of Na_2SO_4, 15 g of Rochelle salt, 30 g of Na_2CO_3, and 20 g of $NaHCO_3$ in 1 liter of deionized water, while solution B contains 180 g of Na_2SO_4 and 20 g of $CuSO_4 \cdot 5H_2O$ in 1 liter of deionized water.)[13] Next, the tube is lightly capped by placing a marble over the mouth, and then it is placed in a boiling water bath or steamer for 1 hr, after which time, the tube is cooled to 4°. Three milliliters of Nelson reagent is then added, and the entire solution is mixed by vortexing. To prepare the Nelson reagent, 50 g of $(NH_4)_6Mo_7O_{24} \cdot 4H_2O$ in 42 ml of concentrated H_2SO_4 is added to 900 ml of water. Then 6 g of $Na_2HAsO_4 \cdot 7H_2O$ in 50 ml of water is added, and the volume is brought up to 1 liter. The Nelson reagent must be incubated at 37° for 24–48 hr prior to use.[14] After mixing, 10 ml of additional water is mixed into the solution, and its optical density is determined at 500 nm. The amount of reducing sugar in the solution (=carbohydrate) is determined from a standard curve using glucose as the standard sugar. The carbohydrate content of each lignocellulose or APPL is calculated as a percentage of the initial weight of sample (50 mg) subjected to the Klason assay. Again, values should be averages of at least three replicate assays. Losses in the amount of total carbohydrate from lignocellulose residues are calculated by comparing the total amount of carbohydrate present in each degraded lignocellulose residue with that of the starting lignocellulose.

Micro-Kjeldahl Nitrogen and Ash Content Assays. The amount of Kjeldahl (amino) nitrogen in the lignocellulose residues or APPLs is determined using the standard micro-Kjeldahl assay.[15] In particular, a nitrogen

[12] S. P. Antai and D. L. Crawford, *Appl. Environ. Microbiol.* **42,** 378 (1981).

[13] M. Somogyi, *J. Biol. Chem.* **195,** 19 (1952).

[14] N. Nelson, *J. Biol. Chem.* **153,** (1944).

[15] J. Kjeldahl, *Z. Anal. Chem.* **22,** 336 (1883).

determination is required for all APPLs due to the possibility of coprecipitation of protein with APPL when APPLs are recovered by acid precipitation. The percentage Kjeldahl nitrogen value obtained by this assay is converted to a percentage crude protein value by multiplying the nitrogen value by 6.25. An example of the potential contamination of APPLs with protein is that of the APPL recovered from submerged cultures of *S. badius*.[4] These APPLs are contaminated with a large amount (20–25%) of protein (A. L. Pometto III and D. L. Crawford, unpublished results).

The ash content of each APPL can be determined by combusting a known amount of APPL (100–200 mg) in a tube furnace at 750° in an oxygen atmosphere. Combustion is carried out until there is complete oxidation of the organic carbon present, and then the weight of the inorganic residue is determined gravimetrically.[1] As discussed previously, the ash content of an APPL is mainly affected by the type of acid used to precipitate the APPL (concentrated HCl vs H_2SO_4).

Lignin Chemistry Characterizations of APPLs

To confirm their lignin origin and to determine the principal types of chemical changes introduced into the lignin by specific ligninolytic organisms, it is important that the lignin chemistry of the APPLs be characterized. While a variety of spectrophotometric, chemical, and other techniques are available,[16] three chemical degradative analysis procedures can by themselves reveal the information needed to adequately analyze the structure of the APPLs. These analysis include acidolysis, permanganate oxidation, and alkaline ester hydrolysis.[1] Acidolysis shows whether the APPLs contain β-aryl ether linked phenylpropane units as would be found in native lignin, and it reveals how much oxidative change has occurred in the phenylpropane side chains of the APPLs as compared to native lignin. The acidolysis procedure is summarized in chapter [24] of this volume. Permanganate oxidation is used to quantify the free phenolic hydroxyl group contents of APPLs and lignins.[1,2,4] Since the β-ether cleavage and aromatic ring demethylation reactions carried out by lignin depolymerizing *Streptomyces* generate new phenolic hydroxyl groups in both residual lignins and in APPLs,[2,4] the permanganate oxidation procedure can be used to help confirm that the APPL produced by a specific microorganism is a β-ether cleavage product, provided the APPL is enriched in phenolic hydroxyls as compared to the native lignin from which it was derived. Alkaline ester hydrolysis is used to quantify the number of aromatic acids

[16] R. L. Crawford, "Lignin Biodegradation and Transformation." Wiley (Interscience), New York, 1981.

naturally esterified to lignins or APPLs.[1] In corn lignins *p*-coumaric acid is the predominant esterified acid, and it can account for up to 5% or more of the lignin dry weight.[1,4] Ester hydrolysis has been used in the characterization of partially degraded lignins after their degradation by fungi or actinomycetes and in the characterization of actinomycete-produced APPLs.[16] Esterified acids are in effect useful as markers to confirm the lignin origin of APPLs. In the hydrolysis procedure, results are based upon comparisons of the hydrolytic release of esterified acids from APPL as compared with a control lignin prepared from the initial lignocellulose using the Björkman procedure[7] (see chapter [3] in this volume). Björkman lignins are often called milled wood lignins (or milled corn lignins, etc.), and they are relatively unaltered in structure from native lignin.[16]

Permanganate Oxidation. In this procedure, Björkman control lignins and APPLs are subjected to oxidation by permanganate after ethylation to protect their free phenolic hydroxyl groups from oxidation.[17] Permanganate oxidation products from lignin are derived from phenolic units containing free hydroxyl substituents para to propanyl side chains. From ethylated lignins, these products are recovered as ethoxylated benzoic acids.[1,4,17] A primary value of this analysis, therefore, is that it can be used to compare the number of free phenolic hydroxyl groups (=ethylatable hydroxyls) between control and decayed lignins or APPLs.

In the procedure, an air-dried control lignin or APPL is first ethylated by a procedure slightly modified from that described by Kirk and Adler.[17] Ethylation is accomplished using diazoethane. To prepare sufficient diazoethane to ethylate six to eight lignin or APPL samples, 10 ml of 10 *M* NaOH and 30 ml of ether are mixed by stirring in a 150-ml beaker placed in a salt–ice bath at 0° in a hood. Three grams of *N*-nitroso-*N*-ethylurea (Fluka Chemical Corp., Hauppauge, New York) is added slowly over an approximate 1-min time period, while the mixture is stirred vigorously *(Caution: mixture is explosive and appropriate protection must be used).* Then the mixture is covered with a watch glass and stirred continuously for 30 min. Next, the dark yellow ether solution is decanted into a beaker containing 3–4 g of KOH pellets, after which it is left standing for 2 hr at 0° with the watch glass in place. The solution is next filtered, and the filtrate is combined with 30–40 ml of dimethylformamide (DMF). This mixture is left uncovered in the hood for 30 min. The resulting solution of diazoethane ($C_2H_4N_2$) in DMF should be used immediately. To ethylate a specific lignin or APPL sample, 5 ml of diazoethane solution is added directly to a known amount (40–100 mg) of control lignin or APPL previously dissolved in 5 ml of DMF and placed in a preweighed 50-ml

[17] T. K. Kirk and E. Adler, *Acta Chem. Scand.* **24,** 3379 (1970).

round bottom flask. This mixture is loosely corked and allowed to stand at room temperature in the dark. Five milliliters of additional diazoethane solution (freshly prepared) is added on days 1, 3, 5, and 7, and after a total of 9 days, the DMF is removed by evaporation at $<50°$. Evaporation will require that a strong vacuum be applied with a vacuum pump, and the use of a rotary evaporator is recommended. The final preparation will be a pale yellow-brown glassy solid or oil. After air drying, the weight of the ethylated residue is determined by reweighing the round bottom flask.

The entire ethylated sample is subjected to the permanganate oxidation procedure. Approximately 10–20 ml of *tert*-butanol–water (1 : 1) is added to the sample to dissolve the ethylated material. After thorough mixing, the solution is drawn off with a pipette. Unethylated material remains behind, and the amount of this undissolved material is determined by reweighing the flask after allowing it to air dry. The difference between the total ethylated residue and final weights is equivalent to the amount of ethylated lignin, or APPL, which dissolved in the *tert*-butanol–water. This solution is added dropwise to 75 ml of vigorously stirred aqueous 1% (w : v) Na_2CO_3 in a 250-ml flask placed on a hot plate and maintained at 80° with continuous stirring. A solution of 5% (w : v) $KMnO_4$ is added to the stirring mixture at a rate sufficient to maintain a purple color over the brown color produced by MnO_2, which is produced as a result of the oxidation. Addition is continued for a period of 3 hr, and the total amount of permanganate solution required will average about 15 ml/sample. After 3 hr, the reaction is halted by addition of a few milliliters of 95% ethanol. Next, insoluble MnO_2 is filtered off. This residue is washed with a few milliliters of hot 1% (w : v) Na_2CO_3. Then the clear washings and filtrate are combined and allowed to cool. The filtrate (pH 10) is washed with ether and then hexane, and then it is neutralized with 2 *M* H_2SO_4. Then the washed filtrate is vacuum evaporated at 50° to a volume of about 30 ml. Na_2CO_3 (0.8 g) and then H_2O_2 (5 ml of a 30% solution) is added, and this mixture is incubated in a 50° water bath for 10 min. Excess peroxide is then destroyed by adding a small amount of additional permanganate (the solution turns purple), and then excess permanganate is destroyed by addition of a small amount of $Na_2S_2O_5$ (the solution turns brown). The solution is then acidified to pH 1–2 with 2 *M* H_2SO_4, at which time it turns clear. This solution is extracted four times with chloroform–acetone (1 : 1) and then once with chloroform only. The chloroform only extract should be washed once with water before it is combined with the chloroform–acetone solution. Finally, the combined extracts are transferred to a preweighed beaker and evaporated to dryness in a hood at room temperature. The final dry product, when derived from lignin or APPL, contains a mixture of ethylated *p*-hydroxybenzoic acid (4-hydroxybenzoic acid), protocatechuic acid (3,4-dihydroxybenzoic acid), vanillic acid (4-hydroxy-3-methoxybenzoic acid), and syringic acid (4-hydroxy-3,5-dimethoxybenzoic acid) in which the hydroxyl group for each compound is replaced with an ethoxyl group. Therefore, the amount of each (calculated as a percentage of initial lignin) can be related to the numbers of structures in the lignin or APPL which contained free phenolic hydroxyl units prior to ethylation.[4,17] The acids are quantified by GLC of trimethylsilyl (TMS) -derivatives as described in chapter [16] in this volume. The underivatized compounds can also be quantified by HPLC as described in chapter [■] in this volume.

By preparing standard curves for each of the ethoxylated benzoic acids and then comparing the quantities of ethylated *p*-hydroxybenzoic acid, protocatechuic acid, vanillic acid, and syringic acid for control lignins versus APPLs, it is possible to determine whether APPLs produced by a specific organism are more or less phenolic than the lignin from which they were derived. APPLs produced by *S. viridosporus,* for example, give higher yields of the aromatic acids upon permanganate oxidation than do control lignins, while APPLs produced by *S. badius* yield much lower amounts.[1,4] The ratios of each of the benzoic acids will also vary markedly depending on whether the initial substrate used is from a softwood, hardwood, or grass lignocellulose.[12]

Alkaline Ester Hydrolysis. Aromatic acids are often esterified to peripheral units of the lignin macromolecule in native lignins.[1,2,16] The quantitatively most important aromatic acids include *p*-hydroxybenzoic acid, vanillic acid, syringic acid, ferulic acid (4-hydroxy-3-methoxycinnamic acid), and *p*-coumaric acid [3-(4-hydroxyphenyl)-2-propenoic acid]. In grass lignins derived from corn, the predominant esterified acid is *p*-coumaric acid.[4] Ligninolytic *Streptomyces* are often able to hydrolyze the ester bonds, liberate the aromatic acids, and then metabolize them for carbon and energy.[5]

A simple ester hydrolysis procedure can be used to quantify the content of esterified acids in lignins and APPLs. A control Björkman lignin or APPL (30 mg) is added to 1.0 ml of 1 *M* NaOH and incubated with shaking for 48 hr at room temperature. Esterified aromatic acids are hydrolyzed under these conditions and are released into the aqueous phase. After incubation, the solutions are acidified to pH 1–2 with concentrated H_2SO_4 and then extracted twice with ethyl ether and once with ethyl acetate. The organic extracts are combined, dewatered with anhydrous sodium sulfate, and transferred to a preweighed beaker. After the ether is removed by evaporation, the beaker is reweighed to determine the quantity of material recovered. Next, the identity and quantity of each aromatic acid are determined by either GLC or HPLC as described in chapter [16] and [17] in this volume.

Studies of New Organisms

The APPL production and characterization procedures described above can be readily applied to the study of other filamentous microorganisms. While it should be possible to use the lignin solubilization assay described in chapter [24] of this volume to identify tentative APPL-producing strains of either actinomycetes or fungi, the techniques discussed in this chapter must be used to confirm that the acid-precipitable products produced by a specific culture are actually lignin derived. Several genera of actinomycetes in addition to *Streptomyces* have now been found to produce APPL-like products as they decompose lignin.[18,19] White rot fungi, such as *Phanerochaete chrysosporium* and *Coriolus versicolor,* also appear to produce small amounts of acid-precipitable polymers when growing on lignocellulose.[3,20,21] However, detailed chemical characterizations of these products have not yet been reported.

[18] A. J. McCarthy and P. Broda, *J. Gen. Microbiol.* **130,** 2905 (1984).
[19] A. J. McCarthy, M. J. MacDonald, A. Peterson, and P. Broda, *J. Gen. Microbiol.* **130,** 1023 (1984).
[20] I. D. Reid, G. D. Abrams, and J. M. Pepper, *Can. J. Bot.* **60,** 2357 (1982).
[21] A. Hutterman, C. Herche, and A. Haars, *Holzforschung* **34,** 64 (1980).

[6] Chemical Synthesis of Lignin Alcohols and Model Lignins Enriched with Carbon Isotopes

By KONRAD HAIDER, HARTMUT KERN, and LUDGER ERNST

Synthetic lignins labeled with carbon isotopes are the best models presently available for research on the biodegradation of lignin,[1,2] the earth's second most abundant plant polymer. Methods to label lignin *in situ* by feeding plants with labeled lignin precursors have been also described.[3,4]

Synthesis of ^{13}C- or ^{14}C-labeled lignin alcohols (coniferyl, *p*-coumaryl or sinapyl alcohol, Scheme 1) can be accomplished from commercially

[1] K. Haider and J. Trojanowski, *Arch. Microbiol.* **105,** 33 (1975).
[2] T. K. Kirk, W. J. Connors, R. D. Bleam, W. F. Hackett, and J. G. Zeikus, *Proc. Natl. Acad. Sci. U.S.A.* **72,** 2515 (1975).
[3] D. L. Crawford, R. L. Crawford, and A. L. Pometto III, *Appl. Environ. Microbiol.* **33,** 1247 (1977).
[4] K. Haider, J. P. Martin, and E. Rietz, *Soil Sci. Soc. Am. J.* **41,** 556 (1977).

available starting materials. Each of the labeled alcohols can be polymerized as such or admixed with the other unlabeled alcohols to obtain dehydrogenation polymers (DHP) similar to the lignin of conifers, deciduous trees, or gramineous plants.[5,6] Polymerization is carried out in aqueous solution, catalyzed by horseradish peroxidase (EC 1.11.1.7) with H_2O_2.[7] Laboratory equipment for manipulating small amounts of material must be available. Synthetic steps using labeled CO_2 need a vacuum line equipped with pumps to evacuate the line to less than 1 Pa. The necessary equipment for working with radioactive compounds must be available if ^{14}C-labeled lignin alcohols or model lignins are to be prepared.

Scheme 1 shows the structure of the three alcohols with designation of the carbons according to "common notation."[8] Methods are described to enrich distinct carbons of the three phenylpropanoid compounds by ^{14}C or ^{13}C with trans configuration of the double bond in the side chain.

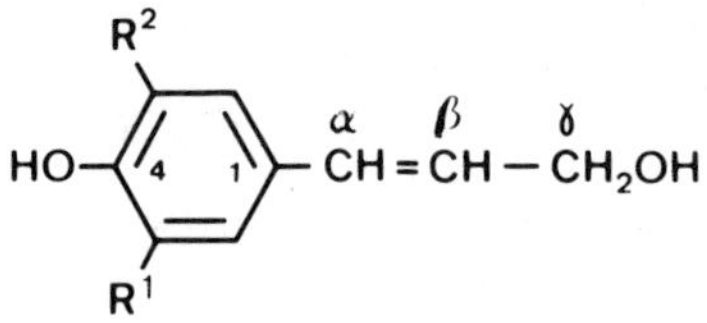

$R^1 = R^2 = H$	p-coumaryl alcohol
$R^1 = OCH_3$, $R^2 = H$	coniferyl alcohol
$R^1 = R^2 = OCH_3$	sinapyl alcohol

SCHEME 1

General Materials and Methods

C-β- or C-γ-labeled alcohols can be obtained by condensation of 4-hydroxybenzaldehyde, vanillin, or syringaldehyde with approximately equimolar amounts of malonic acid, labeled at C-1 or C-2.[9,10] Condensation is carried out in pyridine solution containing a few drops of piperidine as a catalyst (8 hr at 60°, longer than recommended in Ref. 10), and the yield in this Knoevenagel reaction is generally 70–80%. The condensation of the aldehydes with C-1-labeled malonic acid implies the loss of one carboxyl group as CO_2. Several attempts to circumvent these losses of labeled material by condensation of the aldehydes with nitriloacetic acid[9] ($^{14}CN^-$) or with [1-^{14}C]acetic anhydride[11] did not improve the radiometric yields obtained by the Knoevenagel reaction.

[5] K. Freudenberg and H. H. Hübner, *Chem Ber.* **85,** 1181 (1952).
[6] W. Schweers and O. Faix, *Holzforschung* **27,** 208 (1973).
[7] K. Freudenberg, K. Jones, and H. Renner, *Chem. Ber.* **96,** 1844 (1963).
[8] K. V. Sarkanen and C. H. Ludwig (eds.), "Lignins," p. 916. Wiley (Interscience), New York, 1971.
[9] K. Freudenberg and F. Bittner, *Chem. Ber.* **86,** 155 (1953).
[10] K. Kratzl and K. Buchtela, *Monatsh. Chem.* **90,** 1 (1959).
[11] H. M. Balba and G. G. Still, *J. Labelled Compd.* **15,** 309 (1978).

The desired lignin alcohols can be obtained by reduction of the cinnamic acids with $LiAlH_4$ after acetylation of the hydroxy and esterification of the carboxyl group.[5,11] It is also possible to reduce the intermediate acid chlorides directly with $LiAlH_4$,[12] but yields are better if the esters are reduced (Scheme 2). The reduction of coniferylaldehyde with $NaBH_4$ to give coniferyl alcohol has also been described.[2]

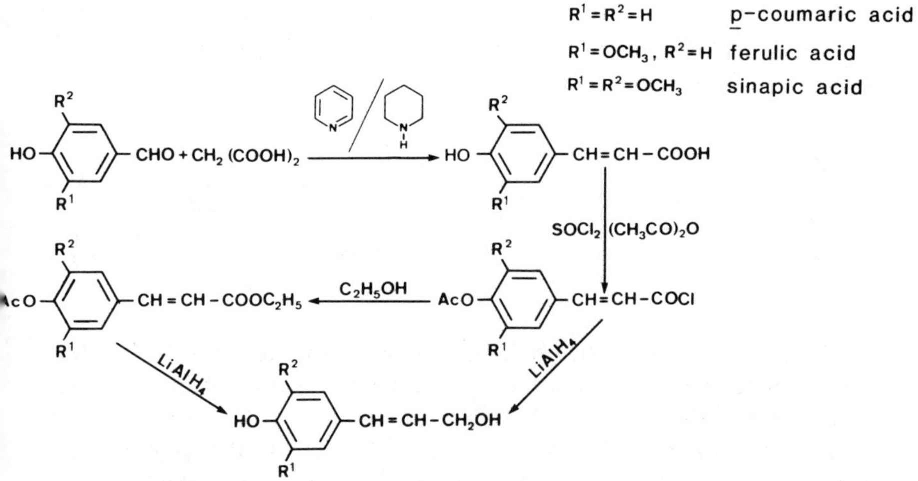

SCHEME 2

The synthesis of benzaldehydes labeled with carbon isotopes in the carbonyl groups can be accomplished by carboxylation of appropriate lithium compounds with labeled CO_2 and subsequent reduction of the carboxyl via its acid chloride to the aldehyde group by Rosenmund reduction.[12,13] (Scheme 3). Condensation of the labeled aldehydes with malonic acid or its monoethyl ester, followed by reduction (Scheme 2) yields the corresponding C-α-labeled alcohols.

The preparation of vanillin or syringaldehyde, enriched with carbon isotopes in the methoxyl groups, utilizes the facile condensation of isotopically enriched CH_3I or $^{14}CH_2N_2$ with methyl 4-*O*-benzyl protocatechuate or with methyl gallate.[14] ^{14}C- or ^{13}C-labeled CH_3I can be obtained from commercial sources, but it is necessary to buy it as fresh batches without traces of iodine. By ester cleavage, the free 4-*O*-benzyl vanillic acid is

[12] K. Kratzl and G. Billek, *Holzforschung* **7,** 66 (1953); K. Kratzl and G. Billek, *Monatsh. Chem.* **85,** 845 (1954).

[13] D. Gagnaire and D. Robert, *Makromol. Chem.* **178,** 1477 (1977).

[14] K. Haider and S. Lim, *J. Labelled Compd.* **1,** 294 (1965).

released and can be reduced to vanillin (Scheme 3). Similarly, methyl trimethoxybenzoate is obtained after reaction of gallate with CH_3I, and the free acid is released by ester cleavage. This can be demethylated in the fourth position by treatment with H_2SO_4 at 40° to give syringic acid in good yields.[15] Direct methylation of protocatechuic or caffeic acid for the preparation of *O*-[*methyl*-^{14}C]vanillic or -ferulic acid also can be carried out by using commercially available *S*-[*methyl*-^{14}C]adenosyl-L-methionine and catechol *O*-methyltransferase from porcine liver (EC 2.1.1.6).[16] Yields of 30–40% relative to the starting ^{14}C material, after purification of the labeled acids, have been reported.[16]

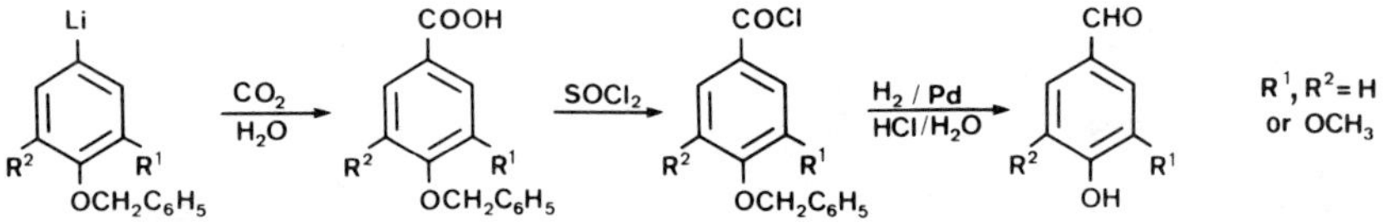

SCHEME 3

Several methods for the preparation of ring-labeled lignin alcohols have been described.[2,17,18] These are more complicated than labeling of the side chains. Ring-labeled lignins, however, are valuable for studying their complete metabolization[2,19] or for following alterations in the chemical environment of ring carbons during the attack of lignolytic microorganisms by means of ^{13}C NMR spectroscopy.[20,21]

The synthesis of coniferyl or 4-coumaryl alcohol uniformly ^{14}C labeled in the aromatic ring starts from commercially available [^{14}C]phenol. This can be formylated directly to give *p*-hydroxybenzaldehyde[17] or it can be transformed, via catechol and guaiacol, into vanillin. The latter synthesis is accomplished by condensation of [^{14}C]phenol with 2-chloro-4-nitrobenzophenone to give 2-phenoxy-4-nitrobenzophenone, introduction of the second OH group by oxidation with H_2O_2, followed by methylation of the free OH group with CH_2N_2. Guaiacol is then liberated by boiling the reaction product with piperidine (Scheme 4). A similar pathway is followed in the synthesis of the two alcohols, enriched with ^{13}C at C-4. This reaction

[15] M. J. Bogert and B. B. Coyne, *J. Am. Chem. Soc.* **51,** 569 (1929).
[16] A. C. Frazer, I. Bossert, and L. Y. Young, *Appl. Environ. Microbiol.* **51,** 80 (1986).
[17] K. Haider, *J. Labelled Compd.* **2,** 174 (1966).
[18] K. Kratzl and F. W. Vierhapper, *Monatsh. Chem.* **102,** 224 (1971); K. Kratzl, F. W. Vierhapper, and E. Tengler, *Monatsh. Chem.* **106,** 321 (1975).
[19] J. P. Martin and K. Haider, *Appl. Environ. Microbiol.* **38,** 283 (1979).
[20] P.-C. Ellwardt, K. Haider, and L. Ernst, *Holzforschung* **35,** 103 (1981).
[21] K. Haider, H. Kern, and L. Ernst, *Holzforschung* **39,** 23 (1985).

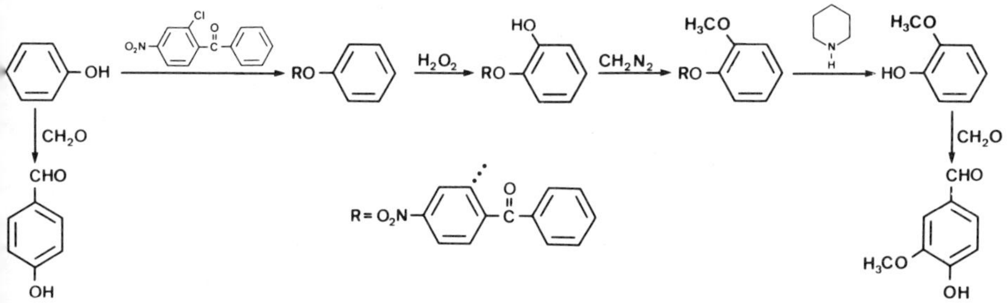

SCHEME 4

starts from commercially available [1-^{13}C]benzoic acid, which is first converted into [1-^{13}C]phenol.[17,20] The best reagent for the introduction of an aldehyde group into phenol or guaiacol was found to be dichloromethyl methyl sulfide with $SnCl_4$ as the catalyst.[22]

Polymerization of coniferyl alcohol or of mixtures of the three alcohols by a peroxidase-catalyzed (EC 1.11.1.7) free radical reaction in phosphate buffer with H_2O_2 yields insoluble ligninlike polymers.[6,7,23,24] Two alternative methods can be applied. (1) A dilute aqueous solution of the alcohol(s) and, separately, a dilute H_2O_2 solution are added slowly and simultaneously to a dilute solution of peroxidase under vigorous stirring. (2) The dilute solution of peroxidase and of the alcohol is added dropwise to the stirred H_2O_2 solution. The first method is more like the process which occurs during lignin formation in plants and yields a higher molecular-weight lignin than the second. Using the latter method with faster addition of the reactants, it is possible to isolate di- and oligolignols.[25]

Preparation of Cinnamic Acids and Cinnamyl Alcohols

The method described follows the procedure of Ref. 5. Similar methods with good yields are also available.[2,12,13] A mixture of equimolar amounts of malonic acid (the sodium salt has to be converted into the free acid) and vanillin, 4-hydroxybenzaldehyde, or syringaldehyde is dissolved in pyridine (1 : 10, w : w) containing a few drops of piperidine. Both solvents must be freshly distilled and dried. The mixture is reacted at 60° for 8 hr and acidified upon cooling with 6 *N* HCl. The crystalline cinnamic acid derivatives are separated and recrystallized from water. Small amounts of the acids can be isolated by ether extraction of the mother liquids. Final yields

[22] H. Gross and G. Matthey, *Chem. Ber.* **97,** 2606 (1964).

[23] G. Brunow and H. Wallin, "The Ekman Days," Vol. 4, p. 128. Swedish Forest Products Research Laboratory, Stockholm, Sweden, 1981.

[24] O. Faix and G. Besold, *Holzforschung* **32,** 1 (1978).

[25] T. Higuchi, F. Nakatsubo, and Y. Ikeda, *Holzforschung* **28,** 189 (1974).

are approximately 80%. When using [1-^{14}C]malonic acid, the evolved $^{14}CO_2$ should be trapped in soda lime.

Reduction of the cinnamic acids to the respective alcohols is carried out by acetylation with acetic anhydride in pyridine and followed by transformation into the acid chloride with $SOCl_2$ (small excess) in benzene (2 hr, 75°). After removal of the solvents, the acetylated acid chloride can be purified by distillation at 1 Pa. This is better than recrystallization from xylene[5] because of its sensitivity to moisture. The acid chlorides can be immediately transformed into their respective ethyl esters by a small excess of absolute ethanol and are isolated after evaporation of the ethanol. In order to label the alcohols at C-α, the esters can be directly prepared by condensation of the labeled aldehydes with malonic acid monoethyl ester.[13]

The esters are dissolved in dry ether (1 g in 25 ml) and slowly reduced with $LiAlH_4$ in ether (0.4 g in 100 ml) at −10° to −15° during 8–10 hr. The LiAl salts of the alcohols are decomposed first with moist ether and then by dropwise addition of water. The free alcohols are exhaustively extracted with ether. After evaporation of the extracting agent, the alcohols can be recrystallized from dichloromethane and petroleum ether or purified by silica gel chromatography with benzene/diethyl ether (9:2, v:v) as the eluent.

Preparation of Vanillin, 4-Hydroxybenzaldehyde, or Syringaldehyde

The preparations of the three aldehydes enriched by carbon isotopes in the carbonyl group follow the procedures in Refs. 12 and 13. For the synthesis of vanillin, 10 mmol of 4-bromoguaiacylbenzyl ether are dissolved in absolute ether in a flat bottom flask with two ground side arms and cooled to −60°. To the cold solution, an ether solution of *n*-butyllithium (6.66 mmol) is added while continuously stirring under N_2. The reaction mixture is stirred until the temperature reaches −10°. Then the flask is connected to a vacuum manifold line[12] and immediately frozen with liquid N_2. The closed flask is evacuated to less than 1 Pa, and CO_2 from isotopically enriched $BaCO_3$ (3.33 mmol) is released by acidification. The CO_2 is condensed into the reaction flask by cooling with liquid N_2 and reacts quantitatively with the lithium compound after reaching a temperature of −60 to −50°. After acidification, the free acid can be extracted with ether and purified by reextraction with $NaHCO_3$ solution and recrystallization from water. The syntheses of the benzyl ethers of 4-hydroxybenzoic or syringic acid follow similar procedures.[12] It is recommended that an excess of unlabeled CO_2 be added immediately to the reaction mixture for preparation of 4-hydroxybenzoic acid benzyl ether when the reaction of the labeled CO_2 has been completed.[26]

The benzyl ethers of the acids (5 mmol) are treated with 8 ml of neat $SOCl_2$ for 2 hr at 70°, and the resulting acid chloride is distilled *in vacuo* after evaporation of the $SOCl_2$. The reduction to the aldehyde is accomplished by the Rosenmund reduction. The acid chloride (5 mmol) is dissolved in 30 ml of absolute *p*-xylene, and after addition of 100 mg of a Pd catalyst (5%) on $BaSO_4$ together with one drop of quinoline-S, the suspension is treated with a vigorous stream of H_2 while boiling under reflux. The course of the reaction can be monitored by collecting the liberated HCl in 0.1 *N* NaOH and by retitration with 0.1 *N* HCl. When the reaction is complete, the catalyst is removed and the xylene evaporated. Cleavage of the benzyl ether is accomplished with boiling HCl (150 ml, 6 *N*) under continuous distillation of the benzyl chloride formed. After neutralization, the aldehyde is exhaustively extracted with ether and sublimed *in vacuo*. For further purification, it is chromatographed on preparative silica gel plates with *n*-hexane/ether (40/60 ml + 0.5 ml of acetic acid).[12,20]

Vanillin or syringaldehyde labeled in the methoxyl groups can be obtained by methylation of methyl 4-*O*-benzyl protocatechuate or of methyl gallate with labeled CH_3I.[14]

The compounds are dissolved in 2-butanone and mixed with stoichiometric amounts of $^{13}CH_3I$ or $^{14}CH_3I$. Dry K_2CO_3 is added, and the mixture is left for 2 hr at ambient temperature. Thereafter, it is boiled under reflux (condenser cooled with methanol/water of −5°) for 20 hr. After filtration, the 2-butanone is removed and the ester bond is cleaved by alkaline hydrolysis. Upon acidification, benzylvanillic or trimethoxybenzoic acid is filtered off and recrystallized from aqueous ethanol. The benzylvanillic acid can be used for the further synthesis of ferulic acid and coniferyl alcohol as described above. The 4-OCH_3 group of trimethoxybenzoic acid can be removed by suspending in concentrated H_2SO_4 (300 mg in 0.8 ml) and stirring at 50° for 24 hr.[15] The syringic acid crystallizes after addition of H_2O and can be recrystallized from H_2O. The preparation of OCH_3-labeled sinapyl alcohol is effected by acetylation of the OH group followed by reduction of the carboxyl group as described above.[5,14]

Methods for the preparation of ring-labeled aromatic aldehydes will be only briefly described. For further details, the reader is referred to Refs. 2, 17, and 18. Phenol, uniformly ^{14}C-labeled or enriched with ^{13}C at C-1 is condensed with 2-chloro-5-nitrobenzophenone,[18] and the resulting 2-phenoxy-5-nitrobenzophenone is transformed in three separate steps, in-

[26] S. Acerbo, R. Kastori, H. Söchtig, H. Harms, and K. Haider, *Z. Pflanzenphysiol.* **69,** 306 (1973).

cluding methylation with CH_2N_2, into 2-(2-methoxyphenoxy)-5-nitrobenzophenone, which is recrystallized from glacial acetic acid. Cleavage of the ether bond by treatment with piperidine at 150° yields guaiacol. This compound can be formylated[22] to give vanillin. By direct formylation of phenol using the same procedure, 4-hydroxybenzaldehyde can be prepared. Both aldehydes can be used for the subsequent preparation of ring-labeled *p*-coumaryl or coniferyl alcohol.

Comments on ^{13}C NMR Spectra of ^{13}C-Enriched Model Lignins

Figure 1A shows the ^{13}C NMR spectrum of a coniferyl alcohol DHP with ^{13}C at natural abundance. It reflects the different types of linkages of the coniferyl alcohol units in the polymer. The signal assignments for the different structural entities follow from Refs. 20, 21, 27, and 28. A relatively intense signal at $\delta = 55.8$ is attributed to methoxy groups. The range $\delta = 60-63$ contains three signals for C-γ, while that of $\delta = 71-88$ contains several signals for C-α and C-β of the side chain. The signals of the aromatic ring appear between $\delta = 110$ and 150 with C-3 and C-4 near the lower field limit. The NMR spectrum of the [γ-^{13}C]DHP (Fig. 1D) shows enhanced signals at $\delta = 70.9$ for C-γ in pinoresinol and at $\delta = 62.9$ in phenylcoumaran units. The most intense signal at $\delta = 61.6$ is assigned to C-γ of coniferyl alcohol units with a free side chain. A double signal at $\delta = 60.1$ and 59.8 indicates C-γ in units connected by β-aryl ether bonds. The aldehyde carbons of cinnamaldehyde side chains absorb at $\delta = 193.9$. The signals for C-β in a DHP prepared from [β-^{13}C]coniferyl alcohol (Fig. 1C) are also significantly enhanced and absorb at $\delta = 128.5$ and 128.0, indicating subunits with free side chains. A group of signals between $\delta = 84.6$ and 81.5 indicate β-carbons in β-aryl or α,β-diaryl ethers. Signals at $\delta = 53.6$ and 53.0 belong to C-β in pinoresinol and in phenylcoumaran units, respectively. A weaker signal at $\delta = 34.3$ is characteristic for C-β in dihydroconiferyl alcohol units. The corresponding [α-^{13}C]DHP is shown in Fig. 1B. Strong signals at $\delta = 128.9$ and 128.5 indicate α-carbons in coniferyl alcohol units with free side chains. Those at $\delta = 87.1$ and 86.9 and at $\delta = 85.1$ and 84.9 show the α-carbons in phenylcoumaran and pinoresinol units, respectively. Signals at $\delta = 79.0$ and 71.7–70.8 are characteristic for the α-carbons in α,β-diaryl ethers and β-aryl ethers, respectively. A weaker signal at $\delta = 31.2$ indicates C-α in dihydroconiferyl alcohol units.

Thus, Fig. 1 clearly demonstrates the usefulness of selectively ^{13}C-labeled DHPs for facilitating the interpretation of lignin ^{13}C NMR spectra.

[27] H. Nimz, J. Mogharab, and H. Lüdemann, *Makromol. Chem.* **175,** 2563 (1974).
[28] C.-L. Chen, M. G. S. Chua, J. Evans, and H.-M. Chang, *Holzforschung* **36,** 239 (1982).

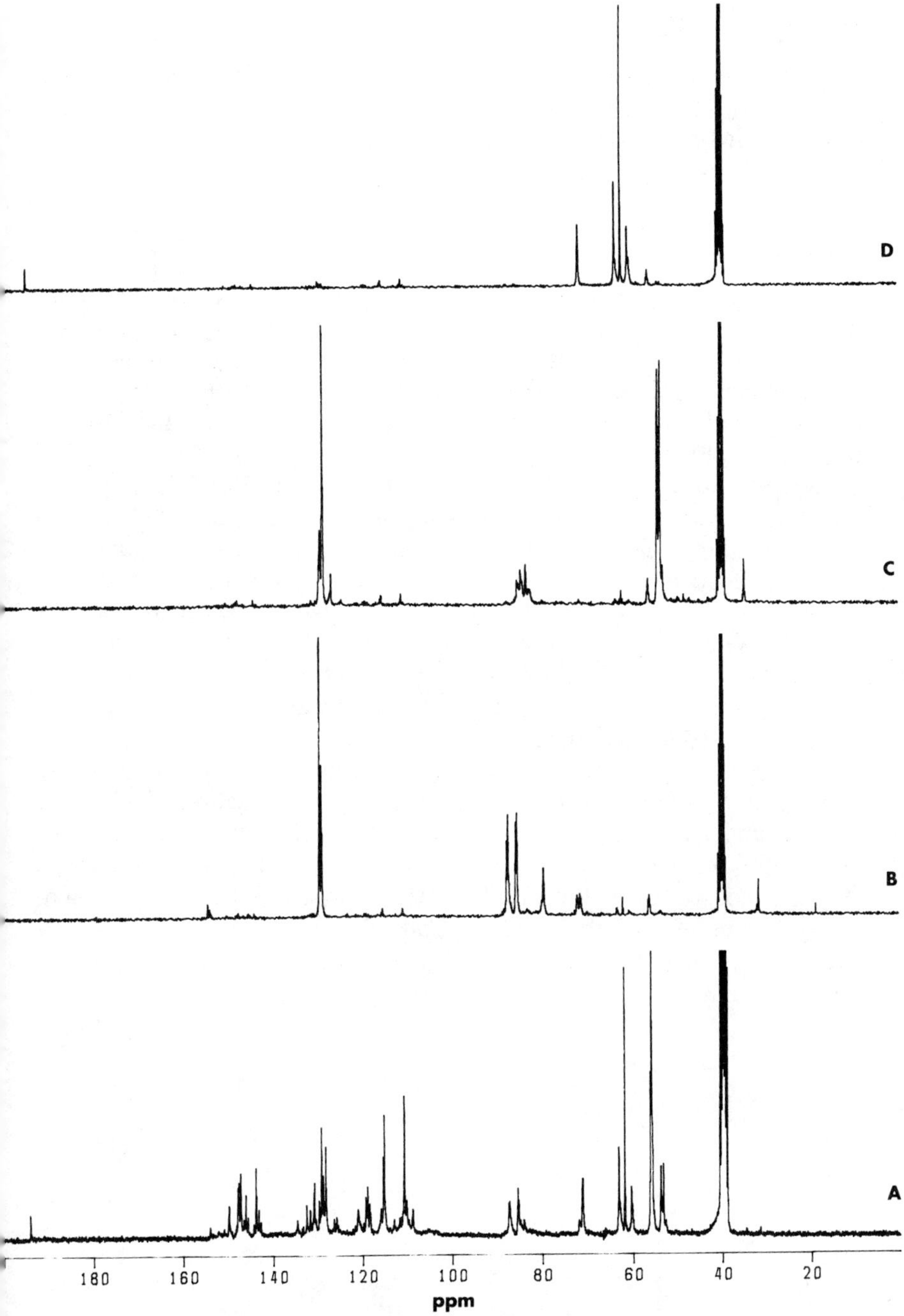

FIG. 1. ^{13}C NMR spectra (75.5 MHz) [DMSO-d_6 (d is deuterium) solutions, ~25 mg/0.6 ml, Bruker AM 300 spectrometer, solvent signal used as reference, $\delta = 39.5$] of DHP from coniferyl alcohol with natural abundance of ^{13}C (A) and ^{13}C-labeled at C-α (B), C-β (C), and C-γ (D), respectively.

Alterations in the chemical environment after attack by lignolytic organisms can be followed by measuring the ^{13}C NMR spectra of ^{13}C-enriched model lignins after incubations and become obvious by new prominent signals.[20,21]

Addendum

Recently, Newman and co-workers[29] reported possible improvements in the syntheses of [β,γ-^{13}C]-, [γ-^{13}C]-, and [α-^{13}C]coniferyl alcohol. The preparation of [β,γ-^{13}C]coniferyl alcohol involves condensation of 4-*O*-ethoxyethylidene-vanillin in dry tetrahydrofuran (THF) with commercially available triethyl [1,2-^{13}C]phosphonoacetate to give the [^{13}C]ethyl ferulate-labeled derivative in 98% yield. This ester was reduced by a complex of diisobutylaluminium hydride and *n*-butyl lithium (*n*-BuLi) in dry toluene.[30] Removal of the protecting group then furnished a 64% yield of labeled coniferyl alcohol after recrystallization. For the preparation of [γ-^{13}C]coniferyl alcohol, 4-*O*-ethoxyethylidene-vanillin was reacted with propyl [1-^{13}C]dimethylphosphonoacetate to give propyl [*carbonyl*-^{13}C] 4-*O*-ethoxyethylidene-ferulate (77% yield) and reduced to [γ-^{13}C]coniferyl alcohol as described. The [^{13}C] phosphonoacetate-labeled derivative can be prepared by reaction of dimethyl methylphosphonate with *n*-BuLi and $^{13}CO_2$, followed by esterification with *n*-propanol.

A proposal by Newman *et al.*[29] could facilitate the synthesis of the [*formyl*-^{13}C]vanillin by avoiding the Rosenmund reduction of the 4-*O*-benzyl-vanillic acid chloride[12,13] (Scheme 3). For this reason, methyl [*carbonyl*-^{13}C]vanillate was reduced by $LiAlH_4$ in dry THF to vanillyl alcohol. This alcohol was treated in dry THF with recrystallized 2,3-dichloro-5,6-dicyanoquinone, and the resulting [*formyl*-^{13}C]vanillin was purified by column chromatography on silica gel (81% yield).

[29] J. Newman, R. N. Rey, G. Just, and N. G. Lewis, *Holzforschung* **40,** 369 (1986).
[30] S. Kim and K. H. Ahn, *J. Org. Chem.* **49,** 1717 (1984).

[7] Synthesis of Lignols and Related Compounds

By FUMIAKI NAKATSUBO

Lignin macromolecules are composed of several lignin substructural units linked by many different stable carbon–carbon and ether linkages. Studies using substructural models are extremely important for the elucidation of the detailed mechanism of lignin biodegradation.

Many synthetic methods for lignin model compounds have been reported.[1] This chapter is concerned with the preparation of two such model oligolignols: guaiacylglycerol-β-guaiacyl ether **(1)** and 1-(4-hydroxy-3,5-dimethoxyphenyl)-2-(4-hydroxy-3-methoxyphenyl) propane-1,3-diol **(2)**. The synthesis of compounds **(1)** and **(2)** is presented in Figs. 1 and 2.

Compound **(1)** has historically been the most important model for the arylglycerol-β-aryl ether (β-O-4′) substructure. This unit is the most common substructure in lignin, comprising about 45% of the interphenylpropane linkages.[2]

Compound **(2)** is a significant model for the 1,2-diarylpropane-1,3-diol (β-1′) substructure. This unit is one of the main substructures in lignin, comprising about 15% of the interphenylpropane linkages.[2] This model is often used for lignin-degrading enzyme assays because its degradation in ligninolytic culture proceeds more smoothly than those of β-O-4′, β-5′, and β-β′ models, and its degradation products are easily detectable by silica gel chromatography.[3]

General Considerations

The procedures outlined here require a reasonable amount of skill in handling organic chemicals, and directions in all footnotes in this chapter should be followed carefully. In particular, care should be taken especially for the anhydrous reactions using lithium diisopropylamide. However, once one becomes accustomed to this method, the synthesis in high yield[1] of other lignin substructural models, e.g., β-5′ and β-β′ models, using any combination of guaiacyl, syringyl, or *p*-hydroxyphenyl derivatives[1] becomes possible.

[1] F. Nakatsubo, *Wood Res.* **67,** 59 (1981).

[2] Y. Z. Lai and K. V. Sarkanen, *in* "Lignins" (K. V. Sarkanen and C. H. Ludwig, eds.), p. 228. Wiley (Interscience), New York, 1971.

[3] T. Higuchi, *in* "Biosynthesis and Biodegradation of Wood Components" (T. Higuchi, ed.), p. 557. Academic Press, Orlando, Florida, 1985.

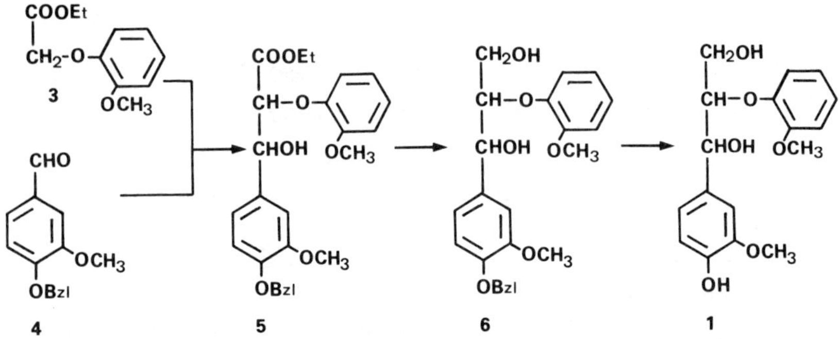

FIG. 1. Synthetic route for guaiacylglycerol-β-guaiacyl ether (**1**).

All chemicals used in these procedures were of laboratory grade or reagent grade and were purchased from Aldrich Co. or Merck & Co., Inc.

Synthesis of Guaiacylglycerol-β-guaiacyl Ether (**1**)[4]

Preparation of Ethyl 2-Methoxyphenoxyacetate (**3**). Ethyl chloroacetate (24.5 g, 0.2 mol), guaiacol (24.8 g, 0.2 mol), finely powdered potassium carbonate (41.5 g, 0.3 mol), and powdered potassium iodide (0.83 g, 0.005 mol) are added to 300 ml of acetone while stirring vigorously with a magnetic stirrer. The reaction suspension is refluxed for 3 hr with continuous stirring. The completion of the reaction is indicated on silica gel thin-layer chromatography (TLC)[5] by the disappearance of guaiacol (R_f values: 0.33 for ethyl 2-methoxyphenoxyacetate and 0.47 for guaiacol in ethyl acetate/*n*-hexane, 1 : 4 by volume). After cooling the reaction mixture to room temperature, the suspension is filtered and the precipitate is washed three times with 40 ml of ethyl acetate. The combined filtrate and washings are evaporated to dryness *in vacuo* to give a slightly yellow oil which is dissolved in 100 ml of ethyl acetate. The ethyl acetate solution is washed with 40 ml of a saturated sodium chloride solution (hereafter referred to as brine), is dried over anhydrous sodium sulfate, and is evaporated *in vacuo* to give a slightly yellow oil. This oil is purified by fractional distillation under reduced pressure (132°/2 mm Hg) to afford the expected compound (**3**) as a colorless viscous oil (30 g, 70% yield).

[4] F. Nakatsubo, K. Sato, and T. Higuchi, *Holzforschung* **29,** 165 (1975).

[5] Merck GMbH (Darmstadt, Federal Republic of Germany), Silica gel 60 plates F-254 (0.25 mm thickness; size, 20 × 20 cm) are cut into small plates (1.5 × 5 cm) and used to check reaction products. Each of the intermediate and product compounds absorbs ultraviolet light, as a result of the aromatic ring (or rings) incorporated into the compound. After the development of a TLC plate with an appropriate solvent, the plate should be irradiated with UV light (254 nm), using an inspection lamp. The various compounds may then be easily observed as dark spots against the light green background.

Preparation of Benzylvanillin (**4**). Vanillin (15.2 g, 0.1 mol) and benzyl chloride (15.2 g, 0.12 mol) are dissolved in 300 ml of tetrahydrofuran (THF) in a 500-ml round bottom flask. Finely powdered anhydrous potassium carbonate (21 g, 0.15 mol) and tetra-*n*-butylammonium iodide (1.85 g, 0.005 mol) are added at room temperature while stirring vigorously with a magnetic stirrer.[6] The reaction suspension is refluxed for 4 hr with continuous stirring. At the beginning of the reaction, the reaction mixture is a thick suspension, but the particle size and suspension viscosity gradually decreases as the reaction proceeds. The reaction is followed by TLC (R_f values: 0.4 for vanillin and 0.6 for benzylvanillin in ethyl acetate/*n*-hexane, 1:2 by volume). After cooling the reaction mixture to room temperature, the suspension is filtered and the precipitate is washed three times with 40 ml of ethyl acetate. The combined filtrate and washings are evaporated to dryness *in vacuo* to give a yellow oil which is dissolved in 100 ml of ethyl acetate. The ethyl acetate solution is washed with 40 ml of brine, is dried over anhydrous sodium sulfate, and is evaporated *in vacuo* to give a yellow oil. The product is crystallized from ethanol (20 ml) and *n*-hexane (80 ml) in a freezer at −20° to afford colorless crystals (22.3 g, 92% yield, mp 61.5–62°).

Preparation of β-Hydroxy Ester (**5**).[7] Diisopropylamine (6.1 g, 0.06 mol)[8] is dissolved in 50 ml of anhydrous THF[9] in a 500-ml three-necked flask equipped with a 100-ml dropping funnel, a drying tube containing anhydrous calcium sulfate (Aldrich Co., Drierite), and a rubber septum. The flask is cooled in a dry ice/acetone bath (−10°) for 30 min while stirring under nitrogen supplied through a Drierite column. To this solution, 38.2 ml of *n*-butyllithium (0.06 mol) in *n*-hexane (1.57 *N*)[10] is

[6] A mechanical stirrer is recommended for this stirring.

[7] This reaction must be carried out under completely anhydrous conditions. All glassware must be dried at 110° in an oven overnight and assembled while still hot. Vacuum grease is spread on all glass joints before assembly.

[8] Diisopropylamine is freshly distilled from sodium metal.

[9] Special grade THF (about 300 ml) is dried over molecular sieve 4A and then distilled from potassium metal (about 1 g) and benzophenone (about 2 g) to obtain an anhydrous solvent.

[10] *n*-Butyllithium is highly corrosive and should be handled with protection from rubber gloves and safety glasses [F. Fieser and L. Fieser, "Reagents for Organic Synthesis," Vol. 1, p. 95 (1967) and Vol. 2, p. 51 (1969). Wiley, New York]. This reagent is titrated in the following manner before use (S. C. Watson and J. F. Eastham, *J. Organomet. Chem.* **9**, 165 (1967). A 5-ml round bottom flask containing a stirring bar is dried in an oven at 110° overnight. After cooling the flask in a desiccator containing P_2O_5, several crystals of *o*-phenanthroline are added and the flask is quickly sealed with a rubber septum. Two milliliters of anhydrous THF and 100 μl of 2-butanol ($d = 0.811$, MW = 74.12) freshly distilled from sodium metal are added and the resulting solution is titrated with commercial *n*-butyllithium at 0–5°. The endpoint of the titration is easily detected by the appearance of a brown color, indicating the production of the lithium chelate of *o*-phenanthroline. The normality of the *n*-butyllithium is $1.1/x$ *(N)*, where x is milliliters of *n*-butyllithium consumed during the titration.

carefully added dropwise from a syringe over a period of 30 min. The stirring is then continued for an additional 30 min at the same temperature. After cooling the reaction mixture below −70° in a dry ice/acetone bath, ethyl 2-methoxyphenoxyacetate **(3)** (10.5 g, 0.05 mol) dissolved in 50 ml of anhydrous THF[11] is added dropwise over a period of 1 hr. After stirring for an additional 30 min, benzylvanillin **(4)** (12.1 g, 0.05 mol) dissolved in 50 ml of anhydrous THF[12] is added dropwise over a period of 1 hr. The temperature of the cooling bath is gradually raised to −50° over a period of 1 hr, and the resulting slightly yellow reaction mixture is diluted with 200 ml of ethyl acetate. The ethyl acetate solution is washed with brine until the washings became neutral, then is dried over anhydrous sodium sulfate and evaporated *in vacuo* to give a slightly yellow oil. This oil is purified by column chromatography (silica gel, 250 g; column size, 5 × 25 cm, eluted with dichloromethane) to give a slightly yellow viscous oil (20 g, 88.5% yield). The product is crystallized from ethanol (40 ml) and *n*-hexane (40 ml). The resulting crystal mass is crushed to a fine powder, which is filtered and washed three times with 20 ml of ethanol/*n*-hexane (1 : 4, by volume) to afford the pure *erythro* form of **(5)** as colorless crystals (12 g, 53% yield, mp 80.5–81°). The mother liquor, consisting of both *erythro* and *threo* isomers (roughly 1 : 1 ratio), gives one spot on a TLC plate developed with ethyl acetate/*n*-hexane (1 : 2, by volume) [R_f values: 0.17 for the expected *β*-hydroxy ester **(5)** and 0.53 both for benzylvanillin and ethyl 2-methoxyphenoxyacetate]. The above mixture of isomers is used for subsequent reactions without separation.

Preparation of Guaiacylglycerol-β-guaiacyl Ether **(1).** Lithium aluminum hydride (LAH, 2.3 g, 0.06 mol) is placed in a 500-ml three-necked flask equipped with a gas inlet tube, a dropping funnel, and a reflux condenser. The latter two are fitted with drying tubes containing Drierite. One hundred milliliters of anhydrous THF is carefully added dropwise. While stirring this solution in a 60° bath, crystalline *erythro-β*-hydroxy ester **(5)** (9.04 g, 0.02 mol) dissolved in 80 ml of anhydrous THF is added dropwise under nitrogen over a period of 2 hr. The stirring is continued for an additional 30 min. The disappearance of the starting material is confirmed by TLC [R_f values: 0.34 for the expected compound **(6)** and 0.5 for the *β*-hydroxy ester **(5)** in methanol/dichloromethane, 5 : 95 by volume;

[11] Anhydrous THF is placed into the dropping funnel first. 2-Methoxyphenoxyacetate is then added from a syringe to give a homogeneous solution.

[12] Benzylvanillin is placed into a 100-ml round bottom flask and dried over P_2O_5 in a vacuum desiccator overnight. After breaking the vacuum by the addition of dry nitrogen, the flask is quickly sealed with a rubber septum. Anhydrous THF is then added and the benzylvanillin is dissolved. The THF solution is transferred into the dropping funnel by the use of a syringe.

0.2 for the expected compound **(6)** and 0.4 for the β-hydroxy ester **(5)** in ethyl acetate/*n*-hexane, 1:1 by volume]. The reaction mixture is then cooled in an ice water bath and 4.32 ml of water (0.24 mol) diluted with 10 ml of THF is carefully added[13] dropwise over a period of 20 min for the decomposition of excess LAH. After the addition of 150 ml of ethyl acetate to this reaction mixture, about 25 ml of water is added dropwise with vigorous stirring until a white precipitate settles to the bottom of the flask.[14] Finely precipitated white inorganic salts are collected by filtration and are washed three times with 30 ml of ethyl acetate. The combined filtrate and washings are washed with brine until the washings become neutral. The organic solution is dried over anhydrous sodium sulfate, is filtered, and is evaporated *in vacuo* to give a colorless crust in quantitative yield. The colorless product obtained is dissolved in 100 ml of ethanol, and 10% palladium charcoal (1 g) is added.[15] The suspension is stirred under hydrogen at room temperature for 6 hr until the hydrogenolysis has proceeded completely. The reaction is followed by TLC [R_f values: 0.34 for the starting material **(6)** and 0.28 for the final product, guaiacylglycerol-β-guaiacyl ether **(1)** in methanol/dichloromethane, 5:95 by volume]. The palladium charcoal is filtered off and washed three times with 30 ml of ethanol. The combined filtrate and washings are evaporated *in vacuo* to give a colorless oil. The product is crystallized from ethyl acetate (15 ml) and *n*-hexane (10 ml) to afford colorless crystals [5.5 g, 85% overall yield from β-hydroxy ester **(5)**, mp 101–102° (triacetate: 110–111°)].

The β-hydroxy ester **(5)** consisting of *erythro* and *threo* isomers (about 8 g) obtained above is also subjected to LAH reduction and subsequent hydrogenolysis with 10% palladium charcoal/hydrogen as described above to give a colorless crust, consisting of a mixture of the *erythro* and *threo* isomers of the final product **(1)** [4.3 g, 76% overall yield from β-hydroxy ester **(5)**].

Synthesis of 1-(4-Hydroxy-3,5-dimethoxyphenyl)-2-(4-hydroxy-3-methoxyphenyl)propane-1,3 diol (2)[16]

Preparation of Benzylacetovanillone **(7)**. Acetovanillone (6.64 g, 0.04 mol) and benzyl chloride (5.1 ml, 0.048 mol) are dissolved in 100 ml of THF. Finely powdered anhydrous potassium carbonate (8.3 g, 0.06 mol)

[13] If the water/THF solution is added quickly, a violent reaction, accompanied by vigorous hydrogen evolution, will occur.

[14] If excess water is added to the reaction mixture, the precipitate dissolves to afford a fine emulsion which is difficult to separate into its organic and water layers.

[15] Before the addition of palladium charcoal, the flask should be purged with nitrogen to avoid ignition.

[16] F. Nakatsubo and T. Higuchi, *Holzforschung* **29**, 193 (1975).

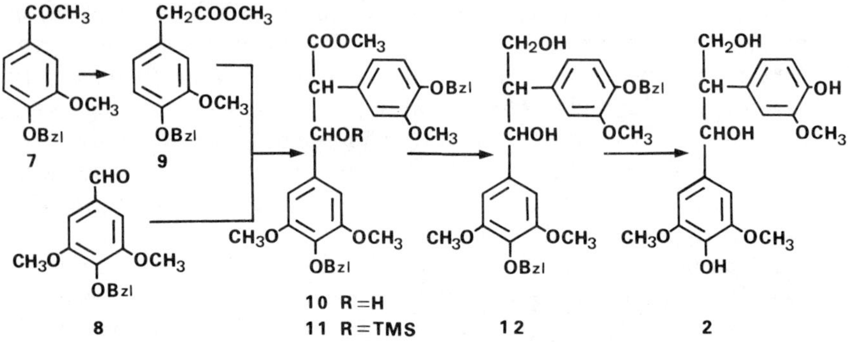

10 R=H
11 R=TMS

FIG. 2. Synthetic route for 1-(4-hydroxy-3,5-dimethoxyphenyl)-2-(4-hydroxy-3-methoxyphenyl)propane-1,3 diol (**2**).

and tetra-*n*-butylammonium iodide (1.5 g, 0.04 mol) are added and the reaction mixture is refluxed for 2 hr with vigorous stirring. The reaction is followed by TLC [R_f values: 0.31 for acetovanillone and 0.55 for the expected compound (**7**) in ethyl acetate/*n*-hexane, 1 : 1 by volume]. After subjecting the reaction mixture to the same workup as that used in the above preparation of benzylvanillin (**4**), slightly yellow crystals are obtained. The product is recrystallized from ethanol (20 ml) and *n*-hexane (80 ml) to give needles (9.7 g, 95% yield, mp 86°).

Preparation of Benzylsyringaldehyde (**8**). Syringaldehyde (5.46 g, 0.03 mol) and benzyl chloride (4.14 ml, 0.036 mol) are dissolved in 100 ml of THF. Finely powdered potassium carbonate (6.22 g, 0.045 mol) and tetra-*n*-butylammonium iodide (1.1 g, 0.03 mol) are added and the reaction mixture is refluxed for 14 hr with vigorous stirring. The reaction is followed by TLC (R_f values: 0.14 for syringaldehyde and 0.46 for benzylsyringaldehyde in ethyl acetate/*n*-hexane, 1 : 2 by volume). After subjecting the reaction mixture to the same workup as that used in the above preparation of benzyl vanillin (**4**), a slightly orange-colored oil is obtained. The product is crystallized from ethanol (10 ml) and *n*-hexane (60 ml) to give needles (7.5 g, 92% yield, mp 61.5–62°).

Preparation of Methyl Benzylhomovanillate (**9**). Benzylacetonvanillone (**7**) (5.1 g, 0.02 mol) is added slowly to a solution of thallium(III) nitrate [$Tl(NO_3)_3 \cdot 3H_2O$, 9.3 g, 0.021 mol][17] and 75% perchloric acid (10 ml) in 50 ml of methanol. The reaction mixture is stirred for 4 hr at room tem-

[17] Thallium(III) nitrate is commercially available from Aldrich Co., but may be easily prepared by dissolving thallium(III) oxide in concentrated nitric acid (d = 1.42) following the method reported by A. McKillop, J. P. Hunt, E. C. Taylor, and F. Kienzle [*Tetrahedron Lett.* **60,** 5275 (1970)].

perature. The reaction is followed by TLC [R_f values: 0.55 for the starting compound **(7)** and 0.68 for the expected compound **(9)** in dichloromethane]. The thallium(I) nitrate precipitate is filtered off and is washed three times with 30 ml of ethyl acetate.[18] The combined filtrate and washings are diluted with 200 ml of ethyl acetate and the solution is washed with a saturated sodium bicarbonate solution until the washings become neutral. This solution is washed once with brine, is dried over anhydrous sodium sulfate, and is evaporated *in vacuo* to give a brown oil. This oil is then dissolved in about 20 ml of dichloromethane and is passed through a silica gel column (silica gel, 10 g, column size, 2.5 × 3.5 cm) to obtain a slightly yellow oil. The product is crystallized from ethanol (20 ml) and *n*-hexane (60 ml) to afford colorless crystals (4.8 g, 84% yield, mp 66–67°).

Preparation of β-Hydroxy Ester **(10)**[7]. This compound is prepared in the same manner as compound **(5)**. Diisopropylamine (2.02 g, 0.02 mol)[8] is dissolved in 70 ml of anhydrous THF[9] in a 200-ml three-necked flask equipped with a dropping funnel, a drying tube containing Drierite, and a rubber septum. The flask is cooled in a dry ice/acetone bath (−10°) for 30 min while stirring under nitrogen supplied through a Drierite column. To this solution, 13 ml of *n*-butyllithium (0.02 mol) in *n*-hexane (1.57 *N*)[10] is added dropwise from a syringe over a period of 30 min. The stirring is then continued for an additional 30 min at the same temperature. After cooling the reaction mixture below −70°, methyl benzylhomovanillate **(9)** (4.29 g, 0.015 ml) dissolved in 20 ml of anhydrous THF is added dropwise over a period of 20 min. The reaction mixture is stirred for an additional 30 min and the solution becomes greenish yellow. Benzylsyringaldehyde **(8)** (4.08 g, 0.015 mol) dissolved in 20 ml of anhydrous THF is added dropwise over a period of 30 min. After addition of the solution, the bath temperature is gradually raised to −50° over a period of 80 min. The reddish yellow reaction mixture is neutralized by the addition of dry ice and turns slightly yellow. The resulting mixture is diluted with 200 ml of ethyl acetate, is washed once with water (100 ml) and three times with brine (50 ml), is dried over anhydrous sodium sulfate, and is evaporated *in vacuo* to yield a slightly yellow oil. Crystallization from ethanol (100 ml) gives colorless crystals consisting of both the *erythro* and *threo* isomers (roughly 1:1 ratio) (7 g, 84% yield, mp 113–115°). The crystals produce two spots on a TLC plate developed with ethyl acetate/*n*-hexane, 1:2 by volume [R_f values: 0.12 for the *threo* isomer of compound **(10)**, 0.20 for the

[18] If the reaction mixture is evaporated for the removal of methanol before partitioning, the expected methyl benzylhomovanillate will be nitrated to give the 6-nitro derivative (mp 161.5–162.5°) in about 80% yield. This compound is easily crystallized and is difficult to dissolve in standard solvents, e.g., ethyl acetate, THF, or dioxane.

erythro isomer of compound **(10)**, 0.50 for the starting compound **(8)**, and 0.61 for the starting compound **(9)**]. Pure *threo* and *erythro* isomers may be effectively obtained by column chromatography at this stage, if necessary.[16]

Preparation of 1-(4-Hydroxy-3,5-dimethoxyphenyl)-2-(4-hydroxy-3-methoxyphenyl)propane-1,3-diol **(2)**. To a stirred solution of β-hydroxy ester **(10)** (5.58 g, 0.01 mol) dissolved in 40 ml of anhydrous THF, triethylamine (8.4 ml, 0.06 mol) distilled from sodium metal is added dropwise under nitrogen. Chlorotrimethylsilane (43.5 ml, 0.03 mol) is then added dropwise over a period of 10 min at 0°. The reaction mixture is brought to room temperature and is stirred for 30 min to complete the formation of the trimethylsilyl (TMS) ether derivative **(11)**.[19]

Lithium aluminum hydride (2.3 g, 0.6 mol) is placed in a 500-ml three-necked flask equipped with a dropping funnel, a reflux condenser with a drying tube containing Drierite, and a rubber septum. Anhydrous THF (50 ml) is slowly added to the flask under nitrogen. The THF solution is compound **(11)** is transferred to the dropping funnel. This solution is added dropwise over a period of 30 min under nitrogen with stirring and gentle reflux (oil bath temperature, 50°). After the addition of the trimethylsilyl ether solution, the stirring is continued for an additional 30 min. After the reaction mixture has been cooled in an ice water bath, 2.16 ml of water (0.12 mol) diluted with 10 ml of THF is carefully added dropwise to the solution for the decomposition of excess LAH.[13] About 16 ml of water is added dropwise to the vigorously stirred solution until finely precipitated white inorganic salts appear.[14] The precipitate is filtered off and is washed three times with 40 ml of ethyl acetate. The combined filtrate and washings are washed successively with a 2 *N* hydrochloric acid solution, brine, a saturated sodium bicarbonate solution and brine, then are dried over anhydrous sodium sulfate. The solvents are evaporated *in vacuo* to give a colorless oil (5.3 g, 100% yield), which gives one spot on TLC developed with ethyl acetate/*n*-hexane, 1 : 1 by volume [R_f values: 0.15 for the expected compound **(12)**, 0.37 and 0.49 for the *erythro* and *threo* isomers of the starting compound **(10)**, respectively]. The colorless compound obtained **(12)** is used for the next reaction without further purification.

Compound **(12)** is dissolved in 100 ml of methanol. To this solution, 10% palladium charcoal (1 g) is added[15] and the reaction mixture is stirred for 4 hr under hydrogen. The reaction is followed by TLC [R_f values: 0.11 for the expected compound **(2)**, 0.30 and 0.34 for the *threo* and *erythro*

[19] The LAH reduction of β-hydroxy ester **(10)** gives several byproducts, such as α,β-cleaved compounds and deformylated products. Therefore, the trimethylsilylation of β-hydroxy ester **(10)** to compound **(11)** is absolutely essential before LAH reduction.

isomers of the starting compound **(12)** in methanol/dichloromethane, 5:95 by volume]. The palladium charcoal is filtered off and is washed three times with 30 ml of methanol. The combined filtrate and washings are evaporated *in vacuo* to give a colorless oil which is dissolved in 20 ml of 20% methanol/dichloromethane. The solution is applied to a silica gel column (silica gel, 10 g; column size, 2.5 × 3.5 cm) and the compound is eluted with 20% methanol/dichloromethane. The eluent is evaporated *in vacuo* to give a colorless oil. The product is crystallized from about 40 ml of ethyl acetate to afford colorless crystals [3.22 g, 92% overall yield from *β*-hydroxy ester **(10)**, mp 157–159°].

[8] Synthetic ^{14}C-Labeled Lignins

By T. KENT KIRK and GÖSTA BRUNOW

One of the keys to progress in lignin biodegradation research has been the development of unequivocal quantitative assays for degradation based on ^{14}C-labeled lignins. The radioactive lignins can be prepared either by labeling specifically the lignin in plant materials (see chapter [3] in this volume), or as described here, by *in vitro* synthesis.[1]

Lignin is synthesized in plant cell walls by the polymerization of radicals generated by the one-electron oxidation of the lignin precursors, which are three *p*-hydroxycinnamyl alcohols: *p*-coumaryl alcohol, coniferyl alcohol, and sinapyl alcohol (see Scheme 1 in chapter [6] of this volume for structures). For a description of the principles of this polymerization, see Adler[2] and Sarkanen[3] (see also chapter [12] in this volume).

Most gymnosperm lignins are essentially homopolymers derived from coniferyl alcohol, with minor proportions of the units derived from *p*-coumaryl and sinapyl alcohols. Angiosperm lignins are largely copolymers of coniferyl and sinapyl alcohols, but they also contain a small proportion of units derived from *p*-coumaryl alcohol. Polymerization takes place primarily by the addition of incoming "monomer" radicals to radicals in the growing polymer within and between the plant cell walls. The *in vivo*

[1] T. K. Kirk, W. J. Connors, R. D. Bleam, W. F. Hackett, and J. G. Zeikus, *Proc. Natl. Acad. Sci. U.S.A.* **72,** 2515 (1975).

[2] E. Adler, *Wood Sci. Technol.* **11,** 169 (1977).

[3] K. V. Sarkanen, *in* "Lignins" (K. V. Sarkanen and C. H. Ludwig, eds.), p. 95. Wiley (Interscience), New York, 1971.

METHODS IN ENZYMOLOGY, VOL. 161

oxidant that generates the radicals is thought to be a phenol-oxidizing peroxidase.[4]

Similarly, synthetic lignins can be prepared in the laboratory by the peroxidase-catalyzed polymerization of the precursor alcohols. The latter are synthesized using organic chemical techniques with commercially available starting materials. For most studies, it is sufficient to use only coniferyl alcohol, which is the most common natural lignin precursor. Synthetic lignin is often referred to in the literature as "dehydrogenative polymerizate," or simply "DHP."

Polymerization of the *p*-hydroxycinnamyl alcohols, alone or in mixtures, is accomplished by the separate and simultaneous addition over a period of several hours of two aqueous solutions (precursor and hydrogen peroxide) to a buffered solution of peroxidase. The insoluble lignin polymer is recovered by centrifugation, is washed, is fractionated by molecular size if desired, and is stored as a frozen aqueous suspension.

Preparation of the Precursors

In the following, we describe the synthesis of ^{14}C-labeled coniferyl alcohol, with the label in the β- and γ-carbons of the side chain, in the methoxyl carbon, or uniformly in the aromatic ring carbons. These coniferyl alcohols permit the synthesis of synthetic gymnosperm-type lignins. As mentioned, these lignins should suffice for most investigations. However, at the end of this section, we briefly summarize methods that have been, or could be, used for synthesizing labeled sinapyl and *p*-coumaryl alcohols, if it is desired to prepare homopolymers from these other precursors or copolymers derived from mixtures. In addition, we briefly reference methods that can be used to label specific carbon atoms in the side chains.

[β,γ-^{14}C]Coniferyl Alcohol

Vanillin methoxymethyl ether is synthesized by condensing the sodium salt of vanillin with chloromethylmethyl ether. The sodium salt of vanillin is prepared by reacting sodium ethoxide with vanillin in toluene.[5,6] Sodium vanillate (16.4 g, 94 mmol) is powdered in a mortar and added to a three-necked 500-ml round bottom flask fitted with a mechanical stirrer and containing 100 ml of benzene. Chloromethylmethyl ether (10.0 g, 125 mmol) is added with stirring. As the reaction takes place, the mixture warms. Stirring is continued for 5 hr. The mixture is then extracted with

[4] J. M. Harkin and J. R. Obst, *Science* **180,** 296 (1973).
[5] H. Pauly and K. Wäscher, *Chem. Ber.* **56,** 603 (1923).
[6] K. Freudenberg and T. Kempermann, *Annalen* **602,** 184 (1957).

1 *M* NaOH until the aqueous layer is colorless and then with water until neutral. Evaporation of the benzene and recrystallization from ether gives a product in excess of 90% (mp 40°). Alternatively, the product can be purified by distillation (bp 135–138°/1.5 mm Hg).

Coniferaldehyde methoxymethyl ether is synthesized by condensing vanillin methoxymethyl ether with [^{14}C]acetaldehyde. Vanillin methoxymethyl ether (5.88 g, 30 mmol) is dissolved in 35 ml of methanol/65 ml of H_2O in a 500-ml three-necked round bottom flask fitted with a mechanical stirrer, addition funnel, and water-cooled condenser. The flask is maintained at 75° in an oil bath. Acetaldehyde (1.45 g, 33 mmol, containing [1,2-^{14}C]acetaldehyde) in 30 ml of methanol/10 ml of H_2O is added dropwise from the addition funnel to the stirred solution over 11 hr. The reaction mixture is maintained at pH 9–9.5 by adding 5 *M* NaOH periodically. The cooled reaction mixture is then extracted with benzene; this solution is washed with water, is dried over Na_2SO_4, and most of the benzene is removed by vacuum evaporation. The product is purified by silica gel column chromatography, with benzene, then with 2% (v:v) ethyl ether in benzene as eluting solvent; vanillin methoxymethyl ether is eluted by the benzene. A suitable silica gel for the column chromatography is Sil-A-200 (Sigma); approximately 500 g (1.25 liter) suffices for this separation. The final yield of coniferaldehyde methoxymethyl ether after recovery from the column and recrystallization from methanol is approximately 3.0 g (45% based on vanillin methoxymethyl ether).

The methoxymethyl group is hydrolyzed to yield coniferaldehyde.[7] For this, the above product is heated for 30 min on a steam bath in 35 ml of 50% aqueous acetic acid containing one drop of concentrated H_2SO_4. The mixture is poured over ice, and the (yellow) coniferaldehyde precipitate is recovered by filtration and is dried. Recrystallization from ethyl ether gives [β,γ-^{14}C]coniferaldehyde (mp 82–83°) in approximately 95% yield.

The coniferaldehyde is reduced to coniferyl alcohol with $NaBH_4$. Coniferaldehyde (1.8 g, 10 mmol), is dissolved in 100 ml of 95% ethanol in a 1-liter Erlenmeyer flask to give a yellow solution, and a solution of $NaBH_4$ (200 mg, 5.4 mmol) in water is added in portions at room temperature. The solution becomes colorless when the reduction is complete. The solution is *carefully*[8] adjusted to pH 6.5–7.0 with 1 *M* H_2SO_4; it is then saturated with NaCl, and is extracted with ethyl ether. The ether solution is dried over Na_2SO_4, and the ether is evaporated; the coniferyl alcohol

[7] H. Pauly and K. Feuerstein, *Chem. Ber.* **62,** 297 (1929).

[8] Coniferyl alcohol is readily polymerized in the presence of acid (to a nonligninlike material), and care must be taken to avoid overacidification, or the use of acid-containing solvents; we have found that chloroform sometimes contains too much acid and have avoided its use.

crystallizes and is obtained in essentially quantitative yield. It can be recrystallized from 1,2-dichloroethane. Coniferyl alcohol is stable for prolonged periods if stored dry, under inert gas, in the dark, and at low temperature.

Because acetaldehyde labeled only in C-1 or C-2 is apparently not available commercially, coniferyl alcohol labeled only in C-β or C-γ cannot be prepared by this synthesis unless one first synthesizes the appropriate [^{14}C]acetaldehyde. However, syntheses of specifically C-β- and C-γ-labeled coniferyl alcohols can be accomplished with [1-^{14}C]- or [2-^{14}C]malonate in a preparation that is essentially that described below for ring-labeled coniferyl alcohol. Coniferyl alcohol labeled in C-α can be made from [^{14}CHO]-vanillin, which is prepared according to Kratzl and Billek.[9]

[ring-U-^{14}C]Coniferyl Alcohol

[ring-U-^{14}C]Guaiacol is prepared from uniformly labeled phenol using published procedures.[10–14] In the first step, phenol (940 mg, = 10 mmol, containing [^{14}C]phenol), 2-chloro-5-nitrobenzophenone (2.62 g, 10 mmol), and powdered KOH (680 mg, 12 mmol; powdered by grinding in a dry, hot mortar) are mixed well in 2 ml of *N,N*-dimethylformamide (DMF) and are heated under reflux (~150°) for 45 min, and then are allowed to cool; the reaction mixture solidifies. The product is broken up with a spatula and is washed with a few milliliters of 0.1 *M* NaOH into a coarse porosity fritted glass funnel. The product is washed with approximately 80 ml of 0.1 *M* NaOH and then with cold distilled water until the filtrate is approximately pH 6. The bright yellow product, 2-[^{14}C]phenoxy-5-nitrobenzophenone, is then dried *in vacuo* over P_2O_5. The yield is approximately 3 g (95%).

In the second step, the product from above is dissolved in 6 ml of concentrated H_2SO_4 with efficient stirring (a large magnetic stirring bar is convenient) in a 125-ml Erlenmeyer flask; dissolution takes approximately 30 min. Glacial acetic acid (39 ml) is added, followed by 8.5 ml of 30% H_2O_2, which is added dropwise with stirring. During addition of H_2O_2, a tan solid begins to separate while the liquid is still very dark; the reaction mixture finally becomes buff colored. The mixture is allowed to stand for

[9] K. Kratzl and G. Billek, *Monatsh. Chem.* **85,** 845 (1954).
[10] K. Freudenberg and H. Hübner, *Chem. Ber.* **85,** 1181 (1952).
[11] D. Gagnaire, C. Lacoste, and D. Robert, *Bull. Soc. Chim. Fr.* **1970,** 1067 (1970).
[12] K. Kratzl and F. Vierhapper, *Monatsh. Chem.* **102,** 224 (1971).
[13] K. Kratzl and F. Vierhapper, *Monatsh. Chem.* **102,** 425 (1971).
[14] J. Okabe and K. Kratzl, *Tappi* **48,** 347 (1965).

3 hr and then is poured over 200 g of ice, and the mixture is held until the ice melts. The bright yellow product, 2-(2-hydroxy[^{14}C]phenoxy)-5-nitrobenzophenone, is recovered by filtration, is washed until the wash water is pH 5–7, and is dried *in vacuo* over P_2O_5. The yield is approximately 2.9 g (93%).

The third step is methylation of the phenolic hydroxyl group with methyl iodide. The above product is dissolved in 40 ml of DMF in a 250-ml glass-stoppered flask equipped with a magnetic stirring bar. Powdered K_2CO_3 (1.4 g) and 1100 μl of CH_3I are added, the flask is stoppered, and the mixture is stirred overnight at room temperature. The mixture is filtered through glasswool and the DMF removed by vacuum evaporation. The residue is dissolved in a few milliliters of dry acetone and is filtered to remove insolubles. Removal of the acetone yields 2-(2-methoxy[^{14}C]phenoxy)-5-nitrobenzophenone as a pale yellow syrup which crystallizes. The yield is essentially quantitative. Alternatively, the methylation can be performed with diazomethane in ether/methanol, which also gives an essentially quantitative yield.[1] (Caution must be used with diazomethane which is explosive.)

In the fourth step, [^{14}C]guaiacol is cleaved from the above product. For this, the product is dissolved in 5 ml of piperidine by stirring overnight, and the solution is sealed in a Pyrex glass tube. This tube in turn is sealed under piperidine in a stainless-steel tube bomb and heated in an oil bath at 150° for 90 min, followed by cooling in cold water. The tube is opened and the contents are transferred to a 125-ml separatory funnel with 25 ml of benzene. The piperidine is removed by three extractions with 1 *M* H_2SO_4 (3 × 25 ml). The aqueous solution is back-extracted with ether and the ether is added to the benzene solution. This organic solution is extracted with three portions (total 100 ml) of 2 *M* NaOH. The alkaline extract is carefully acidified with 5 *M* H_2SO_4 (~20 ml), and the guaiacol is extracted out with chloroform–acetone, 1:1 by volume (3 × 100 ml), followed by chloroform (100 ml). The organic phase is washed with approximately 5 ml of saturated NaCl, is dried over Na_2SO_4, and solvents are removed under reduced pressure at <30°. (Care must be taken to avoid distilling off the guaiacol.) The yield of [ring-U-^{14}C]guaiacol is approximately 850 mg (75%); the product is slightly impure.

Vanillin is synthesized by formylating the impure [^{14}C]guaiacol with dichloromethylthiomethyl ether.[13] Guaiacol (~850 mg, 6.85 mmol) is transferred with and is dissolved in 60 ml of dry CH_2Cl_2 in a 125-ml Erlenmeyer flask, and this is cooled on an ice bath. With stirring, 1.66 ml (2.25 g; 17.2 mmol) of dichloromethylthiomethyl ether[15] is added, fol-

[15] F. Boberg, G. Winter, and J. Moos, *Annalen* **616,** 1 (1958).

lowed by the dropwise addition, still with stirring, of 2.1 ml (17.8 mmol) of $SnCl_4$. Five min after the final addition, the mixture is transferred into a vigorously stirring solution of 16.3 g of $HgCl_2$ in 100 ml of 2 *M* HCl in a 250-ml beaker at 5°. Stirring is continued for an additional 10 min. The black precipitate is removed by centrifugation (in Teflon tubes) and is washed with CH_2Cl_2. The combined (orange) organic phase is separated from the aqueous phase, which is washed several times with CH_2Cl_2. The combined organic phase is washed with a few milliliters of water, is dried over $MgSO_4$, and solvents are removed by vacuum evaporation.

[ring-U-^{14}C]Vanillin is isolated from the reaction mixture by silica gel column chromatography. For this, a 500-ml column containing approximately 220 g of silica gel (Sil-A-200, Sigma) is packed in ethyl acetate : hexanes (2 : 5, v : v). The column is run with the same solvent at approximately 2 ml/min *o*-Vanillin elutes first, followed by vanillin, and then isovanillin; separation of these isomers is essentially complete. The yield of recovered vanillin is approximately 700 mg (67% based on guaiacol).

An alternate procedure for formylating guaiacol involves the use of bis(1,3-diphenylimidazolidinylidene-2).[16] The guaiacol is condensed with an equimolar amount of this reagent in refluxing DMF solution. We have found that the product, without purification, can be hydrolyzed directly to yield vanillin, which is extracted with ether and recrystallized from water. The yield after recrystallization is 38%.

Vanillin is condensed with monoethyl malonate to give ethyl ferulate. The [^{14}C]vanillin from above is combined with 0.75 ml of pyridine, 0.77 ml of monoethyl malonate,[17] one drop of aniline, and one drop of piperidine in a 25-ml boiling flask and is held at 55° for 24 hr (the flask is stoppered after reaching 55°). The cooled reaction mixture is dissolved in 25 ml of ethyl ether; this solution is extracted with 50 m*M* H_2SO_4 (5 × 20 ml), is washed with 1 ml of saturated NaCl, is dried over $MgSO_4$, and solvents are removed *in vacuo*. The product, ethyl [ring-U-^{14}C]ferulate, is purified by column chromatography. A column similar to that used for vanillin above is used with the solvent mixture ethyl acetate : hexanes (3 : 2, v : v). The yield of recovered product is approximately 0.9 g (90%).

The ethyl ferulate is reduced to coniferyl alcohol with bis(2-methoxyethoxy)aluminum hydride (Red-Al, Aldrich Chemical Co., Milwaukee, Wisconsin). Red-Al (4.05 g) is mixed with 40 ml of dry benzene in a dry 250-ml three-necked flask fitted with a condenser with a drying tube. Ethyl ferulate from above in 20 ml of dry benzene is added dropwise from an addition funnel over 20 min with stirring. The reaction mixture is refluxed for 30 min and is cooled to room temperature. Water (20 ml) is then added

[16] H. Giesecke and J. Hocker, *Liebigs Ann. Chem.* **1978,** 345 (1978).
[17] R. E. Strube, *Org. Synth. Coll.* **4,** 417 (1963).

carefully with stirring (dropwise at first), followed by 25 ml of saturated sodium potassium tartrate. With continued stirring, the reaction mixture is adjusted with approximately 1.75 ml of acetic acid to pH 6.5. The organic layer is washed with saturated $NaHCO_3$, and the combined aqueous layers are extracted exhaustively with ethyl ether. The combined organic layers are washed with a few milliliters of water, are dried over $MgSO_4$, and solvents are removed *in vacuo.* The yield of coniferyl alcohol is approximately 600 mg (78%). The coniferyl alcohol can be recrystallized from 1,2-dichloroethane.

[methoxyl-^{14}C]Coniferyl Alcohol

Methoxyl-labeled coniferyl alcohol is prepared by methylating the (unlabeled) intermediate 2-(2-hydroxyphenoxy)-5-nitrobenzophenone in the above synthesis with $^{14}CH_3I$, and then by following the same procedure as above to prepare coniferyl alcohol. The methylation is done in two steps, the first with $^{14}CH_3I$, and the second with excess unlabeled CH_3I to complete the methylation. Alternatively, [^{14}C]diazomethane can be used for this methylation.[1]

Polymerization of Coniferyl Alcohol[18]

Polymerization is done with a total of 1.7 g of coniferyl alcohol (9.45 mmol). Consequently, the total coniferyl alcohol from one of the above syntheses is diluted with unlabeled coniferyl alcohol to give 1.7 g. This sample is dissolved in about 20 ml of acetone and is added with stirring and under N_2 to 400 ml of degassed sodium phosphate buffer (0.01 *M*, pH 6.5) to give a clear solution. To this is added 1200 purpurogallin units of horseradish peroxidase (we have used Sigma Type II, 200 U/mg). A second 400 ml of the degassed buffer contains 9.45 mmol H_2O_2 (the concentration must be accurate; 9.45 mmol is approximately 1.1 ml of 30% H_2O_2). Both solutions are maintained under a stream of N_2. The two solutions are added simultaneously over approximately 20 hr[19] to a stirring solution of approximately 30 mg of vanillyl alcohol or guaiacylglycerol[20,21] in 200 ml of the same degassed phosphate buffer as above; this

[18] The polymerization procedure should be practiced with unlabeled coniferyl alcohol before it is attempted with the labeled material.

[19] We have used peristaltic pumps for this addition. The rates of delivery by each pump (or by each tube) are determined, and the volume of one of the solutions is adjusted if necessary to compensate for the difference; this assures simultaneous delivery of the two solutions.

[20] The vanillyl alcohol or guaiacylglycerol serves as a water-soluble initiator (site of polymerization) and retards precipitation of the growing polymer. Guaiacylglycerol, because it has four hydroxyl groups, is more water-soluble and, therefore, better than vanillyl alcohol (two hydroxyl groups), but it must be synthesized.[21]

[21] E. Adler and S. Yllner, *Acta Chem. Scand.* **7,** 570 (1953).

reaction mixture, in a 2-liter Erlenmeyer flask, also is maintained under a stream of N_2, and it is kept dark by covering the flask with aluminum foil. After final addition of reactants, stirring is continued in the dark under N_2 for 10 hr. The synthetic [^{14}C]lignin is recovered by centrifugation and is washed twice with a total of 150 ml of water. The yield is approximately 1.5 g.

The lignin is stored as a suspension in 100 ml of water at $-20°$ in the dark; it is stable for several years under these conditions. Solid lignin can be recovered by carefully evaporating the water from some of the suspension using a rotary evaporator; the temperature should be maintained below 30°. The dry lignin can be stored in the dark at $-20°$ also, but may not be stable for more than a few months.

The synthetic lignin, recovered by evaporating the water, should be soluble in DMF (a 20% solution is readily achieved); the lignin is conveniently added as a DMF solution to water to give a fine suspension for addition to microbial cultures or to enzyme reaction mixtures.[22] If the sample is not soluble in DMF, it has probably condensed due to the use of too much H_2O_2. The polymerization procedure should then be repeated with special care being taken in making up the H_2O_2. Alternatively, the DMF-insoluble portion of the preparation can be separated by centrifugation from the soluble portion, and the lignin can be recovered from the latter by solvent evaporation. The lignin can be stored as a DMF solution in the dark at $-20°$ for several months.

The synthetic lignin is best characterized by ^{13}C NMR spectroscopy.[23–25] It should be similar but not identical to milled wood lignin or Brauns' native lignin isolated from conifer wood (see chapter [1] in this volume).

The synthetic lignins prepared by this procedure should be essentially free of oligomers smaller than 1000 MW i.e., the lignins should be excluded from Sephadex LH-20 (DMF as solvent).[26] If a preparation contains low-molecular-weight components, these can be removed by preparative gel permeation chromatography,[27] or perhaps by precipitating into ethyl ether from 1,2-dichloroethane–ethanol as described for milled wood lignin (see chapter [1] in this volume).

[22] T. K. Kirk, E. Schultz, W. J. Connors, L. F. Lorenz, and J. G. Zeikus, *Arch. Microbiol.* **117,** 277 (1978).

[23] H. Nimz, J. Mogharab, and H.-D. Lüdemann, *Makromol. Chem.* **175,** 2563 (1974).

[24] D. Gagnaire and D. Robert, *Makromol. Chem.* **178,** 1477 (1977).

[25] H. Nimz, U. Tschirner, M. Stähle, R. Lehmann, and M. Schlosser, *J. Wood Chem. Technol.* **4,** 265 (1984).

[26] W. J. Connors, L. F. Lorenz, and T. K. Kirk, *Holzforschung,* **32,** 106 (1978).

[27] O. Faix, M. D. Mozuch, and T. K. Kirk, *Holzforschung,* **39,** 203 (1985).

Preparation of *p*-Coumaryl and Sinapyl Alcohols

Both *p*-coumaryl and sinapyl alcohols have been prepared using the same methods as for coniferyl alcohol. For preparation of the former,[28] *p*-hydroxybenzaldehyde (12.2 g) is reacted with monoethyl malonate (19.8 g, 1 ml pyridine and 0.2 ml piperidine; 60°; 24 hr) to yield *p*-coumaric acid ethyl ester (mp 137°, 87% yield), after a similar workup as for ethyl ferulate. The product is reduced with Red-Al in the same manner as for coniferyl alcohol. *p*-Coumaryl alcohol crystallizes from dichloromethane (mp 124°, ~80% yield). ^{14}C Labeling of the alcohol in the aromatic nucleus is accomplished by formylating labeled phenol in the same manner as described above for guaiacol or by the newer procedure for Giesecke and Hocker,[16] which employs bis(1,3-diphenylimidazolidinylidene-2) as the formylating reagent. By using labeled malonic acid, the alcohol can be labeled in the propyl side chain.

For preparation of sinapyl alcohol,[29] acetylsyringaldehyde (5 g, prepared by acetylation of syringaldehyde) is condensed with monoethyl malonate (7 g, 15 ml of pyridine and three drops each of piperidine and aniline; 50°; 24 hr); the reaction mixture is diluted with 25 ml of ethanol and is poured into a mixture of 70 ml of 4 *M* HCl and 50 ml of ethanol. After cooling to 0°, the crystalline product is filtered and washed with water. Yield is approximately 5 g (80%, mp 120–121°). The product is reduced with Red-Al in the same manner as for coniferyl alcohol. The yield, however, is lower, and it is often difficult to obtain the alcohol in crystalline form (mp 66–67°), even when it is chromatographically pure. It should be noted that sinapyl alcohol is extremely sensitive to air oxidation. Ring-labeled sinapyl alcohol can be prepared from ring-labeled vanillin via 5-iodovanillin according to Haider,[30] and side-chain-labeled sinapyl alcohol can be prepared by the use of labeled malonic acid.

[28] K. Freudenberg and G. Gehrke, *Berichte* **84,** 443 (1951).
[29] K. Freudenberg and H. Hübner, *Berichte* **85,** 1181 (1952).
[30] K. Haider, *J. Labelled Compd.* **2,** 174 (1966).

[9] Use of Polymeric Dyes in Lignin Biodegradation Assays

By MICHAEL H. GOLD, JEFFREY K. GLENN, and MARGARET ALIC

Historically, various ^{14}C-radiolabeled and unlabeled substrates have been used to screen for ligninolytic activity. However, these assays are relatively slow and cumbersome and often require synthesis of substrates which are not commercially available. Furthermore, these assays are not particularly well suited either for screening organisms for lignin-degrading ability or for isolating mutants defective in the lignin degradative system. The development of assays utilizing polymeric dyes as substrates for the lignin degradative system have facilitated these screening procedures.

The polymeric dyes used in these assays are inexpensive and can be obtained commercially in high purity. They are stable and readily soluble, have high extinction coefficients, and low toxicity toward *Phanerochaete chrysosporium* and other white rot fungi and bacteria tested. *o*-Anisidine and other low-molecular-weight dyes that have been used in similar assays[1,2] could be taken into the cells, whereas polymeric dyes will remain extracellular, at least during the initial stages of degradation, and thus will provide a better model for lignin degradation. Fungal decolorization of polymeric dyes has led to the development of simple, rapid, and quantitative spectrophotometric assays for the lignin degradative system in microorganisms.[3]

A growing body of evidence indicates that the dyes serve as substrates for at least some component(s) of the lignin degradative system and that dye decolorization is correlated with the onset of secondary metabolism and ligninolytic activity. High nitrogen in the medium represses decolorization as well as the ligninolytic system.[3] Oxygen induces lignin degradation and dye decolorization, and inhibitors of lignin degradation likewise inhibit dye decolorization.[3] Recent studies indicate that only lignin-degrading fungi are able to decolorize the dye Poly B-411 and that efficiency of decolorization is correlated with the ability to degrade several lignin model compounds.[4,5]

[1] P. Ander and K.-E. Eriksson, *Arch. Microbiol.* **109,** 1 (1976).
[2] M. H. Gold, M. B. Mayfield, T. M. Cheng, K. Krisnangkura, M. Shimada, A. Enoki, and J. K. Glenn, *Arch Microbiol.* **132,** 115 (1982).
[3] J. K. Glenn and M. H. Gold, *Appl. Environ. Microbiol.* **45,** 1741 (1983).
[4] M. W. Platt, Y. Hadar, and I. Chet, *Appl. Microbiol. Biotechnol.* **21,** 394 (1985).
[5] I. Chet, J. Trojanowski, and A. Hütterman, *Microbios Lett.* **29,** 37 (1985).

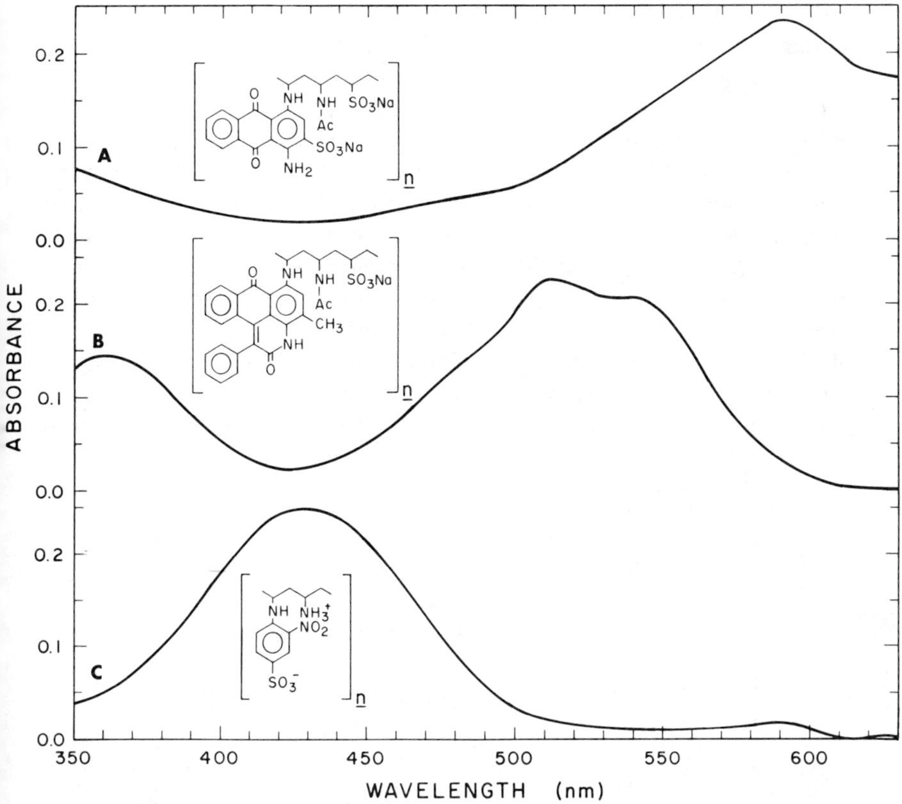

FIG. 1. Chemical structure and visible spectra of the polymeric dyes (A) Poly B; (B) Poly R; and (C) Poly Y. The spectra of the dyes (0.002%) in sodium 2,2-dimethylsuccinate, pH 4.5, were measured with a Cary 15 spectrophotometer.

Assay Methods

Liquid Culture of Phanerochaete chrysosporium. Wild-type and mutant strains are maintained on slants as previously described.[6] Erlenmeyer flasks (250 ml) containing 25 ml of medium [2% glucose, 1.2 m*M* $(NH_4)_2$ tartrate in 20 m*M* sodium 2,2-dimethylsuccinate buffer, pH 4.5, and minimal salts as previously described[7]] are inoculated with 5×10^7 conidia and incubated at 37° under air. After the first 3 days, the cultures are purged periodically with O_2.[3]

[6] M. H. Gold and T. M. Cheng, *Appl. Environ. Microbiol.* **35,** 1223 (1978).

[7] T. K. Kirk, E. Schulz, W. J. Connors, L. F. Lorenz, and J. G. Zeikus, *Arch. Microbiol.* **117,** 227 (1978).

Decolorization Assays in Liquid Medium. The chemical structures and visible spectra of the three polymeric dyes, Poly B-411, Poly R-481, and Poly Y-606, are shown in Fig. 1. The dyes are added to the fungal cultures on day 6, as aqueous solutions, to a final concentration of 0.02%. Decolorization is measured by removing 0.1 ml of extracellular culture medium and diluting 10-fold in water. Adsorption of dye to fungal mycelia reduces its intensity: the drop in absorbance following the addition of Poly R to sodium azide-inactivated cultures of *P. chrysosporium* is shown in Fig. 2. However, under these conditions the absorbance ratio A_{513}/A_{362} remains essentially unchanged with time. Thus with whole cultures, the absorbance ratio is used. Absorbance of Poly B is measured at 593 and 483 nm and Poly Y at 430 and 392 nm (Fig. 1). These wavelengths produce the largest change in the absorbance ratio during dye degradation.

In an alternative assay, the dye is added to aliquots of filtrate from

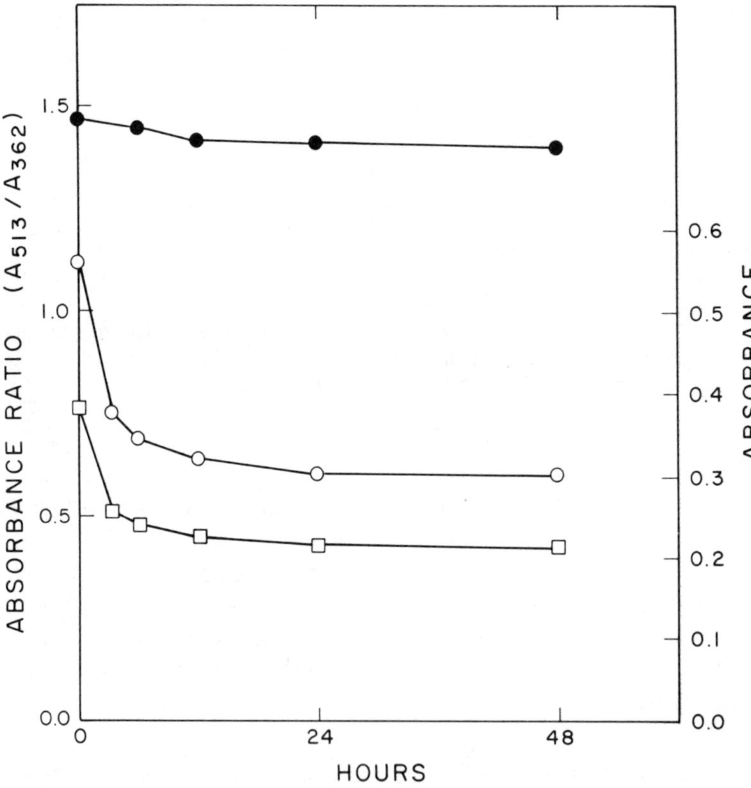

FIG. 2. A_{513} (○), A_{362} (□), and A_{513}/A_{362} (●) of Poly R in cultures grown under 100% O_2 and inactivated with N_3^- before adding the dye as described in the text. Aliquots (0.1 ml) of medium were removed at the times indicated and diluted 10-fold with water.

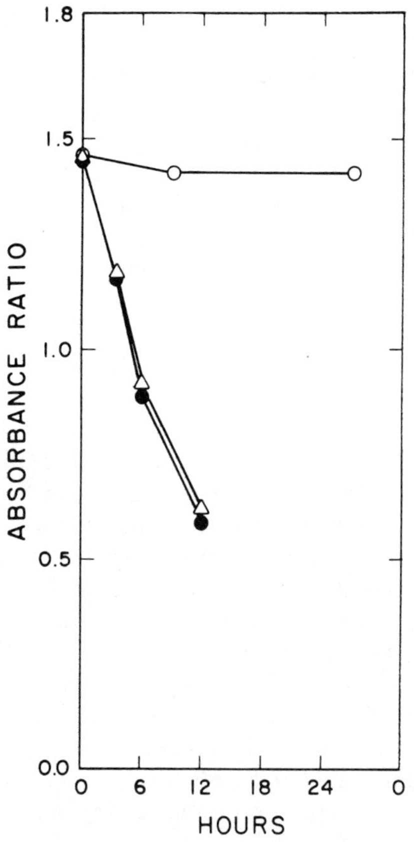

FIG. 3. Dye decolorization by wild-type and mutant strains of *P. chrysosporium*. Poly R was added to 6-day-old duplicate cultures of the wild-type (●), 104-2 (○), and 424-2 revertant (△) strains grown under O_2. Samples were removed and diluted at the times indicated, and absorbance ratios were measured as decribed in the text.

5-day-old cultures in the presence of glucose–glucose oxidase or H_2O_2 Reactions are carried out under O_2 at 38°, and the rate of decolorization is measured as the change in absorbance at one wavelength.[8,9]

It has now been shown that a Mn peroxidase produced during secondary metabolism is responsible, at least in part, for the polymeric dye decolorization.[10]

[8] M. H. Gold, J. K. Glenn, M. B. Mayfield, M. A. Morgan, and H. Kutsuki, *in* "Recent Advances in Lignin Biodegradation Research" (T. Higuchi, H.-M. Chang, and T. K. Kirk, eds.), p. 219. Uni, Tokyo, 1983.

[9] M. Kuwahara, J. K. Glenn, M. A. Morgan, and M. H. Gold, *FEBS Lett.* **169,** 247 (1984).

[10] J. K. Glenn and M. H. Gold, *Arch. Biochem. Biophys.* **242,** 329 (1985).

Fungal decolorization of Poly B and Poly R begins after an initial growth period of about 3 days. Poly B decolorization reaches a peak rate at about day 8 and then declines, whereas decolorization of Poly R continues at a steady rate for at least 11 days. Decolorization of Poly Y is significantly slower and nonexistent in cultures grown in air without O_2 purging.[3] The polymeric dye blue dextran can also be used in these assays.

The liquid decolorization assays can also be used for monitoring the effects of mutations on the lignin degradative system (Fig. 3). As shown here, dye decolorization by a secondary metabolic mutant of *P. chrysosporium* (104-2)[2] and by a phenotypic revertant (424-2) correlates with the ligninolytic capabilities of these strains.

Plate Assays

Whereas the above assays are useful for screening new species for potential ligninolytic capabilities, plate assays can be used for screening mutants defective in lignin degradation and for phenotypic revertants as well as for screening new species. Mutagenized conidia[2] of *P. chrysosporium* are plated on suitable medium containing the dye (0.02%) and are supplemented with 4% sorbose to induce colony formation.[11] After several weeks, wild-type colonies develop a decolorized halo, whereas mutant colonies remain unchanged. Such assays are also useful for selecting recombinants of *P. chrysosporium* from genetic crosses.[12]

Thus, polymeric dye assays are useful, not only for studying the enzymology of the lignin degradative system of *P. chrysosporium,* but also for isolating ligninolytic mutants and for screening other microorganisms for ligninolytic function.

[11] M. H. Gold and T. M. Cheng, *Appl. Environ. Microbiol.* **35,** 1223, (1978).
[12] M. Alic and M. H. Gold, *Appl. Environ. Microbiol.* **50,** 27 (1985).

[10] Ligninolytic Activity of *Phanerochaete chrysosporium* Measured as Ethylene Production from α-Keto-γ-methylthiolbutyric Acid

By ROBERT L. KELLEY

Recent advances in the understanding of lignin degradation by *Phanerochaete chrysosporium* were made concurrent with the development of several new ways of measuring ligninolytic activity. Glenn and Gold[1] showed a correlation between decolorization of polymeric dyes (i.e., Poly R-481, Poly B-411, and Poly Y-606) and ligninolytic activity (defined as $^{14}CO_2$ production from [^{14}C]lignin). Tein and Kirk[2] found a similar correlation between ligninolytic activity and the activity of a hydrogen peroxide (H_2O_2)-requiring, extracellular enzyme (ligninase). Both of these assays are spectrophotometric assays and are mentioned elsewhere in this volume ([9] and [23]). Kelley and Reddy[3] have shown a positive correlation between ethylene production from α-keto-γ-methylthiolbutyric acid (KTBA) and ligninolytic activity.

The basis of this assay is originally thought to be dependent on hydrogen peroxide (H_2O_2)-derived, hydroxyl radical ($\cdot$OH), which is known to react with KTBA to produce ethylene.[4,5] Recently, extracellular enzymes, ligninase, capable of degrading a wide variety of chemical bonds found in lignin have been isolated.[2,6] At least one of these enzymes is also capable of generating ethylene from KTBA.[2] Therefore, the ethylene assay may be a measure of $\cdot$OH and/or ligninase. In either case, ethylene production from KTBA is associated with ligninolytic activity.

Procedure

Organism and Growth Conditions. Phanerochaete chrysosporium (ATCC 34541) is maintained on slants of malt extract (ME) agar. This medium contains per liter: malt extract, 20 g; glucose, 20 g; Bacto-peptone, 1.0 g; and Bacto-agar, 20 g. The pH of the medium is 4.5. For preparing

[1] J. K. Glenn and M. H. Gold, *Appl. Environ. Microbiol.* **45,** 1741 (1983).
[2] M. Tien and T. K. Kirk, *Proc. Natl. Acad. Sci. U.S.A.* **81,** 2280 (1984).
[3] R. L. Kelley and C. A. Reddy, *Biochem. J.* **206,** 423 (1982).
[4] L. J. Forney, C. A. Reddy, M. Tien, and S. D. Aust, *J. Biol. Chem.* **257,** 1455 (1982).
[5] R. L. Crawford and D. L. Crawford, *Enzyme Microb. Technol.* **6,** 434 (1984).
[6] M. H. Gold, M. Kuwahara, A. A. Chiu, and J. K. Glenn, *Arch. Biochem. Biophys.* **234,** 353 (1984).

conidial inoculum, the organism is grown on plates of ME medium for 4–6 days. Ten milliliters of sterile glass distilled water is added to each plate to suspend the conidia. This conidial suspension is passed through sterile glass wool to remove contaminating mycelia. The filtrate is diluted to give 1.25×10^6 conidia/ml. Foam-stoppered 50-ml Wheaton serum bottles containing 5 ml of low-nitrogen or low-carbohydrate medium[3,7] are inoculated with the above condial suspension (0.5 ml inoculum/10 ml medium) and are incubated at 37° without agitation.

Assay

To estimate ligninolytic activity, the Wheaton bottle containing the culture is closed with a sterile, rubber septum and is sealed with an aluminum cap. During this step and all subsequent steps, the disruption to the culture should be minimized because rough handling of the cultures decreases activity. A filter-sterilized solution of KTBA (0.5 ml) is carefully injected through the rubber septum to give a final concentration of 3.3 m*M*. After 3–5 hr of additional incubation at 39°, the reaction is stopped with 1 ml of 3 *M* H_2SO_4. The ethylene from a sample of the head space (20–100 μl) is then analyzed with a Varian 2440 gas chromatograph fitted with a Porapak N (Waters Associates, Milford, Massachusetts) column and a flame-ionization detector. The carrier gas is N_2 at 30 ml/min. The detector gas flow rates are 300 ml of air/min and 30 ml of H_2/min. The chromatograph oven is maintained at 120°, with the injector and detector both set at 140°. A typical chromatogram shows only a ethylene peak with a retention time of 1.32 min, which is quantified with a standard mixture of ethylene and N_2.

In our investigations using this assay, we have found that under nitrogen- or carbohydrate-starved conditions, which promotes ligninolytic activity in *P. chrysosporium,* the pattern of ethylene production is comparable with that of lignin degradation ([^{14}C]lignin → $^{14}CO_2$) (Fig. 1A and B).[3] We have described the isolation and characterization of mutants of *P. chrysosporium* lacking the primary source of H_2O_2, glucose oxidase, which are deficient not only in their ability to produce H_2O_2, but in lignin degradation and production of ethylene from KTBA.[8] With the same fungi under similar growth conditions except under 100% O_2, Faison and Kirk[9] assayed for ·OH with an assay based on the decarboxylation of benzoate

[7] T. K. Kirk, E. Schultz, W. J. Connors, L. F. Lorenz, and J. G. Zeikus, *Arch. Microbiol.* **117,** 277 (1978).

[8] R. L. Kelley, K. Ramasamy, and C. A. Reddy, *Arch. Microbiol.* **144,** 254 (1986).

[9] B. D. Faison and T. K. Kirk, *Appl. Environ. Microbiol.* **46,** 1140 (1983).

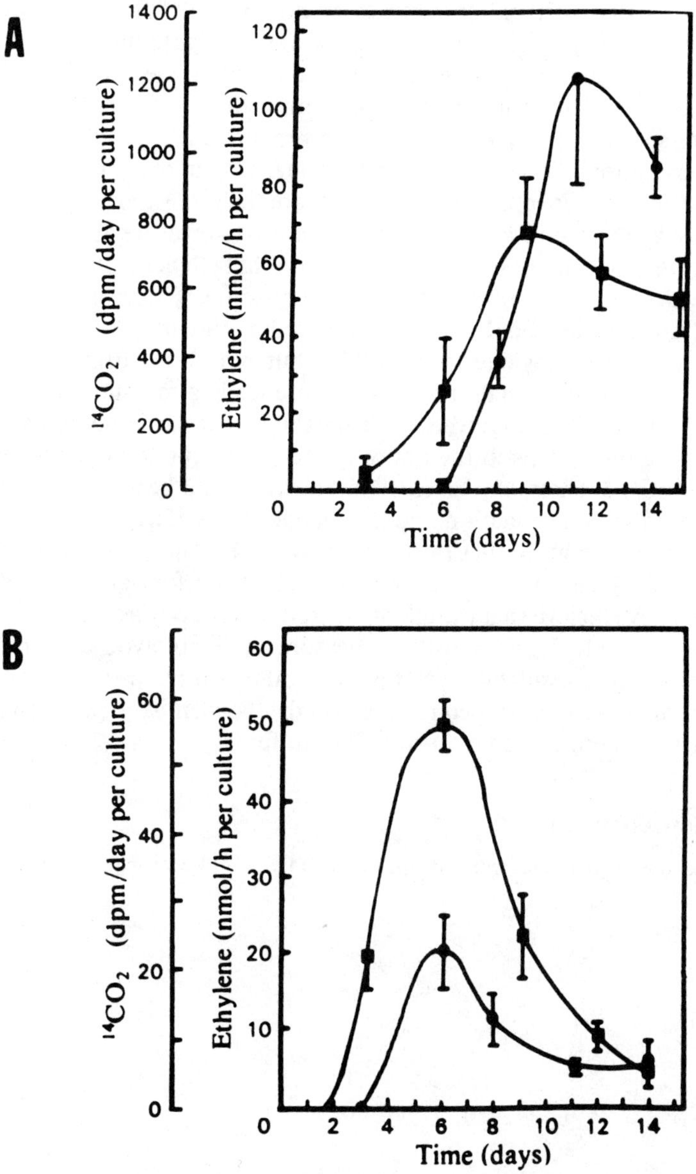

FIG. 1. Ligninolytic activity (^{14}C-labeled synthetic lignin → $^{14}CO_2$) and ·OH production, as determined by the ethylene production from α-keto-γ-methylthiobutryic acid. by *P. chrysosporium*. (A) Grown in the low-nitrogen medium and (B) grown in the low-carbohydrate medium described in the text. (●) $^{14}CO_2$ and (■) ethylene. Bars represent the SD for three replicate cultures. Reproduced from Kelley and Reddy.[3]

and found that neither the kinetics or final activity of ·OH production correlated with ligninolytic activity. However, similar to our findings,[5] they did see an inhibition of lignin degradation with the addition of ·OH scavengers, i.e., salicylate or benzoate. Although they claim that this assay is more specific for ·OH than ethylene production, they also admit that decarboxylation of benzoate may be catalyzed by an enzyme.

The advantages of the ethylene assay are high sensitivity, the ability to measure activity in intact cells, and the ability to measure as small as 5 ml of culture. Possible caveats of this assay are that ethylene production does not always correlate with $^{14}CO_2$ production from [^{14}C]lignin. This point is best illustrated by Gold *et al.*,[10] who found mutant *P. chrysosporium*, which produced ethylene from KTBA, but not $^{14}CO_2$ from [^{14}C]lignin. However, we believe that our assay is more likely a measure of the earlier nonspecific reactions involved in the breakdown of the lignin polymer, and the lack of other steps in the pathway may block the total degradation of lignin to CO_2. Therefore, users of this assay should be aware of the complexity of lignin degradation and how it effects specific assay of this event. Also, we must caution that this assay may not be suitable for some microorganisms which may have a different mechanism for lignin degradation. However, we believe that the ethylene production assay as described in this chapter is a rapid and sensitive measure of ligninolytic activity by *P. chrysosporium*. Availability of this and other simple assays should give further impetus to expanded research on the biochemistry, physiology, and genetics of lignin degradation by white rot fungi.

Acknowledgment

This research was done with the supervision of Dr. C. A. Reddy in his laboratory.

[10] M. H. Gold, J. K. Glenn, M. B. Mayfield, M. A. Morgan, and H. Kutsuki, *in* "Recent Advances in Lignin Biodegradation Research" (T. Higuchi, H.-M. Chang, and T. K. Kirk, eds.). Uni, Tokyo, 1983.

[11] Assays for Extracellular Aromatic Methoxyl-Cleaving Enzymes for the White Rot Fungus *Phanerochaete chrysosporium*[1]

By VAN-BA HUYNH, RONALD L. CRAWFORD, and ANDRZEJ PASZCZYŃSKI

Introduction

Demethoxylation of lignin's aromatic methoxyl groups by wood-rotting fungi is known to be an important degradative transformation, occurring universally during wood decay.[2-6] The most studied white rot fungus, *Phanerochaete chrysosporium,* demethoxylates lignin and lignin model compounds in two distinct ways: (1) demethoxylation of lignin substructures containing phenolic moieties (e.g., 3-methoxy-4-hydroxyphenyl structures) and (2) demethoxylation of substructures bearing methoxyls, but no phenolic groups (nonphenolic lignin).

Recent work has shown that *P. chrysosporium* can demethoxylate nonphenolic lignin model compounds using a novel, extracellular heme protein that has been called ligninase or lignin peroxidase.[7-9] It is not known whether ligninase is responsible for demethoxylation of aromatic methoxyl groups within the lignin polymer, most work having been done with model compounds. Enzymes responsible for demethoxylation of phenolic lignin could include any of numerous phenol-oxidizing enzymes excreted by the fungus, including both peroxidases and laccases.[10] Alternatively, some enzymes responsible for demethoxylation of the lignin polymer may still await discovery.

[1] This work was performed under grant number PCM-8318151 from the United States National Science Foundation.

[2] J. Trojanowski and A. Leonowicz, *Ann. Univ. Maria Curie-Skłodowska, Sect. C* **18,** 441 (1963).

[3] P. Ander and K.-E. Eriksson, *Prog. Ind. Microbiol.* **14,** 1 (1978).

[4] T. K. Kirk, *Holzforschung* **29,** 99 (1975).

[5] T. K. Kirk and H.-M. Chang, *Holzforschung* **28,** 217 (1974).

[6] R. L. Crawford, "Lignin Biodegradation and Transformation." Wiley (Interscience), New York, 1981.

[7] M. Tien and T. K. Kirk, *Science* **221,** 661 (1983).

[8] P. J. Kersten, M. Tien, B. Kalyanaraman, and T. K. Kirk, *J. Biol. Chem.* **280,** 2609 (1985).

[9] A. Paszczynski, V.-B. Huynh, P. Olson, and R. L. Crawford, *Arch. Biochem. Biophys.* **244,** 750 (1986).

[10] B. R. Brown, *in* "Oxidative Coupling of Phenols" (W. I. Taylor and A. R. Battersby, eds.), pp. 167–201. Marcel Dekker, New York, 1967.

In order to detect and study extracellular demethoxylases, quick and sensitive enzyme assays are required. The development of such assays is quite empirical. In this chapter, we describe several spectrophotometric assays that are useful in studying extracellular demethoxylating activities of *P. chrysosporium.*

Assay 1: Demethoxylation of 2,6-Di-tert-butyl-4-methoxyphenol (I), a Phenolic Lignin Substructure Model

The assay is performed in a 1-ml quartz cuvette containing 0.1 m*M* substrate I (commercially available, recrystallized before use; stock solution dissolved in 10% dimethylformamide in water), 100 m*M* sodium tartrate (pH 5.0 or 3.0),[11] 0.1 m*M* $MnSO_4$ (if the reaction is performed using *P. chrysosporium* manganese peroxidase; see this volume, chapter [25]), and 0.1 m*M* H_2O_2. Enzyme and water are added to give a final volume of 1.0 ml. The reaction is initiated by addition of H_2O_2. A reference cuvette contains all components except the enzyme. The amount of enzyme required for oxidation of **I** to produce 1 μmol of product **II** (2,6-di-*tert*-butyl-*p*-benzoquinone) per minute is defined as one unit of

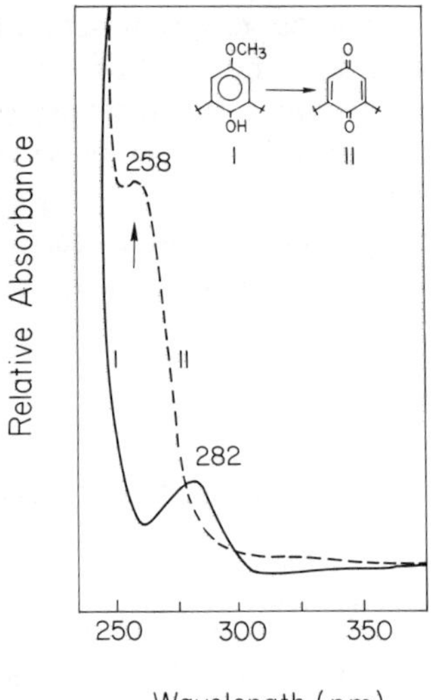

FIG. 1. Ultraviolet-visible spectra of 2,6-di-*tert*-butyl-4-methoxyphenol and its demethoxylation product in sodium tartrate at pH 5.0.

[11] The pH optimum for *P. chrysosporium* Mn^{2+}-dependent peroxidase is pH 5.0, while that of its lignin peroxidase is pH 3.0 (see Refs. 7 and 9).

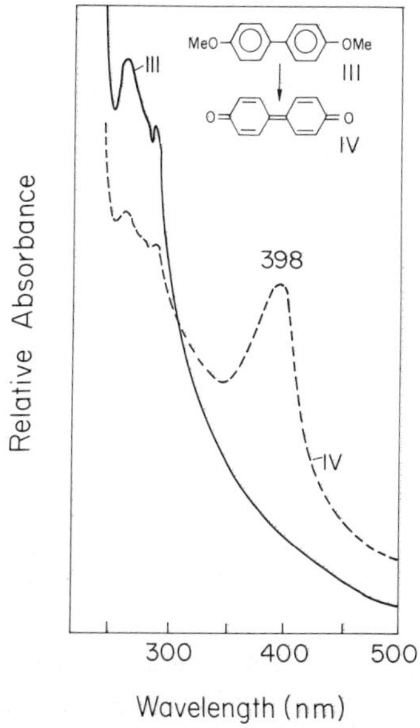

FIG. 2. Ultraviolet-visible spectra of 4,4′-dimethoxybiphenyl and its demethoxylation product in sodium tartrate at pH 3.0.

activity. Formation of product is followed by monitoring the increase in absorbance at 258 nm, which is the wavelength of maximal absorbance for dibutylbenzoquinone ($E = 11{,}500$) (Fig. 1).

Enzyme preparations used are prepared according to methods summarized in chapters elsewhere in this volume (concentrated culture filtrates of *P. chrysosporium* growth media,[12] lignin peroxidase [23], manganese peroxidase [25]. Both lignin peroxidase and manganese peroxidase from *P. chrysosporium* readily catalyze this reaction.

Assay 2: Demethoxylation of 4,4′-Dimethoxybiphenyl (III), a Nonphenolic Lignin Model Compound

Compound **II** is prepared from 4,4′-dihydroxybiphenyl (**III**, commercially available) by methylation with dimethyl sulfate. One gram of dihydroxybiphenyl is dissolved in 10 ml of 4 *N* KOH and 5 ml of dimethyl sulfate is added. The solution is heated at 65° for 90 min, is cooled, and is extracted with ether. Product is recovered on evaporation of the ether, in approximately quantitative yield.

Demethoxylation of **III** is assayed in a 1-ml quartz cuvette containing 0.1 m*M* substrate **III** (a stock solution of **III** is prepared using 10% di-

[12] A. Paszczyński, V.-B. Huynh, and R. Crawford, *FEMS Microbiol. Lett.* **29,** 37 (1985).

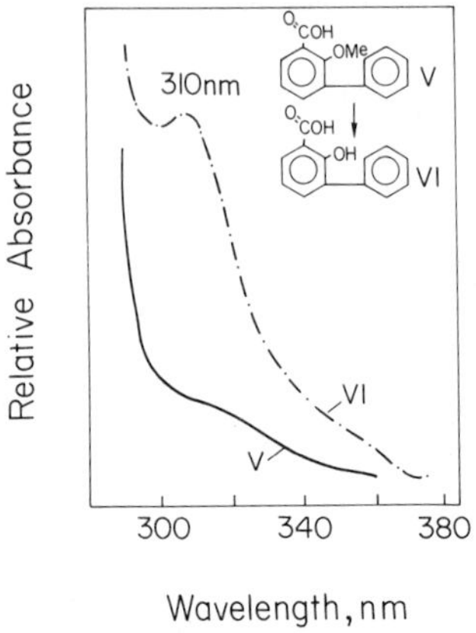

FIG. 3. Ultraviolet-visible spectra of 2-methoxy-3-phenylbenzoic acid and its demethoxylation product 2-hydroxy-3-phenylbenzoic acid in sodium tartrate at pH 5.0.

methylformamide in water), 100 m*M* sodium tartrate (pH 3.0),[11] 0.1 m*M* H_2O_2, plus enzyme and water to 1.0 ml. The reaction is initiated by addition of H_2O_2. A reference cuvette contains all components except the enzyme. The amount of enzyme required to convert 1 μmol of **III** to 1 μmol of **IV** per minute is defined as 1 U of activity. The conversion of **III** to **IV** is followed by monitoring the increase in absorbance of the solution at 398 nm, the maximal wavelength of absorbance of **IV** ($E = 69{,}000$) (Fig. 2). The mechanism of this reaction probably is identical to that described by Kersten *et al.*[8] for the demethoxylation of methoxybenzenes by lignin peroxidase. Manganese peroxidase does not catalyze this reaction.

Assay 3: Demethoxylation of 2-Methoxy-3-phenylbenzoic Acid (**V**)

Phanerochaete chrysosporium excretes an enzymatic activity that demethoxylates substrate **V** forming the free phenol, 2-hydroxy-3-phenylbenzoic acid (**VI**), as an important product.[13] This activity is dependent on addition of H_2O_2 and Mn(II) and probably is catalyzed by a heme peroxidase. However, the activity is unstable and probably is catalyzed by a different enzyme than lignin peroxidase or manganese peroxidase.[13]

Substrate **V** is prepared as described by Huynh and Crawford.[13] Assays are performed in 1-ml quartz cuvettes containing 1 ml of solution. The reaction solution contains 0.1 m*M* substrate **V** (a stock solution of **V** is prepared using 10% dimethylformamide in water), 0.1 m*M* $MnSO_4$,

[13] V.-B. Huynh and R. L. Crawford, *FEMS Microbiol. Lett.* **28,** 119 (1985).

0.05 mM H_2O_2, 100 mM sodium tartrate, plus enzyme (crude culture filtrate[12]), and water to volume. A reference cuvette contains all components except substrate. The reaction is monitored by observing the increase in absorbance at 310 nm [the wavelength of maximal absorbance of product **VI** ($E = 5{,}860$) (Fig. 3)]. One unit of activity is defined as that amount of enzyme that converts 1 μmol of **V** to **VI** per minute.

[12] Lignin Determination

By T. Kent Kirk and John R. Obst

Introduction

Lignin is a natural plastic containing carbon, hydrogen, and oxygen. Composed of phenylpropane units, lignin is heterogeneous and chemically complex. It is intimately associated with, and to some extent covalently bonded to, plant cell wall hemicelluloses. Because of these structural features, lignin is difficult to measure quantitatively. The methods commonly used are imperfect, and the researcher must have a clear understanding of the plant material being analyzed and of the limitations of the analytical method being used. The structure of lignins is discussed extensively by Sarkanen and Ludwig[1] and by Adler,[2] and various lignin determination procedures are discussed in detail by Browning.[3] (As is common in lignin chemistry, we refer to lignin in the singular in the remainder of this chapter; the reader should recognize, however, that within a fundamental structural principle, lignin differs among plant species, among plant parts, and even within plant cell walls.)

The heterogeneity of lignin is the result of its formation via a random free radical coupling of three different-*p*-hydroxycinnamyl alcohols: *p*-coumaryl, coniferyl, and sinapyl alcohols (Fig. 1). Phenoxy radicals are produced in the lignifying cell by the peroxidase-catalyzed one-electron oxidation of the alcohols. The radical from each of these precursors exists in several mesomeric forms, and coupling occurs between almost all of them, leading to over 12 different interunit C–C and C–O–C linkages for each alcohol. Further heterogeneity results by coupling between radicals derived from the three different precursors. Some intermonomer linkage types

[1] K. V. Sarkanen and C. H. Ludwig (eds.), "Lignins." Wiley (Interscience), New York, 1971.
[2] E. Adler, *Wood Sci. Technol.* **11,** 169 (1977).
[3] B. L. Browning, "Methods of Wood Chemistry," Vol. II. Wiley (Interscience), New York, 1967.

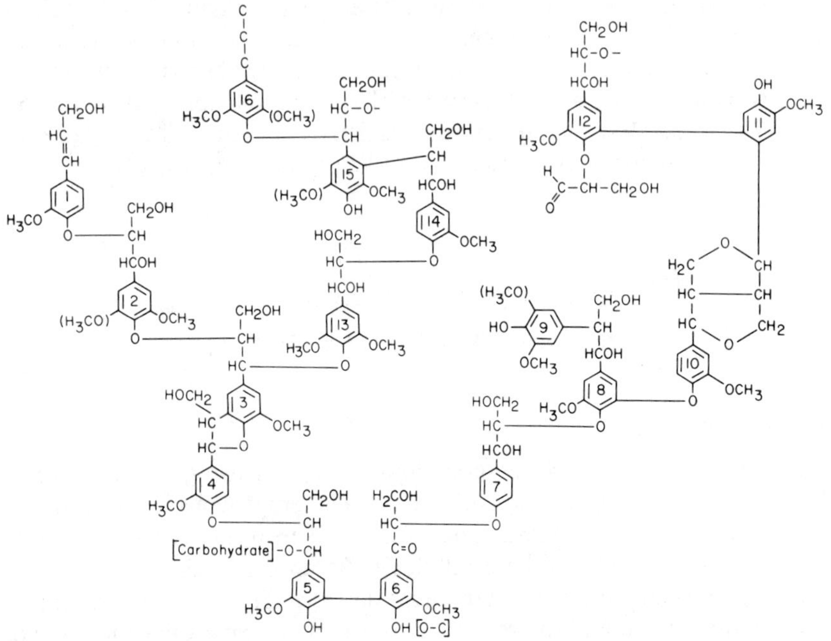

FIG. 1. Schematic structural formula for lignin, adapted from Adler.[2] *This structure is intended to illustrate the various interunit linkages. It is not a quantitatively accurate depiction of the various substructures.*

dominate (those shown in Fig. 1), and consequently, have the greatest influence on the chemical and biochemical reactivities of lignin. Angiosperm lignins typically are formed from all three of the cinnamyl alcohols, whereas most gymnosperm lignins are derived primarily from coniferyl alcohol. *p*-Coumaryl alcohol usually makes only a small contribution to lignin. Figure 1 contains units derived from the three precursors. Units derived from coniferyl alcohol are termed *guaiacyl* and those from sinapyl alcohol are termed *syringyl;* similarly, lignins are referred to as being of a guaiacyl, guaiacylsyringyl, or guaiacylsyringyl-*p*-hydroxyphenyl type.

In the following chapter, we have summarized the most widely used methods for determining lignin. Researchers should always include known lignin samples (see chapter [1] in this volume) or samples of known lignin content as controls in any procedure. The procedures that are described were developed for plant tissues, but should be adaptable to isolated lignins in reaction mixtures.

It is important to keep in mind that microbial or enzymatic modification of lignin will affect the results of qualitative and quantitative analyses.

Consequently, researchers must use caution in interpreting results where such changes have taken place.

We begin by describing some qualitative procedures.

Qualitative Determinations

Through the years, many color stains have been reported for lignin.[3] The two most widely used are the Wiesner (phloroglucinol-HCl) method and the Mäule method. The basis for both color reactions has been elucidated,[4,5] and both are fairly specific for certain structures in lignin. These stains generally provide an indication that lignin is present or absent, but should not be the sole basis for conclusions.

The best methods for determining whether lignin is present in samples are based on chemical degradations to known lignin-derived products. Three such procedures are described in the following sections, and a fourth promising new method is referenced. These procedures can also be used to gain information about the structure of the lignin in the samples. Before describing these procedures, we consider sample preparation, which can of course markedly affect the outcome of analysis.

Sample Preparation

Preparation of wood samples for analysis has been described by Browning.[3] In general, plant materials are air-dried, ground to pass a 40-mesh screen, and then are dried in a vacuum oven at <50°. The investigator must decide what subsequent steps, if any, might be desirable to remove nonlignin contaminants. Wood is usually extracted with organic solvents (see chapter [1] in this volume) prior to analysis. Dried and ground herbaceous samples might be extracted with neutral and/or acid detergent (see Goering/Van Soest procedure below) or treated with proteases.[3] Finally, the air-dry sample is freed of solvents in a vacuum oven at <50°.

Isolated lignin or other lignin-containing samples, too, may first be purified by appropriate procedures if desired (see chapter [1] in this volume).

Nitrobenzene Oxidation.[6] A commonly used procedure for determining the presence of lignin is nitrobenzene oxidation. Vanillin (**1**) is produced from guaiacyl lignins and vanillin plus syringaldehyde (**2**) from guaiacyl/

[4] D. Fengel and G. Wegener, "Wood Chemistry, Ultrastructure and Reactions." de Gruyter, Berlin, Federal Republic of Germany, 1984.

[5] G. Meshitsuka and J. Nakano, *Mokuzai Gakkaishi* **25,** 588 (1979).

[6] B. Leopold, *Acta Chem. Scand.* **6,** 38 (1952).

syringyl lignins. Total yields of these two products may be as much as 25% by weight of the lignin present. From both angiosperm and gymnosperm lignin, several other products are also formed, of which *p*-hydroxybenzaldehyde (**3**) can sometimes be a major one. Production of vanillin (and syringaldehyde) provides a good indication that lignin is present; however, certain lignans, flavanoids, condensed tannins, and perhaps suberins can interfere by giving the same products. Compounds closely related to vanillin and syringaldehyde may also be produced from such nonlignin components and care must be taken in separating and identifying these products.

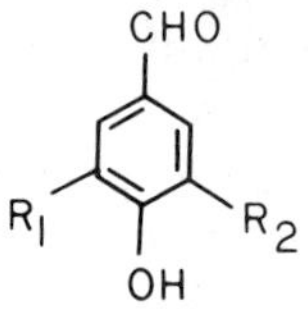

1 $R_1 = OCH_3$, $R_2 = H$

2 $R_1 = R_2 = OCH_3$

3 $R_1 = R_2 = H$

PROCEDURE. The dried and weighed sample, containing approximately 15–25 mg of lignin, is sealed, together with 0.5 ml nitrobenzene and 5 ml 2 *M* NaOH, in a stainless-steel tubing bomb (approximately 1.3 cm i.d. × 3.1 cm long). The tube is tumbled in an oil bath at 170° for 2.5 hr. After cooling, the contents of the bomb are washed with 100 ml of water into a beaker containing an internal standard for gas chromatographic (GC) analysis (approximately 5 mg of accurately weighed *o*-vanillin, 3-hydroxy-4-methoxybenzaldehyde, is convenient).

The alkaline solution is washed twice with 50-ml portions of chloroform in a separatory funnel. The aqueous phase in the funnel is then acidified with concentrated HCl to approximately pH 5 and is extracted three times with chloroform (40 ml each time). The combined chloroform extracts are dried over anhydrous magnesium sulfate or sodium sulfate, are filtered, and are evaporated at reduced pressure to approximately 4 ml.

The major products from hardwood lignins, vanillin (**1**), syringaldehyde (**2**), and a little *p*-hydroxybenzaldehyde (**3**), are conveniently determined quantitatively by GC. We have used a 60-m "DB-5" fused silica capillary column (J & W Scientific, Inc., Rancho Cordova, California). (This column gives efficient separations; less efficient columns might require a different internal standard than *o*-vanillin.)

Methylation–Oxidation.[7] Oxidation of lignin, after methylation, is another widely used procedure. It is more diagnostic for lignins than the nitrobenzene procedure, because it yields more identifiable products characteristic of lignin. The major products are veratric acid from guaiacyl units and tri-O-methylgallic acid from the syringyl units. As described below, the presence of additional products provides very strong evidence that the sample contains lignin. Samples are methylated and then are oxidized sequentially with $KMnO_4$ and H_2O_2, and the resulting substituted benzoic acids are separated and identified as their methyl esters. Pretreatment of the lignin with a pulping reagent or other depolymerizing reagent can be used to increase the yields of products.[8] The procedure is complemented by GC and GC/mass spectrometry (GC/MS).

A variation of the procedure, in which the sample is ethylated instead of methylated,[9] can provide information about microbial or enzymatic modification, namely, ring hydroxylation or demethylation of aromatic methoxyl groups.

PROCEDURE.[7] *i. Methylation.* Methylation with dimethyl sulfate is described in this section. Diazomethane methylation can also be used and has the advantage of creating essentially no residue from which the sample must be separated. (Diazomethane must be used with caution! It is explosive.) However, it is more difficult to methylate lignin quantitatively with diazomethane. For ethylation, diethyl sulfate or diazoethane can be used.[9]

For methylation with dimethyl sulfate, the weighed sample, containing 25–50 mg of lignin, is suspended in 10 ml of a mixture of dioxane–water (5:3) in a 50-ml round bottom flask. The suspension, heated to 65°, is maintained under a slow stream of N_2 and is stirred efficiently with a magnetic stirrer. Over 1 hr, dimethyl sulfate and 25% NaOH are added alternately and dropwise.[10] The solution is maintained above pH 10. Complete methylation is indicated by a negative test with diazotized sulfanilic acid,[11] diazotized *p*-nitroaniline (available commercially as the te-

[7] M. Erickson, S. Larsson, and G. E. Miksche, *Acta Chem. Scand.* **27,** 127 and 903 (and references cited therein) (1973).

[8] The yields of products can be more than doubled by depolymerizing the lignin. One procedure for doing this is acidolysis; acidolysis is described in this chapter, but following acidolysis, the entire sample, not only solvent extractables, must be recovered.[9] Another procedure for depolymerizing lignin is kraft pulping, which is described in chapter [1] of this volume. Its use with methylation–oxidation analysis is described by Erickson *et al.*[7]

[9] T. K. Kirk and E. Adler, *Acta Chem. Scand.* **24,** 3379 (1970).

[10] An improvement in the methylation procedure[7] employs a pH stat connected to magnetic valves controlling base addition. The procedure involves a single addition of dimethyl sulfate at the outset, and it results in more reproducible methylation with use of lower volumes of reagents.

[11] B. N. Ames and H. K. Mitchell, Jr., *J. Am. Chem. Soc.* **74,** 252 (1952).

trafluoroborate salt), or other sensitive phenol reagents.[3] The sample is transferred to a centrifuge tube, is acidified to pH 4 with H_3PO_4, is diluted with an equal volume of water, and is centrifuged. The residue is washed with 1% Na_2SO_4 and again recovered by centrifugation. The supernatant and the washings are combined and extracted three times with chloroform–acetone (1 : 1, v : v); each extraction is with a volume of organic solvent equal to the volume of the aqueous solution. The chloroform–acetone solution is back-washed with a few milliliters of water and is dried over anhydrous granular Na_2SO_4, and solvents are removed by vacuum evaporation. The residue from centrifugation is combined with the extract from the aqueous phase for oxidation.

ii. Oxidation. The methylated samples are oxidized in two steps. The first is with permanganate at high pH, and the second is a milder oxidation with alkaline hydrogen peroxide to convert arylglyoxylic acids formed in the first oxidation to the corresponding aromatic acids. In the permanganate oxidation, a minimum of permanganate is used and is regenerated with periodate; this avoids accumulation of MnO_2, which can interfere with sample oxidation and product recovery.

The combined methylated sample from above is dissolved/suspended in 40 ml of *tert*-butanol : water (3 : 1, v : v) in a 250-ml three-necked round bottom flask. With stirring, 40 ml of 0.5 *M* NaOH is added. Then 20 ml of 0.06 *M* $NaIO_4$ is added, followed by 20 ml of 0.03 *M* $KMnO_4$. Finally, an additional 30 ml of the periodate solution is added. *The order of addition is important.* The flask is heated during about 1 hr to reflux temperature, and the contents are refluxed for 5 hr. When boiling first occurs, 25 ml of additional periodate solution is added. The oxidation is stopped by adding a few milliliters of ethanol, which causes the solution to become brown; boiling is continued for 15 min, and the hot solution is filtered through a layer of Celite. The cooled filtrate is extracted twice with diethyl ether (50 ml each time). The aqueous phase is neutralized with 2 *M* H_2SO_4 (about 10 ml), and the solution is concentrated to about 30 ml by rotary vacuum evaporation. Solid Na_2CO_3 (0.9 g) is added, followed by 5 ml of 30% H_2O_2. The solution is heated at 50° for 10 min. Excess H_2O_2 is destroyed with permanganate solution (added dropwise until the purple color persists), and the permanganate is destroyed with a few crystals of $Na_2S_2O_5$. The solution is acidified to about pH 3 with 2 *M* H_2SO_4 and is extracted with 1 : 1 (v : v) acetone : chloroform (3 × 60 ml) and then with chloroform (30 ml). The combined extracts are back-washed with a few milliliters of saturated NaCl. The solution is dried with anhydrous Na_2SO_4 and solvents are evaporated to about 5 ml. Methanol (approximately 5 ml) is added, and the solution is methylated with excess diazomethane in ether.

Evaporation of solvents leaves a yellow oil, which is dissolved in acetone for GC.

iii. Gas chromatography and identification of products. Samples are analyzed with a nonpolar column, such as a methylated silicone elastomer, or with a capillary column. We have used 5% silicone elastomer OV-101 (Analabs, Inc. North Haven, Connecticut) on Chromosorb G, acid-washed and treated with dimethyldichlorosilane, 80–100 mesh. The lignin-derived aromatic acid methyl esters are tentatively identified from their retention times by comparison with those of authentic standards. Mass spectrometry (GC/MS) is recommended to positively identify the products. The gas chromatographic procedure can be used quantitatively with a flame ionization detector by including a suitable internal standard such as pyromellitic acid tetramethyl ester.[7]

The major products from angiosperm lignins are veratric acid (**4**) and tri-O-methylgallic acid (**5**); with some samples, such as aspen wood and grasses, anisic acid (**6**) is also a major product. Products **4** and **5** are obtained in about 2–3% yield based on the lignin in the sample, and their presence provides good evidence that lignin is present. Strong evidence for lignin is provided by the identification of additional products, especially **10–12** (numbers following structure numbers refer to approximate percentage yield from lignin): isohemipinic acid (**7**, 0.4); metahemipinic acid (**8**, 0.3); 3,4,5-trimethoxyphthalic acid (**9**, 0.2); 5,5′-dehydrodiveratric acid (**10**, 0.3); 3′,4,5-trimethoxy-3,4′-oxydibenzoic acid (**11**, 0.2); and 3′,5′,4,5-tetramethoxy-3,4′-oxydibenzoic acid (**12**, 0.8). Reference standards of compounds **4** and **5** are available commercially; the others must be synthesized (see Ref. 7 and citations therein).

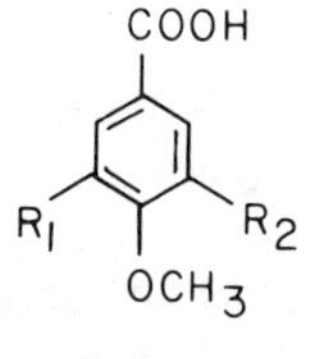

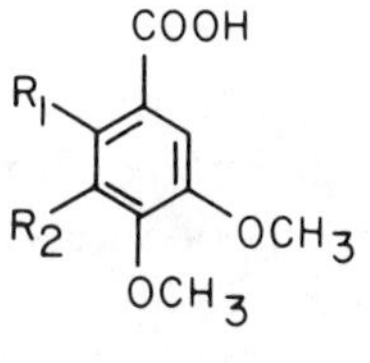

4 $R_1 = OCH_3$, $R_2 = H$

5 $R_1 = R_2 = OCH_3$

6 $R_1 = R_2 = H$

7 $R_1 = H$, $R_2 = COOH$

8 $R_1 = COOH$, $R_2 = H$

9 $R_1 = COOH$, $R_2 = OCH_3$

Acidolysis. Acidolysis provides a relatively easy and reliable procedure for qualitatively determining lignin. Lignin-containing samples are dehy-

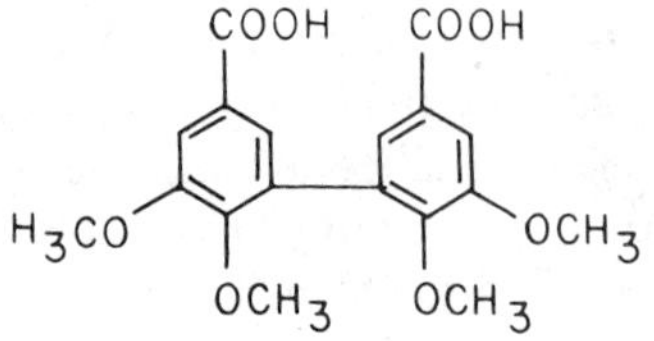

10

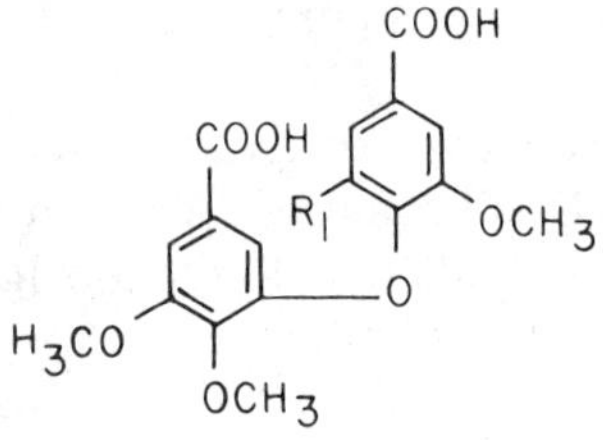

11 $R_1 = H$

12 $R_1 = OCH_3$

drated/hydrolyzed in hot dioxane/HCl solution, which results in formation of two ketols (among several other products) that are diagnostic of the major interunit linkage in lignin (the arylglycerol-β-aryl ether linkage, found between units 1 and 2, 2 and 3, 4 and 5, etc. in Fig. 1). These two products are 3-hydroxy-1-(4-hydroxy-3-methoxyphenyl)-2-propanone **(13)**

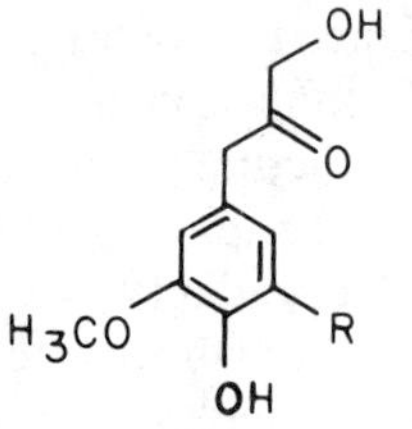

13 R = H

14 $R = OCH_3$

from appropriately linked guaiacyl units and the corresponding 3-hydroxy-1-(3,5-dimethoxy-4-hydroxyphenyl)-2-propanone **(14)** from syringyl units. Yields of these two products vary from 4 to 10% each. The compounds are readily separated and estimated by GC and are identified by MS. It is important to note that the yield of products **9** and **10** is especially sensitive to any structural modification in the units from which they are derived.

PROCEDURE.[12] *i. Acidolysis.* The sample is dissolved or suspended in purified[13] *p*-dioxane–water (9 : 1) containing 0.2 *M* HCl and is refluxed under N_2 for 4 hr. Sufficient sample is required to yield enough of products **13** and **14** for detection by GC (see below). We typically use samples

[12] K. Lundquist, *Acta Chem. Scand.* **27,** 2597 (1973), and references cited therein.

[13] A. I. Vogel, "Practical Organic Chemistry," 3rd Ed. Wiley, New York, 1966.

containing approximately 25 mg of lignin, but much less can be used. A few milliliters of acidolysis reagent are adequate; the amount must only be sufficient to supply HCl in excess. The cooled mixture is diluted with 0.4 *M* $NaHCO_3$ to raise the pH to approximately pH 3 and then is extracted three times with chloroform–acetone (1 : 1, v : v), each time with a volume equal to the volume of the solution. The extract is washed with a small volume of water and is dried over anhydrous Na_2SO_4, and solvents are removed by rotary vacuum evaporation to give a brown-yellow oil. Ketols **13** and **14** are separated—and are quantified if desired—as their trimethylsilyl derivatives by GC. For quantitation, we have used methyl arachidonate as the internal standard.

ii. Derivatization and analysis. A sample of the oil from above is dissolved in 50–100 μl of dioxane containing 10% pyridine, 50 μl of *N,O*-bis(trimethylsilyl)acetamide [or *N,O*-bis(trifluoromethylsilyl)acetamide] is added, and the container is heated briefly (50–60°) and is stoppered. Samples are then analyzed within 3–4 hr using the same gas chromatographic conditions as used for the methyl esters of products **4–12** above. A precolumn filter of fine glass wool is desirable to trap nonvolatiles.

Reference ketols **9** and **10** are not available commercially, but can be synthesized.[12] However, as an alternative, it is suggested that preextracted birch sapwood (see *Sample Preparation* above for *Methylation–Oxidation*) be used as a reference sample; the major monomeric acidolysis products from birch lignin are ketols **9** and **10**; their identity can be confirmed by GC/MS. Note that both ketols form the di- trimethylsilyl derivatives under the conditions used here, but on prolonged reaction with the trimethylsilylating reagent they can form the tri-TMS derivatives.[14]

Two alternative methods have been described for analyzing the acidolysis products: GC or GC/MS, following reduction–acetylation,[15] and HPLC of the nonderivatized products.[16]

Thioacidolysis.[17] This procedure employs dioxane–ethanethiol, 9 : 1, with 0.2 *M* BF_3 etherate to degrade and derivatize uncondensed arylglycerol-β-aryl ether structures (such as units 2 and 7 in Fig. 1), with formation of 1-(4-hydroxy-3-methoxyphenyl)-1,2,3-(tristhioethyl)propane **(15)** from guaiacyl units and the corresponding compound **(16)** from syringyl units. These products are separated and quantitated by HPLC or, as TMS derivatives, by GC. Both analytical methods resolve the threo and erythro

[14] K. Lundquist and T. K. Kirk, *Acta Chem. Scand.* **25,** 889 (1971).

[15] P. Kristersson, K. Lundquist, and A. Strand, *Wood Sci. Technol.* **14,** 297 (1980).

[16] C. Lapierre, C. Rolando, and B. Monties, *Holzforschung* **37,** 189 (1983).

[17] C. Lapierre, B. Monties, and C. Rolando, *J. Wood Chem. Technol.* **5,** 277 (1985); C. Lapierre, B. Monties, and C. Rolando, *Holzforschung* **40,** 47 (1986).

stereoisomers of **15** and **16**. Like acidolysis, thioacidolysis is diagnostic for lignin (with the same slight reservations) and reportedly has the advantages of giving higher yields and fewer products.[17] It has the disadvantage of yielding sugar and other derivatives that must be separated from **15** and **16**

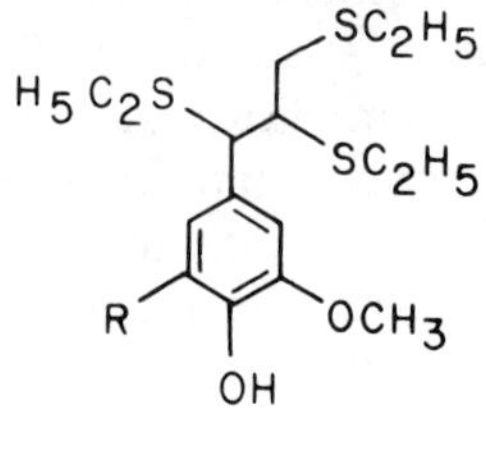

15 R=H

16 R=OCH3

when plant tissues rather than isolated lignins are analyzed. Such separation, however, is apparently accomplished by the procedures already described[17] (C. Lapierre and B. Monties, personal communication).

Quantitative Procedures

Both chemical and physical methods have been described for quantitatively determining lignin. The chemical methods are the best, but as we pointed out initially, they must be used with a full knowledge of the substrates being analyzed and of the limitations of the procedure being used. For measuring the products formed from lignin in enzymatic or microbial degradation, ^{14}C-labeled lignin is often the preferred substrate (see chapters [3] and [8] in this volume).

Chemical Methods

Klason or 72% H_2SO_4 Method. The most widely used method for lignin determination by wood chemists is probably the simplest and overall the most reliable, despite its limitations. Samples are digested with 72% sulfuric acid, then with dilute sulfuric acid, to hydrolyze and solubilize the polysaccharides; the insoluble residue is dried and weighed as lignin. The major disadvantages of the procedure are as follows: (1) other components, including proteins and suberins, may condense and analyze as Klason lignin and (2) some lignins, notably those rich in syringyl residues (angiosperm tissues), are partially solubilized. (Acid-soluble lignin from angiosperms can be estimated from the UV absorbance of the hydrolysate.[3]) The method gives good reproducibility with both gymnosperm and angiosperm tissues.

PROCEDURE.[13,18] The sample should be ground to pass at least a 20-mesh screen and should be freed of extractive components and of solvents before analysis (see description of *Sample Preparation* under **Qualitative Determinations** above and also that under the *Goering–Van Soest Method* below).

Approximately 200 mg of sample is accurately weighed to the nearest 0.1 mg into a shell vial or small beaker. One milliliter of 72% H_2SO_4 (conveniently determined by specific gravity) is added for each 100 mg of sample. The mixture is placed in a water bath at $30 \pm 0.5°$ and is stirred frequently to assure complete solution. After exactly 1 hr, it is diluted and transferred quantitatively to a 125-ml Erlenmeyer flask, using 28 ml of water for each 1 ml of acid. Secondary hydrolysis is in an autoclave at 120° for 1 hr. The hot solution is filtered through a tared Gooch, alundum, or fritted glass crucible, and the Klason lignin residue is washed with hot water to remove the acid. The crucibles containing the samples are then dried to constant weight at 105° and are weighed to the nearest 0.1 mg. Lignin is expressed as a percentage of the original sample. The lignin can be ashed, and the lignin content can be corrected for acid-insoluble inorganics if these are suspected or known to be present. Also, an indication of the purity of the lignin can be obtained by having it analyzed for methoxyl content. Methoxyl contents of Klason lignin from a variety of plant tissues are given by Brauns.[19] Klason lignin contents of many tissues are given by Brauns[19] and also by Pettersen.[20]

The filtrate from the Klason analysis may be used for a total carbohydrate determination. After dilution to a suitable volume, a portion is neutralized with $CaCO_3$ to the methyl orange end point and total sugar content is determined.[21]

Goering–Van Soest Method. This method is also widely used, especially for analysis of cereals, forages, and feed ingredients which often have high contents of protein and other nonlignocellulosic components. In this method, samples are pretreated with acid detergent to remove protein, hemicelluloses, and other components from the cellulose and lignin, and the lignin is determined by the Klason method described above or by

[18] M. J. Effland, *Tappi* **60,** 143 (1977).

[19] F. E. Brauns, "The Chemistry of Lignin." Academic Press, New York, 1952; F. E. Brauns and D. A. Brauns, "The Chemistry of Lignin," Suppl. Vol. Academic Press, New York, 1960.

[20] R. C. Pettersen, *in* "The Chemistry of Solid Wood" (R. M. Rowell, ed.), p. 57. Am. Chem. Soc. Press, Washington, D.C. 1984.

[21] R. C. Pettersen, V. H. Schwandt, and M. J. Effland, *J. Chromatogr. Sci.* **22,** 478 (1984).

difference after removing it from the cellulose via permanganate oxidation–solubilization. In using the latter procedure, care must be taken to assure sample penetration by the permanganate. It is likely, too, that some lignin might be removed in the pretreatments. Nonlignin components, such as condensed tannins and suberin, can analyze as lignin, although correction for the latter can be made, as described below.

PROCEDURE.[22] *i. Acid–detergent fiber preparation.* The sample of known weight (0.3–1 g dry weight basis) and moisture content, ground to pass a 40-mesh screen, is placed in a beaker fitted with an apparatus suitable for refluxing.[18] Acid–detergent solution (100 ml) is added at room temperature, and the mixture is heated to boiling in 5–10 min. [Acid–detergent solution is prepared by dissolving 20 g of cetyl trimethylammonium bromide (CTAB, technical grade) in 1 liter of 0.5 *M* H_2SO_4.] Heat is reduced after boiling begins to avoid foaming, and the sample is refluxed 1 hr. The sample is then filtered in a tared, 50-ml coarse porosity Gooch crucible and is washed thoroughly with hot (90–100°) water. The sample is then repeatedly washed with acetone until no more color is removed. Acetone is allowed to evaporate, and the sample is then freed of solvents in a vacuum oven or is dried overnight at 100–105° and is weighed.

ii. Lignin determination. Lignin can be determined on the above sample by the Klason method described above, with a correction for acid-insoluble ash. If cutin is present, it will analyze as lignin; the following alternative lignin determination method can be extended to estimate and correct for cutin. (Note: We have found that the Klason method underestimates the value of lignin in hardwood samples that have been acid–detergent treated, suggesting that the following method might be more accurate.)

This second method for lignin determination involves removing the lignin by its selective chemical oxidation and extraction. The crucible containing the dry acid–detergent fiber is placed in a shallow pan containing cold water to a depth of about 1 cm; fibers should not be wet. About 25 ml of a 2:1 (v:v) mixture of saturated potassium permanganate and lignin buffer solution is added, and the level of water in the pan is increased to reduce flow of the solution out of the crucibles. [Saturated $KMnO_4$ is prepared by dissolving 50 g of $KMnO_4$ and 0.05 g of Ag_2SO_4 in 1 liter of distilled water. The lignin buffer is prepared by dissolving 6 g of $Fe(NO_3)_3 \cdot 9H_2O$ and 0.15 g of $AgNO_3$ in 100 ml of distilled water, combining with 500 ml of acetic acid and 5 g of potassium acetate, and adding 400 ml of *tert*-butanol.] The sample is maintained at 20–25° for 90 ± 10 min; more mixed permanganate solution is added if needed to keep the sample wet, and a glass rod is used to assure good mixing.

[22] H. K. Goering and P. J. Van Soest, "Forage Fiber Analysis," Agric. Handb. No. 379. ARS/U.S. Dept. Agric., Govt. Printing Office, Washington, D.C. 1970.

The treating solution is removed from the sample by suction filtration, and the crucibles are placed in a clean pan and are filled about half full with demineralizing solution. Care must be taken to avoid foaming. (Demineralizing solution is made by dissolving 50 g of oxalic acid dihydrate in 700 ml of 95% ethanol, adding 50 ml of concentrated HCl and 250 ml of distilled water, and mixing.) After about 5–15 min, the demineralizing solution is removed by filtration, fresh demineralizing solution is added, and the process is repeated until the sample is white. Care must be taken to ensure that all mineral deposits on the glass crucible are treated with demineralizing solution. The sample is washed thoroughly in the crucible with 80% ethanol and then with acetone. Finally, it is dried at 100° and is weighed. The weighed residue is cellulose plus cutin (and ash).

A correction for ash can be made as described above for Klason lignin. If cutin is suspected to be present, a correction for it can be made by treating the sample with the Klason procedure described above to dissolve the cellulose. The Klason "lignin" thus determined estimates cutin. With or without this step, the residue is corrected for ash.

Other Methods. A number of other procedures have been described for quantitatively estimating lignin. Some of the most useful are briefly described in the following sections.

KAPPA METHOD.[23] A method related to the above permanganate procedure is widely used to determine lignin in chemical pulps and might be useful in certain biochemical studies. The procedure involves oxidizing the lignin in a measured amount of pulp with a known quantity of standard permanganate under acidic conditions. The unused permanganate is then determined by titration with sodium thiosulfate solution, and the lignin content (kappa number) is calculated. The method is simple and provides reproducible results.

THIOGLYCOLIC ACID METHOD.[3] This procedure has not been as widely used as the above procedures, but might also have some advantages in certain studies. The procedure involves acid-catalyzed derivatization of lignin with thioglycolic acid to produce base-soluble, acid-insoluble lignin thioglycolate. The method has the advantage over the Klason method that the isolated lignin is much less modified (aside from being derivatized), is soluble, and can be analyzed more readily. Interference by nonlignin components is apparently minimal, but methoxyl and other analyses of the isolated lignin thioglycolates should be used to verify purity. Correction for the thioglycolate moieties is based on a sulfur analysis.

CHLORINE CONSUMPTION METHOD.[3] Lignin reacts readily with chlorine, so that consumption of chlorine by a sample can form the basis for an

[23] Technical Association of the Pulp and Paper Industry, "Kappa Number of Pulp," TAPPI Official Standard OS-76. TAPPI, Atlanta, Georgia, 1976.

estimation of lignin. The method has been especially used in measuring lignin in pulps, but has also been used to measure residual isolated lignin in microbial degradation studies.[24]

METHODS BASED ON CHEMICAL DEGRADATION PRODUCTS. Browning[3] has described a procedure for estimating lignin in samples by quantifying the yield of vanillin or vanillin plus syringaldehyde, following nitrobenzene oxidation (described above as a qualitative procedure). Similarly, the methylation–oxidation procedure can be adapted to lignin determination by using special methylation apparatus[10] and precautions in the methylation and the oxidation procedures.[7] It is possible, too, that the new thioacidolysis method, described above as a qualitative method, can be adapted for quantitative analysis.

Physical Methods

Physical methods for estimating lignin in samples have not yet been developed into generally useful, reliable methods. A method for estimating lignin by ultraviolet (UV) spectroscopy following solubilization by acetylation and dissolution in acetic acid has been reported.[25] The method suffers from two notable disadvantages: (1) it is of course very sensitive to changes in the molar extinction coefficients of the UV chromophores of lignin and (2) it is not specific for lignin [any nonlignin aromatic (or other UV-absorbing) component in the sample will interfere].

Near-infrared (NIR) spectroscopic procedures are being developed for the prediction of digestibility and the estimation of composition of plant materials. At this time, however, these empirical methods have only limited utility for quantifying lignin for research purposes.

A promising technique is ^{13}C NMR. This spectroscopic technique is now widely employed for the study of lignin. The >40 signals in the spectra of milled wood lignins from gymnosperm and angiosperm tissues have been assigned.[26] These assignments are useful in qualitatively identifying an isolated material as lignin and allow classification of the lignin as a guaiacyl-, guaiacylsyringyl, or guaiacylsyringyl-*p*-hydroxyphenyl type.

Research on quantitative ^{13}C NMR of isolated, soluble lignins is now coming to fruition. Data acquisition and processing parameters have been determined which enable precise and accurate quantitative measurements within the lignin spectrum.[27] Solid-state, cross polarization-magic angle spinning ^{13}C NMR has the potential to yield quantitative measurements of

[24] T. Hiroi and K.-E. Eriksson, *Sven. Paperstidn.* **79,** 157 (1976).
[25] D. B. Johnson, W. E. Moore, and L. C. Zank, *Tappi* **44,** 793 (1961).
[26] H. H. Nimz, D. Robert, O. Faix, and M. Nems, *Holzforschung* **35,** 16 (1981).
[27] J. R. Obst and L. L. Landucci, *Holzforschung,* **40,** 87 (Supplement) (1986).

the lignin content of lignocellulosic materials.[28] The resolution of lignin peaks in solid-state ^{13}C NMR is usually inferior to that obtained with lignin solutions. However, advances continue, and the technique has the potential to become a highly sensitive research tool in studies of lignin biochemistry.

[28] J. F. Haw, G. E. Maciel, and W. A. Schroeder, *Anal. Chem.* **56,** 1323 (1984).

[13] Chemical Degradation Methods for Characterization of Lignins

By MITSUHIKO TANAHASHI and TAKAYOSHI HIGUCHI

Introduction

Lignins are complex aromatic polymers which are generally classified into three major groups based on their structural monomer units (Fig. 1, structures **1**–**3**). Guaiacyl-lignin (**2**) which occurs in conifers, lycopods, ferns, and horsetails is composed of a dehydrogenation polymer of coniferyl alcohol (**5**). Guaiacylsyringyl-lignin (**2,3**) which is composed of a mixed dehydrogenation polymer of approximately equal amounts of coniferyl and sinapyl alcohols occurs in angiosperms. Guaiacylsyringyl-*p*-hydroxyphenyllignin (**1**–**3**), which is formed in grasses, is composed of a mixed dehydrogenation polymer of coniferyl (**5**), sinapyl (**6**), and *p*-coumaryl alcohols (**4**), and *p*-coumaric acid (**7**) is esterified with γ-hydroxyl group of the lignin side chains.

The monomeric phenylpropane units in lignins are connected by both ether and carbon–carbon linkages. Arylglycerol-β-aryl ether (**8**) is the most abundant interphenylpropane linkage (45%) in lignins, followed by phenylcoumaran (**9**) (14%), diarylpropane (**10**) (15%), resinol (**11**) (13%), biphenyl (**12**) (25%), and diphenyl ether (**13**) (5%) (Fig. 2).[1] Thus, it is important for the characterization of lignins to determine the ratio of guaiacyl, syringyl, and *p*-hydroxyphenyl units with these lignin substructures.[2]

[1] Y. Z. Lai and K. V. Sarkanen, *in* "Lignins" (K. V. Sarkanen and C. H. Ludwig, eds.), p. 228. Wiley (Interscience), New York, 1971.

[2] Substructure is defined as a dimeric structure containing the principal linkage mode between monomeric phenylpropane units in lignin macromolecules (Fig. 2).

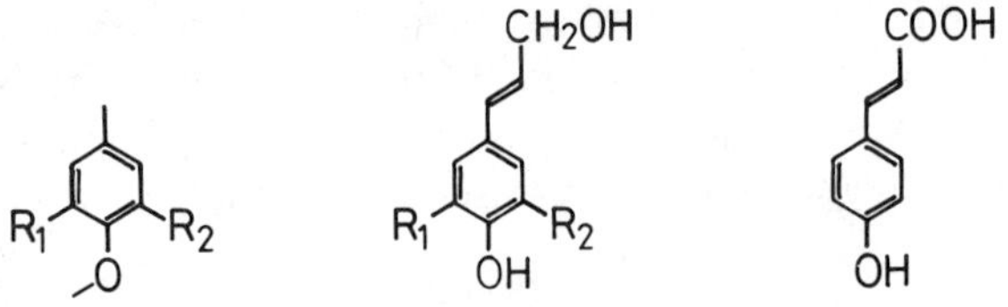

(1) $R_1 = R_2 = H$ (4) $R_1 = R_2 = H$ (7)
(2) $R_1 = H$, $R_2 = OCH_3$ (5) $R_1 = H$, $R_2 = OCH_3$
(3) $R_1 = R_2 = OCH_3$ (6) $R_1 = R_2 = OCH_3$

FIG. 1. Aromatic units of lignins and monolignols.

Hydrolysis

Alkyl–aryl ether linkages are the main intermonomeric bonds in lignins, and their cleavage, which leads to depolymerization of lignin, has been extensively studied in relation to chemical and biological delignification.[3] Ethanolysis,[4] acidolysis,[5] thioacidolysis,[6] and thioacetolysis,[7] which give low-molecular-weight lignin degradation products, have been adopted for characterization of lignin. Among these methods, acidolysis has been used as a common method. However, the yield of lignin degradation products is relatively lower than in thioacidolysis and thioacetolysis due to

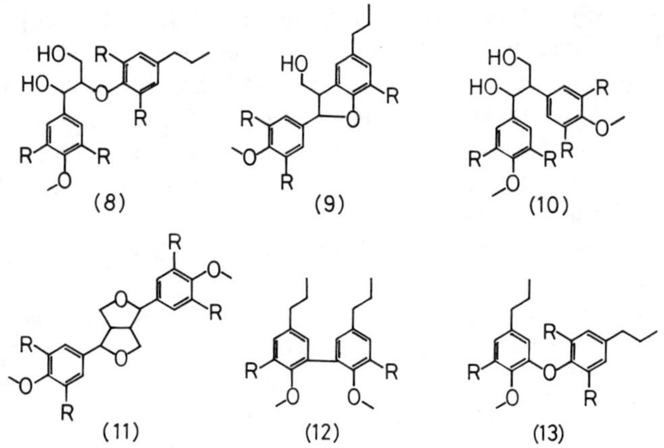

(8) (9) (10) (11) (12) (13)

FIG. 2. Dimeric substructures of lignin.

[3] A. F. A. Wallis, *in* "Lignins" (K. V. Sarkanen and C. H. Ludwig, eds.), p. 345. Wiley (Interscience), New York, 1971.
[4] W. B. Hewson, J. L. McCarthy, and H. Hibbert, *J. Am. Chem. Sco.* **63,** 3041 (1941).
[5] E. Adler, J. M. Pepper, and E. Erikson, *Ind. Eng. Chem.* **49,** 1391 (1957).
[6] C. Lapierre, B. Monties, and C. Rolando, *J. Wood Chem. Technol.* **5,** 277 (1985).
[7] H. Nimz, *Angew. Chem., Int. Ed. Engl.* **13,** 313 (1974).

the condensation reaction of the products in acidic medium. In the latter methods, the condensation is avoidable by conversion of the benzylic cation intermediates to *S*-benzyl derivatives.

Acidolysis

By acidolysis (reflux temperature in dioxane–2 *N* HCl, 9 : 1, v : v), both α- and β-aryl ether linkages **(14)** of lignins are cleaved to form β-oxy-*p*-coumaryl alcohol derivatives **(16)**, which are converted to the corresponding γ-methyl isomers **(17,18)**. When the acidolysis reaction is continued longer, these isomers are converted to the corresponding diketone **(19)** and acetone **(20)** derivatives by oxidation and reduction of the respective isomers (Fig. 3, pathway A, structures **14–20**). These acidolysis monomers, called Hibbert's ketones, are characteristic compounds derived from the arylglycerol-β-aryl ether substructure of lignin. Since the prolonged reaction (24 hr) gives only acetone and diketone derivatives from syringyl, coniferyl, and *p*-coumaryl units of lignins, this method can be used for the determination of arylglycerol-β-aryl ether linkages **(15)** in lignin.

Procedure A. Lignin or extractive-free wood (100 mg) is added to 10 ml of a mixture of dioxane–2 *N* HCl (9 : 1, v : v) in a 20-ml flask fitted with a reflux condenser. The reaction mixture is refluxed for 4 hr under atmospheric nitrogen. When extractive-free wood is used, the reaction mixture is filtered and washed with a small amount of dioxane. The reaction mixture or filtrate is added dropwise into 100 ml of water with stirring and is adjusted to pH 3 with 0.4 *N* $NaHCO_3$ and the stirring is continued for 30 min. The precipitated lignin is filtered and washed with water. The residual lignin collected is dried over P_2O_5 *in vacuo* and is weighed to calculate the yield of unhydrolyzable lignin. The filtrate is then extracted with chloroform (55 ml, three times). The chloroform extract is dried over anhydrous Na_2SO_4 and is evaporated *in vacuo* at 40°. The residue is dried *in vacuo* over KOH and P_2O_5.

Procedure B. Lignin (10 mg) is dissolved in 1.0 ml of a mixture solution

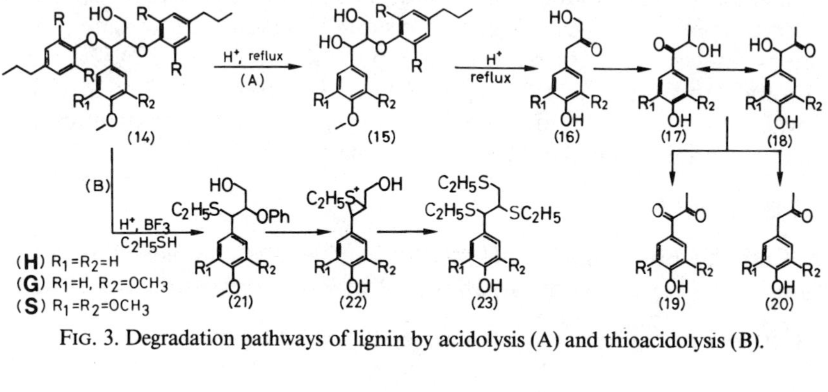

FIG. 3. Degradation pathways of lignin by acidolysis (A) and thioacidolysis (B).

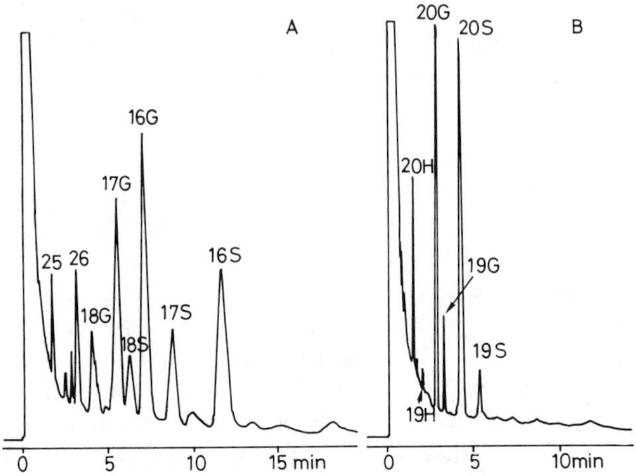

FIG. 4. GC analysis of TMS derivatives of acidolysis monomers from beech lignin [reaction time (A) 4 hr and (B) 24 hr]. The number of the respective peaks denotes the number of the compound in Figs. 3 and 6. H, *p*-Hydroxyphenyl; G, guaiacyl; S, syringyl.

of dioxane–2 *N* HCl (9:1, v:v) in a glass tube. The glass tube is sealed after flashing with nitrogen gas and is heated at 120° for 24 hr. After cooling, the glass tube is opened and the reaction mixture is transferred to a separation funnel. The reaction tube is washed with a small amount of dioxane, and the washing and 2 ml of water are added to the funnel. The solution is adjusted to pH 3 with 0.4 *N* $NaHCO_3$ and is extracted with chloroform (3 × 2 ml). The chloroform extract is dried over anhydrous Na_2SO_4 and is evaporated to dryness *in vacuo,* and the residue is dried *in vacuo* over KOH and P_2O_5.[8]

Quantitative Determination of Acidolysis Products by Gas–Liquid Chromatography. The dried residue is dissolved in pyridine (0.1 ml), and then hexamethyldisilazane (0.1 ml) and trimethylchlorosilane (0.05 ml) are added successively. The reaction mixture is shaken vigorously for 1 min. After 5 min at room temperature, the reaction mixture is evaporated to dryness *in vacuo* over P_2O_5. The residue is dissolved in hexane (5 ml) and is analyzed by gas chromatography (GC) (column: 3% SE-52 on Chromosorb W, 2 m, at 195° or OV-17 on Chromosorb W, 2 m, at 180°). The products could be identified by comparing the retention times and mass spectra to those of authentic compounds (Fig. 4).

Thioacidolysis[6]

Compared to acidolysis, thioacidolysis gives less complex mixtures of monomers **(23)** as shown in pathway B in Fig. 3 (structures **14, 21–23**).

[8] T. Higuchi, M. Tanahashi, and F. Nakatsubo, *Wood Res.* **54,** 9 (1973).

The monomer yields in thioacidolysis from lignin are higher than in acidolysis. This increased yield is particularly evident for hardwood lignins. Therefore, thioacidolysis gives more reliable monomeric composition of hydrolyzable structures of lignin.

Procedure. Lignin (20 mg) is dissolved in 10 ml of the 9:1 (v:v) mixture of dioxane and ethanethiol, containing 0.2 *M* BF_3 etherate in a stainless tube, with a Teflon-lined screw cap. The tube is kept under argon at 100° (oil bath) for 4 hr and is shaken occasionally. The ice-cooled reaction mixture is diluted with 10 ml of water, is adjusted to pH 3–4 with 0.4 *N* $NaHCO_3$ solution, and then is extracted with CH_2Cl_2. One milligram of tetracosane in CH_2Cl_2 as the internal standard is added to the extract, and the extract is dried over Na_2SO_4. The solvent is evaporated *in vacuo,* and the remaining oil is quantitatively transferred to 1 ml of dichloromethane. All these steps should be carried out without interruption. The dichloromethane solution (0.1 ml) is silylated with 1 ml of *N,O*-bis-(trimethylsilyl)trifluoroacetoamide (BSTFA) in a 2-ml reaction vial for 24 hr at room temperature and is analyzed by GC or GC-MS [column: a fused silica capillary column (50 m × 0.25 mm i.d., 1-nm film thickness, CPSIL 5B, Chrompack)]. The temperature is programmed from 180–280° (+5°/min). The products could be identified by comparing the retention times and mass spectra to those of authentic compounds (Fig. 5).

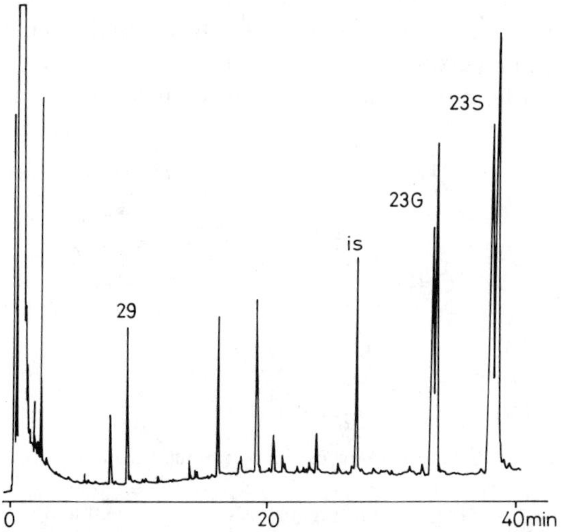

FIG. 5. GC analysis of TMS derivatives of thioacidolysis monomers from poplar lignin. The number of the respective peaks denotes the number of the compound in Fig. 3. G, Guaiacyl; S, syringyl; is, internal standard.[6]

Oxidation

Lignin is susceptible to a wide variety of oxidants[9]: (1) degradation of lignin to aromatic aldehydes and carboxylic acids by nitrobenzene, molecular oxygen, or metal oxides; (2) degradation of aromatic rings of lignin by nitric acid, chlorine, chlorine dioxide, etc.; and (3) selective oxidation of the functional groups by dichlorodicyano-*p*-benzoquinone, periodic and nitrosodisulfonic acid salt, etc.

In the following sections nitrobenzene and permanganate oxidations, which are commonly used for characterization of lignin, are described.

Alkaline Nitrobenzene Oxidation

This procedure, first introduced by Freudenberg, and co-workers (1940),[10] gives a high yield (see Fig. 6, structures **24–28**) of vanillin **(25)** (20–28%) from gymnosperm lignin and both syringaldehyde **(26)** (30–40%) and vanillin (6–12%) from angiosperm lignin. In addition, a considerable amount of *p*-hydroxybenzaldehyde **(24)** is produced from grass lignin and compression wood lignin of gymnosperms. Small amounts of **(27)** and **(28)** are produced from condensed structures, and carboxylic acids are produced by the Cannizzaro reaction from corresponding benzaldehydes. The determination of each aldehyde in the oxidation mixture leads to the chemical characterization of the lignin. This method has been used as an important tool for phylogenetic and taxonomic classification of woody species.

Procedure. A sample (60 mg of extractive-free wood meals or 15 mg of lignin) is placed in a 20-ml stainless-steel bomb, and 2 *N* sodium hydroxide (5 ml) and nitrobenzene (0.5 ml) are added. The bomb is sealed with a Teflon gasket and screw cap and is heated with agitation for 2 hr at 180°.

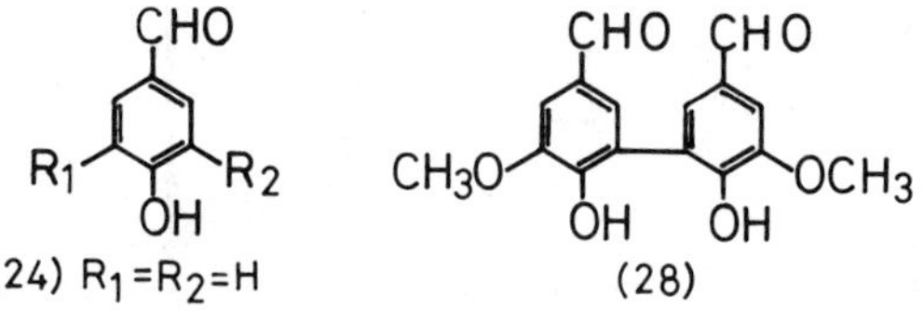

(24) $R_1=R_2=H$ (28)
(25) $R_1=H$, $R_2=OCH_3$
(26) $R_1=R_2=OCH_3$
(27) $R_1=OCH_3$, $R_2=CHO$

FIG. 6. Nitrobenzene oxidation products from lignins.

[9] H.-M. Chang and G. G. Allan, *in* "Lignins" (K. V. Sarkanen and C. H. Ludwig, eds.), p. 433. Wiley (Interscience), New York, 1971.
[10] K. Freudenberg, W. Lautsch, and K. Engler, *Chem. Ber.* **73,** 167 (1940).

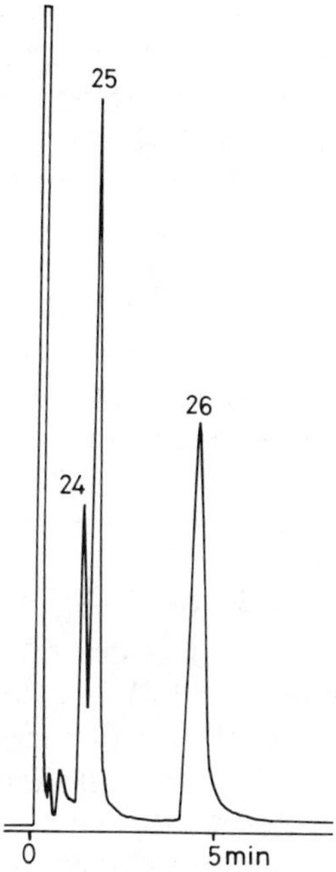

FIG. 7. GC analysis of nitrobenzene oxidation products from barley straw lignin. The number on the respective peaks denotes the number of the compound in Fig. 6.

After cooling the reaction bomb, the contents are filtered, and the residues are washed with a dilute (~0.2*N*) sodium hydroxide solution. The filtrate is then extracted with ether (3 × 50 ml) to remove neutral materials. Then the alkaline solution is acidified to pH 1 with 6 *N* HCl and is extracted again with ether (3 × 50 ml) to obtain the lignin oxidation products. These last ether extracts are dried over Na_2SO_4 and are evaporated to dryness.

Quantitative Determination of Alkaline Nitrobenzene Oxidation Products by Gas–Liquid Chromatography. The oxidation products are dissolved in ethanol (1 ml) and are analyzed by gas–liquid chromatography (column: 2% DC-550 on Chromosorb W, 2 m at 185°). The identification of the products could be made by comparison of the retention times with those of authentic compounds (Fig. 7).

Permanganate Oxidation

This technique, initially developed by Freudenberg *et al.*,[11] was modified by Larsson and Miksche.[12] The products mainly give information

[11] K. Freudenberg, A. Janson, E. Knopf, and A. Haag, *Chem. Ber.* **69,** 1415 (1936).
[12] S. Larsson and G. E. Miksche, *Acta Chem. Scand.* **25,** 674 (1971).

about "condensed type" structures of lignin. Isohemipinic acid **(33)** is formed from the phenylcoumaran structure **(9)**, metahemipinic acid **(35)** from the C–C linkage on position six of the aromatic ring, **(36)** from the 5–5′ unit **(12)**, and **(41, 42)** from the biphenyl ether unit **(13)**, etc. Main degradation products are shown in Fig. 8 (structures **29–46**).

Procedure. PREDEGRADATION BY NaOH–CuO. Lignin (200 mg) is dissolved in 20 ml of 2 *N* NaOH in a small stainless bomb, and 1 g of CuO is added. The reaction mixture is heated at 170° with stirring for 2 hr under atmospheric nitrogen to cleave the ether linkages in the lignin polymer. After cooling, the precipitate of CuO is filtered, and the filtrate is acidified with 2 *N* HCl to pH 2–3. The turbid lignin solution is saturated with NaCl and is extracted with a mixed solution of acetone–chloroform (2:1, v:v). The organic layer is dried over Na_2SO_4 and is evaporated to dryness.

METHYLATION BY DIMETHYL SULFATE. The lignin residue is dissolved in dioxane–water (5 ml/3 ml) and is heated at 65° under atmospheric nitrogen; 25% NaOH solution (33 ml) and dimethyl sulfate (18 ml) are added mutually little by little to the reaction mixture with stirring for 1 hr, and the reaction mixture is kept slightly alkaline. After 8 hr of stirring, the solution is acidified and is extracted three times with chloroform–dioxane (1:1, v:v). The organic layer is washed with a small amount of water, and the solvent is evaporated *in vacuo* to leave methylated lignin.

FIRST OXIDATION BY $KMnO_4$. The residue is dissolved in a small amount of acetone, and 1% KOH (35 ml) is added. Then 5% of potassium

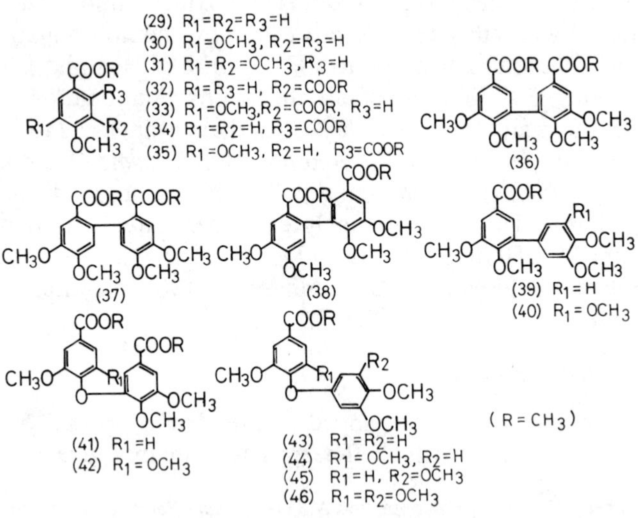

FIG. 8. Permanganate oxidation products from lignin.

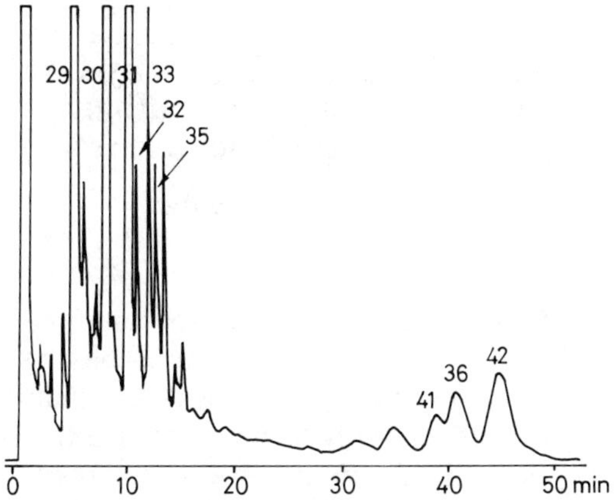

FIG. 9. GC analysis of permanganate oxidation products from bamboo lignin. The number on the respective peaks denotes the number of the compound in Fig. 8.

permanganate solution is added dropwise into the methylated lignin solution at 90–100° at pH 11–12 with stirring. The reaction is continued until the color of $KMnO_4$ does not disappear. The hot reaction mixture is filtered, and the precipitate of MnO_2 is washed with hot KOH solution. The mixed solution of filtrate and washings is cooled and extracted with ether to remove the neutral fraction.

SECOND OXIDATION BY H_2O_2. The aqueous solution is condensed under reduced pressure to 100 ml and is adjusted to pH 10 with 2 *N* NaOH, and then 30% of H_2O_2 (25 ml) is added. The reaction mixture is kept at room temperature for 30 min. The excess of H_2O_2 is degraded with $KMnO_4$ solution, and the reaction mixture is acidified with sulfuric acid. The solution is extracted three times with chloroform–acetone (1 : 1, v : v). The organic layer is dried over Na_2SO_4, and then the solvent is evaporated *in vacuo*.

Analysis by Gas–Liquid Chromatography. A mixture of aromatic carboxylic acids thus obtained from lignin is methylated with diazomethane in methanol and is analyzed by gas–liquid chromatography (column: 5% SE-30 on Chromosorb W, 160–265°, +7.5°/min). The identification of the products could be made to compare the retention times to those of authentic compounds (Fig. 9).

[14] Characterization of Lignin by Oxidative Degradation: Use of Gas Chromatography-Mass Spectrometry Technique

By CHEN-LOUNG CHEN

Lignin is one of the integral components of woody tissues in vascular plants, forming about one-fourth to one-third of these tissues, and is the second most abundant organic material next to cellulose in the plant kingdom.[1-3] It occurs in the woody tissues encrusted to cellulose fibers in the cell wall as well as filling the space between wood cells, i.e., the middle lamella. In addition, lignin is physically and chemically bonded to carbohydrates in the woody tissues, imparting mechanical strength to the wood. In general, lignins are produced *in vivo* by an enzyme-induced dehydrogenative polymerization of phenylpropane monomers, i.e., *p*-hydroxycinnamyl, coniferyl, and sinapyl alcohols. The molar ratio of the phenylpropane monomers depends on the nature of plant species. Thus, gymnosperm (softwood) lignins are derived mostly from coniferyl alcohol, while angiosperm (hardwood) lignins are produced mainly from coniferyl and sinapyl alcohols, and grass lignins from all three alcohols. Unlike most of the other naturally occurring biopolymers, the monomeric phenylpropane units in lignins are linked to each other, not by a single intermonomeric linkage, but by several different ether and carbon-to-carbon linkages. It is, therefore, not surprising that lignins in plant tissues are not extractable by organic solvents. Isolation of lignin from plant tissues by any procedure ultimately results in change of the chemical structure of lignin. Except for technical lignins, it is of primary importance to use a protolignin, i.e., extract-free plant tissues, or a lignin preparation isolated from plant tissues with minimal change in the chemical structure for characterization of lignin by chemical means. Milled wood lignin (MWL) is so far the best lignin preparation for use in the characterization of lignin. However, MWL is usually isolated from plant tissues in a yield of about 30–40% of the total lignin in a plant tissue.[4] Thus, MWL does not repre-

[1] K. Freudenberg, *in* "Constitution and Biosynthesis of Lignin" (A. C. Neish and K. Freudenberg, eds.). Springer-Verlag, New York, 1968.

[2] A. G. Waldrop, *in* "Lignins" (K. V. Sarkanen and C. H. Ludwig, eds.). Wiley (Interscience), New York, 1971.

[3] E. Adler, *Wood Sci. Technol.* **11,** 168 (1977).

[4] Y. Z. Lai and K. V. Sarkanen, *in* "Lignins" (K. V. Sarkanen and C. H. Ludwig, eds.). Wiley (Interscience), New York, 1971.

sent whole lignin occurring in a plant tissue. For characterization of lignin by physical means, such as by ^1H and ^{13}C NMR spectroscopy, MWL is usually used. Plant tissues (protolignin) cannot be directly used for these purposes because of the presence of other plant constituents, such as cellulose and hemicelluloses. This chapter deals with determination of extract, moisture and lignin contents in plant tissues, preparation of MWL from woody plant tissues, characterization of lignin using plant tissues and MWL by means of nitrobenzene oxidation,[5-8] and potassium permanganate–sodium periodate oxidation.[9-12] In addition, the following analytical procedures will also be described. The analysis of oxidation mixtures with high-performance liquid chromatography (HPLC), gas chromatography (GC), and gas chromatography-mass spectrometry (GC-MS) as well as interpretation of mass spectra thus obtained.

The nitrobenzene oxidation of a lignin provides data about the minimal quantities of aromatic ring uncondensed phenylpropane monomeric units present in the lignin. In addition, the method can be conveniently used for classification of lignins, since the procedure is relatively simple. The potassium permanganate–sodium periodate oxidation of a lignin affords data about the minimal quantities of both aromatic ring condensed and uncondensed monomeric phenylpropane monomeric units in the lignin. Moreover, the data lead to elucidation of the nature of the condensed units.

The objectives of gas chromatographic-mass spectrometric analysis of a sample are 2-fold: (1) to confirm the identity of compounds established by GC analysis of a sample or by other chromatographic analysis and (2) to elucidate the possible structure of unidentified components in the sample. Before describing the aforementioned analytical procedures, a short description of the GC-MS technique will be presented.

Gas Chromatography-Mass Spectrometry (GC-MS)

Gas chromatography-mass spectrometry consists of gas chromatograph, mass spectrometer, and computer-based data acquisition systems. The operational conditions of a GC system for GC-MS analysis of a sample

[5] K. Freudenberg and W. Lautsch, *Naturwissenschaften* **27,** 277 (1939).
[6] K. Freudenberg, W. Lautsch, and K. Engler, *Chem. Ber.* **73,** 167 (1940).
[7] B. Leopold, *Acta Chem. Scand.* **6,** 38 (1952).
[8] B. Leopold and I.-L. Malmstrom, *Acta Chem. Scand.* **6,** 49 (1952).
[9] K. Freudenberg, M. Meister, and E. Flickinger, *Chem. Ber.* **70,** 500 (1937).
[10] K. Freudenberg, C.-L. Chen, and G. Cardinale, *Chem. Ber.* **95,** 2814 (1962).
[11] S. Larsson and G. E. Miksche, *Acta Chem. Scand.* **25,** 647 (1971).
[12] M. Erickson, S. Larsson, and G. Miksche, *Acta Chem. Scand.* **27,** 903 (1973).

are determined by those for GC analysis of the sample. It is, therefore, of primary importance to elucidate the optimal conditions for the GC operation, when the GC analysis of the sample is being conducted.

An appropriate amount (usually 5–10 μl) of a sample solution is injected into the GC system of a GC-mass spectrometer with an appropriate capillary column. The effluent emerging from the GC column is continuously introduced into the ion sources of the mass spectrometer with the pressure in the order of 10^{-6} to 10^{-7} mm Hg. In the ion source, the energy of the bombarding electrons can be adjusted between approximately 10 and 100 eV. However, the usual operating energy is 70 eV. The numerous mass spectra thus produced are scanned quickly (approximately 45–80 scans/min depending on the type of GC-mass spectrometer), and the spectral data must be recorded accordingly fast in a computer system. The resulting gas chromatogram is plotted on the basis of total ion counts versus time and is referred to as total ion chromatogram. The mass spectrum of a particular component is produced by retrieval and concomitant edition of the mass spectral data stored in the computer system. The editorial process involves subtraction of the background spectra from the spectrum obtained at a scan number corresponding to the highest point of the peak of interest. The mass spectrum is interpreted to elucidate the possible structure for the compound. Most of GC-MS facilities subscribe to one or more mass spectral data centers where the standard mass spectra of over 10,000 compounds are stored in a computer system. These facilities provide a library research service for finding a matching spectrum through computer search. Occasionally, several matching spectra are found for the spectrum, and sometimes none are found. Thus, the knowledge of interpreting mass spectra is still required.

Determination of Extract and Moisture Contents

Reagents

95% Ethanol–benzene (1 : 2, v/v)
Ethanol
P_2O_5

Air-dried woody tissue is ground in a Wiley mill to pass a 40-mesh screen. The resulting wood meal is extracted continuously with 95% ethanol–benzene (1 : 2, v/v) in a Soxhlet extractor for 48 hr. The extracted wood meal is air-dried and then is dried over P_2O_5 in a vacuum desiccator for a few days. Determine the total weight of the dried extracted wood meal, then store it in a vacuum desiccator using P_2O_5 as the drying agent. Acetone–water (10 : 1, v/v) can be used for the extraction instead of 95%

ethanol–benzene (1 : 2, v/v). Weigh accurately specimen (about 1 g) of dried extracted wood meal into a weighing bottle. Dry the specimen to a constant weight in an oven at 105° (about 4 hr) and weigh. Calculate the moisture content of the extracted wood meal in weight percentage from the weight loss. Remove the solvent from the 95% ethanol–benzene solution and dry the residue in a vacuum oven at 40° for 24 hr and weigh the dried residue (extract). Calculate the extract content in the original plant tissue on an oven-dry basis as follows:

$$\text{Extract/oven-dry original plant tissue (\%)} = \frac{100R}{W(1 - M/100) + R} \quad (1)$$

where R is the weight of dried residue (g), W is the total weight of the extracted wood meal specimen (g), and M is the moisture content of the extracted wood meal specimen (%).

Determination of Total Lignin Content

Acid-Insoluble Lignin (Klason Lignin Content)[13,14]

Reagent

72% H_2SO_4 (24 ± 0.1 N)

To the accurately weighed specimen (about 1 g) of extracted wood meal (dried over P_2O_5) is added 10 ml of 75% H_2SO_4 The mixture is held at 18–20° for 2 hr with occasional stirring then is diluted with water to a volume of 400 ml. Boil the resulting solution under reflux for 4 hr. The acid-insoluble material (Klason lignin) is filtered on a glass crucible. Keep the filtrate for the determination of acid-soluble lignin. The Klason lignin is washed with hot water until free of acid. Dry the crucible with the Klason lignin to a constant weight in an oven at 105° and weigh. Calculate the acid-insoluble lignin content in the original plant tissue on an oven-dry basis as follows:

$$\text{Acid-insoluble lignin/oven-dry original plant tissue (\%)} = \frac{100A_i}{W(1 - M/100)(1 + E/100)} \quad (2)$$

where A_i is the weight of acid-insoluble lignin (g), W is the weight of extracted wood meal specimen (g), M is the moisture content of extracted wood meal specimen (%), and E is the extract content of oven-dry original plant tissue (%).

[13] K. Freudenberg and T. Ploetz, *Chem. Ber.* **73,** 754 (1940).

[14] TAPPI Standard Method T-222 om-83 (1983).

Acid-Soluble Lignin[15,16]

Reagent

0.5 N H_2SO_4

The filtrate obtained from the removal of acid-insoluble lignin [see *Acid-Insoluble Lignin (Klason Lignin Content)]* is diluted with water to a total volume of 480 ml. Measure the UV absorbance at λ 205 nm in a cuvette with a 10-mm light path using 0.5 N H_2SO_4 as a reference solution. If the absorbance is higher than the equivalent of 70% transmittance, dilute the filtrate in a volumetric flask with an appropriate volume of 0.5 N H_2SO_4, so that the absorbance will be in a range equivalent to 20–70% transmittance. Use the diluted filtrate as a test specimen. Calculate the acid-soluble lignin content in the original plant tissue on an oven-dry basis as follows:

$$\text{Acid-soluble lignin content/oven-dry original plant tissue (\%)} = \frac{100 A_s V}{110 \times 1000 W(1 - M/100)(1 + E/100)} \quad (3)$$

where A_s is the absorbance at λ 205 nm, (the average extinction coefficient for acid-soluble lignin is 110 liters/g-cm at λ 205 nm), V is the total volume of the filtrate (ml), W is the weight of extracted wood meal specimen (g), M is the moisture content of extracted wood meal specimen (%), and E is the extract content of oven-dry original plant tissue (%).

Usually, the acid-soluble lignin content in gymnosperms is less than 0.5%, which is negligible. In contrast, the acid-soluble lignin content in angiosperms and grasses is about 5%, which is significant.

Total Lignin Content

The total lignin content in the original plant tissue on an oven-dry basis is calculated as follows:

$$\text{Total lignin content/oven-dry original plant tissue (\%)} = \text{Acid-insoluble lignin contest} + \text{Acid-soluble lignin content} \quad (4)$$

[15] Y. Musha and D. Goring, *Wood Sci.* **7**, 133 (1974).
[16] TAPPI Useful Method UM-250.

Preparation of Milled Wood Lignin (MWL)[17]

Reagents

Fresh purified toluene
Fresh purified dioxane
Glacial acetic acid
1,2-Dichloroethane
Absolute ethanol
Absolute ether
n-Hexane

Apparatus

Vibratory ball mill [vibration model SA 0.6 with two 0.6-liter grinding barrels (Siebtechnik G.m.b.H., Muhlheim, Federal Republic of Germany) (Dealership in the United States: Tema Inc., Cincinnati, Ohio 45242)]

Grinding

Accurately weighed extracted wood meal (about 6 g, dried over P_2O_5) is placed in each of the grinding barrels. The vessels are filled with fresh toluene to cover the steel balls (to about 80% of the volume of the barrel), are purged with N_2 gas, and are closed tightly. The barrels are then placed on the apparatus and are milled for 48 hr at room temperature. The milled wood dispersed in toluene is drained. The milled wood remaining in the barrel is washed with fresh toluene. The toluene is combined with that drained from the sample. Let the milled wood settle down, and decant as much of the supernatant (toluene) as possible. The remainder is centrifuged to separate the milled wood. The collected toluene can be reused without purification. The grinding can be conducted without the presence of toluene (dry grinding). The MWL isolated from the milled wood obtained by the dry grinding usually contains more phenol and α-carbonyl compounds than the corresponding MWL obtained from the milled wood prepared by the grinding in toluene.

Extraction

After drying in a vacuum oven at 40°, the milled wood (about 11 g), in a centrifuge bottle, is dispersed in 100 ml of fresh dioxane–water (96:4, v/v) and is stirred mechanically at room temperature for 24 hr. The aqueous dioxane solution is removed by centrifugation. This operation is

[17] A. Björkman, *Sven. Papperstidn.* **59,** 477 (1956).

repeated three times with a stirring time of 6 hr instead of 24 hr. Combine the aqueous dioxane solutions and filter off any insoluble material, then freeze-dry to obtain crude MWL.

Purification

The crude MWL is dissolved in glacial acetic acid–water (9 : 1, v/v) in the approximate ratio of 1 g of MWL/20 ml of solvent. The resulting solution is added dropwise into 10 times the volume of distilled water in a centrifuge bottle under mechical stirring. The mixture is centrifuged, and the supernatant is removed. To the wet precipitate (MWL) in the centrifuge bottle is added 25 ml of distilled water under mechanical stirring, then water is removed by centrifugation. This operation is repeated three times. The precipitate (MWL) is stirred with 25 ml of distilled water and is freeze-dried to obtain MWL. The MWL is then dissolved in 1,2-dichloroethanol (2 : 1, v/v) in the approximate ratio of 1 g of MWL/25 ml of solvent. The resulting solution is centrifuged, or filtered if possible, to remove any insoluble material. The clear solution is then added dropwise into 10 times the volume of absolute ether in a centrifuge bottle with occasional stirring and is centrifuged. The precipitate is washed twice with 25 ml of fresh ether, and finally, with 25 ml of *n*-hexane by stirring with the solvents and by subsequent centrifugation. After air-drying, the purified MWL is dried in a drying pistol at 50° in a vacuum over P_2O_5 for 48 hr. Weigh the MWL and calculate the yield in percentage by weight per lignin content in the plant tissue. Store dried MWL in a vacuum desiccator with P_2O_5 as the drying agent.

The purified MWL usually contains carbohydrate up to about 5% by weight and is hydrophilic. Consequently, the determination of the carbohydrate content in the MWL should be performed in addition to the elemental analysis (percentage of C, H, and OCH_3) before the characterization of MWL. Also, the specimen of MWL for any experiment should be dried in a drying pistol over P_2O_5 at 50° in a vacuum at least 48 hr beforehand. Table I shows the carbohydrate content, elemental composition, with carbohydrate correction, and C_9-unit formula of some MWLs. The carbohydrate correction is made according to the composition of the carbohydrates in MWL. In general, the C_9-unit formula and the C_9-unit weight of a MWL preparation with the carbohydrate correction are similar to those of the corresponding protolignin within the limit of experimental error. The C_9-unit weight can be, therefore, used in evaluation of experimental results using protolignin, i.e., plant tissue as the starting material. Otherwise, the experimental results should be expressed on the basis of lignin content in the plant tissues.

TABLE I
CARBOHYDRATE CONTENT, ELEMENTAL COMPOSITION, AND C_9-UNIT FORMULA OF MILLED WOOD LIGNIN (MWL) FROM WOOD OF SOME PLANT SPECIES

Plant species	Carbohydrate content (% MWL)	Elemental composition (%)[a] C	H	O	OCH_3	C_9-Unit formula
ɔruce (*Picea glauca)*	2.2	64.30	6.21	29.48	15.72	$C_9H_{7.66}O_2(H_2O)_{0.48}(OCH_3)_{0.94}$
ırch (*Betula papyrifora)*	4.8	61.88	6.32	33.44	22.53	$C_9H_{6.87}O_2(H_2O)_{0.86}(OCH_3)_{1.52}$
ıong-Yang Mu (*Bischofia polycarpa)*	5.6	61.22	5.94	32.38	22.78	$C_9H_{6.58}O_2(H_2O)_{0.89}(OCH_3)_{1.13}$

[a] Corrected for carbohydrate content, based on average of two elemental analyses.

Nitrobenzene Oxidation[6-8]

Reagents

Fresh distilled nitrobenzene
2 *N* NaOH
Meconin (6,7-dimethoxylphthalide)
Chloroform acetonitrile
Acetic acid
Dichloromethane

Nitrobenzene Oxidation

Weigh accurately about 200 mg of dried extracted wood meal (40 mesh) with a known moisture content or 50 mg of dried MWL (see Preparation of MWL) or 50 mg of dried purified technical lignin. The specimen, 7 ml of 2 *N* NaOH and 0.4 ml of nitrobenzene, is placed in a 10-ml stainless-steel bomb. The bomb is sealed tightly with a Teflon gasket and screw cap and is heated at 170° for 2.5 hr with occasional shaking. Allow a 10-min warm-up period for the bomb. The heater is an aluminum heating block with a thermostat and holes fitted to accommodate the bomb or oil bath with a thermostat, both preheated to 170°. After removing and cooling the reaction bomb with water, the reaction mixture is transferred to a chloroform extractor and is extracted continuously for 6 hr with chloroform to remove excess nitrobenzene and its reduction products. The reaction mixture is then acidified to pH 4 with concentrated HCl and is further extracted continuously with chloroform for 48 hr.

Quantitative Determination of the Oxidation Product by High-Performance Liquid Chromatography (HPLC)

The solvent from the second extract is removed under reduced pressure. The residue is dissolved in 5 ml of acetonitrile, transferred into a 10-ml volumetric flask. Make up the solution to 10 ml with the solvent. Place 2 ml of the stock solution and 1 or 2 ml of 97 mg of meconin in 50 ml of acetonitrile (1 ml = 0.01 mmol) as an internal standard into a 25-ml volumetric flask and dilute to 25 ml with acetonitrile–water (1:2, v/v). (The amount of internal standard depends on the concentration of the oxidation products in the solution.)

The sample solution is filtered through Millipore membrane HVLP

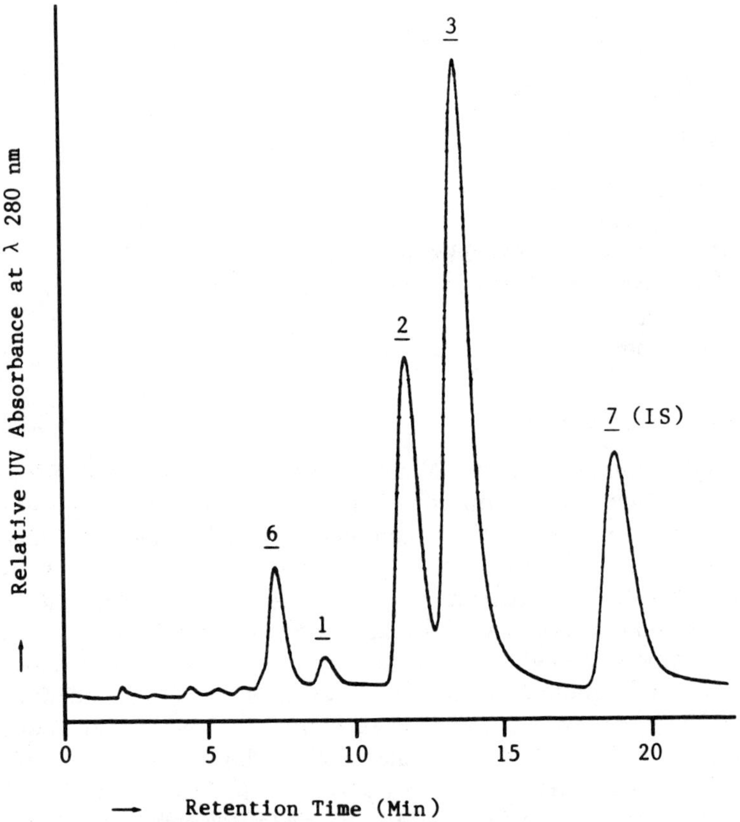

FIG. 1. High-performance liquid chromatogram of nitrobenzene oxidation mixture from lignin in sweetgum wood (*Liquidambar styraciflua* L.). Numbers refer to Scheme 1.

(pore size 0.45 μm) to remove any high-molecular-weight contaminants. Inject 10–15 μl of the sample solution into a HPLC system (chromatography pump, Model M-6000 A, Waters Associates) with a chromega bond C_{18} column (10 μm, 30 cm $\times$ 4.6 mm i.d., ES Industries, Marlton, New Jersey) to determine quantitatively the oxidation products. Acetonitrile–water (1:8, v/v) with 1% acetic acid is used as an eluent (flow rate 2 ml/min). The eluent is monitored with a UV detector (Model 440) at λ 280 nm. Figure 1 shows a high-performance liquid chromatogram of a nitrobenzene oxidation mixture from lignin in sweetgum wood.

Yield of nitrobenzene oxidation products are usually calculated in mmol/100 mg of lignin from Eq. (5) and in mol % from Eq. (6).

$$Y_1(\text{mmol/100 mg of lignin}) = \frac{500CA_1I}{A_2W} \tag{5}$$

where Y_1 is the yield of a product, A_1 is the peak area of a product, A_2 is the peak area of an internal standard (i.e., meconin), I is the amount of an internal standard added (mmol), W is the weight of lignin used (mg), and C is the conversion factor (see Table II).

$$Y_2(\text{mol \%/C}_9\text{-unit}) = Y_1M \tag{6}$$

where Y_1 is the yield of a product [mmol/100 mg of lignin (=mol/100 g of lignin), see Eqs. (5) or (7)]; Y_2 is the yield of a product (mol %/C_9-unit); and M is the C_9-unit weight of lignin (g).

TABLE II
CONVERSION FACTOR (C) AND HPLC RETENTION TIME FOR SOME NITROBENZENE OXIDATION PRODUCTS OF LIGNIN

Compounds[a]	Conversion factor[b]	HPLC retention time (min)
p-Hydroxybenzaldehyde (**1**)	0.4261	8.8
Vanillin (**2**)	0.5507	11.6
Syringaldehyde (**3**)	0.7706	13.4
p-Hydroxybenzoic acid (**4**)	—	6.1
Vanillic acid (**5**)	0.4407	7.0
Syringic acid (**6**)	0.8134	7.5
Meconin (**7**)	—	18.8

[a] The numbers in parentheses refer to Scheme 1.

[b] Calculated from appropriate calibration curve and relative to meconin, the internal standard.

Quantitative Determination of Oxidation Products by Gas Chromatography (GC)

The residue from the second chloroform extraction (see *Nitrobenzene Oxidation*) is dissolved in 5 ml of dichloromethane and is transferred into a 10-ml volumetric flask. Make up the solution to 10 ml with the solvent. Place 5 ml of the stock solution and 2 or 3 ml of 97 mg of meconin in 50 ml of dichloromethane (1 ml = 0.01 mmol) in a 10-ml volumetric flask and dilute to 10 ml with the solvent. Keep the stock solution for GC-MS analysis.

The sample solution is filtered through Millipore membrane HVHP to remove any high-molecular-weight contaminants. It is important to store the purified sample solution in a GC vail with a Teflon gasket and a screw cap. Inject 0.5 μl of the solution into a GC system (Hewlett-Packard Model 5880A) with a DB-5-bonded phase, fused silica capillary column [30 m × 0.32 mm i.d., film thickness, 0.25 μm (J & W Scientific)] and a flame ionization detector (temperature, 270°). The carrier gas is helium (flow ratio, 2 ml/min). The injection part (temperature, 200°) is fitted with a

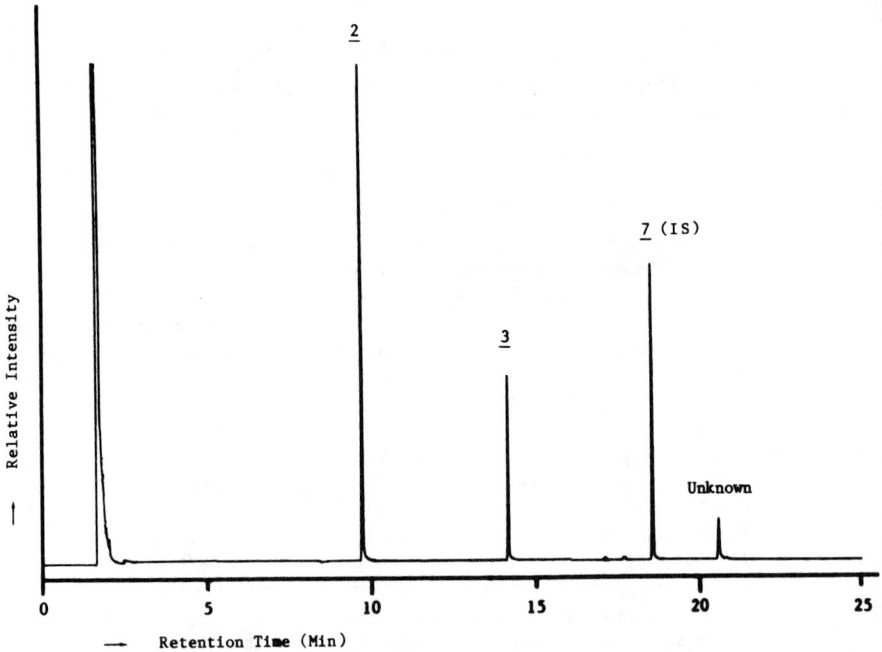

FIG. 2. Gas chromatogram of a nitrobenzene oxidation mixture from milled wood lignin (MWL) prepared from wood of Zhong-Yang Mu *(B. polycarpa)*. Numbers refer to Scheme 1.

split liner and a solvent purge system. Temperature profile for the GC operation is as follows: hold 1 min at 60°, 60–140° (15°/min), 140–250° (5°/min), final temperature 255° (10 min), and postvalue 260° (10 min). Figure 2 shows a gas chromatogram of the nitrobenzene oxidation mixture from lignin in wood of *Bischofia polycarpa,* an exceptional angiosperm wood species grown south of the Yangtse River in China.

Yield of the nitrobenzene oxidation products in mmol/100 mg of lignin is calculated from Eq. (7) and in mol % from Eq. (6).

$$Y_1(\text{mmol/100 mg of lignin}) = \frac{200CA_1I}{A_2W} \tag{7}$$

where Y_1 is the yield of a product, A_1 is the peak area of a product, A_2 is the peak area of an internal standard (i.e., meconin), I is the amount of an internal standard added (mmol), W is the weight of lignin used (mg), and C is the conversion factor (see Table III).

Analysis of Oxidation Mixture by Gas Chromatography-Mass Spectrometry (GC-MS)

The remaining stock solution (5 ml) of nitrobenzene oxidation mixture [see *Quantitative Determination of Oxidation Products by Gas Chromatography (GC)*] is diluted to 10 ml with dichloromethane and is filtered through a Millipore membrane HVHP. The sample solution should be stored in a GC vial with a Teflon gasket and screw cap. Inject about 10 μl of the sample solution into a GC-MS system (Hewlett-Packard Model 5985B) with a DB-5-bonded phase, fused silica capillary column [30 m × 0.32 mm i.d., film thickness 0.25 μm (J & W Scientific)]. The temperature

TABLE III
CONVERSION FACTOR (C) AND GC RETENTION TIME FOR THE MAJOR NITROBENZENE OXIDATION PRODUCTS OF LIGNIN

Compounds[a]	Conversion factor[b]	GC retention time (min)
p-Hydroxybenzaldehyde (**1**)	1.2886	9.0
Vanillin (**2**)	1.3158	9.7
Syringaldehyde (**3**)	1.4302	14.2
Meconin (**7**)	—	18.6

[a] The numbers in parentheses refer to Scheme 1.
[b] Calculated from appropriate calibration curve and relative to meconin, the internal standard.

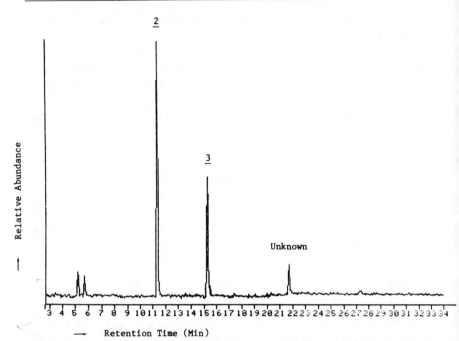

FIG. 3. Total ion chromatogram of nitrobenzene oxidation mixture from milled wood lignin (MWL) prepared from wood of Zhong-Yang Mu *(B. polycarpa)*. Numbers refer to Scheme 1.

profile for GC operation is as follows: hold 1 min at 60°, then 60–250° (10°/min). The MS conditions are EI mode (70 eV) and scanning rate of approximately 45 scans/min. Figure 3 shows a total ion chromatogram of the nitrobenzene oxidation mixture of lignin in wood of *B. polycarpa*.

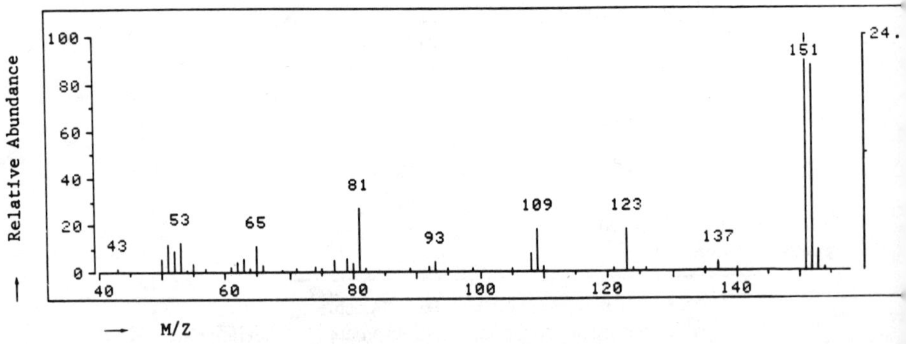

FIG. 4. Mass spectrum of vanillin [Scheme 1 **(2)**].

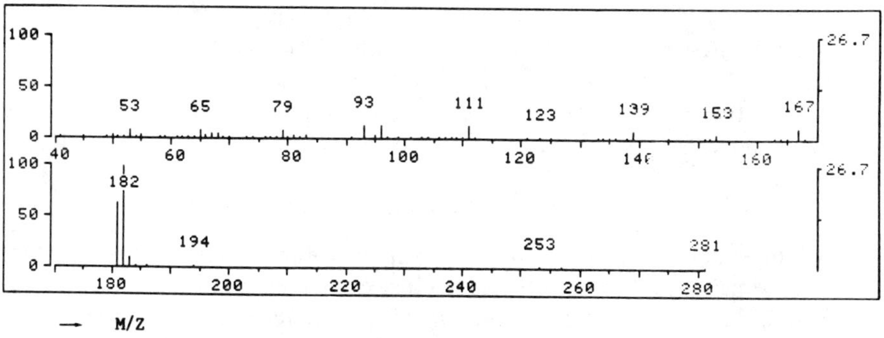

FIG. 5. Mass spectrum of syringaldehyde [Scheme 1 (3)].

Interpretation of Mass Spectra

The GC-MS analysis of the nitrobenzene oxidation mixture of lignin in wood of *B. polycarpa* results in obtaining mass spectra of the major components with GC retention time 11.3 and 15.2 min as shown in Figs. 4 and 5. The analysis of the spectra shows that the compounds are vanillin and syringaldehyde, respectively. The major mass spectral fragmentation pattern of *p*-hydroxybenzaldehyde [Scheme 1 (**1**)], vanillin [Scheme 1 (**2**)], and syringaldehyde [Scheme 1 (**3**)] is loss of aldehyde hydrogen from the molecular ion (M) to give the (M − 1) ion species. An elimination of CO from the (M − 1) species affords the (M − 29) ion species. The M and (M − 1) ion species are the most predominant ions in the spectra with relative intensity of about 90–100%. Thus, the spectrum of vanillin shows that the M and (M − 1) ions at *m/z* 152 ($C_8H_8O_3$) and 151 have relative intensities of 87 and 100% (base peak), respectively. Similarly, M ion of syringaldehyde at *m/z* 182 ($C_9H_{10}O_4$) has a relative intensity of 100% (base peak), while that of (M − 1) ion at *m/z* 181 is 63%. Although the mass spectrum of *p*-hydroxybenzaldehyde is not shown here, the spectrum also

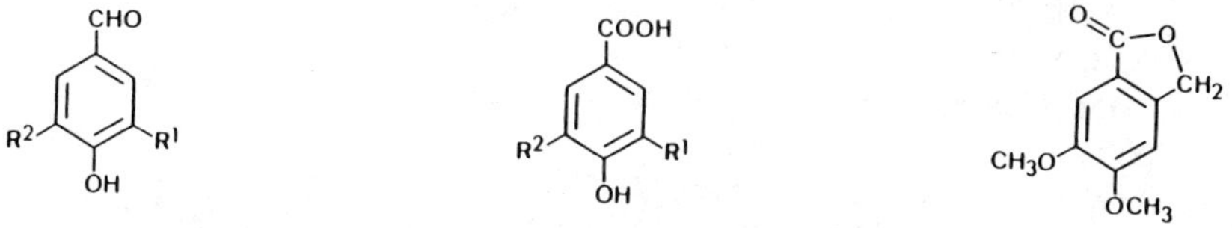

1, $R^1 = R^2 = H$
2, $R^1 = OCH_3$; $R^2 = H$
3, $R^1 = R^2 = OCH_3$

4, $R^1 = R^2 = H$
5, $R^1 = OCH_3$; $R^2 = H$
6, $R^1 = R^2 = OCH_3$

7

SCHEME 1.

exhibits the M and (M − 1) ions at *m/z* 122 ($C_7H_6O_2$) and 121 with relative intensities of 100% (base peak) and 98%, respectively. Except for syringaldehyde, the (M − 29) ion species are moderately abundant, with the relative intensity in the range of 20–30%. The (M − 29) ion from syringaldehyde is insignificant, with the relative intensity of only about 5%.

Chemical Meaning of Results from Nitrobenzene Oxidation

In general, lignins of gymnosperms produce predominantly vanillin on nitrobenzene oxidation, whereas lignins of angiosperms afford mostly vanillin and syringaldehyde, and lignins of grasses afford *p*-hydroxybenzaldehyde, vanillin, and syringaldehyde. Moreover, the yield of these aldehydes depends on the nature of the plant species.

Table IV shows yields of the aromatic aldehydes from lignins of some plant species on nitrobenzene oxidation. The results indicate the type of lignin and the minimal quantity of aromatic condensed phenylpropane monomeric units present in a lignin. Thus, lignin in wood of spruce *(Picca glauca)* is a guaiacyl-type lignin and contains at least about 33 aromatic uncondensed guaiacylpropane units per 100 phenylpropane units. In contrast, lignin in wood of birch *(Betula papyrifera)* is a guaiacylsyringyl-type lignin and contains at least about 15 and 35 aromatic ring uncondensed guaiacyl- and syringylpropane units per 100 phenylpropane units, respectively. The results do not mean that the spruce lignin contains about 67 aromatic ring condensed phenylpropane units per 100 C_9 units, or that birch lignin contains about 49 aromatic ring uncondensed phenylpropane units per 100 C_9 units. However, the results do provide a relative measure of the extent of condensation in aromatic moieties of lignins.

Potassium Permanganate–Sodium Periodate Oxidation[11,12]

Reagents

Cupric oxide
2 *N* NaOH
N_2 gas
Concentrated HCl
NaCl
Acetone
Chloroform
1,2-Dimethoxyethane
Methanol
5% KOH
2 *M* H_3PO_4

TABLE IV

YIELDS OF AROMATIC ALDEHYDES FROM WOODS AND MILLED WOODS LIGNINS (MWLS) OF SOME PLANT SPECIES ON NITROBENZENE OXIDATION

Wood species	Specimen	Total lignin content (%)	C_9-Unit weight[b]	Yield (mol%/C_9-unit)[a]			
				p-Hydroxybenzaldehyde	Vanillin	Syringaldehyde	Total aldehyde
Spruce	Wood	27.6		+	33.4	+	33.4
(Picea glauca)	MWL	97.8	185.63	+	33.9	+	33.9
Birch	Wood	22.6		–	14.5	36.2	50.7
(Betula papyrifera)	MWL	95.2	209.68	–	14.1	34.2	49.5
Zhong-Yang Mu	Wood	33.3		–	26.7	10.4	37.1
(Bischopia polycarpa)	MWL	94.4	197.83	–	26.8	11.8	38.6

[a] +, Trace amount; –, not detected.

[b] C_9 formula (see Table I).

tert-Butanol
0.5 *N* NaOH
0.06 *M* $NaIO_4$
0.03 *M* $KMnO_4$
SO_2 gas
1% Na_2CO_3
30% H_2O_2
Ether
Pyromellitic acid tetramethyl ester
About 2% ethereal diazomethane

Cupric Oxide Pretreatment[18]

Weigh accurately about 200 mg of extracted wood meal (oven-dry base) or 50 mg dried MWL or 50 mg dried and purified technical lignin. The specimen, 7 ml of 2 *N* NaOH and 1 g of CuO, is placed in a 10-ml stainless-steel bomb then is flashed with N_2 gas. The bomb is sealed tightly with a Teflon gasket and screw cap and is heated at 170° for 2 hr with occasional shaking. Allow about a 10-min warm-up period for the bomb. The heater is an aluminum heating block with thermostat and holes fitted to accommodate the bomb or oil bath with thermostat, both preheated to 170°. After removing and cooling the bomb with water, the reaction mixture is filtered to remove any insoluble materials. Wash the insoluble materials twice with 5 ml of 2 *N* NaOH solution. The filtrate and washing are combined, are acidified with concentrated HCl to pH 2–3, and are saturated with NaCl. The resulting mixture is extracted four times with 50 ml of acetone–chloroform (2 : 1, v/v). The organic solutions are combined, and the solvent is removed under reduced pressure at 50°. Dry the residue.

Methylation

Caution: Dimethyl sulfate is a liquid (bp 188.5°) without odor. Both vapor and liquid dimethyl sulfate are highly poisonous. The reagent should be only used in a hood with a good draft. Rubber or vinyl gloves should be worn when handling dimethyl sulfate. Inhalation of dimethyl sulfate vapor may lead to giddiness or even more serious results. If the liquid is accidentally splashed upon the skin, wash immediately with an abundant amount of concentrated ammonia solution in order to hydrolyze the compound

[18] J. M. Pepper, B. W. Casselman, and J. C. Karapally, *Can. J. Chem.* **45,** 3009 (1967).

before it can be absorbed through the skin, then rub gently with a wad of cotton wool soaked in ammonia solution.

The residue from the cupric oxide pretreatment (see the previous section) is dissolved in 10 ml of 1,2-dimethoxyethane : methanol : water (3.5 : 3.5 : 3.0, v/v/v) and is placed in a 100-ml three-necked flask. The solution is adjusted to pH 11 with 5% KOH, to which 1 ml of dimethyl sulfate is added with mechanical stirring at room temperature under a nitrogen atmosphere. The pH of the solution is maintained at pH 11 throughout the course of the reaction by addition of 5% KOH. The reaction is terminated after KOH consumption levels off (about 18–24 hr). The reaction mixture is acidified to pH 4 with 2 *M* H_3PO_4 and is transferred into a separatory funnel. The three-necked flask is washed thoroughly three times with 15 ml of acetone–water (6 : 1, v/v). The washings are combined and are added to the reaction mixture in the separatory funnel. The solution is then shaken with 30 ml of $CHCl_3$. The aqueous layer is shaken three times with 60 ml of acetone–$CHCl_3$ (1 : 1, v/v). The organic phases are combined and are washed with 5 ml of water, and the solvent is removed under reduced pressure at 50° to obtain methylated lignin preparation as the residue.

Potassium Permanganate–Sodium Periodate Oxidation

The residue is dissolved in 40 ml of *tert*-butanol–water (3 : 1, v/v) and is transferred into a 250-ml round bottom flask with a condenser. To this solution are added 40 ml of 0.5 *N* NaOH, 100 ml of 0.06 *M* $NaIO_4$, and 7 ml of 0.16 *M* $KMnO_4$. The solution is heated on a thermostatically controlled water bath at 85° for 6 hr with mechanical stirring. After cooling, the reaction mixture is transferred in an ether extractor and is extracted continuously with ether for 4 hr to remove any neutral substances. The aqueous solution in the extractor is bubbled with SO_2 gas to dissolve MnO_2. The pH of the solution is about 1.8 at the end point. The acidic solution is extracted continuously with ether for 48 hr. After removal of the solvent from the ether solution under reduced pressure, the residue is dissolved in 30 ml of 1% Na_2CO_3 and then is adjusted to pH 11 with concentrated NaOH. After adding 5 ml of 30% H_2O_2, the solution is heated at 50° for 10 min. A small amount of $KMnO_4$ (about 100 mg) is added to decompose any excess of H_2O_2. SO_2 gas is bubbled through the solution to dissolve MnO_2. The solution is extracted continuously with ether for 48 hr. After removal of the solvent from the ether solution under reduced pressure, 3.1 or 6.2 mg pyromellitic acid tetramethyl ester (3.1 mg = 0.01 mmol) in 10 ml of methanol is added to the residue as an internal standard. To the resulting solution is added 15 ml of about 2% ethereal diazomethane solution. After 4 hr, the solvent is removed under reduced pressure. The residue is dissolved in 15 ml of $CHCl_3$. The solution is dried over anhydrous $MgSO_4$, and the solvent is removed under reduced pressure.

Quantitative Determination of Oxidation Products by Gas Chromatography (GC)

The dried sample is dissolved in 2 ml of $CHCl_3$, and the solution is filtered through a Millipore membrane HVHP to remove any high-molecular-weight contaminants. The sample solution is stored in a GC vial with a Teflon gasket and screw cap. Inject 0.5 μl of the sample solution into a GC system (Hewlett-Packard Model 5880A) with a DB-5-bonded phase, fused silica capillary column [30 m × 0.32 mm i.d., film thickness 0.25 μm (J & W Scientific)] and a flame ionization detector (temperature 280°). The carrier gas is He (flow rate, 2 ml/min). The injection port (temperature 200°) is fitted with a split liner and a solvent purge system. Temperature profile for the GC operation is as follows: hold 1 min at 60°, 60–140° (15°/min), 140–155° (5°/min), final temperature 260° (10 min) and post value 270° (10 min). Figure 6 shows a gas chromatogram of $KMnO_4$–$NaIO_4$ oxidation mixture from MWL of spruce *(P. glauca)*.

Yields of potassium permanganate–sodium periodate oxidation products in mmol/100 mg lignin are calculated from Eq. (8) and in mol % from Eq. (6).

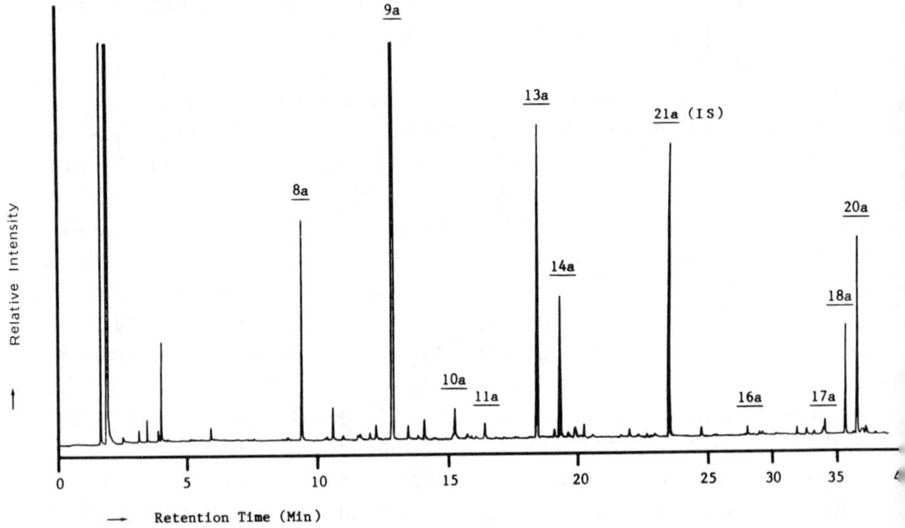

FIG. 6. Gas chromatogram of potassium permanganate–sodium periodate oxidation mixture from lignin in wood of spruce *(Picea glauca)*. Numbers refer to Schemes 1 and 2.

$$Y_1 \text{ (mmol/100 mg lignin)} = \frac{100CA_1I}{A_2W} \tag{8}$$

where Y_1 is the yield of a product, A_1 is the peak area of a product, A_2 is the peak area of an internal standard (i.e., pyromellitic acid tetramethyl ester), I is the amount of an internal standard added (mmol), W is the weight of lignin used (mg), and C is the conversion factor (see Table V).

Analysis of Oxidation Mixture by Gas Chromatography-Mass Spectrometry (GC-MS)

Inject about 5–10 μl of the sample solution (see *Quantitative Determination of Oxidation Products by Gas Chromatography (GC)* section) into a GC-Ms system (Hewlett-Packard Model 5985B) with a DB-5-bonded phase, fused silica capillary column [30 m × 0.32 mm i.d., film thickness 0.25 μm (J & W Scientific)]. The temperature profile for GC operation is as follows: hold 1 min at 60°, 60–140° (15°/min), 140–260° (5°/min). The

TABLE V
CONVERSION FACTOR (C) AND GC RETENTION TIME FOR SOME POTASSIUM PERMANGANATE–SODIUM PERIODATE OXIDATION PRODUCT OF LIGNIN

Compound[b]	Conversion factor[b]	GC retention time (min)
p-Anisic acid methyl ester **(8a)**	1.2050	9.4
Veratric acid methyl ester **(9a)**	1.2152	13.0
3,4,5-Trimethoxybenzoic acid methyl ester **(10a)**	1.2206	15.4
4-Methoxyisophthalic acid dimethyl ester **(11a)**	1.2298	16.5
4-Methoxyphthalic acid dimethyl ester **(12a)**	1.3116	17.3
Isohemipinic acid dimethyl ester **(13a)**	1.2358	18.5
m-Hemipinic acid dimethyl ester **(14a)**	1.2538	19.4
3,4,5-Trimethoxyphthalic acid dimethyl ester **(15a)**	1.2576	20.1
2,3,3′,4′Tetramethoxybiphenyl-5-carboxylic acid methyl ester **(16a)**	—	28.4
2,2′,3-Tetramethoxybiphenyl-5,5′-dicarboxylic acid dimethyl ester **(17a)**	1.1218	34.3
2,2′,3-Trimethoxydiphenylether-4′,5-dicarboxylic acid dimethyl ester **(18a)**	1.0314	35.9
2,2′,3,6′-Tetramethoxydiphenylether-4′,5-dicarboxylic acid dimethyl ester **(19a)**	1.1618	37.7
Dehydrodiveratric acid dimethyl ester **(20a)**	1.1938	36.9
Pyromellitic acid tetramethyl ester **(21a)**	—	23.6

[a] The numbers in parentheses refer to Scheme 2.

[b] Calculated from appropriate calibration curve and relative to pyromellitic acid tetramethyl ester, the internal standard.

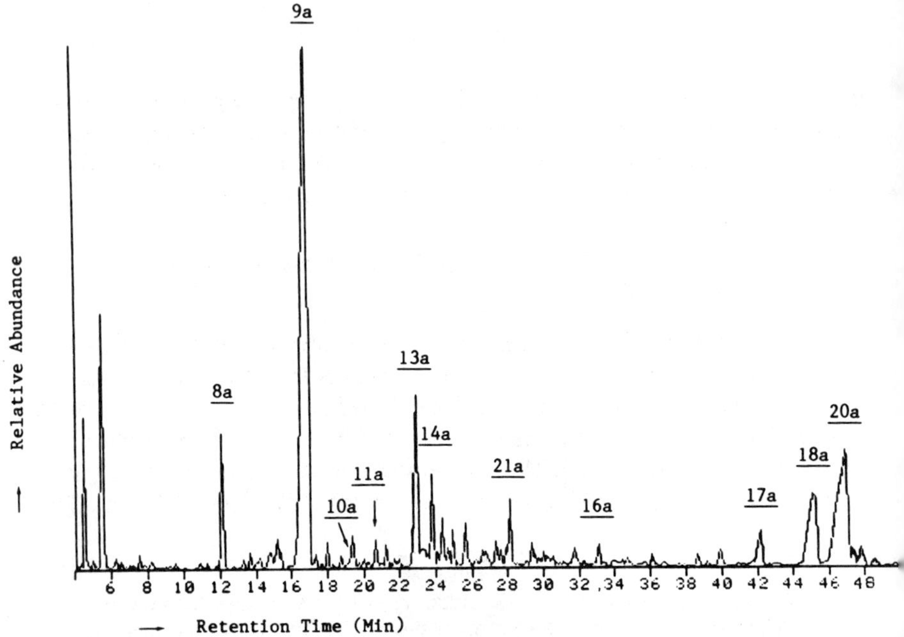

FIG. 7. Total ion chromatogram of potassium permanganate–sodium periodate oxidation mixture from lignin in wood of spruce *(Picea glauca)*. Numbers refer to Schemes 1 and 2.

MS operation is as follows: EI mode (70 eV) and scanning rate of approximate 45 scans/min. Figure 7 shows a total ion chromatogram of the sodium periodide–potassium permanganate oxidation mixture from lignin in wood of spruce *(P. glauca)*.

Interpretation of Mass Spectra

The GC-MS analysis of the potassium permanganate–sodium periodide oxidation of lignin in the spruce wood results in obtaining mass spectra of 22 components with relative abundance of more than 1%. The analysis of these spectra leads to identification of 10 aromatic acid methyl esters [Scheme 2 **(8)–(21a)**]. These compounds include *p*-anisic acid methyl ester **(8a)**; veratic acid methyl ester **(9a)**; 3,4,5-trimethoxybenzoic acid methyl ester **(10a)**; 4-methoxyisophthalic acid dimethyl ester **(11a)**; isohemipinic acid dimethyl ester **(13a)**; *m*-hemipinic acid dimethyl ester **(14a)**; 2,3,3′,4′-tetramethoxybiphenyl-5-carboxylic acid methyl ester **(16a)**; 2,2′,3-trimethoxybiphenyl-5,5′-dicarboxylic acid dimethyl ester **(17a)**; 2,2′,3-trimethoxydiphenyl ether-4′,5-dicarboxylic acid dimethyl ester **(18a)**; and dehydrodiveratric acid dimethyl ester **(20a)**. Among these com-

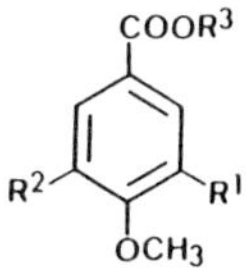

8, $R^1 = R^2 = R^3 = H$

8a, $R^1 = R^2 = H$; $R^3 = CH_3$

9, $R^1 = OCH_3$; $R^2 = R^3 = H$

9a, $R^1 = OCH_3$; $R^2 = H$; $R^3 = CH_3$

10, $R^1 = R^2 = OCH_3$; $R^3 = H$

10a, $R^1 = R^2 = OCH_3$; $R^3 = CH_3$

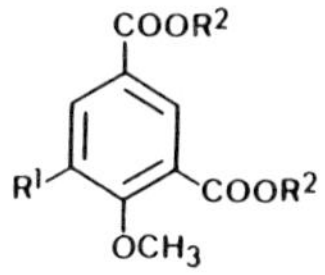

11, $R^1 = R^2 = H$

11a, $R^1 = H$; $R^2 = CH_3$

13, $R^1 = OCH_3$; $R^2 = H$

13a, $R^1 = OCH_3$; $R^2 = CH_3$

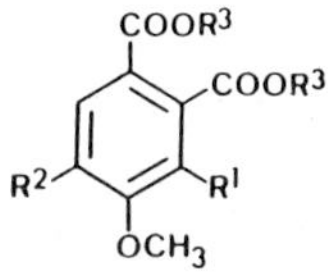

12, $R^1 = R^2 = R^3 = H$

12a, $R^1 = R^2 = H$; $R^3 = CH_3$

14, $R^1 = R^3 = H$; $R^2 = OCH_3$

14a, $R^1 = H$; $R^2 = OCH_3$; $R^3 = CH_3$

15, $R^1 = R^2 = OCH_3$; $R^3 = H$

15a, $R^1 = R^2 = OCH_3$; $R^3 = CH_3$

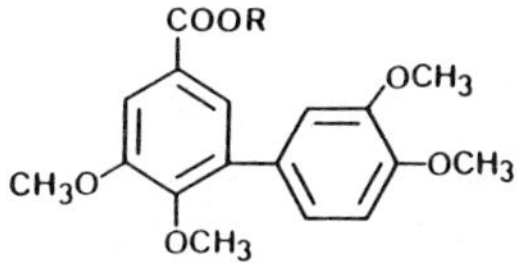

16, $R = H$

16a, $R = CH_3$

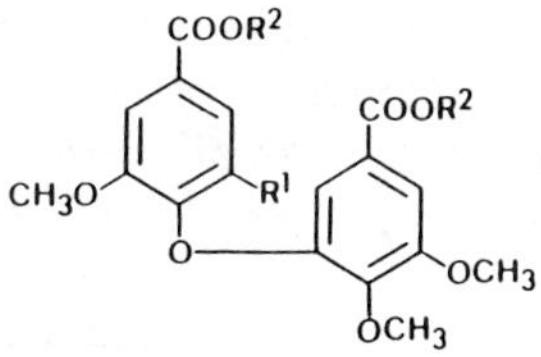

18, $R^1 = R^2 = H$

18a, $R^1 = H$; $R^2 = CH_3$

19, $R^1 = OCH_3$; $R^2 = H$

19a, $R = OCH_3$; $R = CH_3$

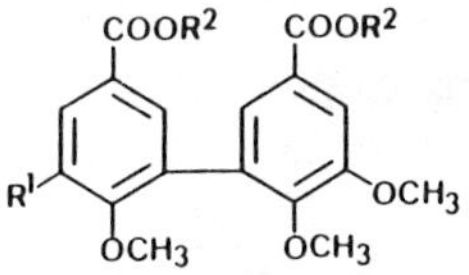

17, $R^1 = R^2 = H$

17a, $R^1 = H$; $R^2 = CH_3$

20, $R^1 = OCH_3$; $R^2 = H$

20a, $R^1 = OCH_3$; $R^2 = CH_3$

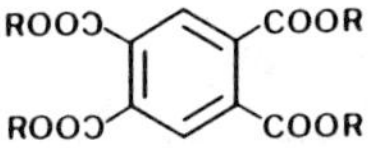

21, $R = H$

21a, $R = CH_3$

SCHEME 2.

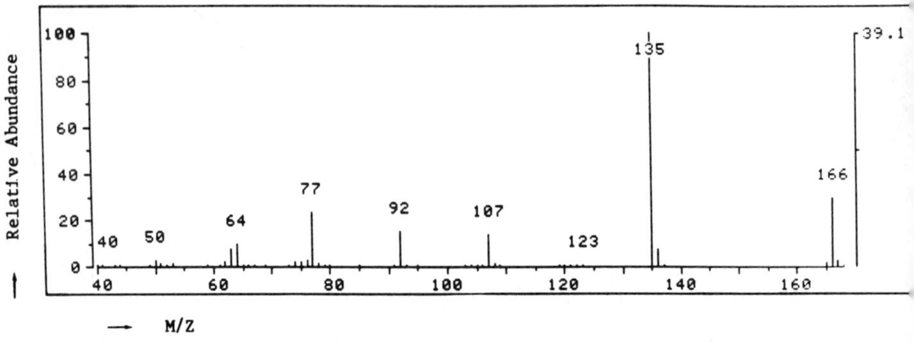

FIG. 8. Mass spectrum of anisic acid methyl ester **(8a)**.

pounds, **(9a)**, **(13a)**, **(14a)**, **(18a)**, and **(20a)** are the major $KMnO_4$–$NaIO_4$ oxidation products of spruce lignin.

Figures 8–14 show the mass spectra of some representative $KMnO_4$–$NaIO_4$ oxidation products of lignin in spruce wood. The major mass spectral fragmentation pattern of all aromatic acid methyl esters is loss of methyl ester $CH_3O^{\cdot}$ from the molecular ion (M) to give the (M − 31) ion species. Thus, the mass spectrum of compound **(8a)** (Fig. 8) exhibits a rather insignificant M ion at *m/z* 166 ($C_9H_8O_3$; relative intensity, 30%) and a predominant (M − 31) ion at *m/z* 135 (base peak). In contrast, the mass spectrum of compound **(9a)** (Fig. 9) shows the predominant M and (M − 1) ions at *m/z* 196 (base peak; $C_{10}H_{12}O_4$) and 165 with relative intensity of 100 and 93%, respectively. The mass spectrum of compound **(10a)** (Fig. 10) exhibits a rather predominant (M − 15) species at *m/z* 211 (relative intensity, 74%) produced from the M ion by loss of methoxyl $^{\cdot}CH_3$, in addition to the M and (M − 1) ions at *m/z* 226 (base peak;

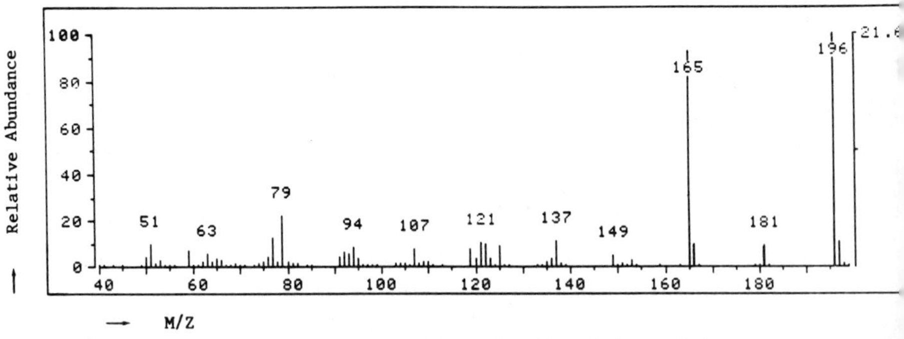

FIG. 9. Mass spectrum of veratric acid methyl ester **(9a)**.

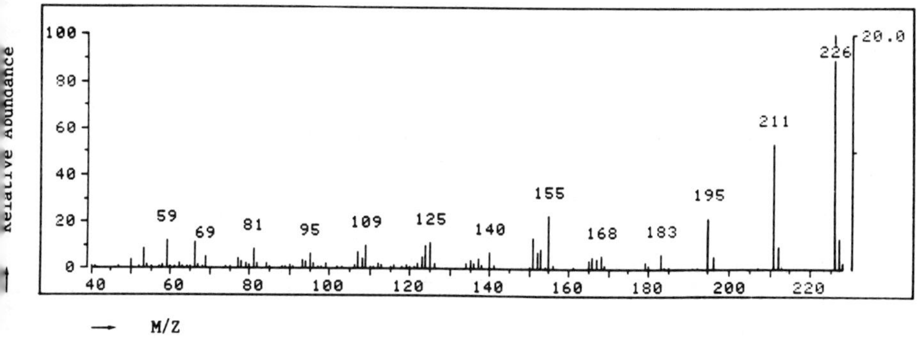

FIG. 10. Mass spectrum of 3,4,5-trimethoxybenzoic acid methyl ester **(10a)**.

$C_{11}H_{14}O_5$) and 195 with relative intensity of 100 and 22%, respectively. The mass spectrum of compound **(13a)** (Fig. 11) shows a predominant (M – 33) ion at *m/z* 221 (base peak), in addition to the M and (M – 31) ions at *m/z* 254 ($C_{12}H_{14}O_6$) and 223 with relative intensity of 61 and 96%, respectively. The (M – 33) ion is produced by successive losses of CO and 2H˙ from the M ion due to *ortho*-effects involving methyl ester group at C-3 and methoxyl group at C-4. In contrast, the mass spectrum of compounds **(14a)** (Fig. 12), an isomer of compound **(13a)**, exhibits a less predominant M ion at *m/z* 254 ($C_{12}H_{14}O_6$; relative intensity, 50%) and a predominant (M – 31) ion at *m/z* 223 (base peak). The predominance of the latter ion is due to the *ortho*-effects of the two vicinal methyl ester groups at C-1 and C-2. The mass spectra of compounds **(11a)** and **(12a)** are analogous to those of compounds **(13a)** and **(14a)**, respectively. As shown in Figs. 13 and 14, the mass spectra of other compounds show the M ion (base peak) predominant over the (M – 31) ion.

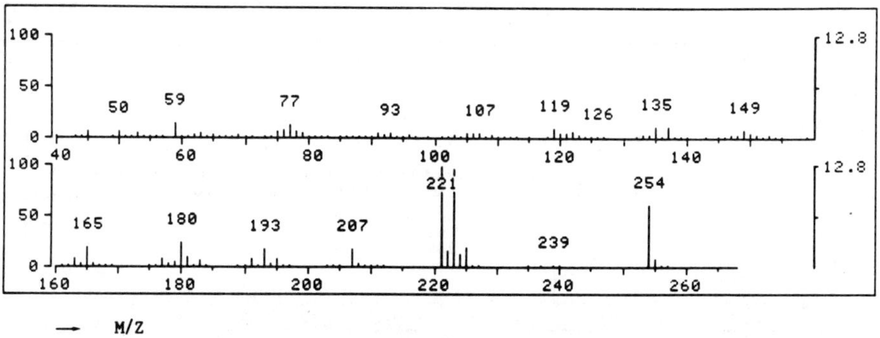

FIG. 11. Mass spectrum of isohemipinic acid dimethyl ester **(13a)**.

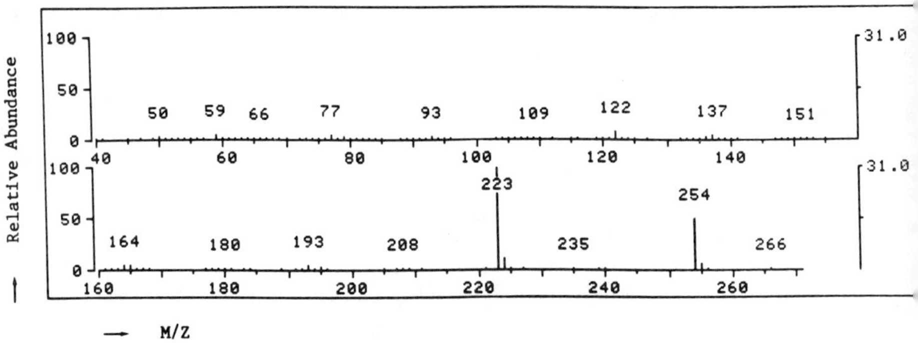

FIG. 12. Mass spectrum of *m*-hemipinic acid dimethyl ester **(14a)**.

Chemical Meaning of Results from Potassium Permanganate–Sodium Periodate Oxidation

On $KMnO_4$–$NaIO_4$ oxidation, lignins of gymnosperms generally give aromatic acids derived from guaiacylpropane units, e.g., veratric acid **(9)**; isohemipinic acid **(13)**; *m*-hemipinic acid **(14)**; 2,2′,3-trimethoxy-diphenylether-4′,5-dicarboxylic acid **(18)**; and dehydrodiveratric acid **(20)**. Veratric acid is the major product. In addition to these acids, lignins of angiosperms also produce aromatic acids derived from syringylpropane units with veratric acid **(9)** and 3,4,5-trimethoxybenzoic acid **(10)** as the major products. The oxidation products from lignins of grasses are qualitatively similar to those from lignins of angiosperms, except for the presence of aromatic acids derived from *p*-hydroxyphenylpropane units in appreciable amounts. Yield of these acids depends on the nature of plant species.

Table VI shows the major $KMnO_4$–$NaIO_4$ oxidation products from woods and MWLs of spruce and birch. In general, yields of the major

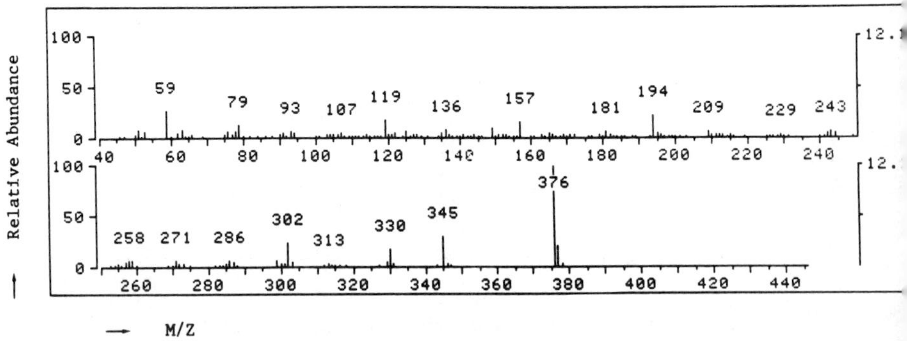

FIG. 13. Mass spectrum of 2,2′,3-trimethoxydiphenylether-4′5-dicarboxylic acid dimethyl ester **(18a)**.

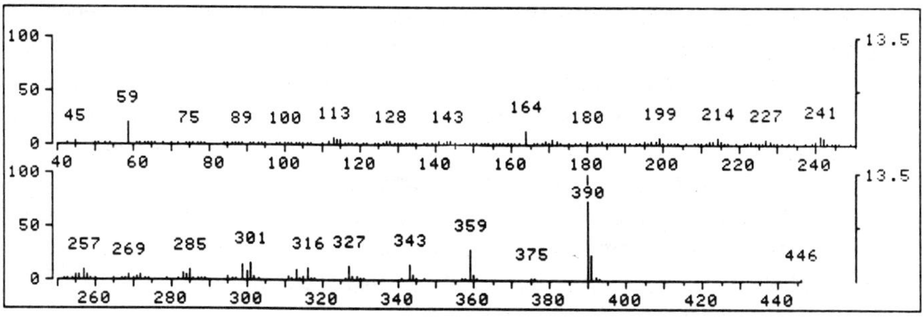

FIG. 14. Mass spectrum of dehydrodiveratric acid dimethyl ester (**20a**).

products, e.g., veratric acid (**9**) and/or 3,4,5-trimethoxybenzoic acid (**10**), are smaller than those of the corresponding aromatic aldehyde obtained from the same specimen on nitrobenzene oxidation. However, $KMnO_4$–$NaIO_4$ oxidation of a lignin provides information not only about a relative measure of aromatic ring condensed and uncondensed phenylpropane units in the lignin but also about types of linkage between aromatic moieties in the condensed phenylpropane units.

Preparation of Ethereal Diazomethane[19]

Caution: Diazomethane is not only very toxic but also explosive. Therefore, the preparation of diazomethane should be carried out only in a

TABLE VI

YIELD OF MAJOR AROMATIC ACIDS FROM WOODS AND MILLED WOOD LIGNINS (MWL'S) OF SPRUCE AND BIRCH ON POTASSIUM PERMANGANATE–SODIUM PERIODATE OXIDATION

		Yield (mol%/C_9-unit)[a,b]									
Wood species	Specimen[c]	(8)	(9)	(10)	(13)	(14)	(15)	(18)	(19)	(20)	Total
Spruce	Wood	0.8	28.6	0.5	4.2	0.8	+	1.2	+	3.4	39.5
(*Picea glauca*)	MWL	0.9	29.2	0.5	4.0	0.7	+	1.1	+	3.2	39.6
Birch	Wood	+	10.9	20.4	1.2	0.3	0.3	0.4	1.1	0.9	35.5
(*Betula papyrifera*)	MWL	+	11.2	20.2	1.3	0.3	0.2	0.2	1.2	1.0	35.6

[a] +, Trace amount; –, not detected.

[b] Number column headings refer to structures in Schemes 1 and 2.

[c] C_9 Formula (see Table I).

[19] A. Vogel, "A Text Book of Practical Organic Chemistry" (revised by B. S. Furniss, A. J. Hannaford, V. Rogers, P. W. G. Smith, and A. R. Tatchell), 4th Ed., p. 291. Longmans, London, 1978.

hood provided with a powerful exhaust system. The use of a screen of safety glass is strongly recommended, in addition to wearing safety spectacles. The use of glassware without sharp edges or ground glass joints is recommended and the use of boiling chips should be avoided. Aldrich Chemical Company, (Milwaukee, Wisconsin 53233) offers a Diazald Kit, a glassware set for the preparation of ethereal diazomethane up to 100 mmol. Clear-seal glass joints are used for the glassware in this kit in order to prevent the accidental explosion of diazomethane caused by friction of ground glass joints. These joints also do not require grease even for vacuum application.

Reagents

N-Methyl-*N*-nitroso-*p*-toluenesulfonamide (Diazald, Aldrich Chemical Company)
Ether
96% Ethanol
Potassium hydroxide

To a solution of 10 g of potassium hydroxide in 15 ml of water is added 50 ml of 96% ethanol. Place this solution in a 200-ml distillation flask equipped with a dropping funnel and an efficient double surface condenser. Connect the condenser to two conical flasks of 50 and 100 ml, respectively, to act as receivers. Place 40 ml of ether in the smaller flask and arrange in such a manner that the inlet tube of the smaller receiver dips below the surface of the ether. This will serve to trap any uncondensed diazomethane–ether vapor escaping from the first receiver. Cool both receivers in an ice–salt mixture. Heat the distillation flask in a water bath at 60–65°. Place a solution of 43 g *N*-methyl-*N*-nitroso-*p*-toluenesulfonamide in about 250 ml of ether in the dropping funnel and introduce it into the flask over a period of 45 min. Adjust the rate of addition so that it is about equal to the rate of distillation. When the dropping funnel is empty, add about 30 ml of ether gradually until the ether distilling over is colorless. The combined ethreal solutions in the receivers contain about 5.9–6.1 g of diazomethane.

For smaller quantities of diazomethane, the use of a dropping funnel is unnecessary. Dissolve 4.28 g of *N*-methyl-*N*-nitroso-*p*-toluenesulfonamide in 60 ml of ether, cool in ice, and add a solution of 0.8 g of potassium hydroxide in 20 ml of 96% ethanol. If a precipitate forms, add more ethanol until it just dissolves. After 5 min, distill the ethereal diazomethane solution from a water bath at 60–65°. The ethereal diazomethane contains about 0.6–0.7 g of diazomethane.

[15] Characterization of Lignin by ^{1}H and ^{13}C NMR Spectroscopy

By CHEN-LOUNG CHEN and DANIELLE ROBERT

Since 1951, rapid progress has been made in both the theory and experimental aspects of nuclear magnetic resonance (NMR). This is traceable to the fact that, unlike other forms of spectroscopy, the interpretation of NMR spectra in most cases is straightforward in terms of the fundamental parameters, such as chemical shifts, coupling constants, signal areas (intensities), and relaxation times. These parameters observed in NMR spectral experiments provide the necessary information for solving a wide variety of problems in the chemical and biological sciences.[1-5]

Most of the early work in NMR was focused on ^{1}H NMR spectroscopy. The major reason for this is that the ^{1}H nucleus, i.e., the proton, is the one most sensitive to NMR detection among the nuclei that give a NMR spectrum. However, the recent advances in Fourier transform NMR (FT-NMR) techniques lead to rapid development of the technique for obtaining natural abundance ^{13}C NMR spectra. Thus, ^{13}C NMR spectroscopy has become comparable to ^{1}H NMR spectroscopy in terms of importance as a tool for structural elucidation of organic substances, in spite of the low natural abundance of ^{13}C (1.1 at. %).

There are several advantages of ^{13}C NMR spectroscopy over ^{1}H NMR spectroscopy for the structural determination of organic compounds. In ^{13}C NMR, spectral data are obtained from the "backbone" of the molecule rather than from the exterior of the molecule as in ^{1}H NMR. Consequently, ^{13}C NMR spectrum of a compound provides information about the nature of all carbons in the molecules. In contrast, the corresponding ^{1}H NMR spectrum does not give information about the nature of quaternary carbons in the molecule, such as the carbons in R″-C(R′)= and >C=O groups. The second advantage is that ^{13}C NMR spectra are not complicated by spin–spin coupling. The probability of having two ^{13}C

[1] J. M. Emseley, J. Feenay, and L. H. Sutcliffe, "High Resolution Nuclear Magnetic Resonance Spectroscopy," Vol. I. Pergamon, London, England, 1967.

[2] E. Breitmaier, and W. Voelter, "Carbon-13 NMR Spectroscopy," 3rd Ed. VCH, Weinheim, Federal Republic of Germany, 1987.

[3] D. E. Leyden, and R. H. Cox, "Analytical Applications of NMR." Wiley, New York, 1977.

[4] R. J. Abraham, and P. Loftus, "Proton and Carbon-13 NMR Spectroscopy." Heyden & Son, Sussex, England, 1979.

[5] R. K. Harris, "Nuclear Magnetic Resonance Spectroscopy." Pitman, Marshfield, Massachusetts, 1983.

nuclei adjacent to each other in the same molecule is so low (1/10,000) that the possibility of $^{13}C-^{13}C$ coupling can be ignored. Moreover, ^{13}C NMR spectra are usually obtained with broadband noise, proton decoupling,[6] so that only single signals are observed for each ^{13}C resonance. The third advantage is that the ^{13}C NMR chemical shift range of the majority of diamagnetic organic compounds is about 240 ppm (in δ scale) in comparison with about 12 ppm for ^{1}H NMR. All these imply that there is better resolution and less overlap of signals in ^{13}C NMR spectra of organic compounds, particularly of polymeric natural products, such as lignin, than the corresponding ^{1}H NMR spectra. In addition, there is a greater probability of observing individual carbon resonances. One of the disadvantages of ^{13}C NMR spectroscopy over ^{1}H NMR spectroscopy is that routine ^{13}C NMR spectra are not quantitative. The area under each signal in a routine ^{13}C NMR spectra is not proportional to the number of the corresponding ^{13}C nuclei giving rise to these signals because of the nuclear Overhauser effect (NOE) and the different relaxation times of the different carbon. In order to obtain a quantitative ^{13}C NMR spectrum of an organic compound, a specific pulse sequence is required. The other disadvantage is that the ^{13}C nucleus is significantly less sensitive than the ^{1}H nucleus toward magnetic field, relative sensitivity of 0.00018 versus 1 in a constant magnetic field at natural isotopic abundance.

Among the various physical and chemical methods for characterization of lignins, ^{1}H NMR[7–10] and ^{13}C NMR[11–16] spectroscopy has been shown to be among the most reliable and comprehensive techniques. The characterization of lignin by ^{13}C NMR spectroscopy, in particular, furnishes rather comprehensive data about the nature of all carbons in lignin in terms of chemical structure. By contrast, the other physical and chemical analytical methods only provide incomplete information on the chemical structure of lignin (see chapter [14] in this volume). However, several difficulties are still encountered in the interpretation of the ^{1}H NMR and ^{13}C NMR spectra of lignins, e.g., the assignment of signals, because of the intensive

[6] R. R. Ernst, *J. Chem. Phys.* **45,** 3845 (1966).
[7] C. H. Ludwig, B. L. Nist, and J. L. McCathy, *J. Am. Chem. Soc.* **86,** 1186 (1964).
[8] C. H. Ludwig, B. L. Nist, and J. L. McCathy, *J. Am. Chem. Soc.* **86,** 1196 (1964).
[9] A. Klemola, *Suom. Kemistil. B* **41,** 99 (1968).
[10] K. Lundquist, *Acta Chem. Scand., Ser. B* **35,** 497 (1981).
[11] H.-D. Lüdemann and H. Nimz, *Makromol. Chem.* **175,** 2409 (1974).
[12] H. Nimz, I. Mogharab, and H.-D. Lüdemann, *Makromol. Chem.* **175,** 2563 (1974).
[13] H. Nimz, D. Robert, O. Faix, and M. Nemr, *Holzforschung* **35,** 16 (1981).
[14] C.-L. Chen, M. G. S. Chua, J. Evans, and H.-M. Chang, *Holzforschung* **36,** 239 (1982).
[15] C. Lapierre, J. Y. Lallemand, and B. Monties, *Holzforschung* **36,** 275 (1982).
[16] K. P. Kringstad and R. Mörck, *Holzforschung* **37,** 237 (1983).

overlap of signals for ^{1}H and ^{13}C nuclei in lignin present in similar, but nonidentical chemical environments. Some of these difficulties can be circumvented by the application of more sophisticated ^{13}C NMR pulse sequences[13] such as the attached proton test (APT) experiment,[17,18] and the distortionless enhancement by polarization transfer (DEPT) sequence.[18–22] The DEPT pulse sequence is particularly suitable for characterization of lignins by ^{13}C NMR spectroscopy, overcoming the loss of information caused by proton decoupling.[23,24] This pulse sequence is a one-dimensional pulse sequence, involving spin echo phenomenon to observe separately signals for ^{13}C nuclei of CH, CH_2, and CH_3 groups; hence, the multicity of ^{13}C NMR signals can be unambiguously determined. [In addition, the pulse sequence significantly induces the enhancement of signal intensity due to the polarization transfer involving a spin population interchange between the more sensitive nucleus (^{1}H) and the less sensitive nucleus (^{13}C) to the benefit of the latter.] This chapter deals with the characterization of lignin by ^{1}H and ^{13}C NMR spectroscopy. The latter includes techniques for obtaining routine, quantitative, and DEPT ^{13}C NMR spectra of lignins. Before describing the aforementioned procedures, the principle of NMR spectroscopy will be presented.

Pulse Fourier Transform NMR Techniques[1–5]

One of the important limitations of NMR spectroscopy is the low sensitivity of the method as compared to other spectroscopic techniques, such as infrared and ultraviolet-visible spectroscopy. The major cause for the low sensitivity is traceable to the very small magnitude in the energy changes (in the order of about 10^{-2} cal/mol) involved in NMR transitions. Moreover, the low natural abundance of some nuclei, such as ^{13}C and ^{35}Cl, in nature makes observation of NMR signals for these nuclei on a routine NMR procedure, e.g., continuous wave (CW) technique, using natural abundance samples, even more difficult. One of the most effective methods to overcome this difficulty is the pulsed Fourier transform (FT) NMR technique.

[17] D. L. Robenstein and T. T. Takashima, *Anal. Chem.* **51,** 1465A (1979).
[18] S. L. Patt and J. N. Shoolery, *J. Magn. Reson.* **46,** 535 (1982).
[19] D. M. Doddrell, D. T. Pegg, and M. R. Bendall, *J. Magn. Reson.* **48,** 323 (1982).
[20] D. M. Doddrell, D. T. Pegg, and M. R. Bendall, *J. Chem. Phys.* **77,** 2745 (1982).
[21] O. W. Sorensen and R. R. Ernst, *J. Magn. Reson.* **51,** 477 (1983).
[22] R. Benn and H. Gunther, *Angew. Chem., Int. Ed. Engl.* **22,** 350 (1983).
[23] C. Lapierre, B. Monties, E. Guittet, and J. Y. Lallemand, *Holzforschung* **38,** 333 (1984).
[24] M. Bardet, M.-F. Foray, and D. Robert, *Makromol. Chem.* **186,** 1495 (1985).

Nuclear Magnetic Resonance

When a nucleus with spin quantum number $I \neq 0$ is placed in a constant magnetic field B_0, the magnetic moment takes up one of the allowed orientations of an angle θ to the direction of the magnetic field, as shown in Fig. 1. The magnetic moment μ will then experience a torque L tending to align it parallel to the field. However, since the nucleus is spinning, the torque causes μ to precess about the magnetic field B_0. According to Newton's law, the rate of change of angular momentum p with time is equal to the torque L.

$$dp/dt = L \tag{1}$$

From magnetic theory

$$L = \mu B_0 \tag{2}$$

Substituting Eq. (2) into Eq. (1), then

$$dp/dt = \mu B_0 \tag{3}$$

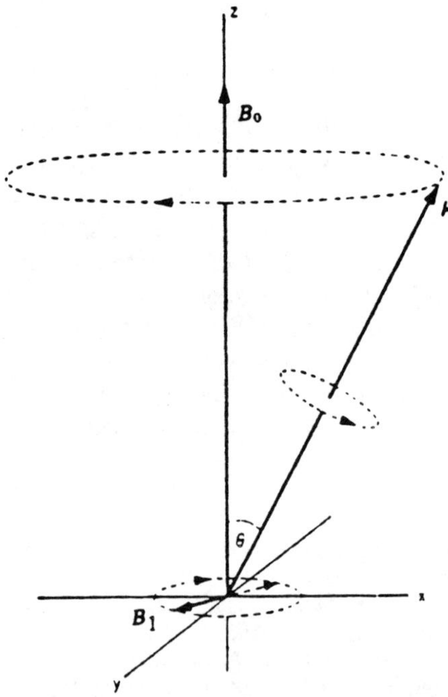

FIG. 1. Classical Larmor precession.[1] (Reproduced by permission of Pergamon, Oxford, England.)

Since $\gamma = \mu/p$ where γ is the magnetogyric ratio, Eq. (3) becomes

$$dp/dt = \gamma p B_0 \tag{4}$$

This equation of motion describes the precession of p about B_0 with angular frequency ω_0 defined by

$$dp/dt = p\,\omega_0 \tag{5}$$

hence,

$$\omega_0 = \gamma\, B_0 \tag{6}$$

Since $\omega_0 = 2\pi\nu_0$ by definition, Eq. (5) can be rewritten in terms of the precession frequency ν_0

$$\nu_0 = (\gamma/2\pi)\, B_0 \tag{7}$$

This equation is called the Larmor equation. An important point regarding this equation is that the frequency ν_0 is independent of the angle of inclination of the nuclear axis to the direction of the field.

If a secondary smaller magnetic field B_1 is applied perpendicular to the direction of the magnetic field B_0, i.e., $x-y$ plane in the stationary frame of reference, and rotating about B_0 in the same direction as μ, then interactions between B_1 and μ occur. In practice, the rotating field B_1 is obtained by passing an alternating current through a coil. The coil is mounted perpendicular to the direction of B_0 in order to produce a magnetic field oscillating along the x-axis in the stationary frame of reference. An application of voltage to the coil at frequency $\omega = 2\pi\nu$ produces two equal counterrotating fields, having vectors in the magnitudes (B_1 cos ωt + B_1 sin ωt) and (B_1 cos ωt − B_1 sin ωt), as shown in Fig. 2. When the frequency of the rotating field B_1, $\nu = \omega/2\pi$, is equal to the precession frequency ν_0 of the magnetic moment μ, the nucleus absorbs energy from B_1, causing the magnetic moment to change the orientation with respect to the direction of the magnetic field B_0. This transition of energy states is called NMR phenomenon. At a magnetic field of 46,974 gauss, the Larmor

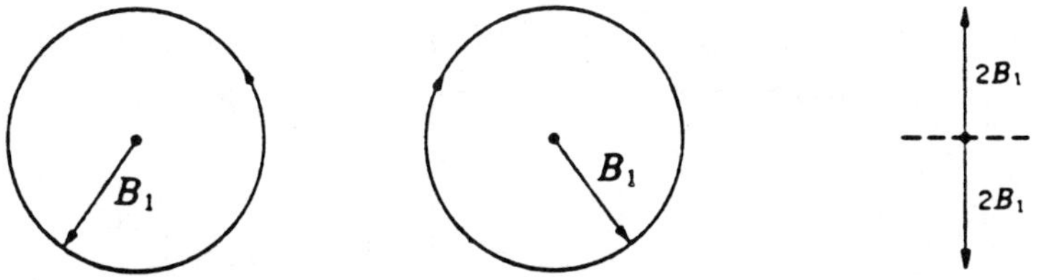

FIG. 2. A schematic illustration of the rotating magnetic field B_1.[1] (Reproduced by permission of Pergamon, Oxford, England.)

frequencies for ^1H and ^{13}C nuclei are in the range of 200 and 50.28 MHz, respectively.

Nuclear Energy Levels and Boltzmann Distribution

The energy E for a magnetic moment μ of a nucleus with spin quantum number I in a constant magnetic field B_0 is given by

$$E = m\gamma(h/2\pi)B_0 \tag{8}$$

where m is magnetic quantum number and has values $-I, -I+1, \ldots, I-1, I$. Thus, the magnetic moment has $2I+1$ energy levels.

For an assembly of identical nuclei with $I = \frac{1}{2}$. e.g., ^1H and ^{13}C nuclei, there are two allowed orientations, i.e., energy levels (Fig. 3), for their magnetic moments with respect to the direction of a constant magnetic field, when the nuclei are placed in the field. From Eq. (8), E for the magnetic moment at the lower energy level $m = -\frac{1}{2}$ (α state) is

$$E_\alpha = -\gamma(h/4\pi)B_0 \tag{9}$$

and E for the magnetic moment at the upper energy level $m = \frac{1}{2}$ (β state) is

$$E_\beta = \gamma(h/4\pi)B_0 \tag{10}$$

The selection rule allows the transition of energy only when $m = \pm 1$. Thus, at the NMR condition, the transition energy ΔE is

$$\Delta E = E_\alpha - E_\beta = \gamma(h/2\pi)B_0 \tag{11}$$

Moreover, according to the selection rule $\Delta m = \pm 1$, there are two allowed transitions: (1) from the α state to the β state with $\Delta m = 1$, which corresponds to an absorption of energy, and (2) from the β state to the α state with $\Delta m = -1$, which corresponds to induced emission, as shown in Fig. 3. Since the coefficients for absorption and induced emission of energy are

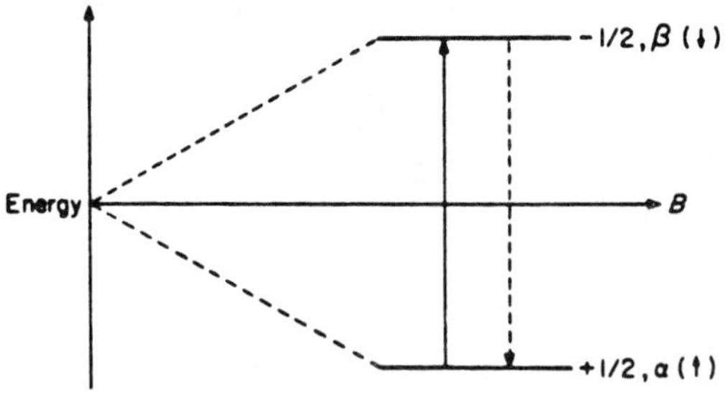

FIG. 3. Energy level for a nucleus with $I = \frac{1}{2}$ in a magnetic field.[4] (Reproduced by permission of Wiley, New York.)

the same at the NMR state, there would be no net absorption of energy from the radio-frequency (rf) radiation to the nuclei if the population of the nuclei in the two states were equal. Therefore, no NMR signal would be obtained.

However, in the thermal equilibrium state, the population of the nuclei in the lower energy level (N^-) is slightly in excess of that in the upper energy level (N^+) according to the Boltzmann equation [Eq. (12)],

$$N^-/N^+ = \exp(-\Delta E/T) \simeq 1 - \Delta E/\kappa T = 1 - 2\mu B_0/\kappa T \qquad (12)$$

where κ is Boltzmann constant and T is the absolute temperature. In general, the energy difference E is in the order of 10^{-2} cal/mol at a magnetic field in the range of 14,000–60,000 gauss. Consequently, the excess population of the nuclei in the lower energy level is in the order of 1 in 10^5 at 25°. This slight excess in the population of the nuclei in the lower energy level gives rise to a resultant magnetization vector $\mathbf{M}_0$ along the direction of B_0, i.e., z-axis in the stationary frame of reference, as shown in Fig. 4.

Relaxation Time

A finite period of time is required for the Boltzmann distribution to be established when an assembly of identical nuclei is placed in a strong constant magnetic field B_0. The nuclei undergo thermal motions and interact with their surroundings (lattice), including the magnetic field B_0.

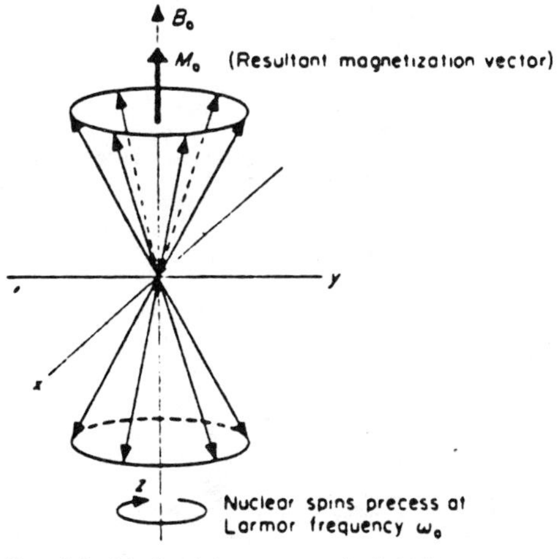

FIG. 4. Motion of nuclei with $I = \frac{1}{2}$ in a magnetic field.[4] (Reproduced by permission of Wiley, New York.)

This interaction involves energy transfer between the spin system and the lattice, resulting in transitions of the nuclei between allowed energy levels. This process is called spin–lattice or longitudinal relaxation and is a nonradiative first order process. The rate of the relaxation is inversely proportional to a relaxation time of T_1. The magnitude of T_1 depends on the physical state of the sample, the temperature, and the type of nucleus under observation. Nuclei of the same species with different chemical environments have different values of T_1. For liquids, T_1 values are generally in the range of $10^{-2}-10^2$ sec for nuclei with $I=\frac{1}{2}$. Another relaxation process is the spin–spin or transverse relaxation, the rate of which is inversely proportional to a spin–spin relaxation time T_2. This process involves exchange of spin orientation between neighboring nuclei by interacting their magnetic moments. The process does not result in a change in the total energy of the system. The magnitude of T_2 is, in general, smaller than that of T_1, i.e., $T_1 \geq T_2$. The line width of the signal at midheight is equal to $1/2\pi T_2$ or $1/2\pi T_2^*$.

Motion of Magnetization Vector in Rotating Coordinate System

During NMR experiments, if the coordination system, i.e., the stationary frame of reference, is rotating about the z-axis at the angular velocity $\omega = 2\pi\nu$ of the rotating field B_1, then the path of the magnetization vector **M** subjected to the effects of magnetic fields B_0 and B_1 can be simplified. The rotating coordinate system is called the rotating frame of reference. The axes of the system are denoted as x', y', and z', as shown in Fig. 5, with the rotating unit vectors $\mathbf{i}'$, $\mathbf{j}'$, and $\mathbf{k}'$ for the components of **M** along x'-, y'-, and z'-axis, respectively.

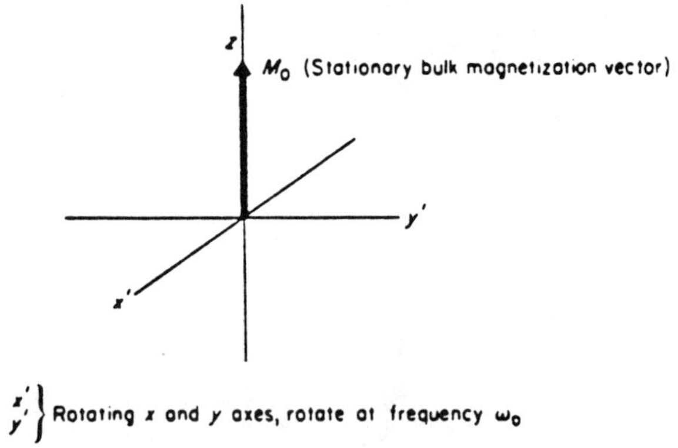

FIG. 5. Motion of nuclei with $I=\frac{1}{2}$ in a magnetic field rotating at the Larmor frequency ω_0 (the rotating frame reference system).[4] (Reproduced by permission of Wiley, New York.)

At NMR state, the magnetic moments of the nuclei with $I = \frac{1}{2}$ change their orientations with respect to the direction of the magnetic field B_0. For an assembly of nuclei with magnetic moment μ, the magnetization vector **M** is the resultant vector sum of the magnetic moments. The direction of **M** is now time dependent, and the change of **M** with time in the rotating frame of reference can be expressed by

$$\begin{aligned}(\partial \mathbf{M}/\partial t)_{\text{rot}} &= \gamma \mathbf{M}(B + \omega/\gamma) \\ &= \gamma \mathbf{M} B_{\text{eff}}\end{aligned} \tag{13}$$

where B is the total magnetic field resulting from the constant magnetic field B_0 and the rf field B_1, i.e., $B = B_0\mathbf{k}' + B_1\mathbf{i}'$, and B_{eff} is the effective magnetic field in the rotating frame of reference, defined by equation

$$\begin{aligned}B_{\text{eff}} &= (B + \omega/\gamma) \\ &= (B_0 + \omega/\gamma)\mathbf{k}' + B_1\mathbf{i}'\end{aligned} \tag{14}$$

Thus, the effect of rotating the coordinate system is to change the effective magnetic field by a term ω/γ resulting from the rotation.

Pulsed NMR Experiment in the Rotating Frame of Reference

In the absence of B_1, the vector **M** has its equilibrium value M_0 along the z'-axis. **M** is thus time-invariant in the rotating frame of reference, so that

$$(\partial \mathbf{M}/\partial t)_{\text{rot}} = \gamma \mathbf{M} B_{\text{eff}} = 0 \tag{15}$$

since $\mathbf{M} = M_0 \neq 0$ and $B_1\mathbf{i}' = 0$, combination of Eqs. (14) and (15) gives

$$B_{\text{eff}} = (B_0 + \omega/\gamma)\mathbf{k}' = 0 \tag{16}$$

Consequently, $(\omega/\gamma)\mathbf{k}' = -B_0\mathbf{k}'$. Thus, the rotational field ω/γ opposes $B_0\mathbf{k}'$ along the z'-axis in the rotating frame of reference as shown in Fig. 6. Equation (14) can then be rewritten

$$B_{\text{eff}} = (B_0 - \omega/\gamma)\mathbf{k}' + B_1\mathbf{i}' \tag{17}$$

Substituting Eq. (6) into Eq. (17), then

$$B_{\text{eff}} = 1/\gamma(\omega_0 - \omega)\mathbf{k}' + B_1\mathbf{i}' \tag{18}$$

If the frame of reference rotates about the z-axis at angular frequency ω matching the Larmor frequency ω_0 of the identical nuclei, then Eq. (18) becomes

$$B_{\text{eff}} = B_1\mathbf{i}' \tag{19}$$

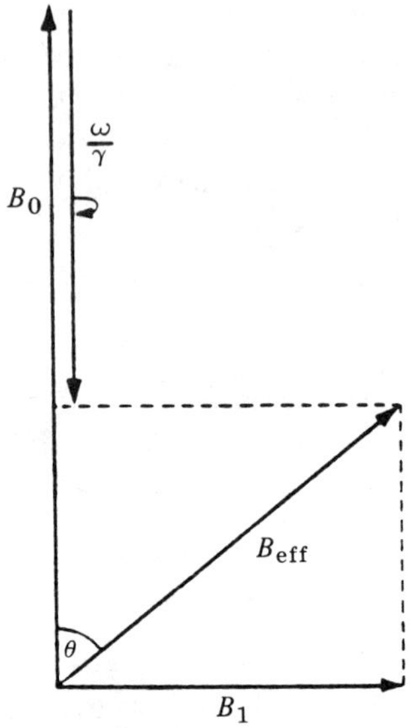

FIG. 6. The effective magnetic field in a rotating coordinate system.[1] (Reproduced by permission of Pergamon, Oxford, England.)

Substituting Eq. (19) into Eq. (13), then

$$(\partial \mathbf{M}/\partial t)_{\mathrm{rot}} = \gamma \mathbf{M} B_1 \mathbf{i}' \tag{20}$$

This equation describes that, at NMR state, the magnetization vector **M** precesses about the field vector $B_1\mathbf{i}'$ of the radio-frequency. Since the coordinate system and the rf field B_1 are chosen to rotate about z'-axis at the same frequency, the direction of B_1 is always along the rotating x'-axis. Consequently, the angular precession frequency ω_1 of **M** about x'-axis also follows the Larmor equation [Eq. (6)].

$$\omega_1 = \gamma B_1 \tag{21}$$

Because angular frequency equals angular velocity, an application of a constant rf field B_1 for a short time $\mathbf{t}_\mathbf{p}$ along the x'-axis causes the vector **M** to precess about the x'-axis from along the z'-axis toward the y'-axis by an angle θ, as shown in Fig. 7.

$$\theta = \omega_1\, \mathbf{t}_\mathbf{p} = \gamma B_1\, \mathbf{t}_\mathbf{p} \qquad \text{(radian)} \tag{22}$$

where $\mathbf{t}_\mathbf{p}$ is pulse width (PW) and θ is flip angle. A rf pulse causing the vector **M** to precess $\theta°$ is called a $\theta°$ pulse, e.g., 30° pulse, 90° pulse, 180° pulse, etc.

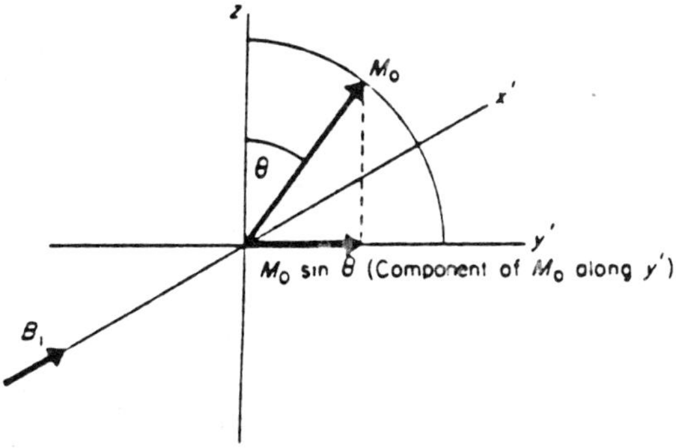

FIG. 7. Effect of a rf pulse with a frequency ω_0 for a time t (sec) on magnetization vector M_0.[4] (Reproduced by permission of Wiley, New York.)

Since the terms $(\omega_0 - \omega)\mathbf{k}'$ and $B_1\mathbf{i}'$ in Eq. (18) are components of vectors with respect to z'- and x'-axes in the rotating frame of reference, as shown in Fig. 6, the magnitude of the effective magnetic field B_{eff} is

$$|B_{eff}| = 1/\gamma[(\omega_0 - \omega)^2 + (\gamma B_1)^2]^{1/2} \qquad (23)$$

If ω_0 is the angular precession frequency of all nuclei of the same nuclear species, and B_1 is chosen such that

$$\omega_0 - \omega \ll \gamma B_1 \qquad (24)$$

then the term $(\omega_0 - \omega)$ can be neglected, and $B_{eff} \simeq B_1$. Since $(\omega_0 - \omega) = \Delta\omega = 2\pi\Delta\nu$, Eq. (24) becomes

$$2\pi\Delta\nu \ll \gamma B_1 \qquad (25)$$

where $\Delta\nu$ is the spectral width (SW), i.e., the chemical shift range of the nuclei to be observed in the unit of Hz. Consequently, if B_1 is chosen large enough, the magnetization vector of all the nuclei having the Larmor frequency within the spectral width $\Delta\nu$ will precess about the x'-axis in the rotating frame of reference as a result of applying a rf field B_1 along the x'-axis.

In order to ensure that the magnetization vector of all the nuclei is rotated by the same angle, i.e., in phase, the pulse width $\mathbf{t}_p$ of the rf pulse applied must be considerably shorter than the relaxation times T_1 and T_2 of the nuclei, so that relaxation is negligible during the pulse. For a 90° pulse $(\theta = \pi/2)$, the limiting condition for $\mathbf{t}_p$ can be estimated by substitution of $\theta = \pi/2$ and Eq. (22) into Eq. (25).

$$\mathbf{t}_{90} \ll 1/(4\Delta\nu) \quad \text{sec} \qquad (26)$$

In general, the pulse width for a 90° flip angle is in the range of 10–30 μsec for ^1H and ^{13}C nuclei. The magnitude of the pulse width depends on the magnetogyric ratio γ of the nucleus being observed, on the amplitude of the rf field [Eq. (22)] and on operational conditions of the NMR spectrometer. In practice, the frequency of rf pulse (carrier frequency) is adjusted in the middle of the spectral width. Compared to the single phase detection, where the carrier frequency is placed at one edge of the spectral width, i.e., slightly outside the sweep width, it has the advantage of a gain for the *S/N* ratio, by preventing the fold over of the base-line noise. In addition, it meets the requirement for a more uniform power level of the exciting rf pulse across the whole frequency range.

According to Eqs. (21) and (22), the application of a rf pulse of width $\mathbf{t_p}$ results in rotation of the magnetization vector M_0 along the z'-axis toward the y'-axis by an angle $\omega_1\mathbf{t_p} = \theta$, as shown in Fig. 7. Consequently, immediately after the rf pulse, the component of M_0 along the y'-axis has the magnitude

$$\mathbf{M}_{y'} = M_0 \sin \omega_1\mathbf{t_p} = M_0 \sin \theta \quad (27)$$

$\mathbf{M}_{y'}$ is also called the transverse magnetization. Thus, the signal induced in the receiver coil having its axis along the y' direction increases with $\mathbf{t_p}$, reaching a maximum for $\omega_1\mathbf{t_p} = \pi/2$, a 90° pulse. For pulse width larger than 90°, the induced signal decreases and becomes zero for $\omega_1\mathbf{t_p} = \pi$, a 180° pulse. In practice, a sample with a strong signal is used, and the pulse width is adjusted so that no signal is detected. One-half of this value is taken to be a 90° pulse.

Immediately after a $\theta°$ pulse, the transverse magnetization vector $\mathbf{M}_{y'}$ lies along the y'-axis. When the rf pulse is turned off, the vector $\mathbf{M}_{y'}$ decays exponentially to zero through spin–spin relaxation with the time constant $1/T_2$. Since the carrier frequency ω_1, i.e., the angular frequency of the rf pulse, is slightly off resonance, the vector $\mathbf{M}_{y'}$ rotates relative to the rotating frame of reference. At a time t after the pulse has been turned off, the vector $\mathbf{M}_{y'}$ has a phase shift of ωt relative to the y'-axis, where $\omega = (\omega_1 - \omega_0)$. Consequently, the vector $\mathbf{M}_{y'}$ does not precess with a constant phase shift of $\pi/2$ relative to the vector of B_1. Thus, the vectors $\mathbf{M}_{y'}$ and $\mathbf{B}_1'$ periodically rephase and dephase in the rotating frame of reference. As a result, a flux is induced in the receiver coil which alternates sign, and decays exponentially with time to zero when the Boltzmann equilibrium state is reestablished. The signal detected in the receiver is in the form of a beat pattern modulated by the difference in frequencies between the carrier frequency and the absorption frequency of the nuclei. In addition, the signal is also modulated with a frequency of J (Hz), where J is the coupling constant, if the nuclei being observed are spin coupled to another type of nuclei species. A plot of the signal versus time as the nuclei return to the Boltzmann equilibrium state after the rf pulse is called the free-induced decay (FID) signal or time domain function $F(\mathbf{t})$. For a sample containing several nonidentical nuclei of the same nuclear species, the beat pattern is very complex, as shown in Fig. 8. The FID signal must be Fourier transformed to obtain frequency domain function $F(\omega)$ or spectrum.

For time average purposes, rf pulses are applied repeatedly at a constant interval with each FID being acquired, added, and stored in a time-averaging device. The time required for acquisition of the FID is called data acquisition time (AQ). The time between two pulses is referred to as the repetition time if it is a constant throughout the experiment; if not, it is called the pulse interval. The time between the end of data acquisition and the next pulse is called the pulse delay (RD), i.e., the difference between the repetition time and the data acquisition time. The number of the FID acquired during a NMR experiment is called the number of scans (NS). Accumulation of FIDs in a digital computer with concomitant noise averaging is known as the computer averaged transients (CAT) method. The signal/noise (S/N) ratio increases with the number of scans n according to Eq. (28).

$$(S/N)_n = (S/N)_1\ (n)^{1/2} \tag{28}$$

For acquisition of FID data, each FID analog signal must be converted into digital form by an analog-to-digital computer (ADC). The FID is then recorded digitally as a series of several thousand data points, the number of which depends on the available data points N (memory capacity) of a digital computer. N is usually a power of 2, e.g., $2^{14} = 16{,}364$ (16K).

In order to obtain a true NMR spectrum after Fourier transformation, sufficient data points of each FID must be collected by the digital computer. The sampling time required to collect the sufficient data points depends on the spectral width $\Delta\nu$ (Hz). According to information theory,[25] two data points per cycle must be collected from each incoming signal. Thus, at least $2\Delta\nu$ data points per second must be collected for a spectrum with a spectral width $\Delta\nu$ (Hz). The maximal sweeping time per one data point, i.e., the dwell time (DW) $\mathbf{t}_{dw}$, must then satisfy Eq. (29).

$$\mathbf{t}_{dw} = (1/2\Delta) \qquad \text{sec/point} \tag{29}$$

For a spectral width of $\Delta\nu = 15{,}000$ Hz usually used in ^{13}C NMR spectra of lignin preparation at the NMR frequency for ^{13}C nuclei in the range of 60–63 MHz, the dwell time is 33.3 μsec.

[25] R. B. Blackman and J. W. Tudey, "The Measurement of Power Spectra." Dover, New York, 1958.

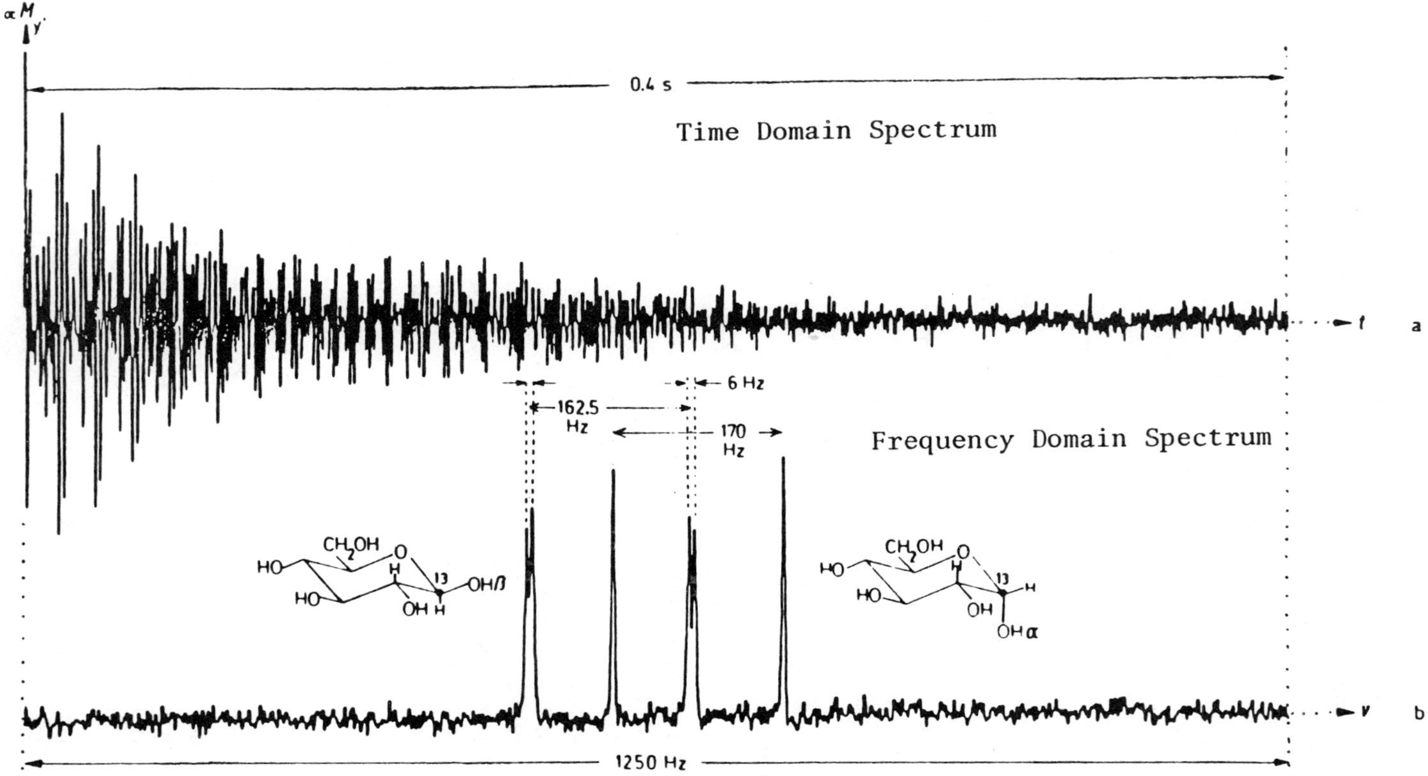

FIG. 8. (a) FID signal of mutarotated D-[1-^{13}C]glucose (60% ^{13}C) without proton decoupling. (b) Fourier transform of (a).[2] (Reproduced by permission of VCH Verlagsgesellschaft, Weinheim, Federal Republic of Germany.)

For a computer with the available data points N, the time required to fill up the memory capacity is

$$T_{aq} = Nt_{dw} = (N/2\Delta) \quad \text{sec} \tag{30}$$

where T_{aq} is the maximal data acquisition time (AQ) for recording a FID. For $\Delta\nu = 15{,}000$ Hz and $N = 16{,}394$ (16K), the maximal data acquisition is 0.55 sec. When several FIDs must be accumulated in order to improve the signal/noise ratio, T_{aq} is the minimal repetition time between two pulses.

Sample Preparations

Except for technical lignins, characterization of lignin by ^{1}H and ^{13}C NMR spectroscopy requires that the lignin should be isolated from plant tissues with minimal change in the structure. Milled wood lignin (MWL) is the best lignin preparation for this purpose (for the procedure for preparation of MWL from plant tissues, see chapter [14] in this volume). It must be mentioned that MWLs usually contain up to about 5% of carbohydrates, mainly hemicelluloses, and a small amount of fatty acids as contaminants. Since signals for ^{1}H and ^{13}C nuclei of these contaminants, particularly of carbohydrates, would appear at the aliphatic region of ^{1}H and ^{13}C NMR spectra of lignin, respectively, the contaminants should be removed from MWLs during purification as much as possible (for the procedure for purification of MWL, see chapter [14] in this volume). Technical lignins also need purification before using these lignins for the characterization. In order to ensure the purity of the lignin preparation, the following analyses of the lignin should be conducted before using it for the structural analysis by ^{1}H and ^{13}C NMR spectroscopy: (1) elemental analysis including methoxyl content, (2) moisture content (for the procedure, see chapter [14] in this volume), (3) total lignin content (for the procedure, see chapter [14] in this volume), and (4) carbohydrate analysis by a suitable procedure either according to Borchardt and Piper[26] or Fengel *et al.*[27] If the carbohydrate content of a lignin preparation is appreciably more than 5% or a carbohydrate-free lignin preparation is required, the lignin preparation should be purified again according to the procedure of Lundquist *et al.*[28]

[26] L. G. Borchardt and C. V. Piper, *Tappi* **53,** 257 (1970).
[27] D. Fengel, G. Wegener, A. Heizmann, and M. Przklenk, *Holzforschung* **31,** 65 (1977).
[28] K. Lundquist, B. Ohlsson, and R. Simonson, *Sven. Papperstidn.* **80,** 143 (1977).

Characterization of Lignin by ^1H NMR Spectroscopy

Reagents

Acetic anhydride
Pyridine
Deuterochloroform ($CDCl_3$) containing 1% tetramethylsiline (TMS)

Preparation of Acetylated Lignin

To a solution of 400-mg lignin preparation, either MWL or technical lignin, in 6 ml of pyridine is added 3 ml of acetic anhydrate. The mixture is kept at room temperature for 48 hr. Centrifuge the reaction mixture to remove insoluble materials, if any. The reaction mixture is then poured into about 40 g of crushed ice in a 100-ml beaker to precipitate acetylated lignin. Adjust the resulting mixture to pH 3 by adding concentrated HCl dropwise, and keep the mixture overnight at room temperature. The precipitate is centrifuged off, is stirred with about 30 ml of distilled water in the centrifuge bottle for about 30 min, and is again centrifuged off. The precipitate is washed once more and then is suspended in about 20 ml of distilled water and is freeze-dried. The final product is dried in a drying pistol over P_2O_5 at 50° under vacuum for 48 hr. The product should be free from pyridine and acetic acid.

^1H NMR Spectrum of Acetylated Lignin

Weigh about 35 mg of dried acetylated lignin. Dissolve the specimen in 0.5 ml of $CDCl_3$ containing 1% TMS (v/v), TMS being the internal reference. The concentration of the sample is about 7% (w/v). Filter off insoluble materials, if any, with beaker filtering (filter disk, 10 mm, i.d.; porosity, 25–50 μm). The solution is transferred to a 5-mm (o.d.) sample tube. The sample solution is then placed in the probe of a ^1H NMR spectrometer operating at 90, 100, 200 or 250 MHz, preferably one of the latter two, and the spectrum is recorded at 25° (298°K) or room temperature. When Fourier transformation (FT) mode is used with ^2H nucleus in $CDCl_3$ as an internal lock for the spectrometer field frequency, then a ^1H NMR spectrum of the acetylated lignin is obtained with pulse width corresponding to flip angle in the range of 60–90°, data acquisition time of about 2–3 sec, pulse delay of 0–2 sec, and the number of scans being about 20. The selection of the operational parameter depends on the nature of ^1H NMR spectrometer. Figure 9 shows a ^1H NMR spectrum of acetylated MWL prepared from sapwood of Zhong-Yang Mu *(Bischofia polycarpa)* recorded at ν ^1H = 250 MHz with a Bruker WM 250 NMR spectrometer.[29]

[29] D. Robert, D. Tai, and C.-L. Chen, *Holzforschung,* submitted for publication.

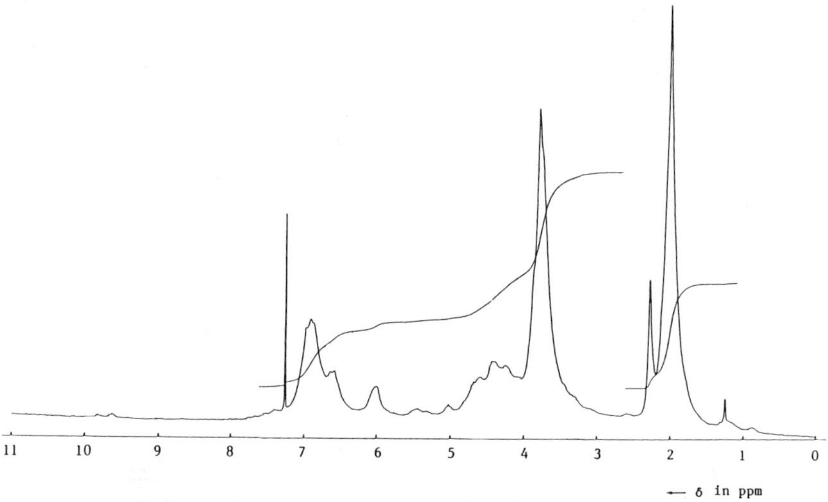

FIG. 9. ¹H NMR spectrum of acetylated milled wood lignin (acetylated MWL) from wood of *B. polycarpa*.

The spectrum is obtained in $CDCl_3$ at 25° in the FT mode with the pulse width 10.5 sec (90° pulse), the data acquisition time about 2.7 sec without a pulse delay, and the number of scans being 16.

Considerable line broadening would be observed in a NMR spectrum if solid materials and paramagnetic and/or ferromagnetic impurities are present in the sample solution. In addition to removal of any insoluble materials from the sample solution, care must be taken to ensure that these impurities are eliminated from the sample solution. For example, after a sample tube is cleaned with a cleaning solution, the tube must be washed thoroughly with distilled water until the tube is free from Cr(III), Cr(II), and other heavy metal ions, then must be rinsed thoroughly with acetone, and finally with carbon tetrachloride. The tube is then dried in a drying pistol or an oven under vacuum at 50° for at least 2 hr to remove any trace of the solvents.

Interpretation of ¹H NMR Spectra of Acetylated Lignins

As shown in Fig. 9, ¹H NMR spectra of acetylated lignins are not only complex, but are also not well resolved. This is also true for ¹H NMR of underivatized lignins using DMSO-d_6 as a solvent. These disadvantages are attributable to the polymeric and complex nature of lignin in terms of the chemical structure, in addition to the rather narrow ¹H chemical shift range. The ¹H nuclei in lignins present in similar, but not identical, chemical environments give rise to intensive overlap of signals in the spectra. In addition, the overlap is enhanced by the multiplication of the signals due to *J* scalar couplings. The presence of carbohydrates in lignin preparations would make the spectra even more complex. Thus, interpretation of the spectra entirely depends on the spectral data obtained from the ^{1}H NMR spectroscopic study of the lignin model compounds. Table I summarizes chemical shifts of ^{1}H nuclei present in the major substructures of acetylated lignins and possible contaminants.

TABLE I

CHEMICAL SHIFTS OF ^{1}H NUCLEI IN SUBSTRUCTURES OF ACETYLATED LIGNINS AND POSSIBLE CONTAMINANTS

Spectral region	Chemical shift range of (δ in ppm)	Types of ^{1}H nucleus
1	9.00–12.00	**Strongly Deshielded Region**
	10.00–12.00	Hydrogens in carboxylic acid groups in acetylated substructu and in *O*-alkylated uronic acid moieties in acetylated pentosa and fatty acids (minor contaminants)
	9.00–10.00	Hydrogens in aldehyde groups in cinnamaldehyde moieties in a tylated substructures (minor components)
2	6.25– 7.90	**Aromatic Region**
	7.80– 7.90	Aromatic hydrogens ortho to carbonyl group in acetylated a 4-*O*-alkylated *p*-hydroxyphenylpropane moieties with α—C= in acetylated substructures
	7.23– 7.80	Aromatic hydrogens ortho to —C=O group in acetylated and 4- alkylated guaiacylpropane moieties with α—C=O group in a tylated substructures. Also, hydrogens on C-α of cinnamaldeh and cinnamic acid moieties in acetylated substructures (mi components)
	7.23– 7.30	Aromatic hydrogens ortho to carboxyl group in acetylated a 4-*O*-alkylated syringylpropane moieties with α—C=O grou acetylated substructures
	6.80– 7.20	Aromatic hydrogens in acetylated and 4-*O*-alkylated *p*-hydro phenylpropane moieties of acetylated substructures
	6.35– 7.25	Aromatic hydrogens of acetylated and 4-*O*-alkylated guaiacyl pane moieties of acetylated substructures. Also, hydrogens on of cinnamaldehyde and cinnamic acid moieties in acetyla substructures (minor components)
	6.25– 6.70	Aromatic hydrogens in acetylated and 4-*O*-alkylated syringyl pane moieties of acetylated substructures. Also, hydrogens C-α of arylvinylene group in cinnamyl alcohol moieties of ac lated substructures (minor components)
3	5.75– 6.25	**Noncyclic Benzylic Region**
	6.10– 6.25	Hydrogens on C-β of arylvinylene group in cinnamyl alcohol m ties of acetylated substructures (minor components)
	5.75– 6.25	Hydrogens on C-α of acetylated β-*O*-4, β-1, and arylglycerol structures (the latter two, minor components)

TABLE I *(continued)*

ꓒectral egion	Chemical shift range of (δ in ppm)	Types of ^{1}H nucleus
4	5.20– 5.75	**Cyclic Benzylic Region**
	5.20– 5.75	Hydrogens on C-α in moiety A of acetylated β-5 and α-*O*-4 substructures
	5.20– 5.50	Hydrogens on C-3 of acetylated pentosans (minor contaminants)
5	2.50– 5.20	**Methoxyl and Major Aliphatic Region**
	4.50– 5.20	Hydrogens on C-β of acetylated β-*O*-4 substructures, on C-γ of cinnamyl alcohol acetate moieties in acetylated substructures, and on C-α of acetylated β–β substructures (the latter two, minor components). Also, hydrogens in C-1 and C-2 of acetylated pentosans and on C-2 and/or C-3 of acetylated hexosans (both minor contaminants)
	3.95– 4.50	Hydrogens on C-γ of acetylated β-*O*-4, β-5, β-1, and equatorial hydrogens on C-γ of β–β substructures (the latter two, minor components). Also, hydrogens on C-4 of acetylated pentosans and on C-1, C-4, and C-6 of acetylated hexosans (both minor contaminants)
	3.55– 3.95	Methoxyl hydrogens in aromatic moieties of acetylated substructures, hydrogens on C-β of β-5 and axial hydrogens on C-γ of β–β substructures. Also, hydrogens on C-5 of acetylated pentosans and hexosans (minor contaminants)
	2.50– 3.55	Hydrogens on C-β in acetylated β-1 and β–β substructures (both minor components). Also, methoxyl hydrogen in acetylated pentosans, hydrogens on $—CH_2COOH$ moieties of fatty acids (all three, minor contaminants)
6	2.20– 2.50	**Aromatic Acetoxyl Region**
	2.20– 2.50	Acetoxyl hydrogens in aromatic moieties of acetylated substructures, except for acetylated 5–5 substructures
7	1.60– 2.20	**Aliphatic Acetoxyl Region**
	1.60– 2.20	Acetoxyl hydrogens in aliphatic moieties of acetylated substructures and in aromatic moieties of acetylated 5–5 substructures
8	0.75– 1.60	**Nonoxygenated Aliphatic Region**
	1.10– 1.60	Hydrogens on $—CH_2—CH_2—CH_2—$ and $—(CH_2)_2—CH—$ moieties of fatty acids and similar aliphatic compounds (all minor contaminants)
	0.75– 1.10	Hydrogens on terminal $CH_3—$ groups in fatty acids and similar aliphatic compounds (all minor contaminants)

The spectrum of acetylated MWL of *B. polycarpa* is analyzed according to Table I.[29] In addition, the major substructure and functional groups are determined semiquantitatively. The results are given in Table II. The lignin contains about 5.6% of carbohydrates, of which about two-thirds are pentosans and the remainder hexosans. Moreover, the lignin has a C_9 unit

TABLE II

RESULTS FROM ^{1}H NMR SPECTRUM OF ACETYLATED MILLED WOOD LIGNIN (MWL) PREPARED FROM SAPWOOD OF ZHONG-YANG MU (*Bischofia polycarpa*)[a]

Chemical shift range (δ in ppm)	Integral	Number of hydrogens	Number of hydrogens after carbohydrate correction	Type of hydrogens
9.58–9.86	Negligible	0.02	0.02	*C*HO in Ar—CH=CH—CHO
7.23–7.90	2.1	0.14	0.14	Ar—H in Ar—COR
6.25–7.23	35.2	2.34	2.34	Ar—H in Ar—R H-α in Ar—CH=CH—CHO H-β in Ar—CH=CH—CHO H-α in Ar—CH=CH—CH_2OAc
5.75–6.25	6.0	0.40	0.40	H-α with α-O-Ac in β-*O*-4 and β-1 H-β in Ar—CH=CH—CH_2OAc
5.20–5.75	2.8	0.19	0.14	H-α with α-O-Ac in β-5 (0.09/C_9) H-α with α-O-Ac in β-*O*-4 and β-1 (0.05/C_9)
4.50–5.20	11.2	0.74	0.58	H-β in β-*O*-4 H-γ in Ar—CH=CH—CH_2OAc H-α in β–β
3.95–4.50	26.5	1.76	1.60 (0.80)	H-γ in β-*O*-4, β-5, β-1, and β–β
3.55–3.95	55.0	3.65	3.52	Ar—O—CH_3, (3.39/C_9) H-β in β-5 (0.09/C_9) H-γ in β–β (0.04/C_9)
2.50–3.55	8.4	0.56	0.56	H-β in β-1, β–β, and others
2.20–2.50	9.0	0.60	0.60	H in Ar—OAc except for 5–5 unit
1.50–2.20	63.8	4.23	3.69	H in Aliph—OAc and Ar—OAc in 5–5 units
1.10–1.50	0.4	0.03	0.03	—
0.75–1.10	+		+	—

formula $C_9H_{6.59}O_2(H_2O)_{0.89}(OCH_3)_{1.13}$ (C_9 unit weight: 197.83) after correction of the carbohydrate content (for analytical data, see chapter [14], Table I, in this volume). This means that one C_9 unit weight (g) of the lignin is associated with about 11.74 g of carbohydrates. Assuming that the mean monomeric unit for pentosans and hexosans are $C_{5.6}H_{8.8}O_{4.4}$ (unit weight: 146.53) and $C_6H_{10.2}O_{5.1}$ (unit weight: 162.94), respectively, then the carbohydrates consist of about 7.83 g (0.053 mol) of pentosans and about 3.91 g (0.024 mol) of hexosans. As shown in Table I, the aromatic methoxyl region (3.55–3.95 ppm) of ^{1}H NMR spectra of lignins also contains signals for ^{1}H nuclei on C-β of acetylated phenylconmaran (β-5) substructures, axial ^{1}H nuclei on C-γ of pinoresinol-type (β–β) substructures, and ^{1}H nuclei on C-5 of acetylated pentosans and hexosans. The C_9 unit formula of the MWL from *B. polycarpa* indicates that the lignin is rather close to MWLs from woods of gymnosperms. The lignin of this type usually contains about 0.09–0.12 and 0.02 U/C_9 unit of β-5 and β–β substructures,[24] respectively, in addition to up to about 5% of carbohydrates. Thus, the area of signals in the aromatic methoxyl region (3.55–3.95 ppm) of the spectrum should correspond to about 3.65 ^{1}H nuclei ($= 3 \times 1.13 + 0.09 + 2 \times 0.02 + 2 \times 0.053 + 0.024$). Since the integral for the area is 55, the integral for one ^{1}H nucleus corresponds to 15.07. The integrals of the other chemical shift ranges are divided by the integral for one ^{1}H nucleus, i.e., 15.07, with a correction due to carbohydrates and other contaminants, if any, to estimate the number of ^{1}H nuclei due to the acetylated MWL in these ranges.

The total aromatic hydrogens in the acetylated MWL from *B. polycarpa* is estimated to be about 2.39–2.42/C_9 units from the aromatic region (6.25–7.90 ppm) of the spectra, taking account of the possible presence of Ar—CH=CH—CHO and Ar—CH=CH—CH_2OH, each about 0.02–0.03 U/C_9 unit in the lignin. The aromatic region also contains signals for ^{1}H nuclei on C—α and C—β of Ar—CH=CH—CHO and on C—α of Ar—CH=CH—CH_2—OAc. Since the MWL has 1.13 OCH_3 groups/C_9 unit, the lignin could consist of guaiacylpropane and syringylpropane units in the approximate molar ratio of 0.87:0.13. If the lignin did not contain any condensed aromatic moiety, the total aromatic hydrogens of the lignin would be 2.87/C_9 unit ($= 3 \times 0.87 + 2 \times 0.13$). The deficiency in the total aromatic hydrogens between the assumed and estimated values, about 0.45–0.48/C_9 unit ($= 2.87 - 2.42$ or 2.39), is the degree of condensation involving aromatic moieties of the lignin. The number of aromatic acetoxyl groups, excluding those of biphenyl (5–5) substructures, are estimated to be about 0.20/C_9 unit from the aromatic acetoxyl region (range, 2.20–2.50 ppm), while the number of aliphatic acetoxyl groups, including those of 5–5 substructures, is determined to be

1.23/C_9 unit. An inspection of the other regions of the spectrum indicates that the acetylated MWL contains a total of about 1.18/C_9 unit aliphatic acetoxyl groups, about 0.38/C_9 unit on C—α (range, δ 5.75–6.25 ppm), about 0.02/C_9 unit of C—γ of Ar—CH=CH—CH_2—OAc (range, δ 4.50–5.20 ppm), and about 0.78/C_9 unit of C—γ of β—O—4, β—5, and β—1 substructures (range, δ 3.95–4.50 ppm). Thus, the number of aromatic acetoxyl groups in 5–5 substructures is estimated to be about 0.05/C_9 unit (= 1.23 − 1.18). Therefore, the aliphatic and phenolic hydroxyl contents in the MWL are determined to be about 1.18/C_9 unit and 0.25/C_9 unit, respectively. The spectrum also indicates that the lignin contains about 0.50 U/C_9 unit of β—O—4 substructures (range, δ 4.50–5.20 ppm), and, probably, 0.08–0.09 and 0.05–0.06 U/C_9 unit of β—5 and α—O—4 substructures, respectively. However, the quantity of other substructures cannot be estimated even in first approximation, mainly, because of line broadening caused by the overlap of signals.

Characterization of Lignin by ^{13}C NMR Spectroscopy

Reagents

Hexadeuterodimethyl sulfoxide (DMSO-d_6)
Hexadeuteroacetone (acetone-d_6)–deuterium oxide (D_2O) (9 : 1, v/v)

Samples and Solvents

Purified MWLs, technical lignins, and acetylated lignin preparations are usually used as samples for obtaining ^{13}C NMR spectra (for purification of lignin preparation, see *Sample preparation* in this chapter, and for the procedure for the preparation of acetylated lignin, see *Preparation of Acetylated Lignin*). Either DMSO-d_6 or acetone-d_6–D_2O (9 : 1, v/v) is used as solvent for MWLs and technical lignins. Usually, DMSO-d_6 is employed as the solvent for these lignin preparations because of greater solubility. For acetylated lignins, acetone-d_6–D_20 (9 : 1, v/v) is used as the solvent.

Preparation of Sample Solution

Weigh about 300–400 mg of dried lignin preparation. Dissolve the specimen in 2 ml of suitable solvent, TMS being the internal reference. Filter off insoluble materials, if any, with a beaker–filter (filter disk, 10 mm, i.d.; porosity, 25–50 μm). The concentration of the specimen is about 15–20% (w/v). The solution is transferred into a 10-mm (o.d.) tube, preparatory to running the ^{13}C NMR spectrum of the lignin. The sample tube must be thoroughly cleaned (for details, see *^1H NMR Spectrum of Acetylated Lignin*).

Routine ^{13}C NMR Spectra

^{13}C NMR spectrum of a lignin preparation in a suitable solvent is usually obtained with a ^{13}C NMR spectrometer operating in a pulse Fourier transform (FT) mode at NMR frequency for ^{13}C nucleus (ν ^{13}C) more than 50 MHz. JEOL FX60 and CX100 NMR spectrometers operate at ν ^{13}C of 14.9 and 100.4 MHz, respectively, while Bruker MW 200, 250, and 400 NMR spectrometers operate at ν ^{13}C of 50.3, 62.9, and 100.4 MHz, respectively. A sample solution of a lignin preparation is placed in the probe of a NMR spectrometer, and ^{2}H nucleus of the solvent, i.e., the ^{2}H nucleus in either DMSO-d_6 or acetone-d_6 is used as the internal lock for the spectrometer field frequency. Routine ^{13}C NMR spectrum of the lignin preparation is then obtained, in general, at 25–50° with a pulse width corresponding to a flip angle in the range of 30–60°, acquisition time about 0.5–1.0 sec, a pulse delay about 0.5–2 sec, and a number of scans about 10,000–20,000. Tetramethylsilane (TMS) is usually used as chemical shift reference, i.e., chemical shifts of signals are expressed in δ values (δ in ppm). The optimal parameters for operation of a spectrometer are determined on the basis of experimental data and the nature of the spectrometer. The broadband noise proton decoupler is turned on during the experiment as shown in Fig. 10.

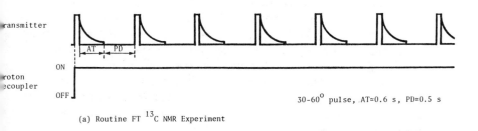

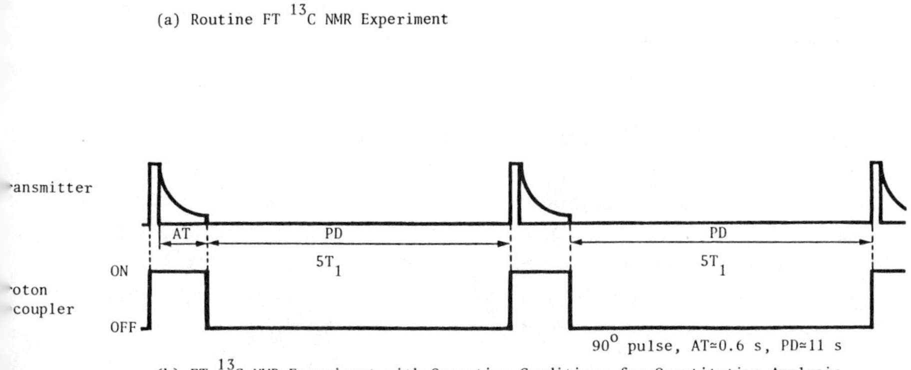

FIG. 10. Routine and inverse gated decoupled FT ^{13}C NMR experiments.

Figure 11 shows a ^{13}C NMR spectrum of MWL from wood of birch *(Betula papyrifera)* recorded at ν $^{13}C = 14.9$ MHz with a JEOL FX 60 NMR spectrometer. The spectrum is obtained in DMSO-d_6 at 50° with the pulse width 20 μsec (60° pulse), the data acquisition time about 1 sec, the pulse repetition time 2 sec, and the number of scans about 20,000. The broadband noise proton decoupler is applied during the operation. Figure 12 shows a ^{13}C NMR spectrum of MWL from wood of birch *(Betula verrucosa)* recorded at ν $^{13}C = 62.9$ MHz with a Bruker WM 250 NMR spectrometer. The spectrum is obtained in DMSO-d_6 at 50° with the pulse width 22 sec (90° pulse), the acquisition time 0.573 sec, the pulse delay 0.527 sec, and the number of scans being 37,000. The broadband noise proton decoupler is applied during the experiment.

A comparison of the spectrum of birch MWLs reveals that the spectrum recorded on a Bruker WM 250 spectrometer (ν $^{13}C = 62.9$ MHz) has a better resolution and more stable baseline, in fact a better *S/N* ratio than the spectrum recorded on a JEOL FX 60 spectrometer (ν $^{13}C = 14.9$ MHz). The resolution depends, in fact increases, on the instrument frequency and on the memory size of the analog-to-digital computer (ADC), respectively, 16,384 versus 8,192 data points for the Bruker and JEOL spectrometers. The signal-to-noise ratio, which defines the sensitivity, depends on B_0 to a power of 3/2, that means, increases by a factor of about eight, on going from 60 MHz (1.4092 *T*) to 250 MHz (5.872 *T*). This clearly demonstrates advantages of operating at a higher field.

Proton decoupling increases the intensity of ^{13}C signals because the intensity of all multiple lines in a ^{1}H- and ^{13}C-coupled spectrum are accumulated in one singlet signal in the decoupled spectrum. However, the intensity of a ^{13}C signal for CH group increases more than twice on proton decoupling. This additional sensitivity enhancement is known as the nuclear Overhauser effect (NOE). The maximal NOE enhancement factor $F_A(X)$ for the A signal for AX group in a X nuclei decoupling experiments depends on the magnetogyric ratios of A and X.[30]

$$f_A(X) = \gamma_X/2\gamma_A \tag{31}$$

Since γ for $^{13}C = 6{,}726$ rad sec^{-1} gauss^{-1} and γ for $^{1}H = 26{,}752$ rad sec^{-1} gauss^{-1}, the maximal NOE enhancement factor for CH groups is about 1.99. For CH_3 and CH_2 groups, the maximal NOE enhancement increases proportional to the number of attached protons. In contrast, quaternary carbons do not undergo the NOE enhancement. Thus, this is one reason why the area of ^{13}C signals in a proton decoupled ^{13}C NMR spectrum is not proportional to the number of the corresponding ^{13}C nuclei.

[30] J. H. Noggle and R. E. Schirmer, "The Nuclear Overhauser Effect: Chemical Applications." Academic Press, New York, 1971.

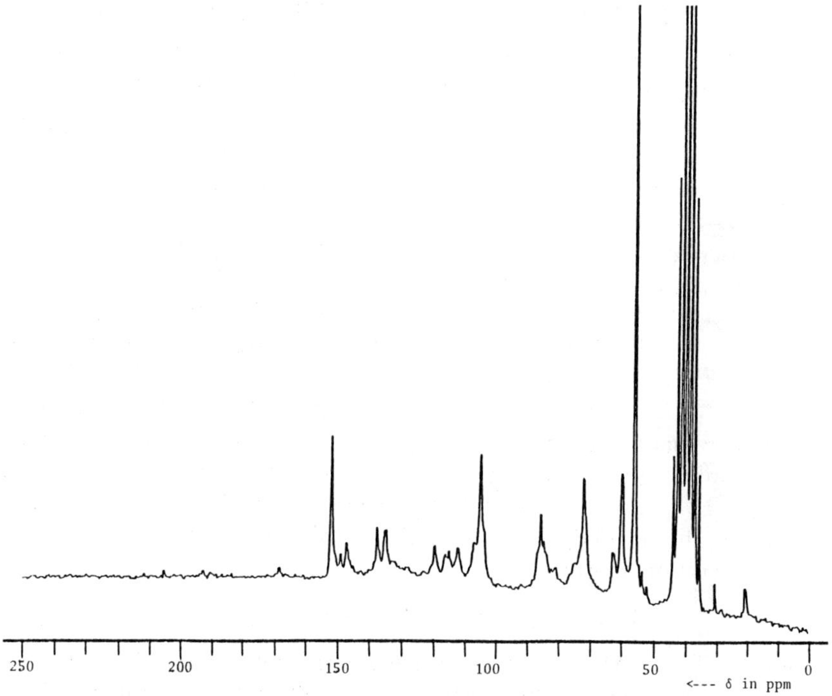

FIG. 11. Routine ^{13}C NMR spectrum of MWL from wood of birch *(B. papyrifera)*.

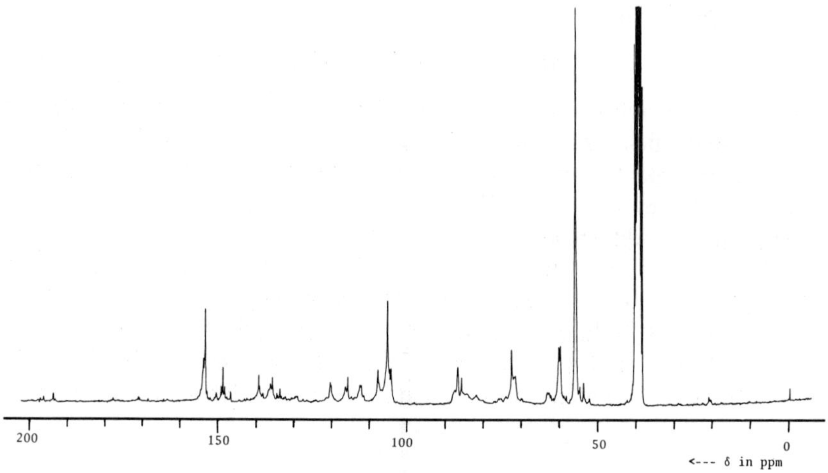

FIG. 12. Routine ^{13}C NMR spectrum of MWL from wood of birch *(B. verrucosa)*.

Quantitative ^{13}C *NMR Spectra: Inverse Gated Decoupling Sequence*

In order to obtain a quantitative ^{13}C NMR spectrum of a lignin preparation, the following factors must be considered: (1) pulse width, (2) time required to reestablish the Boltzmann equilibrium state after the rf pulse, (3) elimination of the NOE sensitivity enhancement due to proton decoupling, and (4) the number of scans in order to obtain a reasonable signal-to-noise ratio.

As described previously, NMR spectrometers only detect signals along the y'-axis in the rotating frame of reference. This requires that a 90° pulse should be applied, since the pulse produces the maximal transverse magnetization vector $\mathbf{M}_{y'}$ along the y'-axis immediately after the pulse according to Eq. (27), i.e., $\mathbf{M}_{y'(0)} = M_0 \sin 90° = M_0$, $\mathbf{M}_{x'(0)} = 0$, $\mathbf{M}_{z'(0)} = M_0 \cos 90° = 0$. After the 90° pulse, the magnetization vectors relax back to the Boltzmann equilibrium state via first order spin–lattice and spin–spin relaxation processes along the z'-axis and the $x'y'$ plane with time constants $1/T_1$ and $1/T_2$, respectively. Without considering the spin–spin relaxation, the signal detected along the y'-axis decays away via the spin–lattice relaxation according to

$$\mathbf{M}_{y'(t)} = \mathbf{M}_{y'(0)} \exp(-t/T_1) = M_0 \exp(-t/T_1) \tag{32}$$

where $\mathbf{M}_{y'(0)}$ and $\mathbf{M}_{y'(t)}$ are the magnetization vector along the y'-axis at time 0 sec and t sec after the application of a 90° pulse. The magnitude of $\mathbf{M}_{y'(0)}$ corresponds to that of M_0 along the z'-axis at the Boltzmann equilibrium state. After a time T_1 (sec), the vector $\mathbf{M}_{y'(0)}$ decays to $\mathbf{M}_{y'(T_1)} = 0.3679\ M_0$; and after a time $5T_1$ (sec), it decays to $\mathbf{M}_{y'(5T_1)} = 0.0067\ M_0$, essentially zero. At the same time, the vector $\mathbf{M}_{y'(0)}$ also decays via the spin–spin relaxation according to

$$\mathbf{M}_{y'(t)} = \mathbf{M}_{y'(0)} \exp(-t/T_2) = M_0 \exp(-t/T_2) \tag{33}$$

The spin–spin relaxation may be accelerated due to inhomogeneities in the magnetic field, so that $(1/T_2^*)_{\text{inhomo}} \leq 1/T_2 \leq 1/T_1$. If $T_1 \simeq T_2^*$, the pulse can be repeated after a pulse delay of about $3T_2^*$ without loss in signal intensity. However, if $T_1 \ll T_2^*$, a pulse delay of about $5T_1$ is required to reestablish the Boltzmann equilibrium state.

The elimination of the NOE sensitivity enhancement can be accomplished by application of the inverse gated proton decoupling sequence. As shown in Fig. 10,[31] the proton decoupler is turned off prior to the 90° pulse so that, when the pulse is applied, the magnetic moments of the ^{13}C nuclei are at the Boltzmann equilibrium state. The decoupler is turned on at the

[31] D. Robert, *Proc. Can. Wood Chem. Symp. 1982* p. 63 (1982).

same time the pulse is applied. Since the sample undergoes proton decoupling during data acquisition time, a full spectrum with no signal enhancement is obtained. The proton decoupling during data acquisition time perturbs the equilibrium state of the system. Consequently, when the decoupler is turned off immediately after the data acquisition, the system is no longer at the Boltzmann equilibrium state. In order to reestablish the equilibrium state, a pulse delay of $5T_1$ sec is required.

The T_1 of all signals in the ^{13}C NMR spectrum of MWL from birch wood *(B. verrucosa)* has been determined by the inversion recovery method.[32-33] Among the T_1s observed for the signals investigated, the longest T_1 is chosen as the standard to estimate the pulse delay required. At operating frequency ν $^{13}C = 50.32$ and 62.29 MHz, the optimal pulse delay is determined to be about 10 sec for MWLs and about 12 sec for acetylated lignin preparation.[31,33] Thus, the operational condition to obtain quantitative ^{13}C NMR spectra of lignin preparation is determined to be the inverse gated decoupling (IGD) sequence with the 90° pulse, the pulse delay about 10 sec, and the number of scans about 10,000. For acetylated lignin preparation, the pulse delay is about 12 sec.[34,35] The quantitative nature of the ^{13}C NMR spectra thus obtained has been verified.[31,33] However, the spectra could still have an error of about ±5% with respect to the proportionality between the area under a signal and the number of the ^{13}C nuclei giving rise to the signal.

Figure 13 shows a quantitative ^{13}C NMR spectrum of MWL from wood of Zhong-Yang Mu *(B. polycarpa)* recorded at 50° in DMSO-d_6 with a Bruker WM 200 NMR spectrometer operating at ν $^{13}C = 50.32$ MHz. The spectrum is obtained by the IGD sequence with the pulse width of 15 μsec (90° pulse), the data acquisition time of 0.7 sec, the pulse delay equal to 10 sec, and the number of scans being 9000. Figure 15d shows a quantitative ^{13}C NMR spectrum of MWL from wood of birch *(B. verrucosa)* recorded at the same experimental conditions for obtaining the spectrum of the MWL from *B. polycarpa.*

Distortionless Enhancement by Polarization Transfer (DEPT) Sequence

The DEPT sequence is a one-dimensional pulse sequence, which involves spin echo phenomenon, and allows separate observation of the ^{13}C NMR signals for the CH, CH_2, and CH_3 groups. In addition to clearly

[32] R. Freeman and H. D. W. Hill, *J. Chem. Phys.* **53,** 4103 (1971).
[33] D. Robert and D. Gagnaire, *Ekman Days, Int. Symp. Wood Pulp. Chem.* **1,** 86 (1981).
[34] D. Robert and G. Brurow, *Holzforschung* **38,** 85 (1984).
[35] D. Robert, M. Bardet, G. Gellerstedt, and E. L. Lindfords, *J. Wood Chem. Technol.* **4,** 239 (1984).

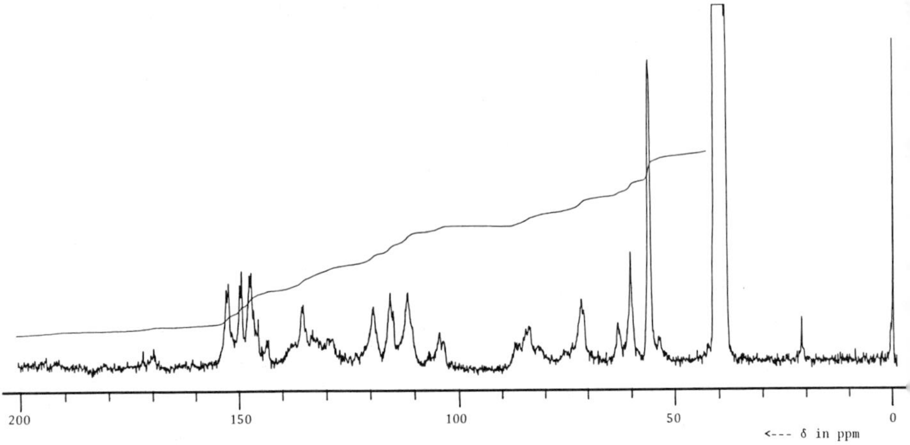

FIG. 13. Guantitative ^{13}C NMR spectrum of MWL from *B. polycarpa.*

revealing the multiplicity of the carbon atoms, there is a large signal intensity enhancement due to the polarization transfer. This polarization transfer relies on a spin population interchange between the more sensitive nucleus (^{1}H) and the less sensitive nucleus (^{13}C) to the benefit of the latter. As in the INEPT sequence,[36] the polarization transfer in the DEPT sequence is accomplished nonselectively through modulations of the transverse magnetization of the more sensitive nucleus (^{1}H) via its coupling to the less sensitive nucleus (^{13}C) by application of 180°x pulses in the ^{1}H and ^{13}C frequency regions, as indicated in the DEPT sequence scheme shown in Fig. 14.[22] The advantage of the DEPT sequence over the INEPT sequence is 2-fold, i.e., the DEPT sequence not only provides the same intensity enhancement for each line of a multiplet without phase distortion, but also is nearly independent of the assumed $^{1}J(^{13}C-^{1}H)$ values.

Figure 14 shows the DEPT pulse sequence and vector diagram of an AX system corresponding to a CH system, where A and X are ^{1}H and ^{13}C nuclei, respectively.[22] After the first 90° (**x**) pulse in the $A(^{1}H)$ region (a), the transverse magnetization of $A(^{1}H)$ is modulated by coupling to the $X(^{13}C)$ nucleus, resulting in a doublet with coupling constant $J(^{13}C-^{1}H)$. After time $t = \frac{1}{2}J$ sec, a phase difference of 180° exists between the vectors of the doublet (b). A 180° (**x**) pulse in the $A(^{1}H)$ region is used to refocus phase error caused by inhomogeneity of the magnetic field. At the same time, a 90° (**x**) pulse in the $X(^{13}C)$ region results in the transverse magnetization of $X(^{13}C)$. Since the magnetization vector of neither $A(^{1}H)$ or $X(^{13}C)$

[36] G. A. Morris and R. Freeman, *J. Am. Chem. Soc.* **101,** 760 (1979).

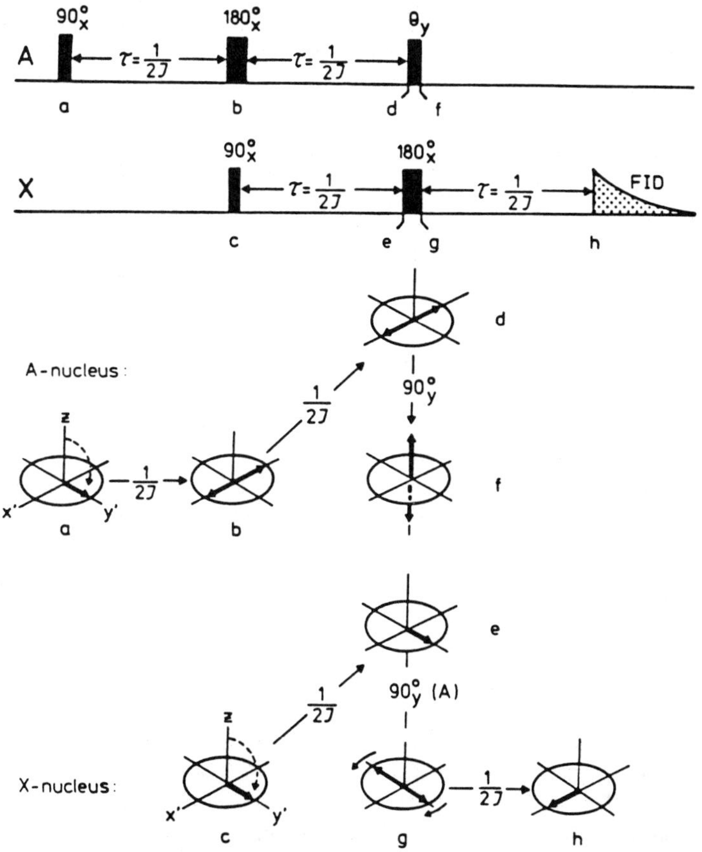

FIG. 14. DEPT pulse sequence and vector diagram for an *AX* system.[22] (Reproduced by permission of VCH Verlagsgesellschaft, Weinheim, Federal Republic of Germany.)

is present along the z'-axis in the double rotating frame of reference, $A(^{1}H)$ and $X(^{13}C)$ are practically decoupled. During the next $\frac{1}{2}J$-sec period, both vectors are stationary in their rotating frames of reference (d and e, respectively). At $t = 2(\frac{1}{2}J)$ sec, a $90°(y)$ pulse $(\theta = 90°)$ in the $A(^{1}H)$ region polarizes the magnetization of A(f). At the same time, this causes polarization of the magnetization of $X(^{13}C)$ (g). The reestablishment of magnetization along the z'-axis in the $A(^{1}H)$ region leads to refocusing of the vectors of $X(^{13}C)$ via spin–spin coupling in the last $\frac{1}{2}J$-sec period (h). The magnetization of $X(^{13}C)$ can be detected with uniform phase as a doublet with coupling constant $J(^{13}C-^{1}H)$ or as singlet with simultaneous ^{1}H decoupling at $t = 3(\frac{1}{2}J)$ sec until the end of the data acquisition time. After a recycling time D_1 sec, the pulse sequence is repeated.

The DEPT experiments for lignin preparation are conducted at 50° in DMSO-d_6 solution using the microprogramming facilities of the ASPECT 2000 pulse programmer,[37] provided with a Bruker WM 200 (or WM 250) NMR spectrometer, according to the following sequences (also see Fig. 16) [Note: ^{1}H decoupler is turned on at the time $t = D_2$, i.e., during data acquisition time (FID), and turned off immediately after FID.]:

$$\begin{array}{ll} ^1\text{H} & 90°(\mathbf{x}) - D_2 - 180°(\mathbf{x}) - D_2 - \theta°(\mathbf{y}) - D_2 \\ ^{13}\text{C} & \qquad\qquad\qquad 90°(\mathbf{x}) - D_2 - 180°(\mathbf{x}) - D_2 \\ & \qquad\qquad\qquad\qquad - \text{FID} - D_1 \end{array}$$

where $D_2 = 1/(2J_{\text{CH}})$ and D_1 = recycling time. For a lignin preparation, D_2 is obtained from an average value of $^1J(^{13}\text{C}-^1\text{H}) = 150$ Hz, e.g., $D_2 = 1/300$ sec. D_1 must be chosen so that $D_1 > T_1$ of ^{1}H, i.e., $D_1 = 3$ sec. Spectra are obtained from three different values of $\theta°$(y) pulse: $\theta_1 = \pi/4$ (45°), $\theta_2 = \pi/2$ (90°), and $\theta_3 = 3\pi/4$ (135°).[20,35] The CH, CH_2, and CH_3 subspectra are edited by linear combination of the spectra obtained with the three values of ($\theta°$, y) pulse as follows:

$$\text{CH subspectrum} \quad \theta_2 - z(\theta_1 + x\theta_3) \tag{34a}$$

$$\text{CH}_2 \text{ subspectrum} \quad \tfrac{1}{2}(\theta_1 + x\theta_3) \tag{34b}$$

$$\text{CH}_3 \text{ subspectrum} \quad \tfrac{1}{2}(\theta_1 + x\theta_3) - y\theta_2 \tag{34c}$$

where θ_1, θ_2, and θ_3 denote the spectra obtained with the three values of $\theta°$(y) pulse, $\theta_1 = \pi/4$, $\theta_2 = \pi/2$, and $\theta_3 = 3\pi/2$, respectively. Theoretically, $x = 1$, $y = 0.71$, and $z = 0$.[20] However, the parameters x, y, and z are experimentally determined by obtaining the optimal cancellation of unwanted signals in the subspectra using the corresponding IGD spectrum of the lignin as reference.

The ^{13}C 90°pulse has a pulse width of 14.5 μsec using the ASPECT pulse programmer. As the ^{1}H $\theta°$(y) pulses are never homogeneous, to calibrate the ^{1}H 90° pulse, the pulse width is adjusted so that the CH_2 signals of the sample are nullified with the $\frac{1}{2}J = D_2$ sec time periods. The calibration usually results in a pulse width of 31–36 sec for the ^{1}H 90° pulse. The pulse widths for the ^{1}H 45 and 135° pulses are directly obtained from the ^{1}H 90° pulse.

Figures 15 and 16 show CH, CH_2, and CH_3 subspectra of MWLs from birch and *B. polycarpa* obtained by the DEPT technique under the aforementioned operational conditions, respectively. Figure 17 shows a quater-

[37] M. R. Bendall, D. T. Pegg, D. M. Doddrell, and W. E. Hull, "DEPT Bruker Information Bulletin." Bruker Analytische Messtechnik, Karlsruhe, Federal Republic of Germany, 1982.

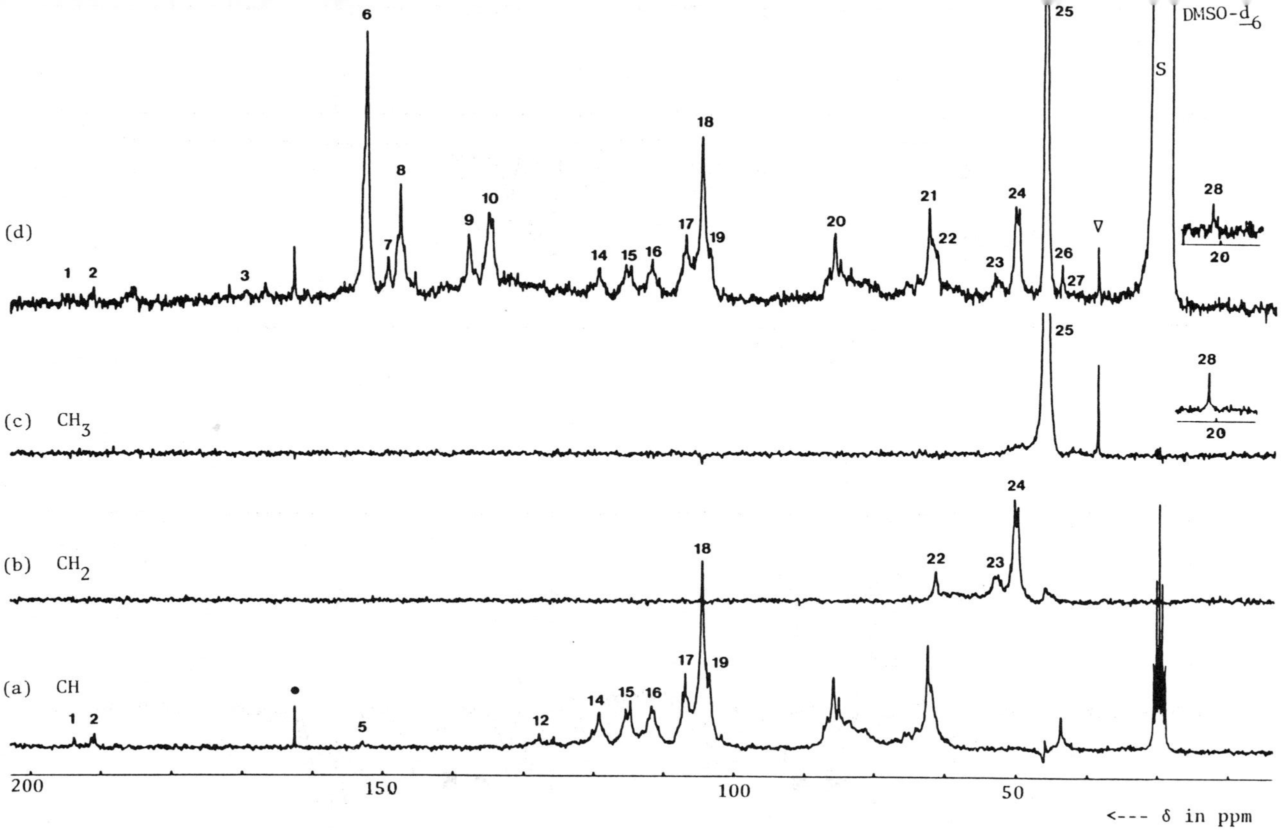

FIG. 15. DEPT-edited ^{13}C NMR spectra of MWL from birch *(B. verrucosa).* (a) CH signals, (b) CH_2 signals, (c) CH_3 signals, and (d) inverse gated decoupled spectrum (● and ∇) Contaminant.[24] (Reproduced by permission of Huthig & Wepf Verlag, Basel, Switzerland.)

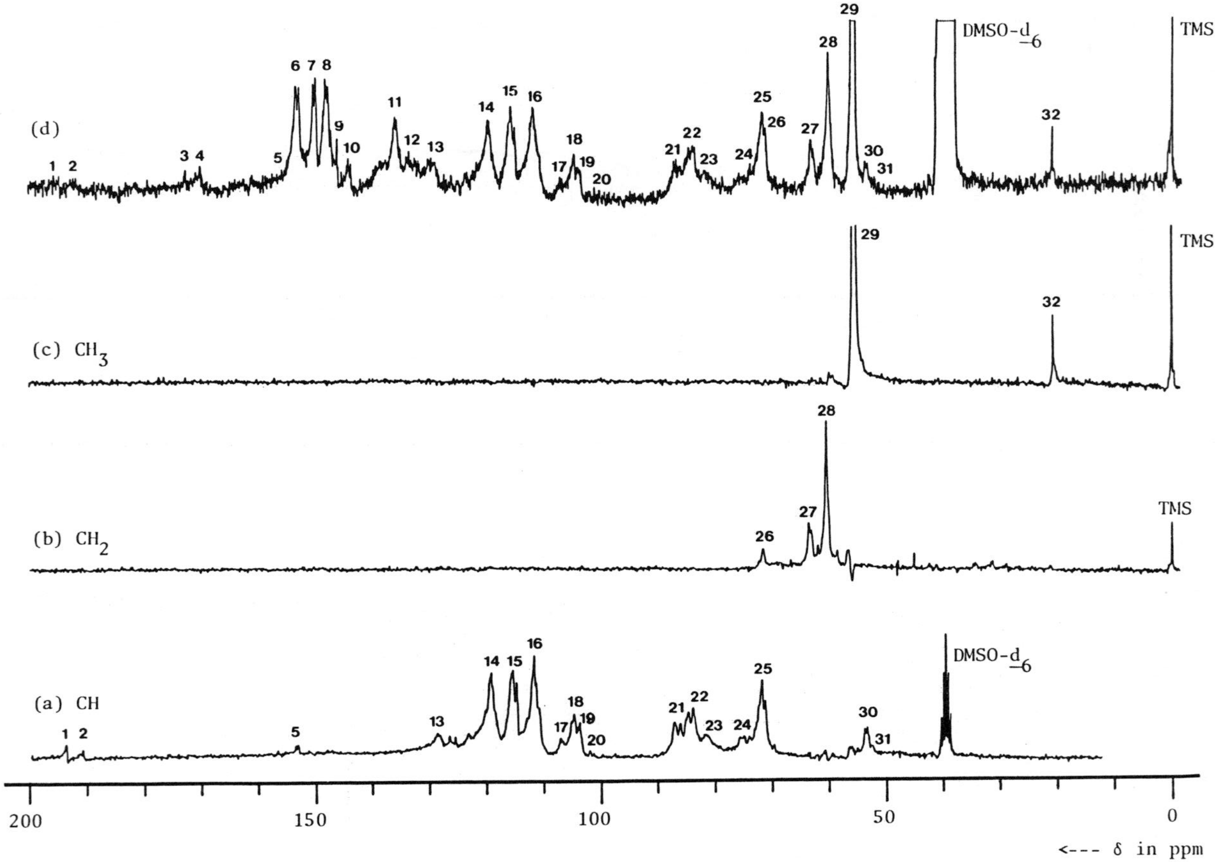

FIG. 16. DEPT-edited ^{13}C NMR spectra of MWL from *B. polycarpa*. (a) CH signals, (b) CH_2 signals, (c) CH_3 signals, and (d)

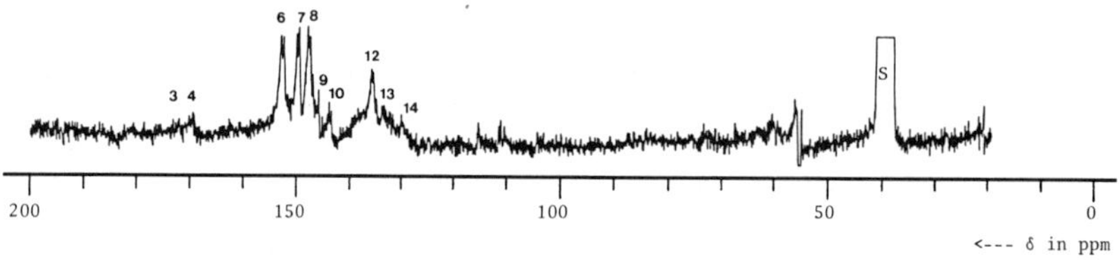

FIG. 17. Difference spectrum of MWL by subtracting the DEPT-edited subspectra for CH, CH_2, and CH_3 signals from the inverse gated decoupled spectrum.

nary C subspectrum obtained by subtraction of Σ DEPT (CH + CH_2 + CH_3) subspectra from the IGD ^{13}C NMR spectrum of the MWL from *B. polycarpa.*

Interpretation of ^{13}C NMR Spectra of Lignins

As compared to the corresponding ^1H NMR spectra, both routine and quantitative ^{13}C NMR spectra of lignins have considerably larger spectral widths, i.e., larger chemical shift range, and better resolution of signals due to proton decoupling. Thus, these spectra provide much more useful information for characterization of lignins than the corresponding ^1H NMR spectra. However, interpretation of the spectra depends entirely on the ^{13}C NMR spectra data of lignin model compounds. Table III summarizes chemical shifts of ^{13}C nuclei in the major lignin substructures and possible contaminants.

The ^{13}C NMR spectra of MWL from birch *(B. verrucosa)* including the DEPT subspectra have been analyzed.[24,31,33] The results are given in Table IV. In this section, the spectra of MWL from wood of *B. polycarpa* will be interpreted semiquantitatively. As described previously, the lignin contains about 5.6% carbohydrate and has a C_9 unit formula of $C_9H_{6.59}O_2(H_2O)_{0.89}(OCH_3)_{1.13}$ (C_9 unit weight, 197.83) after correction of carbohydrates (for analytical data, see chapter [14], Table I, in this volume). An examination of the quantitative ^{13}C NMR spectrum of MWL (Fig. 13) reveals that the MWL indeed contains carbohydrates as evidenced by the presence of signals 4, 20, 24, and 32 at δ 169.6, 101.6, 76–73, and 20.9 ppm corresponding to acetyl C=0, C—1 of xylan, C—2 ~ C—4 of xylan, and acetyl CH_3, respectively. The spectrum shows that the lignin is of the guaiacylsyringyl type. The presence of guaiacylpropane structures is evidenced by signals 14, 15, and 16 at δ 119.6, 115.8, and 112.2 ppm corresponding to C—6, C—5, and C—2 of guaiacyl group, respectively. In contrast to the rather weak signal 9 at δ 145.6 ppm corresponding to C—4 of guaiacyl group, signals 7 and 8 at δ 149.9–149.4 and 147.7–147.2 ppm are very strong. Since signals 7 and 8 correspond to C—3 and

TABLE III
CHEMICAL SHIFTS FOR ^{13}C NUCLEI IN LIGNINS[a]

Chemical shift (δ in ppm)	Type of ^{13}C nucleus
194	C=O in Ar—CH=CH—CHO and Ar—COR
192–191.5	C=O in Ar—CHO
172.1	COOH in aliphatic acid
169.5–169.0	Acetyl C=O in acetylated xylose and in xylan
167.2	COOH in Ar—COOH
165.2	C=O in *p*—hydroxybenzoate
162.0	C-4 in *p*—hydroxybenzoate
153–152	C-3 in etherified biphenyl (5–5)
	C-3/C-5 in S β-*O*-4 (S etherified)
	C-α in Ar—CH=CH—CHO
149.2	C-3 in G β-*O*-4 (G etherified)
147.5–147	C-3 in G nonetherified
	C-4 in G β-*O*-4 (G etherified)
	C-3/C-5 in S nonetherified
145.3	C-4 in G nonetherified
143.3	C-4 in phenylcoumaran (β-5)
	C-4 in 5–5 etherified
141.5	C-1/C-4 in 5–5 nonetherified
137.9	C-4 in S nonetherified
135–134	C-1 in G and S etherified
134–132	C-1 in G and S nonetherified
131.4	C-2/C-6 in *p*—hydroxyphenylbenzoate
129–128	C-β in Ar—CH=CH—CHO
	CH=CH in Ar—CH=CH—CH_2OH
120.5	C-1 in *p*—hydroxybenzoate
120–119	C-6 in G etherified and nonetherified
115.5–115	C-5 in G etherified and nonetherified
112–111	C-2 in G etherified and nonetherified
107–106	C-2/C-6 in S with α-C=O
104.5–103.5	C-2/C-6 in S
102–101	C-1 in xylose unit of xylan
86–85	C-β in β-*O*-4
83.6	C-β in β-*O*-4 with α-C=O
72.2	C-α in β-*O*-4
69.7	C-5 in xylose unit of xylan
63.2	C-γ in β-*O*-4 with α—C=O
62.7	C-γ in β-5
60.1–59.6	C-γ in β-*O*-4
55.5–56	OCH_3 in Ar—OCH_3
53.8	C-β in β–β
53.4	C-β in β–5

[a] G, Guaiacyl; S, syringyl.

TABLE IV
ASSIGNMENTS OF SIGNALS IN ^{13}C NMR SPECTRUM OF MWL FROM BIRCH (*Betula verrucosa*)[a]

gnal	δ (ppm)	Assignment
1	194	C=O in Ar—CH=CH—CHO
2	191.6	C=O in Ar—CHO
3	165–172	COOH in aromatic and aliphatic esters
5	152.6	C-α vinylic in cinnamaldehyde
6	152.1	C-3/C-5 in β-*O*-4 (S etherified)
7	149.2	C-3 in G β-*O*-4 (S etherified)
8	147.4	C-3 in G; C-4 in G β-*O*-4; C-3/C-5 in S nonetherified
9	138	C-1/C-4 in S etherified
10	134.4	C-1 in G etherified; C-4 in β–β
12	129.3	C-β vinylic in cinnamaldehyde
4	120.4	C-6 in G etherified and nonetherified
5	115.1	C-5 in G etherified and nonetherified
6	111.7	C-2 in G etherified and nonetherified
7	106.8	C-2/C-6 in S with α-C=O
8	104.4	C-2/C-6 in S β-*O*-4 and with CHOH
9	103.6	C-2/C-6 in S
0	85.8–82	C-β in β-*O*-4
1	72.2	C-α in β-*O*-4
2	71.8	C-γ in syringaresinol and/or pinoresinol
3	62.9	C-γ in β-*O*-4 with β-C=O; in β-5, in β-1
4	60.1–59.6	C-γ in β-O-4
5	55.9	$—OCH_3$
6	53.4	C-β in syringaresinol unit
7	Sh. 52.7	C-β in phenylcoumaran unit
8	21.0	γ-CH_3 adjacent to a C-β-OH and/or acetyl groups in xylan

[a] G, Guaiacyl; S, syringyl; Sh, shoulder. From Bardet *et al.*[24]

C—4 of 4—*O*—alkylated guaiacyl group, the guaiacyl moieties must be present in the lignin in the form of 4—*O*—alkyl ethers. Moreover, the 4—*O*—alkaylated guaiacyl moieties seem to be present in the lignin predominantly in the form of β—*O*—4 substructures because of the presence of relatively strong signals 22, 25, and 28 at δ 84.6, 71.8, and 60.2 ppm. These signals correspond to C—β, C—α, and C—γ of β—*O*—4 substructure, respectively. The presence of a syringylpropane structure is evidenced by signals 18 and 19 at δ 104.9 and 103.7 ppm, both corresponding to C—2/C—6 of syringl group; in addition to signal 6 at δ 152.6–152.3 ppm corresponding to C—3/C—5 of 4—*O*—alkaylated syringl group. The presence of the signal 21 at δ 87.0 ppm in addition to the signals 25 and 26 indicates further that 4—*O*—alkaylated syringlpropane units are present in the lignin mostly in the form of β—*O*—4 substructure. Signals 30 and 31 at δ 53.4 and 53.4 ppm indicate the presence of β–β and β—5 substructures in the lignin, respectively. The assignment of signals are given in Table V.

The aromatic region of the spectrum is chosen as standard to analyze the spectrum quantitatively, since the region does not contain the signals of carbohydrate contaminants. The DEPT CH subspectrum (Fig. 16a) indi-

TABLE V

ASSIGNMENTS OF SIGNALS IN ^{13}C NMR SPECTRUM OF MWL FROM WOOD OF *Bischofia polycarpa*[a]

Signal	δ (ppm)	Assignment
1	194.0	C=O in Ar—CH=CHO
2	191.6	C=O in Ar—CHO
3	172	COOH in aliphatic acid
4	169.6	Acetyl C=O in acetylated xylose unit in xylan
5	152.9	C-α in Ar—CH=CH—CHO
6	152.9–152.3	C-3 in etherified biphenyl (5–5) C-3/C-5 in S β-*O*-4 (S etherified)
7	149.2–149.4	C-3 in G β-*O*-4 (G etherified)
8	147.7–147.2	C-3 in G nonetherified C-4 in G β-*O*-4 (G etherfied C-3/C-5 in S nonetherified) C-3/C-5 in S nonetherified
9	145.6	C-4 in G nonetherified
10	143.2	C-4 in phenylcoumaran C-4 in 5–5 etherified
11	135.6	C-1 in G etherified
12	135–131	C-1 in G and S nonetherified
13	130–129	CH=CH in Ar—CH=CH—CH_2OH: C-β in Ar—CH=CH—CH
14	119.0	C-6 in G etherified and nonetherified
15	115.8	C-5 in G etherified and nonetherified
16	112.2	C-2 in G etherified and nonetherified
17	106.5	C-2/C-6 in S with α—C=O
18	104.7	C-2/C-6 in S
19	103.7	C-2/C-6 in S
20	101.6	C-1 in xylose unit of xylan
21	87.0	C-β in S β-*O*-4
22	84.6	C-β in G β-*O*-4
23	81.2	—
24	76–73	C-2/C-3/C-4 in xylose unit of xylan
25	71.6	C-α in β-*O*-4
26	71.0	C-γ in G and S β–β
27	62.9	C-γ in β-*O*-4 with α—C=O; in β-5, in β-1
28	60.2	C-γ in β-*O*-4
29	55.8	—OCH_3 in Ar—OCH_3
30	53.8	C-β in β–β
31	53.4	C-β in β-5
32	20.4	

[a] G, Guaiacyl; S, syringyl. From Robert *et al.*[29]

cates the presence of cinnamaldehyde and cinnamyl alcohol structures as evidenced by signals 1, 5, and 13 at δ 194.0, 152.9, and 129.7–129.4 ppm, respectively. From the integral, the quantity of cinnamaldehyde and cinnamyl alcohol is estimated to be each in the order of about 0.02 U/1 aromatic ring. Consequently, the integral of the aromatic region should correspond to 6.08 carbons. Since the total integral of the region is 56, the integral for one carbon is 56/6.08 = 9.21. The DEPT CH subspectrum also shows that the tertiary aromatic carbon region is δ 128–103 ppm excluding 0.08 vinylic carbons. The region δ 128–103 ppm is the syringyl C—2/C—6 region. The quaternary carbon region is δ 156–128 ppm including 0.08 vinylic carbons. The total integral of each region is then divided by the factor for one carbon, i.e., 9.21, to obtain the total number of carbons per one aromatic ring in the region. The results are given in Table VI.

It is obvious that the total number of carbons/benzene ring in the methoxyl region corresponds to the number of methoxyl groups per C_9 unit. Thus, the methoxyl content is 1.15/C_9 unit. This is a good agreement with 1.13/C_9 unit obtained from the elemental analysis. Since one syringyl group always has two C—2/C—6 carbons, the approximate ratio of syringylpropane units/C_9 unit can be obtained by dividing the total number of carbons in the spectral region by 2, i.e., 0.36/2 = 0.18. Assuming that no *p*-hydroxyphenylpropane units are present in the lignin, then the ratio of syringylpropane units to guaiacylpropane units in the lignin is 0.18:0.82/C_9 unit. The number of methoxyl groups/C_9 unit would then be $2 \times 0.18 + 0.82 = 1.18$/$C_9$ unit. The methoxyl content thus obtained is somewhat higher than the experimental value of 1.13/C_9 unit and the value 1.15/C_9 unit obtained from the methoxyl region. The excess value proba-

TABLE VI

INTEGRAL OF SPECTRAL REGIONS IN ^{13}C NMR SPECTRUM OF MWL FROM *Bischofia polycarpa*[a]

Spectral region	Chemical shift range [δ (ppm)]	Integral	Number of carbons per one benzene ring
aternary C.[b]	156–128	33.0	3.58
rtiary C	128–103	23.0	2.50
ringyl C-2/C-6	108–103	3.3	0.36
e chain[c]	90–57.5	28.6	3.11
thoxy	57.5–54.5	10.6	1.15
3 of β–β and β-5	54.5–53.0	0.8	0.09

[a] Factor for one carbon = 56/6.08 = 9.21. From Robert *et al.*[29]

[b] Includes four vinylic carbons.

[c] Excludes C-β of β–β and β-5 substructures.

bly is due to the presence of diphenyl ether (4—*O*—5) substructure. The chemical shift of C—2 and C—6 in 5—aroxyguaiacylpropane is in the chemical shift range δ 110–103 ppm. Thus, the number of 4—*O*—5 substructures would be $1.18-1.15=0.03/C_9$ unit. On the basis of $KMnO_4-NaIO_4$ oxidation, the amount of 4—*O*—5 substructure in lignin has been estimated to be in the order of $0.03-0.05/C_9$ unit.[38] Thus, the value for 4—*O*—5 substructure obtained here is rather reasonable, and the methoxyl content is about $1.15/C_9$ unit. The lignin then consists of guaiacylpropane and syringylpropane units in the molar ratio of 0.85:0.15. Assuming that the lignin does not contain aromatic ring condensed units, then the number of tertiary carbons must be $0.85\times3+0.15\times2=2.85/C_9$ unit. From the tertiary carbon spectral region, a value of $2.50/C_9$ unit is obtained. The degree of condensation for aromatic ring is then $2.85-2.50=0.35/C_9$ unit.

The signal 6 has integral 4.6 corresponding 0.5 carbons/C_9 unit. Since the signal arises from C—3/C—5 of 4—*O*—alkylated syringyl group and C—3 of 4,4′—*O*—dialkylated biphenyl (5–5) substructure, the number of carbons for the *O*-dialkylated 5–5 substructure can be estimated by the difference between the number of carbons under the signal 6 and the number of C—3/C—5 carbons in the syringyl units. Assuming that about 90% of the latter is involved in β—*O*—4 substructure, the number of carbons for C—3 of the 4,4′-*O*-dialkylated biphenyl (5–5) substructure is $0.5-0.15\times2\times0.9=0.23/C_9$ unit. Since the 5–5 substructure is a symmetric dimer, the number of the etherified 5–5 substructure is about $0.11/C_9$ unit. Similarly, the integral for signal 7 is 4.0 corresponding to 0.43 carbons/C_9 unit. Since the signal is characteristic of the C—3 of 4—*O*—alkaylated guaiacylpropane unit, the number of the unit is about $0.43/C_9$ unit. These units are involved in either β—*O*—4 or α—*O*—4 substructures; usually about 80% of these units are present in the form of β—*O*—4. Thus, the number of etherified guaiacyl type β—*O*—4 substructures would be about $0.34/C_9$ unit and that of α—*O*—4 substructures would be about $0.09/C_9$ unit. The number of syringyl type β—*O*—4 substructures would be about $(0.30\times0.90)/2=0.13/C_9$ unit.

[38] M. Erickson, S. Larsson, and G. Miksche, *Acta Chem. Scand.* **27**, 903 (1973).

[16] Gas–Liquid Chromatography of Aromatic Fragments from Lignin Degradation

By ANTHONY L. POMETTO III and DON L. CRAWFORD

When polymeric lignin is chemically or biologically degraded, numerous aromatic fragments are generated. The specific structures of these low-molecular-weight aromatic compounds depend on the type of lignin that is degraded. Softwood, hardwood, and grass lignins are each composed of a combination of three principal phenylpropanoid subunit structures derived from coumaryl (4-hydroxyphenylpropane), guaiacyl (3-methoxy-4-hydroxyphenylpropane), and syringyl (3,5-dimethoxy-4-hydroxyphenylpropane) precursors.[1] The three lignin types vary from one another primarily in the relative abundance of each precursor incorporated into the polymer and in the types and amounts of other aromatic acids which may be esterified to the lignin.[1] When lignins are degraded, each of these groups may be released, either as a complex structure consisting of two or more covalently linked aromatic rings or as single-ring aromatic compounds. The single-ring aromatic fragments typically consist of substituted benzenoid compounds, such as benzyl alcohols, benzaldehydes, or benzoic acids, and often single-ring phenylpropanoid alcohols, aldehydes, or acids are also released.[1-3] The most dominant linkage in lignin is the β-aryl ether bond, which covalently links the propane side chain of one phenylpropane moiety to the aromatic ring of an adjacent phenylpropane moiety.[1] Biological degradation of this ether bond can result in the oxidative substitution of the propane side chain resulting in the introduction of oxygen atoms as carbonyl and/or hydroxy groups into the lignin side chains.[2] When such oxidatively modified monomers are cleaved from the polymer during its biological or chemical degradation, a variety of unique, but structurally related compounds are released.

Researchers interested in identifying and characterizing these complex mixtures of aromatic lignin fragments have largely relied on gas–liquid chromatography (GLC) as their principal analytical tool. Gas–liquid chromatographic instruments equipped with flame ionization detectors (FID), which are extremely sensitive, require a volatile sample.[2-5] In order to

[1] R. L. Crawford, "Lignin Biodegradation and Transformation." Wiley, New York, 1981.
[2] C.-L. Chen and H.-M. Chang, *Holzforschung* **36,** 3 (1982).
[3] D. L. Crawford, *Biotechnol. Bioeng. Symp.* **11,** 275 (1981).
[4] T. K. Kirk, *Acta Chem. Scand.* **24,** 3379 (1970).
[5] K. Lundquist and T. K. Kirk, *Acta Chem. Scand.* **25,** 889 (1971).

convert nonvolatile aromatic compounds into volatile ones, they must be derivatized with a volatile compound such as *N,O*-bis(trimethylsilyl)acetamide (TMSA). Upon heating to high temperature, the now volatile trimethylsilylated aromatic lignin fragments will interact with the stationary liquid phase of the GLC column, migrate with the gaseous mobile phase, and become partitioned into separated compounds. Specific compounds are identifiable by their retention times in the column. Gas–liquid chromatography is a good technique for separating, identifying, and quantifying monomeric to trimeric aromatic lignin degradation fragments. However, because GLC alone provides only a retention time for each compound, other analysis are required to support the identity of the aromatic fragment. These other methods include high-performance liquid chromatography (HPLC), thin-layer chromatography, and gas–liquid chromatography-mass spectroscopy (GLC-MS). Another problem is that identification and quantification of these aromatics by GLC hinge on either the commercial purchase or laboratory synthesis of standard compounds, since pure standards are required to determine the retention time and peak area unit quantity of each aromatic fragment.

Sample Preparation Prior to Gas–Liquid Chromatography

Prior to quantitative analysis of lignin-derived aromatic fragments, the fragments must be concentrated from acidified aqueous solutions by extraction into organic solvent followed by evaporation of the solvent to recover the compounds. The specific weight of recovered extractives is then determined, and a specific amount of the sample is then derivatized for GLC. Although the source of the sample to be analyzed can be aromatic fragment mixtures resulting from either biological or chemical degradation of lignin, generally all samples are treated the same.

Biological Samples

Submerged culture systems are often used for studies of the biodegradation of lignin or lignin model compounds by microorganisms. Ligninolytic cultures may be grown in shake flasks, as stationary cultures, and/or as bubbler tube cultures[6] (see chapter [24] in this volume). Some actinomycetes and fungi, however, may be grown on lignocellulosic substrates using solid-state fermentations[6,7] (see chapter [5] in this volume). In addition, researchers often run cell-free enzymatic reactions where lignin or

[6] A. L. Pometto III and D. L. Crawford, *Appl. Environ. Microbiol.* **51,** 171 (1986).

[7] D. L. Crawford, A. L. Pometto III, and R. L. Crawford, *Appl. Environ. Microbiol.* **45,** 898 (1983).

lignin substructure models are subjected to enzymatic degradation. When submerged cultures are harvested or when enzymatic reactions are stopped, any cell debris or other insoluble material is first removed by filtration for filamentous microorganisms or centrifugation for nonfilamentous bacteria or enzymatic reactions. When solid-state fermentations are harvested, aromatic lignin fragments are usually extracted from the dampened, partially degraded lignocellulosic residues with water. For example, 100 ml of water/g of initial lignocellulose is typically added to the residue, the solution is heated for 1 hr in a boiling water bath, and then the solution is filtered to recover the aqueous phase, which contains the low-molecular-weight aromatic compounds[7] (see chapter [5] in this volume). Alternatively, residues may be extracted with 1 *N* NaOH or NH_4OH to promote complete recovery of lignin degradation intermediates in smaller volumes.[6] In most cases, the quantities of aromatic fragments extracted are small, and their concentrations in the aqueous extracts are such that they must be concentrated. In addition, they must usually be separated from other soluble extracellular material, which may include protein, carbohydrates, and possibly polymeric lignin fragments.[7] Concentration and separation from contaminating materials are easily accomplished by extracting the fragments from the aqueous solutions with organic solvents.

Each aqueous solution is first acidified to pH 1–2. This will precipitate proteins and any polymeric lignin fragments that are present (see chapter [5] in this volume). Next, the acidified supernatant is extracted twice with anhydrous ethyl ether and then once with ethyl acetate. The organic layers from each extraction are combined and dewatered by addition of enough sodium sulfate to make the solution clear.[6] The solvent is then decanted and/or filtered into a preweighed beaker and is placed in a hood where it is evaporated to dryness. The weight of extracted compounds is determined by reweighing the beaker and by substracting the beaker's initial weight from the value obtained. Next, a solution of 95% ethanol–ethyl acetate (1 : 1, v/v) is added to give a concentration of 10 mg of extractives/ml of 95% ethanol–ethyl acetate solution. To a preweighed 1- to 3-ml glass vial, 0.3 ml of the solution is added, and then the solvent is removed by evaporation in the hood. The weight of the extractives in the vial is then determined to ± 0.1 mg accuracy (a final weight of about 3 mg is expected). The sample is now ready for derivatization and chromatographing.[6,8]

Chemical Samples

Chemical degradative characterizations of lignin are commonly used to examine the structure of partially degraded lignins and to compare the

[8] A. L. Pometto III, J. B. Sutherland, and D. L. Crawford, *Can. J. Microbiol.* **27**, 636 (1981).

chemistry of microbially modified lignin to that of native lignin.[1] In research involving studies of lignocellulose degradation by microorganisms, degradative analysis of the lignin component of degraded residues requires that the lignin be purified from the cellulosic components by the methods such as that described by Björkman[9] (see chapter [3] in this volume). Purified Björkman lignins can then be analyzed chemically and/or spectrophotometrically.[1] Acidolysis and permanganate oxidation are the most commonly employed degradative chemical analysis used to characterize and compare the chemistries of decayed and nondecayed Björkman lignins. Acidolysis provides information on changes in the basic integrity of the lignin polymer that result from its degradation by microorganisms[10] (see chapter [24] in this volume), while permanganate oxidation provides information on microbially mediated changes in the number of free phenolic hydroxyl groups in the lignin polymer[11] (see chapter [5] in this volume). In both of these procedures, single aromatic ring degradation products are generated and quantified by GLC, and generally one must also synthesize standards of the expected lignin fragments.[5]

With permanganate oxidation, the aromatic products are ethoxylated benzoic acids. In chapter [24] (this volume), the procedure for recovering these products is described, with the final step involving their extraction into chloroform–acetone (1 : 1, v/v). After this extraction, the solvents containing the extractives are transferred into a preweighed beaker and are evaporated to dryness in the hood. The weight of the extractives is then determined as above, and then sufficient 95% ethanol–ethyl acetate (1 : 1, v/v) solution is added to give a concentration of 10 mg of extractives/ml of 95% ethanol–ethyl acetate solution. As above, 0.3 ml of the mixture is pipetted into a preweighed glass vial and is evaporated to dryness in the hood. Then the weight of the extracted residue is determined to ± 0.1 mg accuracy. The sample is now ready for derivatization and chromatographing.

With acidolysis, the products generated are a combination of substituted benzoic acids and aldehydes along with a mixture of phenylpropanoid compounds such as ketol I [1-hydroxy-3-(4-hydroxy-3-methoxyphenyl)-2-propane] (see chapter [24] in this volume). Some of the compounds are O_2 sensitive. The procedure for recovering the acidolysis products while protecting them from autoxidation is described in chapter [24] in this volume. As with permanganate oxidation, the final recovery steps involve their extraction into chloroform–acetone (1 : 1, v/v). From

[9] A. Björkman, *Sven. Papperstidn.* **59,** 477 (1956).

[10] A. L. Pometto III and D. L. Crawford, *Appl. Environ. Microbiol.* **49,** 879 (1985).

[11] D. L. Crawford, M. J. Barder, A. L. Pometto III, and R. L. Crawford, *Arch. Microbiol.* **131,** 140 (1982).

this point, using the method of Pometto and Crawford,[10] the extracted material is placed into a preweighed beaker, the solvents are removed by evaporation, and the weight of the extractives is determined as described above. The extractives are then solubilized in 95% ethanol–ethyl acetate (1 : 1, v/v) to a concentration of 10 mg of extractives/1 ml of 95% ethanol–ethyl acetate solution. Then, 0.3 ml of the solution is pipetted into a predesiccated, preweighed glass vial, and the solvents are removed by evaporation. The vials are next desiccated for 72 hr under a nitrogen atmosphere to avoid oxidation of acidolysis products and to remove any residual water. The weight of the sample in the vial is then determined to ±0.1 mg accuracy, and at this point, the sample is ready for derivatization and chromatographing.

Aromatic Compound Derivatization

General Considerations

As discussed previously, successful GLC requires that the compounds chromatographed be volatile and stable to heat. They should have a molecular weight of less than 500. Also, the mobile phase is a gas (N_2 or He), and partitioning of compounds by GLC depends primarily on compound interactions with a stationary liquid phase. Volatile forms of the aromatic lignin degradation fragments are formed by derivatization with TMSA, a volatile compound, trimethylsilylates free phenolic hydroxyl and carboxylic acid groups, both of which are very common in lignin-derived aromatic compounds. Aromatic aldehyde groups, which are also common, are not derivatized, but aldehyde-containing compounds are generally volatile enough by themselves to be chromatographed without derivatization. For most analysis, a good internal standard for GLC is veratraldehyde (3,4-dimethoxybenzaldehyde), which does not form a TMS derivative: it is rarely present in extractives as a lignin degradation product, but it chromatographs with a good retention time. By using such an internal standard, relative retention times can be calculated for each unknown compound (relative retention time equals the actual retention time of the compound of interest divided by the retention time of the internal standard). In addition, when an internal standard is used, small changes in retention times caused by interactions of compounds present as mixtures can be adjusted, and this allows for more reliable identification of unknown compounds.

Accurate quantitative determinations of the concentration of a particular compound present in a mixture requires that one know the exact weight of the sample of residue being derivatized, the total volume in the derivatized mixture, and the volume of sample to be injected. In addition, one must also have a pure standard of the compound. Many of the needed standard aromatic compounds can be purchased commercially, but some must be chemically synthesized. Peak area units under each chromatographic peak are usually calculated using a chromatographic integrator. For the quantification of specific aromatic fragments, a value for the number of peak area units per milligram of compound is determined by chromatographing known amounts of standards at concentrations above and below their expected concentrations in extractive mixtures. This standard relationship can then be used to determine the concentration of the compound in the sample residue. Examples of relative retention times and average peak area units per milligram of compound for some common aromatic lignin fragments and lignin substructure model compounds are presented in Table I.[12-14]

Derivatization Procedure and GLC Conditions

A vial containing 1–3 mg of residue is prepared as described above. TMSA derivatization is next accomplished by the addition of 100 μl of dioxane, 10 μl of pyridine, and 50 μl of *N,O*-bis(trimethylsilyl)acetamide (TMSA) (Sigma Chemical Co., St. Louis, Missouri) to the sample (total volume added is 160 μl). For internal standard, one should have previously added 0.1% (w/v) of veratraldehyde (Sigma Chemical Co.) to the dioxane used. The vials are next incubated in a heating block at 35° for exactly 2 hr from the time of TMSA addition. Then samples are injected into the chromatograph. Care must be taken to ensure injection of exactly 2-hr derivatives, and this requires the staggering of TMSA addition to each vial (at least 30 min between additions). A 10-μl GLC syringe is used. The tip is first filled with dioxane, and then the syringe barrel is pulled back to the 1 μl mark. This places 1 μl of air space in the syringe. Next, 1 μl of derivatized sample is taken as the syringe barrel is pulled out to the 2 μl mark. By pulling the sample back into the barrel of the syringe, one can visually determine the exact sample volume by reading it off the barrel. Sequentially, one should, at this point, have 1 μl of sample in the syringe, followed by 1 μl of air space, followed by a syringe tip volume of dioxane, followed by the syringe plunger. The use of this procedure ensures accurate sample volumes and total injection of the sample into the GLC.

[12] M. T. Bogart and J. Erlick, *J. Am. Chem. Soc.* **41,** 801 (1955).

[13] K. Elbs and H. Lerch, *J. Parkt Chem.* **201,** 1 (1976).

[14] E. Adler, B. O. Lindgren, and U. Saedén, *Sven. Papperstidn.* **55,** 245 (1952).

TABLE I

RELATIVE RETENTION TIMES, PEAK AREA UNITS PER MILLIGRAM OF STANDARD COMPOUND, AND SOURCES OF SOME COMMON AROMATIC FRAGMENTS DERIVED FROM LIGNIN AS DETERMINED BY CAPILLARY GLC[a]

Compound	Source[b]	Relative retention Time (min)[c]	area (U/mg) (×10⁶)
Catechol	Eastman	0.854	414.724
trans-Cinnamic acid	U.S. Biochemical	1.167	497.003
p-Coumaric acid	Sigma	1.544	89.602
3,4-Diethoxybenzoic acid	Synthesized[12]	1.397	97.960
Dehydrodivanillin	Synthesized[13]	2.543	22.282
4-Ethoxybenzoic acid	Synthesized[12]	1.186	306.715
4-Ethoxy-3-methoxybenzoic acid	Synthesized[12]	1.361	101.778
4-Ethoxy-3,5-dimethoxybenzoic acid	Synthesized[12]	0.685	98.903
Ferulic acid	Sigma	1.695	64.657
Guaiacol	Aldrich	0.700	713.819
p-Hydroxybenzaldehyde	Sigma	0.931	810.009
p-Hydroxybenzoic acid	Aldrich	1.244	606.773
Ketol I [1-hydroxy-3-(4-hydroxy-3-methoxyphenyl)-2-propane]	Synthesized[5]	1.536 and 1.689	123.159
Protocatechuic acid	Sigma	1.454	268.746
Syringaldehyde	Aldrich	1.311	273.321
Syringic acid	Aldrich	1.505	211.576
Vanillic acid	Aldrich	1.395	427.950
Vanillin	Aldrich	1.142	454.293
Veratraldehyde	Aldrich	1.00	223.771
Veratryl alcohol	Aldrich	0.582	295.966
Veratrylglycerol-β-guaiacyl ether	Synthesized[14]	2.13	13.618

[a] Capillary GLC was performed using a stationary liquid phase of RSL-150 (column, 30 m × 0.25 mm) (Alltech Associates, Inc.). The temperatures used were as follows: injector, 240°; detector, 280°; initial column oven temperature, 120° for 2 min followed by a 20°/min increase to 260°, followed by maintenance of 260° for 3–7 min. The carrier gas was helium at a flow of 20 ml/min, with 1–2 ml/min flow through the column; the excess gas was split off before reaching the column. Sample vials containing a specific amount of compound (1–3 mg) were derivatized by the addition of 100 μl of dioxane containing 0.1% (w/v) veratraldehyde as internal standard, 10 μl of pyridine, and 50 μl of *N,O*-bis(trimethylsilyl)acetamide (Sigma Chemical Co.), giving a total volume of 160 μl. Derivatization was carried out for 2 hr at 35° prior to injection.[6]

[b] Commercial sources were as follows: Aldrich Chemical Co., Milwaukee, Wisconsin; Eastman Kodak Co., Rochester, New York; Sigma Chemical Co., St. Louis, Missouri; United States Biochemical Corp., Cleveland, Ohio. Chemically synthesized compounds are referenced as to the published method for synthesis.

[c] Relative retention times are based on the retention time of the compound of interest divided by that of the internal standard (veratraldehyde). Actual retention time for veratraldehyde was 6.06 min.

The 1-μl sample is next injected into the GLC. Two types of columns can be used, a packed column, which utilizes a solid support to hold the stationary liquid phase, or a capillary column, which has the stationary liquid phase coated on the inside walls of the column. For a packed column (200 × 0.30 cm), the solid support is usually 80/100-mesh acid-washed and dimethyldichlorosilane-treated Chromosorb G coated with 5% (w/w) OV-1 as the stationary liquid phase, and the temperatures used are as follows: injector, 240°; detector, 280°; and column oven, 220°. The carrier gas is N_2 or He at a flow of 30 ml/min.[8] Capillary columns (30 m × 0.25 mm) give much better results than packed columns. With these columns, an OV-1 equivalent such as RSL-150 (Alltech Associates, Inc., Deerfield, Illinois) is used, and the temperatures used are as follows: injector, 240°; detector, 280°; initial column oven temperature, 120° for 2 min, followed by a 20°/min increase to 260°, which is then held for the final 3–10 min. The carrier gas is He, and the flow rate is 20 ml/min, with 1–2 ml/min flow through the column. The excess gas is split off before reaching the column.[6] A capillary column is mandatory for GLC-MS studies, where a very small amount of sample is split from the end of the column such that part of each peak goes to the FID and part goes to the mass spectrometer.

Typically, the order of elution for the aromatic lignin fragments will always be the same under identical or similar conditions, and GLC is tremendously sensitive (Table I). However, the compounds of interest must be heat stable. Heat instability can deleteriously effect quantification and detection sensitivity. Some of the aromatic fragments acidolysis of lignin[4] are heat sensitive, and GLC results with these products must be interpreted with care. Derivatization of acidolysis products is also tricky. It is conducted as above except that the TMSA reaction mixture is chromatographed once every hour for a 3-hr period beginning from the time derivatization is started. During that 3-hr period, the maximum recorded concentration for each of the different compounds will be observed at different derivatization times. Ketol I, the quantitatively most important acidolysis product recovered from native lignin, will show its highest chromatographic concentration after 2 hr of derivatization and always as two major peaks (see Table I), while some of the minor peaks will show their highest concentrations after 1 or 3 hr.[10] Clearly, interpretation of such complex data is difficult because of the potential for data variability due to heat sensitivity and/or quantitative variations associated with derivatization time. In the future, therefore, HPLC may replace GLC for many of these analysis. Lignin-derived aromatic fragments can be readily separated by HPLC, and they can be adequately detected using ultraviolet detectors and spectrally characterized with diode array detectors (see chapter [17] in this volume). HPLC is also nondestructive, which enables purification and further examination of chromatographed fragments. However, for the present, quantification of nanogram amounts of compounds, which is too low a concentration for current HPLC ultraviolet detectors, GLC with FID is still required.

[17] High-Performance Liquid Chromatography of Aromatic Fragments from Lignin Degradation

By ANTHONY L. POMETTO III and DON L. CRAWFORD

Lignin is a complex aromatic polymer composed of phenylpropane subunits including coumaryl, guaiacyl, and syringyl moieties which are covalently linked together by a variety of bonds, but mainly by the β-aryl ether bond.[1] The chemical and/or biological degradation of lignin results in the release of a myriad of low-molecular-weight aromatic fragments which are structurally related to each other, yet unique in specific structure (see chapter [16] in this volume). The physical and spectral properties of these aromatic fragment are such that they may be readily analyzed by high-performance liquid chromatography (HPLC). Usually, the weakest component of HPLC is its detector, with refractive index and ultraviolet (UV) absorption detectors being most commonly used. Both are not very sensitive when compared to the high sensitivity of the flame ionization detector used in gas–liquid chromatography (GLC). However, the aromatic structure of many lignin fragments provides a good chromophore that strongly absorbs UV irradiation in the 190- to 355-nm wavelength range. Often, specific absorption maxima also exhibit fairly high extinction coefficients. Furthermore, many of these aromatic fragments are soluble in aqueous (water) and/or organic (methanol and acetonitrile) polar solvents, an ideal situation for partitioning by reversed-phase column chromatography. In reversed-phase HPLC, porous microparticulates (5–10 μm) of silica are chemically bonded with aliphatic saturated carbon chains of 8 or 18 carbons. The most common mobile phase used with these reversed-phase columns consists of water (acidic, neutral, or buffered) and an organic solvent (methanol or acetonitrile).[2] The partitioning of specific compounds is determined by how they interact with the bonded phase and the mobile

[1] R. L. Crawford, "Lignin Biodegradation and Transformation." Wiley, New York, 1981.
[2] N. A. Parris, *J. Chromatogr. Libr.* **27**, (1984).

phase. The mobile phase can be run isocratically, as a gradient mixture, or as a combination of both.

Aromatic compounds derived from lignin can be readily separated by reversed-phase HPLC. Also, since each unique aromatic fragment has a specific UV absorption spectrum, if one utilizes a chromatograph equipped with a computer-coupled diode array detector (DAD), it is possible to obtain and record the UV absorption spectrum for each chromatographic peak. A unique UV absorption spectrum is a fingerprint for a specific compound, and when both the spectrum and specific column retention time can be determined by HPLC and compared to those of pure standards, then HPLC becomes an excellent technique for identifying unknown lignin degradation aromatic fragments.

High-performance liquid chromatography is also valuable because the detection method is nondestructive. This makes it possible to collect individual compounds from chromatographic peaks for further analysis. For example, if mass spectroscopy is used in conjunction with HPLC for analysis of unknown compounds, researchers obtain not only a specific retention time and UV spectrum (with DAD) for each compound, but also a mass ion. These addition data are of tremendous value to researchers attempting to specifically identify the numerous possible aromatic fragments that might be derived from lignin.

Another area of interest to the lignin researcher is high-performance size-exclusion chromatography, a technique with great potential for characterizing polymeric lignin fragments.[3] This is a research area that thus far has received little attention. The amorphous structure of the lignin polymer and its tendency to form aggregation complexes make lignin molecular-weight determinations quite difficult.[3] Though the subject is not specifically covered in this chapter, we note that the potential for using size-exclusion HPLC in the study of polymeric lignin degradation intermediates such as those generated by *Streptomyces*[4,5] is great.

Sample Preparation

Filtered, aqueous samples containing aromatic lignin fragments generated by biological or chemical degradation of lignin can be removed from reaction mixtures and chromatographed directly, or they can be chromatographed after prior concentration by solvent extraction. However, for

[3] J. Pellinen and M. Salkinoja-Salonen, *J. Chromatogr.* **328,** 299 (1985).

[4] D. L. Crawford, A. L. Pometto III, and R. L. Crawford, *Appl. Environ. Microbiol.* **45,** 898 (1983).

[5] J. R. Borgmeyer and D. L. Crawford, *Appl. Environ. Microbiol.* **49,** 273 (1985).

samples concentrated by solvent extraction, the specific weight of extractives recovered from a given sample volume must be known in order to do quantitative calculations later. For HPLC, particular caution must also be taken to remove all particulate matter before injection of samples into the column, and it is mandatory that a guard column with the same packing material as the separation column be used.

Direct Sampling of Biological Mixtures

Supernatant samples are taken directly from microbial cultures growing on lignocellulose in submerged cultures (e.g., shake flask cultures, liquid stationary cultures, or bubbler tube cultures[6]; see also chapter [5] in this volume). The insoluble cells and residual lignocellulose are then removed by filtration (Whatman No. 1) or by centrifugation (> 16,000 *g* for 30 min) prior to direct analysis by HPLC. If the microorganism has been grown instead on lignocellulose in solid-state fermentations (see chapter [5] in this volume), a different sampling procedure is used. The culture must be harvested by addition of 100 ml of distilled water for each gram of initial lignocellulose substrate used. This mixture is heated for 1 hr in a boiling water bath, and then microbial cells and the lignocellulose residue are removed by filtration or centrifugation[4] just prior to analysis by HPLC. Alternatively, 20 ml of 1 *N* NaOH or NH_4OH/g of initial lignocellulose can be added to the culture. This suspension is mixed, filtered, or centrifuged, and the filtrate is retained.[6] The basic solution is then neutralized prior to analysis by HPLC.

For HPLC studies of low-molecular-weight lignin model compound biodegradation, microbes are grown in the presence of 0.05 – 0.1% (w/v) of model compound, usually in submerged cultures. Periodically during incubation, samples (1 – 2 ml) are taken aseptically and are filtered or centrifuged to remove cell mass.[7] Then these samples are analyzed directly by HPLC to monitor product formation and substrate disappearance over time.

For analysis, samples free of insoluble debris are mixed with an equal volume of the HPLC grade organic solvent to be used in the HPLC run. This solution is again filtered, this time through a 5-μm membrane filter (a polycarbonate filter must be used for samples containing acetonitrile).[6] The mixing and filtration will prevent any solvent-precipitable matter (e.g., protein) from collecting in the column. The sample is now ready for HPLC.

[6] A. L. Pometto III and D. L. Crawford, *Appl. Environ. Microbiol.* **51,** 171 (1986).

[7] D. L. Crawford, T. M. Petty, B. M. Thede, and L. A. Deobald, *Biotechnol. Bioeng. Symp.* **14,** 241 (1984).

Biological Samples Requiring Solvent Extraction

In both submerged cultures and in solid-state fermentations of lignocellulose, the quantities of low-molecular-weight aromatic lignin fragments released into the aqueous phase during lignin degradation are often small.[8] Therefore, one may have to concentrate the fragments and separate them from other extracellular material by organic solvent extraction.[6,8] This is accomplished by first acidifying the culture supernatant (or aqueous extract in the case of solid-state fermentations) to pH 1–2 with concentrated H_2SO_4.

Acidification results in the precipitation of extracellular proteins and polymeric lignin fragments (see chapter [5] in this volume). The acidified supernatant or aqueous extract is next extracted twice with anhydrous ethyl ether and then once with ethyl acetate. The organic phases from each extraction are combined and dewatered with sodium sulfate. The anhydrous solvent layers are next decanted into a preweighed beaker (or round bottom flask) and evaporated to dryness in a hood (or by using rotary vacuum evaporation). The weight of the extracted material is then determined by reweighing the beaker or flask. Then a solution of 95% ethanol and ethyl acetate (1 : 1, v/v) is added to give a final concentration of 10 mg of extractives/ml of 95% ethanol–ethyl acetate solution. To a preweighed glass vial (1–3 ml), 0.3 ml of this solution is added, and the solvents are allowed to evaporate overnight in the hood. The weight of vial plus extractives is then determined to ±0.1 mg accuracy, and the exact quantity of extractives is calculated by subtracting the original weight of the vial from the value obtained. A final extractives weight of 2–3 mg is expected. The residue is next dissolved in 0.5 ml of the HPLC organic solvent to be used as the mobile phase during HPLC, and 0.5 ml of water is added (giving a final 1 : 1 ratio of solvent to water). This solution is filtered through a 5-μm membrane filter (polycarbonate filter for acetonitrile) prior to analysis by HPLC.

Analysis of Fragments Produced by Chemical Degradation of Lignin

Lignin samples are often analyzed by chemical degradation in studies concerned with lignin biodegradation chemistry.[1] Single aromatic ring products released upon chemical degradation of biodegraded lignins are identified, quantified, and compared to results obtained from native (control) lignins. These comparisons allow researchers to determine what chemical changes have occurred in the structure of lignin as a result of microbial degradation. Generally, chemical analysis of the lignin compo-

[8] D. L. Crawford, *Biotechnol. Bioeng. Symp.* **11,** 275 (1981).

nent of lignocelluloses first involves the separation of the lignin component from the cellulosic components, using the lignin purification method of Björkman[9] (see chapter [3] in this volume). This purified lignin is then chemically degraded to analyze its structural properties.[1] Acidolysis and permanganate oxidation are the most commonly used degradative chemical analyses. Acidolysis provides information on changes in the basic phenylpropanoid integrity of the lignin polymer[10] (see chapter [24] in this volume), and permanganate oxidation can be used to quantify changes in the numbers of free phenolic hydroxyl groups in the lignin polymer[11] (see chapter [5] in this volume). In both procedures, the quantification of the degradation products has usually been done by GLC of trimethylsilyl derivatives. However, HPLC is a good alternate for analysis of these fragments. Some examples of chemical degradation-derived fragments are given in Table I.

The procedure for collection and preparation of the samples for each analysis is identical to that described for GLC (see chapter [16] in this volume) up through the step where one obtains the weighed sample of extractive residue. This residue is solubilized into 0.5 – 1.0 ml of the HPLC organic solvent, and then water is added to bring the final solvent to water ratio up to 1 : 1. Next, the solution is filtered through at least a 5-μm filter (polycarbonate for acetonitrile). At this point, the sample is ready for HPLC.

Preparation of Standard Plots for Quantitative Analysis

In order to quantify specific aromatic fragments by HPLC, a standard plot is constructed for each compound of interest at the specific wavelength being used for the chromatogram. The wavelength used should be at or close to an absorption maximum for each of the compounds of interest. The λ_{max} for some common aromatic lignin fragments is given in Table I. For quantification of compounds, the plots can be readily prepared using pure standards. Each available standard is dissolved, at concentrations ranging between 0.1 and 0.001% (w/v), in the most soluble of the solvents being used for HPLC (water or organic). Usually, the HPLC grade organic mobile phase solvent gives the best results, but some dimeric ring compounds or ethoxylated aromatic fragments may require the addition of a small amount of tetrahydrofuran to ensure solubility. Using a typical UV detector, the absorbance unit values should range from 5 times the background noise (0.05 – 0.1 absorbance units) to 2.5 absorbance units. Prior to

[9] A. Björkman, *Sven. Papperstidn.* **59**, 477 (1956).
[10] T. K. Kirk, *Acta Chem. Scand.* **24**, 3379 (1970).
[11] T. K. Kirk and E. Adler, *Acta Chem. Scand.* **24**, 3379 (1970).

TABLE I
RETENTION TIMES AND ABSORPTION MAXIMA FOR SOME COMMON LIGNIN-DERIVED AROMATIC FRAGMENTS AS DETERMINED BY MICROBORE HPLC[a]

Compound	Retention time (min)[b]	λ_{max} (nm)[c]
Catechol	2.76	276
trans-Cinnamic acid	8.34	278
p-Coumaric acid	5.26	308
3,4-Diethoxybenzoic acid	8.78	262
Dehydrodivanillin	4.90	280
4-Ethoxybenzoic acid	8.67	256
4-Ethoxy-3-methoxybenzoic acid	4.80	276
4-Ethoxy-3,5-dimethoxybenzoic acid	8.12	264
Ferulic acid	5.76	322
Gentisic acid	1.05	254
Guaiacol	6.05	274
p-Hydroxybenzaldehyde	4.01	284
p-Hydroxybenzoate	2.61	256
Ketol I [1-hydroxy-3-(4-hydroxy-3-methoxyphenyl)-2-propane]	2.63	280
Protocatechuic acid	6.87	276
Syringaldehyde	5.38	306
Syringic acid	3.84	274
Vanillic acid	3.38	260
Vanillin	4.93	280
Veratraldehyde	6.87	280
Veratryl alcohol	4.73	276
Veratrylglycerol-β-guaiacyl ether	8.08	276

[a] The exact conditions for HPLC are described in the text. The commercial source or the synthesis method for each of the compounds is described in Table I of chapter [16] in this volume.
[b] Retention times are based on a chromatogram with the pilot wavelength set at 276 nm.
[c] λ_{max} were determined from the UV absorption spectra recorded for the chromatographic peaks using a DAD.

chromatography, each solution is filtered. The retention time of each standard is recorded, and when the solutions of varying concentration are chromatographed, the number of peak area units for each is determined by use of a chromatographic integrator. The plot of peak area units against compound concentration should yield a straight line, and the formula for the line can be calculated with a linear regression. Some of the most common aromatic fragments observed as lignin degradation intermediates, their retention times, and their specific absorption maxima are given in Table I. The commercial source and/or the synthesis method for each of the compounds can be found in chapter [16] in this volume.

High-Performance Liquid Chromatography Conditions

High-Performance Liquid Chromatography Column

A reversed-phase column having a nonpolar stationary phase bonded to a solid support (usually silica) is used for partitioning chromatography. Partitioning is dependent on compound interaction with the bonded phase of the column as well as the mobile phase, which consists of a mixture of a polar solvents (water, acetonitrile, or methanol). The most commonly used reversed-phase column contains octadecylsilane (C_{18}) covalently bonded to a silica solid support of 5- or 10-μm pore size. Each supplier of C_{18} reversed-phase columns produces a product that is a little different from that of its competitors; therefore, results will be column specific. However, the sequence of elution of specific compounds should be the same on every column. Differences between columns will be primarily those of retention times and in the quality of resolution of compounds.

A controlled column oven temperature of at least 40° is desirable, because it will make the mobile phase less viscous. Column temperature control also provides the consistent environment needed for obtaining repeatable chromatograms no matter what the temperature of the laboratory. Reversed-phase columns can withstand a maximum temperature of about 80°. One should always include a guard column with the same packing material ahead of the resolving column in order to protect the resolving column from permanent damage. If possible, a microbore column should be employed. Such a column requires very little mobile phase and gives better resolution of aromatic lignin fragments.

High-Performance Liquid Chromatography Solvents

Several criteria are used in the selection of solvents. High purity grade solvents should always be used, especially with UV detectors (=HPLC grade from most suppliers). Solvents typically used for HPLC include water, methanol, acetonitrile, and tetrahydrofuran (do not use dioxane or 2-propanol). The solvents used must dissolve the compounds of interest, and for these aromatic lignin-derived fragments, acetonitrile and water work very well. Ionized carboxyl groups (e.g., in buffered solutions near neutrality water) have no affinity for the C_{18} component of the solid phase, while nonionized carboxyls do. Therefore, to produce symmetrical chromatographic peaks, acidic water is used. It is preferable to use an inorganic acid, such as sulfuric or perchloric acid (never use hydrochloric acid, which damages stainless steel, or glacial acidic acid, which interacts with the aromatic lignin fragments and results in very poor separation). Aromatic lignin fragments also tend to be very soluble in acetonitrile, so it is generally the organic solvent of choice. Water and acetonitrile give good separations, generate symmetrical chromatographic peaks, and are unreactive with the compounds chromatographed. Buffered systems can be used and may give excellent results, but care must be taken to maintain the HPLC pumps by periodically checking valves for problems with precipitated buffer. Further, silica-based supports tend to deteriorate when operated above pH 7–8. A gradient system, described below, gives the best separation of aromatic lignin fragments.

Detector Wavelengths

The detector wavelength used will depend on the spectral properties of the specific compounds of interest. Table I gives examples of common lignin-derived aromatic fragments, along with their respective λ_{max}. The pilot wavelength for recording chromatograms should be one that is absorbed most strongly by all of the compounds of interest. The most commonly used wavelengths are 258, 276, or 310 nm. If a DAD is used, all three wavelengths can be monitored at the same time. This allows for quick identification of compounds by their signal ratios. The DAD also allows one to record a complete UV absorption spectrum for each peak as it moves through the detector.

High-Performance Liquid Chromatography Procedure

In our laboratory (and in obtaining the data for Table I), HPLC is performed on a Hewlett-Packard 1090A high-pressure liquid chromatograph, using a HP-1040A DAD. Pilot chromatograms are run at 276 nm, and spectra (250–350 nm) are stored on disk, for the front, apex, and back side of each chromatographic peak. A 100-mm Hewlett-Packard microbore reversed-phase column (2.1 mm i.d.) of Hypersil ODS (C_{18}) with a 5-μm particle diameter is used, and the column temperature is controlled at 40°. We employ a 5-μl sampling loop and a 0.4 ml/min flow rate. A gradient is employed for solvent delivery, using pH 2.8–3.2 water (one drop of concentrated H_2SO_4 per liter of water) containing 10% acetonitrile initially for 2 min, followed by an increase in acetonitrile concentration to 50% over the next 8 min. The concentration is held at 50% for 2 min and then is returned to 10% acetonitrile over the next 3 min. One complete run takes 15 min.

[18] Conventional and High-Performance Size-Exclusion Chromatography of Graminaceous Lignin–Carbohydrate Complexes

By WOLFGANG ZIMMERMANN, ALISTAIR PATERSON, and PAUL BRODA

Introduction

Graminaceous lignocellulose from cereal straws and corn stovers is produced in agriculture in large quantities and constitutes an important source of renewable biomass. Lignin–carbohydrate complexes (LCC) are a major structural component of grass lignocelluloses, but our understanding of the nature of these polymers has been limited because of their complex and heterogeneous chemical composition and because of the difficulties in their isolation in a native form. The fact that the lignin and carbohydrate components of LCC cannot be separated from each other by fractionation procedures suggests the presence of covalent linkages between these moieties. Lignin is an irregular polymer composed of phenylpropane units, while the carbohydrates in LCC mainly consist of hemicelluloses. Phenolic acids, such as *p*-hydroxycinnamic acid and ferulic acid, are also constituents of grass LCC structures and are thought to be linked to lignin and the carbohydrates, respectively. The high proportion of *p*-hydroxyphenyl residues distinguishes grass lignins as GSH lignins (guaiacylsyringyl-*p*-hydroxyphenyl lignins) from the GS lignins (guaiacylsyringyl lignins) of dicotyledonous plants. Graminaceous LCC and lignins are partially solubilized by mild alkaline treatment at ambient temperature or by simple extraction with hot water.

The study of the chemical and polymeric properties of LCC and their degradation products requires information on molecular weights and on their distribution. Size-exclusion chromatography is a convenient and rapid method for the fractionation of lignins and LCC. Chromatographic separations are, however, complicated by chemical and physical interactions between sample components and the matrix of column packings, which can result in artifacts that disturb the relationship between the elution volume and the size-exclusion process.

Isolation of Lignin–Carbohydrate Complexes

Due to the chemical nature of lignins, their physical location in the plant cell wall, and their close association with cell wall polysaccharides, the isolation of native and homogeneous lignins and LCC without introducing severe changes in their chemical structure and composition *in situ* requires considerable care.

Several methods have been described for the isolation of lignins and LCC.[1,2] In our experience, the preferred method is a modification of that described by Björkman[3] for the isolation of milled wood lignins, which consists of grinding the material in a vibratory ball mill. This treatment progressively reduces the particle size and liberates lignin and lignin–carbohydrate fragments, which are extracted from the resulting fine powder with aqueous dioxane solutions.

Preparation of Milled Straw Lignin from Barley

Barley straw is chopped and ground in a Kek or Wiley mill. To remove lipids and waxes, the coarse powder is successively extracted with acetone and acetone–water (9:1, v/v) in a Soxhlet apparatus. After drying in a vacuum desiccator over phosphorus pentoxide, the extracted material is ground in a vibratory ball mill[4] under nitrogen for 80–100 hr. Lignin–carbohydrate complexes and lignins are extracted from the ball-milled material with a mixture of dioxane and water (9:1, v/v), followed by extraction with a 1:1 (v/v) mixture of dioxane and water. While both extractions yield lignins and LCC, a higher percentage of LCC and carbohydrates is extracted by the more polar dioxane–water (1:1) mixture. The aqueous dioxane solutions are evaporated to dryness under reduced pressure. The milled straw lignin obtained is dried in a vacuum desiccator over phosphorus pentoxide. Other suitable solvents, such as dimethylformamide (DMF) and dimethyl sulfoxide (DMSO), can also be used for the extraction of LCC from ball-milled material.[5,6]

The lignin and lignin–carbohydrate components obtained can be further purified and fractionated by liquid–liquid extraction, dialysis, and precipitation procedures.[7-9]

To obtain carbohydrate-less lignin and water-soluble LCC fractions by liquid–liquid extraction, the milled straw lignin obtained by aqueous dioxane extraction is dissolved in pyridine–acetic acid–water (9:1:4,

[1] Y. Z. Lai and K. V. Sarkanen, *in* "Lignins: Occurrence, Formation, Structure and Reactions" (K. V. Sarkanen and C. H. Ludwig, eds.), p. 165. Wiley, New York, 1971.

[2] D. Fengel and G. Wegener, "Wood: Chemistry, Ultrastructure, Reactions," p. 49. de Gruyter, Berlin, Federal Republic of Germany, 1984.

[3] A. Björkman, *Sven. Papperstidn.* **59,** 477 (1956).

[4] Siebtechnik GmbH, Mülheim, Federal Republic of Germany.

[5] A. Björkman, *Sven. Papperstidn.* **60,** 243 (1957).

[6] I. M. Morrison, *Phytochemistry* **12,** 2979 (1973).

[7] K. Lundquist, B. Ohlsson, and R. Simonson, *Sven. Papperstidn.* **80,** 143 (1977).

[8] J. Azuma, N. Takahashi, and T. Koshijima, *Carbohydr. Res.* **93,** 91 (1981).

[9] A. Scalbert and B. Monties, *Holzforschung* **40,** 119 (1986).

v/v/v), and the solution is extracted with chloroform. The aqueous layer containing water-soluble LCC is dialyzed against water and is lyophilized. The organic layer containing carbohydrate-less lignin and LCC can be further purified by precipitation in ether to remove low-molecular-weight components.

Lignin-carbohydrate complexes can also be obtained from ball-milled straws or milled straw lignin preparations by extraction with alkaline solutions or with hot water.[6,8-11] Alkali-extracted LCC are prepared by stirring the powdered material in an aqueous sodium hydroxide solution. After removal of the residue by filtration, the extracted material is obtained by acid precipitation of the alkaline solution.

Water-soluble LCC are obtained by stirring the powdered material (30 ml/g) in distilled water in a boiling water bath for several hours or by autoclaving with distilled water for 20 min at 120°. After removal of the residue by centrifugation, the water-soluble LCC fraction obtained in the supernatant can be further purified by precipitation in ethanol or by dialysis against water.

Liquid Chromatography of Lignin-Carbohydrate Complexes: Choice of Solvents and Columns

Size-exclusion chromatography (SEC) can be used to separate LCC on the basis of differences in their molecular sizes. If appropriate compounds with known molecular weights are used for calibration of the column, the molecular weights of the samples can be assessed. Narrowly dispersed polystyrene and sulfonated polystyrene standards are commonly used in organic and aqueous solvents, respectively, for the calibration of columns. However, since calibration compounds should have a similar chemical structure and composition to the sample in question to give reliable information on molecular-weight distributions, the use of lignin and lignin-carbohydrate model compounds of known molecular weight is preferable for column calibrations.[12-18]

Mixtures of both linear and branched molecules with widely different

[10] Y. Vered, O. Milstein, H. M. Flowers, and J. Gressel, *Eur. J. Appl. Microbiol. Biotechnol.* **12,** 183 (1981).

[11] K. Atsushi, J. Azuma, and T. Koshijima, *Holzforschung* **38,** 141 (1984).

[12] J. J. Kirkland, *J. Chromatogr.* **125,** 231 (1976).

[13] B. A. Wagner, T. To, D. E. Teller, and J. L. McCarthy, *Holzforschung* **40** (Suppl.), 67 (1986).

[14] R. Concin, E. Burtscher, and O. Bobleter, *J. Chromatogr.* **198,** 131 (1980).

[15] K. Lundquist and B. Wesslen, *Acta Chem. Scand.* **25,** 1920 (1971).

[16] J. Pellinen and M. Salkinoja-Salonen, *J. Chromatogr.* **328,** 299 (1985).

[17] W. J. Connors, L. F. Lorenz, and T. K. Kirk, *Holzforschung* **32,** 106 (1978).

[18] P. Kristersson, K. Lundquist, R. Simonson, and K. Tingsvik, *Holzforschung* **37,** 51 (1983).

sizes and molecular weights are produced by the isolation process, where the polymeric network in the plant cell wall is broken at random sites. Isolated lignins, therefore, show a pronounced polydispersity in molecular weight varying, in part, according to species of origin and location in the plant.[19] The molecular weights of differently isolated straw lignins have been reported to be in the range of 1000–5000.[9,20]

Lignin–carbohydrate complexes have several characteristic properties which, in a complex way, affect their chromatographic behavior. Water-soluble LCC, which exhibit ionizable groups on the surface of the molecule, show a polyelectrolytic behavior that can lead to ionic interactions, polymer expansion, and adsorption effects on matrices of dextran, agarose, and silica gels.[21,22] As a result, exclusion of the sample from the gel or retardation of small sample molecules can occur and consequently lead to an inaccurate estimation of molecular sizes. These artifacts can be effectively reduced or eliminated by the addition of an electrolyte to the mobile phase.[21,22]

Hydrophobic effects, intermolecular orbital interactions, and hydrogen bonding can give rise to solvent–sample interactions, adsorption on the gel surface, and the formation of intermolecular association complexes, which influence the chromatographic behavior of LCC in aqueous and organic solutions.[21–25] These effects can be diminished by using mobile phases with low ionic strength or by adding association-breaking compounds or organic modifiers to dissociate micelles and aggregates.[11,22,25] Adsorption and association effects can also be reduced by derivatization of the sample by acetylation.[26,27]

Generally, the solvent used should be able to readily and completely dissolve the sample. The polarity of the mobile phase must be in a range high enough to allow for a sufficient swelling of the gel material but not to give rise to association and adsorptive effects.[27,28]

[19] D. A. I. Goring, *in* "Lignins: Occurrence, Formation, Structure and Reactions" (K. V. Sarkanen and C. H. Ludwig, eds.), p. 695. Wiley, New York, 1971.

[20] S. Sarkanen, D. C. Teller, J. Hall, and J. L. McCarthy, *Macromolecules* **14,** 426 (1981).

[21] B. Stenlund, *Adv. Chromatogr.* **14,** 37 (1976).

[22] H. G. Barth, *J. Chromatogr. Sci.* **18,** 409 (1980).

[23] S. Sarkanen, D. C. Teller, E. Abramowski, and J. L. McCarthy, *Macromolecules* **15,** 1098 (1982).

[24] F. Yaku, S. Tsuji, and T. Koshijima, *Holzforschung* **33,** 54 (1979).

[25] W. J. Connors, S. Sarkanen, and J. L. McCarthy, *Holzforschung* **34,** 80 (1980).

[26] O. Faix, W. Lange, and E. C. Salud, *Holzforschung* **35,** 3 (1981).

[27] H. L. Chum, D. K. Johnson, M. P. Tucker, and M. E. Himmel, *Holzforschung* **41,** 97 (1987).

[28] W. W. Yau, J. J. Kirkland, and D. D. Bly, "Modern Size-Exclusion Liquid Chromatography." Wiley, New York, 1979.

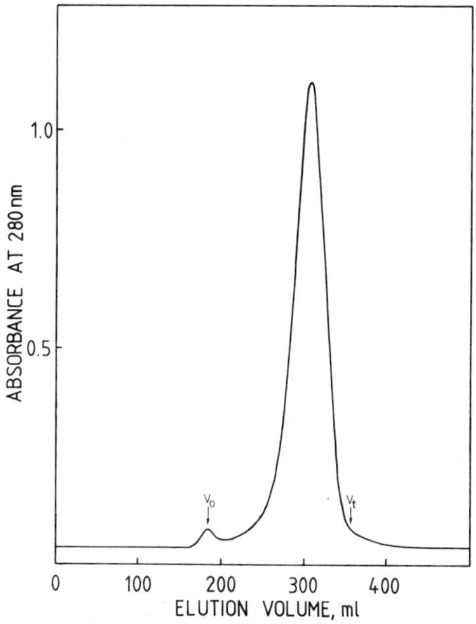

FIG. 1. Elution profile of a milled straw lignin sample from barley obtained by chromatography on Sepharose CL-6B [mobile phase, dioxane–water (7 : 3, v/v)].

Size-Exclusion Chromatography of Lignin–Carbohydrate Complexes with Conventional Organic-Based Packings

The cross-linked dextran gels, Sephadex and Sephacryl, have been used with water, aqueous buffers, aqueous sodium hydroxide solutions, dioxane–water, and DMSO–water mixtures as mobile phases for the analysis of LCC from wheat, ryegrass, and sugar cane bagasse.[6,8,10,11,20,29,30] Sepharose CL,[31] a cross-linked agarose gel, has a good structural and chemical stability in organic solvents and a range of pore sizes, which makes it suitable for the separation of lignins and LCC. We used Sepharose CL-6B (bead diameter, 40–165 μm) with aqueous dioxane as mobile phase for the analysis of LCC from barley straw. Analysis of LCC from sugar cane bagasse with Sepharose 4B has also been described.[8,11] Since such gels have a small number of residual charged groups, adsorption of the sample on the gel can be possible.[13,32] By using dioxane–water mix-

[29] I. M. Morrison, *Biochem. J.* **139,** 197 (1974).

[30] H. Janshekar, T. Haltmeier, and C. Brown, *Eur. J. Appl. Microbiol. Biotechnol.* **14,** 174 (1982).

[31] Pharmacia Fine Chemicals, Uppsala, Sweden.

[32] T. K. Kirk, W. Brown, and E. B. Cowling, *Biopolymers* **7,** 135 (1969).

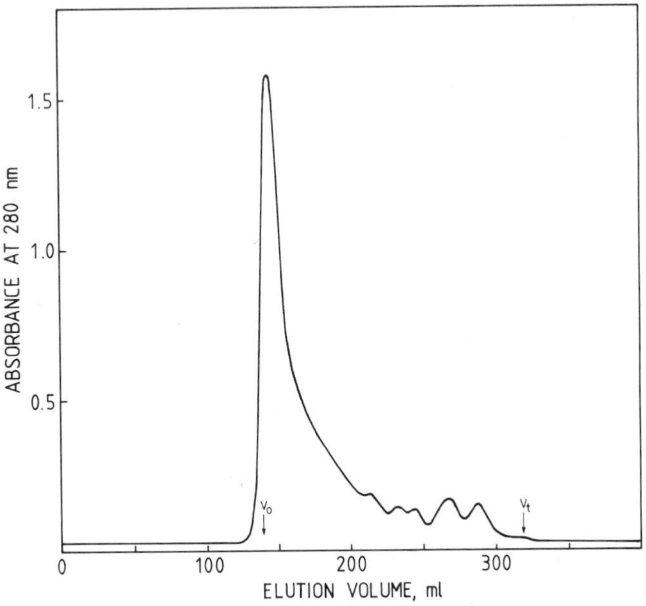

FIG. 2. Elution profile of a milled straw lignin sample from barley obtained by chromatography on Sephadex LH-20 [mobile phase, DMF–acetic acid (200:1, v/v)].

tures as mobile phase, association and adsorption effects are minimized.[32] Also, dioxane–water mixtures are good solvents for lignins and LCC, allowing complete solvation of the sample and sufficient swelling of the gel matrix.

An elution profile of a milled straw lignin sample from barley obtained by chromatography of Sepharose CL-6B, using dioxane–water (7:3, v/v) as mobile phase, is shown in Fig. 1.

The alkylated dextran gels, Sephadex LH-20 and LH-60,[31] have been used for the separation of different graminaceous LCC and lignin model compounds.[14,15,17,25,30,33] We used Sephadex LH-20 with DMF–acetic acid (200:1, v/v) as mobile phase for the chromatography of milled straw lignins. This system has been described to separate lignin–carbohydrates and lignin model compounds according to their molecular weights.[18] Figure 2 shows the elution profile of a milled straw lignin sample from barley obtained by chromatography on Sephadex LH-20 with DMF–acetic acid (200:1, v/v) as solvent. A high proportion of the sample is excluded from the gel under these conditions. Sephadex LH-20 with DMF as solvent has an exclusion limit of 1700 MW[17] and can be used for the fractionation of the low-molecular-weight components in lignin and LCC samples.

[33] W. J. Connors, *Holzforschung* **32,** 145 (1978).

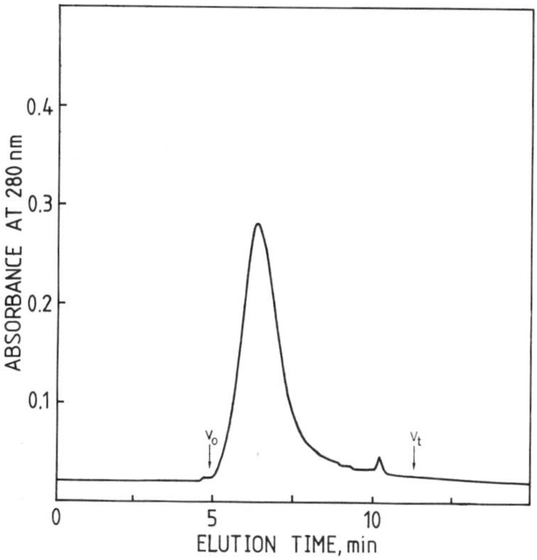

FIG. 3. Elution profile of a water-soluble LCC from barley obtained by chromatography on Zorbax PSM 60 (mobile phase, 0.05 *M* sodium sulfate with 40% methanol).

The chromatographic system used for conventional SEC in organic or partly organic solvents consisted of a borosilicate glass tube (25 × 1000 mm) with solvent-resistant flow adaptors.[31] Flow rates between 25 and 30 ml/hr were maintained. One hundred to five hundred microliters of a 1% (w/v) sample solution was usually applied. The interstitial or void volume, defined as V_o, and the total permeation volume, defined as V_t, were determined with blue dextran and acetone, respectively. Compounds were detected by monitoring the absorption of the column eluent at 275 nm.

High-Performance Size-Exclusion Chromatography of Lignin–Carbohydrate Complexes

High-performance size-exclusion chromatography (HPSEC) of lignins and LCC has been made possible with the development of small, rigid, and porous particle column packings that allow high pressures and flow rates and result in fast separations. High-performance size-exclusion chromatography has been increasingly applied for the analysis of lignin model compounds and graminaceous LCC.[9,15,16,34,35]

[34] A. J. McCarthy, M. J. MacDonald, A. Paterson, and P. Broda, *J. Gen. Microbiol.* **130,** 1023 (1984).

[35] M. E. Himmel, K. K. Oh, D. W. Sopher, and H. L. Chum, *J. Chromatogr.* **267,** 249 (1983).

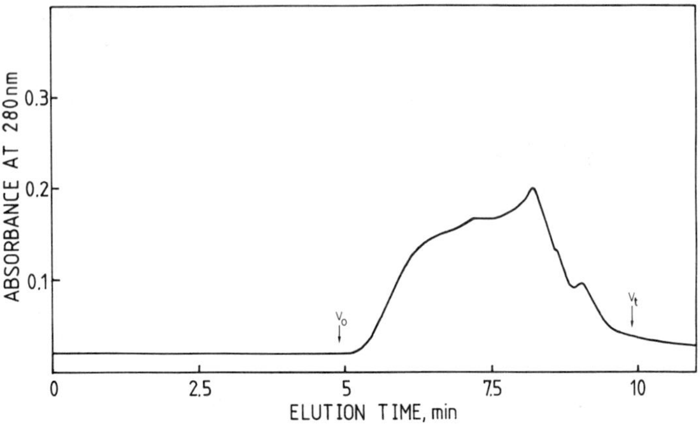

FIG. 4. Elution profile of an acetylated milled straw lignin sample from barley obtained by chromatography on Zorbax PSM 60 (mobile phase, THF).

We used a silica-based HPSEC column material for the chromatography of LCC from straws. The packing consists of small porous silica microspheres (particle size, 5–7 μm; pore size, 60 Å), which provides a homogeneous structure, and has high mechanical strength and chemical resistance.[12] The column has a molecular-weight separation range of 10^2–10^4, determined with polystyrene standards. The separation range can be increased by coupling columns of one or more pore sizes in series or by using columns with mixed particles of two or three discrete pore sizes.[36]

A partially aqueous solvent system was used for the chromatography of water-soluble LCC, and a nonaqueous solvent was used for chromatography of acetylated milled straw lignins. In HPSEC on unmodified rigid silica particles, adsorption effects can be quite severe.[21] These effects are minimized by adding a salt to the mobile phase and by using a polar solvent.[22,28,37] A 0.05 *M* soldium sulfate buffer containing 40% methanol was used as mobile phase. Chromatography of a purified water-soluble LCC obtained by aqueous extraction of milled straw lignin from barley is shown in Fig. 3.

For HPSEC of milled straw lignin with an organic mobile phase, we used tetrahydrofuran (THF) as solvent for the chromatography of the sample, which has been derivatized by acetylation. Tetrahydrofuran has been widely used as a mobile phase for HPSEC of lignins on both silica and styrene–divinylbenzene copolymer gels.[9,15,16,26,27,38] It is, however, not a

[36] W. W. Yau, C. R. Ginnard, and J. J. Kirkland, *J. Chromatogr.* **149,** 465 (1978).

[37] A. R. Walsh and A. G. Campbell, *Holzforschung* **40,** 263 (1986).

[38] R. H. Marchessault, S. Coulombe, and H. Morikawa, *Can. J. Chem.* **60,** 2372 (1982).

good solvent for underivatized lignins. Acetylation strongly increases the solubility and also prevents adsorption effects due to hydrogen bonding between sample and solvent.[26,27] Chromatography of an acetylated milled stray lignin sample using THF as mobile phase is shown in Fig. 4.

A LDC Constametric III high-performance liquid chromatograph[39] with a data system and a UV absorbance detector was employed. A Zorbax PSM 60 HPSEC column[40] was used, packed with porous silica microspheres in a stainless-steel tube (250 × 6.2 mm). Both untreated and surface-deactivated packings are available. The untreated version can be used with both solvent systems described. An inlet filter (2-μm porosity) was fitted between the injection valve and the column to prevent plugging of the column and inlet frit.

Acetylation of the samples was carried out in pyridine–acetic anhydride (1 : 1, v/v) for 20 hr at room temperature. The products were precipitated in water, were collected by centrifugation, and were dried over phosphorus pentoxide. The HPLC-grade solvents were mixed, filtered, and deaerated before experiments. The samples were dissolved in the eluent and were filtered. Typically, 20 μl of a 0.1% (w/v) sample solution was applied to the column. Flow rates between 0.5 and 1.0 ml/min were maintained at room temperature. Compounds were detected by monitoring the absorption of the column eluent at 280 nm. V_o and V_t were determined with high-molecular-weight polystyrene[41] or sulfonated polystyrene[42] standards and acetone, respectively.

Acknowledgments

This work was part of a research program supported jointly by the British Petroleum Venture Research Unit and the Agriculture and Food Research Council.

[39] Liquid Data Control, Riviera Beach, Florida.
[40] Du Pont, Wilmington, Delaware.
[41] Polymer Laboratories, Church Stretton, England.
[42] Polysciences Inc., Warrington, Pennsylvania.

[19] Analysis of Lignin Degradation Intermediates by Thin-Layer Chromatography and Gas Chromatography-Mass Spectrometry

By TOSHIAKI UMEZAWA and TAKAYOSHI HIGUCHI

Introduction

Thin-layer chromatography (TLC) is nowadays an indispensable technique in laboratories dealing with organic compounds including lignin and is used as a rapid, simple, and easy method of analytical and preparative separation of the compounds. TLC has the following important features. (1) Apparatuses are simple and not expensive. (2) Periods required for the analysis are short. (3) All the components can be observed on the TLC plate, different from elution analyses, such as gas chromatography and liquid chromatography. However, identification of compounds by TLC is essentially tentative and not conclusive. Accordingly, unequivocal identification requires the analyses of the compounds purified by preparative TLC by means of other techniques, such as nuclear magnetic resonance spectroscopy (NMR) and mass spectrometry (MS).

In the degradation of lignin and lignin model compounds by chemical or biological reactions, various compounds are produced. The analyses of such a complex mixture including lignin degradation products are usually conducted through two processes: (1) separation of the mixtures into their components (chromatography), followed by (2) identification of the components by means of NMR, MS, etc.

In this chapter, analytical and preparative TLC of degradation products of a model dimer of *β-O*-4 lignin substructure, which is the most frequent interphenylpropane substructure in lignin, by a lignin-degrading basidiomycete is described. Identification of the separated components by gas chromatography-mass spectrometry (GC-MS) is also discussed.

Analytical Thin-Layer Chromatography

Analytical TLC is usually used as a rapid, simple, and easy method of preliminary analysis and is useful for making decisions about the conditions for preparative TLC, column chromatography, and high-performance liquid chromatography. Compounds are identified tentatively by the cochromatography with standards on a TLC plate.

Thin-Layer Chromatography Plate

Many types of TLC plates are commercially available. As an adsorbent, silica gel, alumina, cellulose, and polyamide are coated on glass plates, aluminum sheets, or plastic plates. Adsorbent containing a fluorescent indicator is also commercially available and is recommended in the analysis of lignin-related aromatic compounds. In the authors' laboratory, precoated silica gel glass plates containing a fluorescent indicator (Merck, Kieselgel 60 F_{254}; thickness, 0.25 mm) are usually cut into small plates (1.5×5 cm) and are used for analyses of lignin-related compounds.

Sample Application

When a silica gel layer is used, the sample is dissolved in less polar, volatile solvents such as dichloromethane and is spotted on the TLC plate with a glass capillary. Less volatile and polar solvents, such as water, dioxane, dimethylformamide (DMF), and dimethyl sulfoxide (DMSO), should be removed from the sample before spotting. Water, dioxane, and DMF can be removed under high vacuum. Dimethylformamide and dimethyl sulfoxide are removed by washing the sample solution in organic solvent (diethyl ether for DMF and ethyl acetate for DMSO) several times with water. The organic layer is then washed with a saturated aqueous NaCl solution, is dried over anhydrous Na_2SO_4, and is evaporated to give DMF- or DMSO-free sample.

Developing Solvent and Development

Since many compounds sometimes give practically the same R_f values, developing a sample with at least two different solvent systems is recommended. Ethyl acetate–*n*-hexane systems and methanol–dichloromethane(–*n*-hexane) systems are usually used as developing solvents for the silica gel TLC analysis of fungal and enzymatic degradation products of lignin substructure models.[1] Highest resolution is usually obtained in the R_f value of about 0.3–0.5. If R_f values are too high using ethyl acetate–*n*-hexane (1 : 2, by volume), the ratio of *n*-hexane is increased 2- to 4-fold and vice versa. If R_f values are too low using dichloromethane, 1–8% (by volume) of methanol is added. If R_f values are too high using dichloromethane, dichloromethane–*n*-hexane (2 : 1–1 : 2, by volume) can be used. Besides, ethyl acetate–benzene, ethyl acetate–cyclohexane,

[1] Lignin biodegradation is reviewed in T. Higuchi (ed.), "Biosynthesis and Biodegradation of Wood Components." Academic Press, Orlando, Florida, 1985.

acetone–benzene, acetone–cyclohexane, diethyl ether–petroleum ether, chloroform–methanol, toluene–ethyl formate–formic acid, as well as ethyl acetate–*n*-hexane and dichloromethane–methanol systems are used for the analysis on the silica gel layer of chemical degradation products of lignin[2] by $KMnO_4$ oxidation, catalytic hydrogenolysis, reductive cleavage with sodium in liquid ammonia, thioacetolysis, hydrolysis, and acidolysis and of synthetic intermediates of lignin substructure models.[3]

Ionization of acidic substances is controlled by the addition of a small amount of acetic acid. Lignin-related phenolic acids, such as protocatechuic acid, vanillic acid, ferulic acid, and caffeic acid, are separated on a silica gel layer with benzene–dioxane–acetic acid (90:25:4, by volume) and benzene–methanol–acetic acid (90:16:8, by volume).[4]

Multiple development may increase the resolution of components. Following a single development, the TLC plate is removed from the chamber and is dried gently with a hair drier. The plate is then placed in the chamber containing the same solvent or a different solvent. The procedure may be repeated several times.

Detection

Separated substances are detected by nondestructive methods (observation of fluorescent layers under UV light) and by the reactions with specific reagents and nonspecific reagents (such as H_2SO_4 and phosphomolybdic acid). The detection with iodine vapor is usually nondestructive. Since lignin-related aromatic compounds absorb UV light, the substances separated on the fluorescent layer are easily detected under UV light. Since the method is not destructive and is rapid and clean, it is extremely useful for the TLC analysis of lignin-related compounds. Compounds having specific functional groups, such as phenols, aldehydes, and ketones, are detected by spraying specific color-producing reagents. Table I contains a selection of specific color-producing reagents which are often used in the field of chemistry of lignin biodegradation. (For details of color-producing reagents, see Refs. 5 and 6.)

[2] For details of the solvents, the original articles which are cited in the following books and review articles should be referred to: K. V. Sarkanen and C. H. Ludwig (eds.), "Lignins." Wiley, New York, 1971; E. Adler, *Wood Sci. Technol.* **11,** 169 (1977); A. Sakakibara, *Wood Sci. Technol.* **14,** 89 (1980); and A. Sakakibara, *in* "Lignin no Kagaku" (J. Nakano, ed.; in Japanese), p. 98. Uni Publ., Tokyo, 1979.

[3] F. Nakatsubo, *Wood Res.* **67,** 59 (1981).

[4] G. Pastuska, *Z. Anal. Chem.* **179,** 355 (1961).

[5] E. Stahl (ed.), "Thin-Layer Chromatography," 2nd Ed. Springer-Verlag, Berlin and New York, 1969.

[6] B. Fried and J. Sherma, "Thin-Layer Chromatography: Techniques and Applications." Dekker, New York, 1982.

TABLE I
COLOR-PRODUCING REAGENTS

Compound	Reagent	Procedure	Results
Aldehydes and ketones	2.4-Dinitrophenylhydrazine	Dissolve 1 g of reagent in 1000 ml ethanol and 10 ml concentrated HCl; spray	Yellow to orange spots on pale orange-yellow background
Phenols	$FeCl_3$	(1) 2% $FeCl_3$ in H_2O, (2) 2% $K_3Fe(CN)_6$ in H_2O, and mix 1 ml (1) and 1 ml (2) before use; spray	Blue to purple spots
p-Hydroxybenzyl alcohols	2,6-Dichloroquinonechloroimide	(1) Dissolve 1 g of reagent in 100 ml ethanol, (2) 1 *N* NaOH; spray with (1), then (2)	Blue spots
Cinnamaldehydes	Phloroglucinol–HCl (Wiesner reagent)	(1) Dissolve 1 g reagent in 50 ml ethanol and (2) concentrated HCl; mix 2 ml (1) and 1 ml (2) before use; spray	Red-purple spots
Organic compounds	H_2SO_4–HCHO	Mix 90 parts concentrated H_2SO_4 with 10 parts 35% solution of HCHO in water; spray and heat at 170° on a hot plate for 30 sec	Various colors
Organic compounds	Phosphomolybdic acid	Dissolve 0.5 g reagent in 10 ml ethanol; dip the plate into the solution, dry, and then heat at 170° on a hot plate for 30 sec	Various colors
Organic compounds	I_2	Place the plate in a closed tank containing a few iodine crystals	Brown spots on yellow background

Preparative Thin-Layer Chromatography

Preparative separations are usually accomplished by means of TLC or column chromatography. Column chromatography is generally used for separation or purification of relatively large amounts of substances, while TLC is used for fine separation of small amounts of substances (less than 200 mg/20 × 20 cm plate).

Thin-Layer Chromatography Plates

Precoated silica gel plates (20 × 20 cm, thickness 0.25, 0.5, and 2 mm) are used for preparative separation of samples. Applicable amounts of samples, dependent on the nature of the samples, are as follows: thickness 0.25 mm, <10–15 mg; 0.5 mm, <20–30 mg; 2 mm, <100–200 mg. Thin-layer chromatography plates are also prepared manually in laboratories. Devices such as the Stahl-type apparatus are generally used for this purpose. (For details concerning devices and methods for layer preparation, see Refs. 5 and 6.)

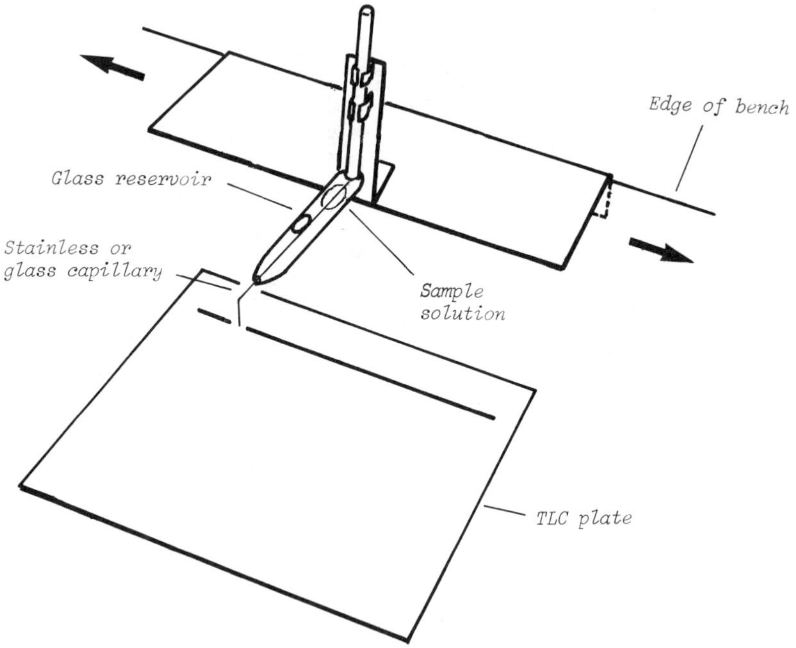

FIG. 1. An apparatus for streak application of sample solutions. The apparatus is moved back and forth along the edge of the bench. This type of apparatus is available from Kontes Co., Vineland, New Jersey.

Sample Application

The sample is dissolved in a small amount of a less polar solvent such as dichloromethane and is applied onto the TLC plate with a special apparatus shown in Fig. 1. Less volatile and polar solvents should be removed from the sample as in the analytical TLC.

Detection, Collection, and Extraction of the Separated Zones

In preparative TLC, procedures which do not harm the separated substances are preferable to identify the zones. Accordingly, the use of fluorescent layers is recommended. When the fluorescent layer is used, separated zones are visualized under UV light and marked with a pencil. When the fluorescent layer is not used or the substances do not absorb UV light, both sides (about 2 cm in width) of the TLC plate are cut off, and the strips thus obtained are submitted to the destructive methods of detection, such as the use of H_2SO_4 or phosphomolybdic acid, by which the zone of the substances on the plate is located (Fig. 2). The marked zones are then scraped or scratched off. The substances are eluted from the silica gel with methanol/dichloromethane (20:80, by volume) and are recovered by solvent evaporation. The process from detection to evaporation should be completed as soon as possible.

Identification of Substances Separated by Preparative TLC

Pure substances after separation or purification by preparative TLC are identified by NMR, MS, and/or IR. Substances of samples purified partially by preparative TLC are repurified by TLC or (especially in cases where the amounts of substances are small) are analyzed by GC-MS to

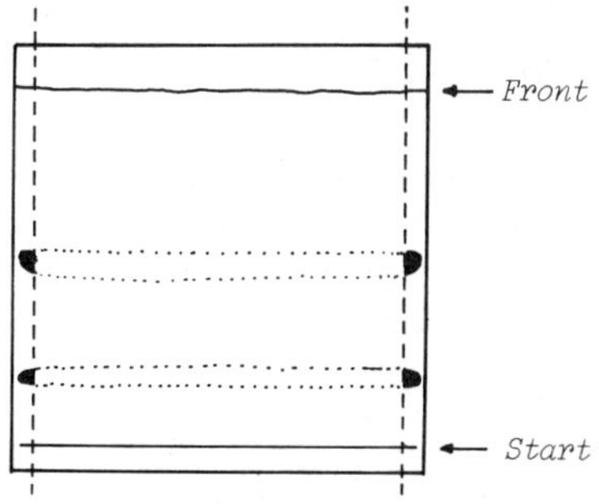

FIG. 2. Location of separated bands. Both sides of the plate are cut off. The strips thus obtained are submitted to the destructive methods of detection. Cares should be taken for the "edge effect": R_f value near the edge of a TLC plate is sometimes a little higher than that in the center.

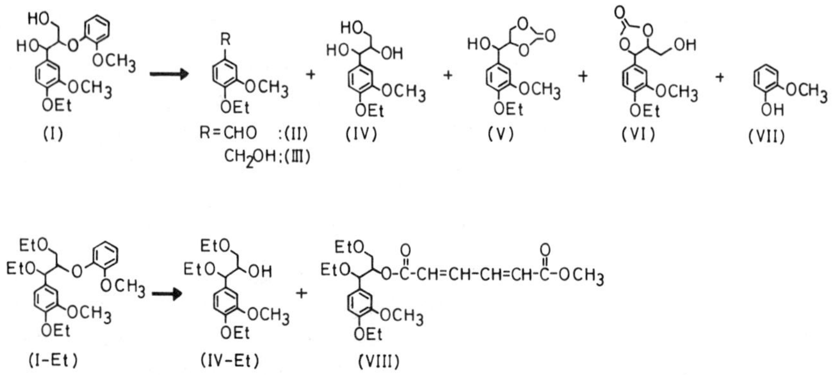

FIG. 3. *β-O*-4 lignin substructure models and their degradation products (**I–VIII**) by intact cells of *Phanerochaete chrysosporium* and *Coriolus versicolor* and by lignin peroxidase of *P. chrysosporium*.

identify the structure of the substances. Many products of degradation of lignin model dimers by lignin-degrading basidiomycetes, *Phanerochaete chrysosporium* and *Coriolus versicolor*, and by lignin peroxidase secreted by *P. chrysosporium* were identified by TLC separation of the degradation products followed by spectrometric analysis or by direct GC-MS analysis without TLC preseparation[1]: when a *β-O*-4 lignin model dimer, 1-(4-ethoxy-3-methoxyphenyl)-2-(2-methoxyphenoxy)-1,3-propanediol **(I)**, was degraded by the lignin peroxidase, a product of propyl side-chain cleavage [4-ethoxy-3-methoxybenzaldehyde **(II)**], a product of *β-O*-4 bond cleavage [1-(4-ethoxy-3-methoxyphenyl)-1,2,3-propanetriol **(IV)**], and products of aromatic ring cleavage [1-(4-ethoxy-3-methoxyphenyl)-1,2,3-propanetriol-2,3-cyclic carbonate **(V)** and -1,2-cyclic carbonate **(VI)**] were identified (Fig. 3).[7] All the products except for **(II)** were also formed in the degradation of **(I)** by intact cells of *P. chrysosporium*; 4-ethoxy-3-methoxybenzyl alcohol **(III)**, reduction product of **(II)** by intact cells, was identified instead of **(II)** (Fig. 3).[8–11] In the degradation of another *β-O*-4 lignin model dimer, 1,3-diethoxy-1-(4-ethoxy-3-methoxyphenyl)-2-(2-methoxyphenoxy) propane (**I**-Et), by the enzyme, an immediate product of aromatic ring cleavage, methyl muconate of 1,3-diethoxy-1-(4-ethoxy-3-methoxyphenyl)-2-propanol **(VIII)**, was identified by alumina TLC separation of the reaction

[7] T. Umezawa, M. Shimada, T. Higuchi, and K. Kusai, *FEBS Lett.* **205,** 287 (1986); and T. Umezawa and T. Higuchi, *FEBS Lett.* **205,** 293 (1986).

[8] T. Umezawa and T. Higuchi, *FEBS Lett.* **182,** 257 (1985).

[9] T. Umezawa and T. Higuchi, *FEMS Microbiol. Lett.* **26,** 123 (1985).

[10] A. Enoki, G. P. Goldsby, and M. H. Gold, *Arch. Microbiol.* **129,** 141 (1981).

[11] T. Umezawa, S. Kawai, S. Yokota, and T. Higuchi, *Wood Res.* **73,** 8 (1986).

products followed by GC-MS analysis with synthesized authentic sample (Fig. 3).[12]

Recently, direct analysis of TLC spots by fast atom bombardment (FAB) mass spectrometry and secondary ion mass spectrometry (SIMS) were reported.[13,14] Since FAB and SIMS are indispensable for the mass spectrometric analysis of relatively nonvolatile, thermolabile molecules, the TLC-MS should be useful for the analyses of lignin-related, nonvolatile, thermolabile molecules, such as oligolignols and lignin–carbohydrate complexes (LCC).

Experimental Procedures

Analytical Thin-Layer Chromatography

Sample. Degradation products of a *β-O*-4 lignin substructure model dimer, 1-(4-ethoxy-3-methoxyphenyl)-2-(2-methoxyphenoxy)-1,3-propanediol (**I**), by *P. chrysosporium* are acetylated with acetic anhydride–pyridine (1 : 1, by volume), and the acetylated products (2 mg) are used as a sample.[8,9] The sample is found to contain trace amounts of guaiacol (**VII**) and 4-ethoxy-3-methoxybenzyl alcohol (**III**) (major product).[10]

Thin-Layer Chromatography Plate. Silica gel plates (Merck, Silica gel 60 F_{254}, 0.25 mm thickness, 20 × 20 cm size) are cut into small plates (1.5 × 5 cm) with a glass cutter.

Sample Application. The sample is dissolved into 0.4 ml of dichloromethane. The solution is spotted on the TLC plate as shown in Fig. 4A with a glass capillary (~0.2 mm, i.d.) and is dried. Then, the solution of a standard, acetate of 4-ethoxy-3-methoxybenzyl alcohol, in dichloromethane is also spotted.

Developing Solvent and Development. Ethyl acetate/*n*-hexane (1 : 4, by volume) and dichloromethane are used as developing solvents. One milliliter of each solvent is added to weighing bottles (i.d., 3.5 cm; height, ~6.5 cm). Then, the TLC plates are placed in the bottles which are covered with lids. After 3 or 4 cm of development (requiring a few minutes), the plates are taken out and dried with a hair drier.

Detection. The substances on the plates are visualized under UV light (254 nm). All substances which absorb UV light in this region are observable distinctly as dark spots on the green fluorescing background. The chromatograms are shown in Fig. 4B and C. Then the plates can be

[12] T. Umezawa and T. Higuchi, *Agric. Biol. Chem.* **51,** 2281 (1987).
[13] T. T. Chang, J. O. Lay, Jr., and R. J. Francel, *Anal. Chem.* **56,** 109 (1984).
[14] Y. Kushi and S. Handa, *J. Biochem. (Tokyo)* **98,** 265 (1985).

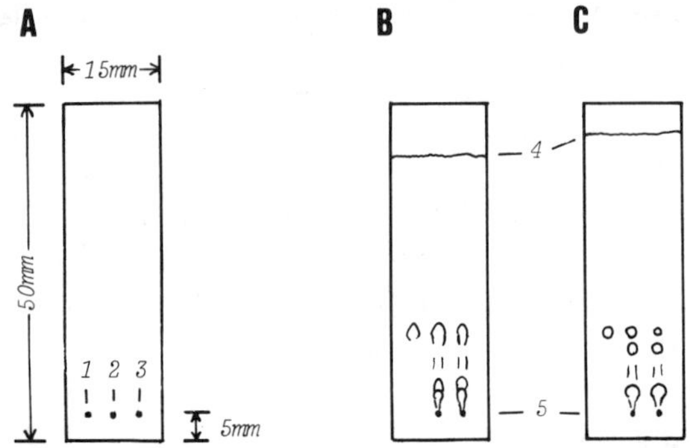

FIG. 4. Sample application and thin-layer chromatograms of degradation products (acetate) of 1-(4-ethoxy-3-methoxyphenyl)-2-(2-methoxyphenoxy)-1,3-propanediol **(I)** by *P. chrysosporium.* (A) A silica gel plate for analytical TLC and location of spotting. 1, Standard (acetate of 4-ethoxy-3-methoxybenzyl alcohol); 2, mixture of the standard and the sample; 3, sample. (B and C) Chromatograms: developing solvent, dichloromethane (B) and ethyl acetate/*n*-hexane (1 : 4, by volume) (C); detection: UV light (254 nm). 4, Front; 5, start.

submitted to color reactions. The spot which has the same R_f value as the standard, acetate of 4-ethoxy-3-methoxybenzyl alcohol, is observed in both the chromatograms after developing by both the solvent systems. The result suggests the presence of 4-ethoxy-3-methoxybenzyl alcohol in the fungal degradation products of **(I)**.

Preparative Thin-Layer Chromatography and GC-MS Analysis of Separated Substances

Sample. The sample is the same as in analytical TLC.

Thin-Layer Chromatography Plates. A silica gel plate (Merck, Silica gel 60 F_{254}, 0.5 mm thickness, 20 × 20 cm size) is predeveloped in ethyl acetate and is cut into two pieces (10 × 20 cm).

Sample Application. As shown in Fig. 5, in order to prevent contamination from the standard, a part of the silica gel layer is removed with a spatula to form a narrow groove (about 2 mm in width). A special apparatus shown in Fig. 1 is used to apply a large volume of samples uniformly as a narrow band onto a plate. The sample solution added into the glass reservoir is streaked on the TLC plate as illustrated, and the plate is dried gently with a hair drier. Then, a standard compound, acetate of guaiacol, is spotted on the narrow alley of the silica gel layer as in analytical TLC.

Development. The plate is then placed into a glass chamber (10 × 23 × 23 cm) containing 50 ml of ethyl acetate/*n*-hexane (1 : 10, by volume). After 15–18 cm of development (it takes about 30–60 min), the

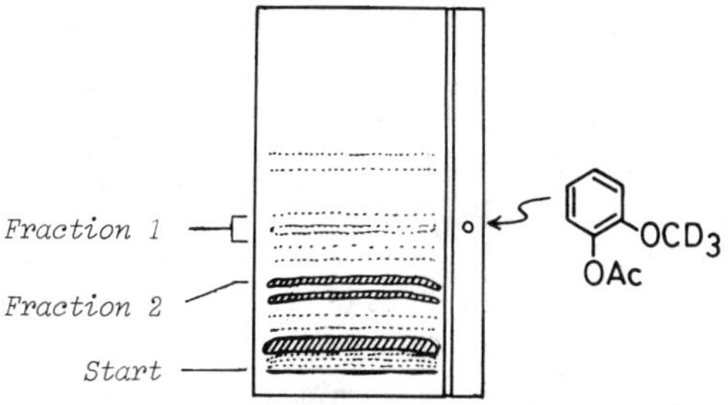

FIG. 5. Thin-layer chromatogram of fungal degradation products (acetate) of 1-(4-ethoxy-3-methoxyphenyl)-2-(2-methoxyphenoxy)-1,3-propanediol (**I**). Separated zones after developing twice are visualized under UV light (254 nm). Deuterated ($D = {}^2H$) standard (acetate of [OC^2H_3]guaiacol) is used. Even if guaiacol as metabolite (unlabeled) would be contaminated for some unexpected reason with the standard compound, GC-MS analysis of Fraction 1 is not affected, since the molecular weight of the standard (MW = 169) is different from that of unlabeled acetate of guaiacol (MW = 166).[9]

plate is taken out and dried gently with a hair drier. The process is repeated again.

Detection, Collection, and Extraction of the Separated Zones. Separated zones are visualized under UV light (254 nm) and the visualized bands are marked with a pencil. The chromatogram is shown in Fig. 5. The marked

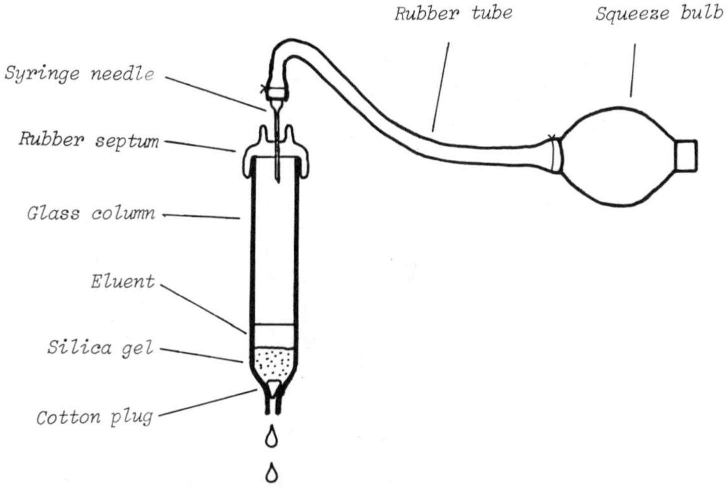

FIG. 6. Device for elution of substances on silica gel. Squeezing the bulb attached to the rubber septum gives rapid elution. Both the ends of the rubber tube should be fasten to the squeeze bulb and the syringe needle with wires. If the amount of silica gel is small, a Pasteur pipette and a rubber bulb can be used instead of the column, septum rubber, and the squeeze bulb.

zone is then scraped and/or scratched off onto a sheet of paper with a small spatula and is crushed to powder with it. Special care should be taken to prevent the sample from contamination by the standard. The powdered gel is then put into a small column with a small plug of cotton as shown in Fig. 6. The substances are eluted from the column with four portions of 1–5 ml of methanol/dichloromethane (20:80, by volume). The solvent is then evaporated off to give the substances (Fractions 1 and 2 in Fig. 5).

Mass Spectrometric Identification of Substances Separated by Preparative Thin-Layer Chromatography. Fraction 1 is analyzed by GC-MS. Figure 7 shows the total ion chromatogram of the fraction. The retention time and mass spectrum of the peak at 3.2 min [MS: *m/z* (%), 166(M^+, 6), 124(100), 109(65)] are identical to those of the standard, acetate of guaiacol. Fraction 2 is analyzed by ^1H NMR. The spectrum obtained [^1H NMR ($CDCl_3$): δ (ppm), 1.47(3H, triplet, $J = 7.0$, Ph-O-C-CH_3), 2.11(3H, singlet, CH_3CO), 3.87(3H, singlet, OCH_3), 4.09(2H, quartet, $J = 7.0$, Ph-O-CH_2-), 5.04(2H, singlet, Ph-CH_2-), 6.9(3H, multiplet, aromatic), *J* (coup-

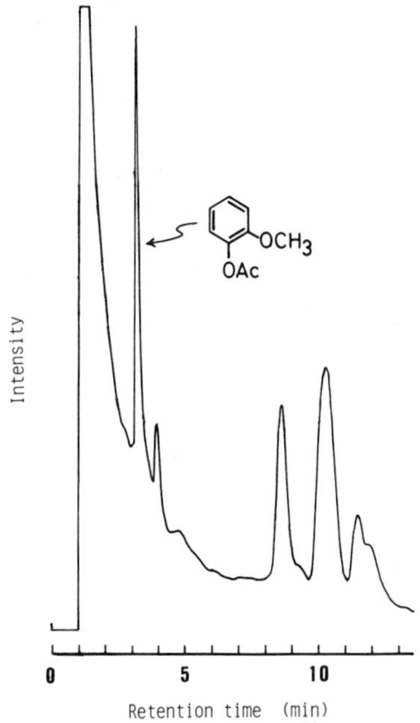

FIG. 7. Total ion chromatogram of Fraction 1. The peak at 3.2 min is identified as acetate of guaiacol by comparison of the mass spectrum and the retention time with those of the standard. Conditions: 1% OV-1 on Chromosorb W (AW DMCS, 80–100 mesh); glass column, 1 m × 0.3 cm (i.d.); column temperature, 127°; Shimadzu-LKB 9000 gas chromatograph-mass spectrometer.[9]

ling constant in Hz)] is identical to that of the standard, acetate of 4-ethoxy-3-methoxybenzyl alcohol. The result shows the presence of guaiacol and 4-ethoxy-3-methoxybenzyl alcohol in the fungal degradation products of **(I)**.

[20] Preparation and Characterization of DNA from Lignin-Degrading Fungi

By UTE RAEDER and PAUL BRODA

Introduction

We describe a set of simple DNA-related experiments that we have used to characterize basic genomic and genetic properties of the ligninolytic basidiomycete *Phanerochaete chrysosporium*.[1-4] The set consists of (1) rapid DNA preparation and construction of genomic libraries; (2) genome size estimation by dot-blot hybridization; (3) use of CsCl–bisbenzimide gradients to separate mitochondrial, ribosomal, and chromosomal DNA and to estimate percentage GC; (4) investigation of DNA sequence variation between different strains and use of restriction fragment length polymorphisms (RFLPs) for spore analysis and genetic mapping of cloned sequences. Similar approaches might be generally useful with other fungi to provide a basis for specific molecular genetics studies. Basic procedures, such as restriction digests, agarose gel electrophoresis, Southern blotting, and nick translation to produce radioactively labeled probes, are described in detail in the cloning manual by Maniatis *et al.*[5]

Cultivation, Spore Formation, and Harvesting for DNA Preparation

As lignin degradation only occurs in response to nutrient depletion,[6,7] ligninolytic cultures are normally grown in defined "low-nitrogen" media.

[1] U. Raeder and P. Broda, *Curr. Genet.* **8,** 499 (1984).
[2] U. Raeder and P. Broda, *Lett. Appl. Microbiol.* **1,** 17 (1985).
[3] U. Raeder and P. Broda, *EMBO J.* **5,** 1125 (1986).
[4] Methods described here sometimes vary slightly from those originally published, due to improvements.
[5] T. Maniatis, E. F. Fritsch, and J. Sambrook, "Molecular Cloning: A Laboratory Manual." Cold Spring Harbor Laboratory, Cold Spring Harbor, New York, 1983.
[6] P. Keyser, T. K. Kirk, and J. G. Zeikus, *J. Bacteriol.* **135,** 790 (1978).
[7] J. A. Buswell, B. Mollet, and E. Odier, *FEMS Microbiol. Lett.* **25,** 295 (1984).

Higher fungal yields for DNA preparation are obtained on 1.5–2% malt extract (Oxoid). The usual and optimum incubation temperature is 37°, but the fungus can also be grown at other temperatures between 7 and 45°.[8]

Growth on malt extract agar leads to the formation of a dense lawn of asexual (see below) mycelium-attached conidiospores which can be stored at room temperature on sealed plates or slopes or scraped off in H_2O for further inoculation. The isolate ME446[9] can be induced to fruit on Walseth cellulose,[10] resulting in formation of sexual homokaryotic (see below) basidiospores that can be collected from the lid of the Petri dish. Cultures from single conidiospores or basidiospores can be obtained by spreading diluted spore suspensions on malt extract agar plates. After 12–14 hr incubation, plate surfaces are examined under the plate microscope, and (as soon as visible) individual growth areas are cut out with a scalpel and transferred to fresh plates. Alternatively, spores can be plated on sorbose-containing agar which induces colony formation.[11]

For DNA preparation (which starts from freeze-dried material, see below), 21 conical flasks (with 400 ml of 1.5–2% malt extract) are inoculated with conidiospore suspensions (from 1 plate for 4–8 flasks) and are incubated at 37° under agitation (if possible with linear shaking, otherwise at 100 rpm) for 20–24 hr. This will yield ~0.5 g mycelial dry weight or 0.5–0.7 mg of DNA per 400 ml of culture. For preparation of smaller amounts of DNA (e.g., 50 μg) from many cultures in parallel, it is convenient to grow them for 2 days in stationary liquid culture in Petri dishes, where mycelial mats are formed. To harvest and freeze dry material from agitated cultures, fungal pellets are collected in a sieve, are transferred to 20 m*M* EDTA, and are collected and dried on filter paper by vacuum filtration. The cake is peeled off, is folded, and is immersed into liquid nitrogen using forceps. Mycelial mats from stationary cultures are taken off by hand, are squeezed using absorbent paper, and are frozen as above. When parallel cultures are to be harvested, it is convenient to use a liquid nitrogen-containing plastic beaker and an ice cube holder to store different samples separately, everything being kept in a polystyrene box with a thin film of liquid nitrogen at the bottom. Frozen samples are lyophilized under vacuum below 1 Torr for sufficient evaporative cooling to maintain the frozen state. Dry samples can be stored at room temperature.

[8] J. A. Staplers, *Stud. Mycol.* **24,** 18 (1984).

[9] *P. chrysosporium* Novobranova, ATCC 24725 (also called *Sporotrichum pulverulentum*), *P. chrysosporium* Burds. Lombard ME446, ATCC 34541, *P. chrysosporium* Burds. Nicot Elphick, CMI 74691, *P. chrysosporium* Burds. White, CMI 110120.

[10] M. H. Gold and T. M. Cheng, *Arch. Microbiol.* **121,** 37 (1979).

[11] M. H. Gold and T. M. Cheng, *Appl. Environ. Microbiol.* **35,** 1223 (1979).

DNA Preparation and Construction of Genomic Libraries

We describe a rapid method for isolation of DNA from filamentous fungi. The procedure starts from freeze-dried material and gives high yields (0.1–0.15% of mycelial dry weight) of high-molecular-weight DNA (≥50 kb) suitable for restriction and ligation.[2] It can be used for processing of many samples in parallel in microfuge tubes. Except for the RNase treatment, all steps are done at room temperature.

Pulverize freeze-dried material using a dry mortar and pestle; this opens the cells. Transfer the ground material with a stiff and circular piece of paper to a dry, phenol- and chloroform-resistant screw cap tube or—for small scale preparations up to 70 mg—to a 1.5-ml microfuge tube and immediately add extraction buffer[12] (~1 ml/0.1 g). (For parallel preparations wipe mortar and pestle with a dry paper towel. The ground material must not be left without extraction buffer.) Suspend homogeneously by manual stirring with a pipette (do not vortex). Add to the slurry ~0.7 volumes of phenol,[13] mix gently but to homogeneity for 2 min, then add 0.3 volumes of chloroform and mix again. Centrifuge for 1 hr at >10,000 *g* in a fixed-angle rotor (e.g., at 12,000 rpm in a Sorvall SS34 rotor with 50-ml polypropylene tubes or in a microfuge) to remove debris, denatured protein, some of the pigment, and to slowly sediment high-molecular-weight polysaccharide. Transfer the upper aqueous phase to a fresh tube and incubate with 1/20 volume of 20 mg/ml RNase A[14] at 37° for 20 min. Extract with ~ 1 volume of chloroform and centrifuge as above, but only for 10 min. Transfer the upper aqueous phase into a sterile tube, add 0.5–0.6 volumes of 2-propanol and shake vigorously. Due to the high DNA concentration (because of the small extraction volume) and its high molecular weight, DNA precipitates quantitatively into a visible aggregate that is allowed to settle in the tube. Immediately afterward, remove liquid carefully by decanting or pipetting (do not centrifuge). Rinse the DNA aggregate and the whole tube with (1) 50% 2-propanol and decant and (2) 70% ethanol and decant. Remove all remaining liquid (which may now be collected by brief centrifugation) with a drawn out Pasteur pipette, dry DNA briefly under vacuum, and resuspend in sterile 10 m*M* Tris-HCl, pH 8, and 1 m*M* EDTA (0.2 ml for DNA prepared from 0.1 g dry myce-

[12] Extraction buffer: 200 m*M* of Tris-HCl, pH 8.5, 250 m*M* of NaCl, 25 m*M* of ethylene diamine tetraacetic acid (EDTA), 0.5% of sodium dodecyl sulfate (SDS).

[13] It is essential to use good quality phenol, which is colorless. We use phenol from BDH (AnalaR) in the following way. Melt crystals at ~50° and immediately equilibrate by mixing with ~0.6 volume 0.1 *M* Tris-HCl, pH 8. Allow phases to separate and store aliquots under 0.1 *M* Tris-HCl, pH 8, at −20°. Always protect from light.

[14] For example, Sigma #4875 in 10 m*M* Tris-HCl, pH 7.5, 15 m*M* NaCl.

lium). After homogeneous resuspension, the DNA concentration is estimated spectrophotometrically.[5]

The DNA is suitable for the construction of gene libraries in standard cloning vectors, e.g., the λ replacement vectors EMBL3 and EMBL4.[15] Both vectors have *Bam*HI cloning sites into which fragments of 9–22 kb can be cloned. In many cases, gene libraries are constructed from size-fractionated partial *Sau*3A digests of total DNA. Conditions for partial restriction to yield fragments predominantly in the 15- to 22-kb range have to be found empirically as outlined, for example, by Karn *et al.*[16] It is easiest to prepare size-fractionated partial digests from large amounts of DNA (e.g., 400 μg) as this minimizes problems due to DNA losses during the preparation. This does not entail the use of large quantities of enzyme, as it is only necessary for *Sau*3A to cut at every 50–100 restriction sites (e.g., we used 1 U of enzyme in a 6 min incubation at 37° to prepare a partial digest of 400 μg of DNA). In our experience, size fractionation using agarose gels[5] is more convenient and gives better resolution and lower losses than sucrose gradients. We obtained good results using preparative horizontal 0.6% low-gelling-temperature agarose gels (Sigma) in Tris-borate buffer[5] prepared and run at 4° with 3–5 V/cm distance between electrodes. The agarose piece with DNA of $\geq$15-kb size (as judged from marker DNA running in a parallel track) is cut out, and DNA can be recovered by phenol extraction. We use the following procedure. Melt the agarose piece at 70°, mix with about 4 volumes of gel buffer (also at 70°), and bring to 0.5 *M* NaCl. Then extract with 1 volume of phenol (equilibrated with 0.1 *M* Tris-HCl, pH 8, and 0.5 *M* NaCl) and separate the phases by centrifugation for 15 min at >10,000 *g* at 4°. Take off ~80% from the upper aqueous layer and precipitate DNA with 2 volumes of ethanol.

To obtain reliably ligatable DNA, we routinely remove any residual agarose-derived contaminants on a small DE52 column.[17] Apply DNA in 25 m*M* of sodium acetate, pH 7 (DNA binding), and wash off contaminants with 2–3 column volumes of 25 m*M* sodium acetate. Elute DNA with 2 column volumes of 2 *M* sodium acetate, pH 7, precipitate with 2 volumes of ethanol, and use for vector ligation and *in vitro* packaging as described elsewhere.[15,18]

[15] A. M. Frischauf, H. Lehrach, A. Poustka, and N. Murray, *J. Mol. Biol.* **170,** 827 (1979).

[16] J. Karn, S. Brenner, and L. Barnett, this series, Vol. 101, p. 3.

[17] Whatman DE-52 swollen overnight and stored in 25 m*M* of sodium acetate, pH 7, and 1 m*M* of EDTA. For ~50-μg size-fractionated DNA, prepare a 500-μl column in syringe or blue pipette tip with siliconized glasswool at the bottom; liquid flow can be regulated by pressure applied with a Pasteur pipette teat.

[18] B. Hohn, this series, Vol. 68, p. 299.

With a haploid genome size for *P. chrysosporium* of 44,000 kb (see below), 8,000 15-kb clones are required for a library that has a probabilitiy of 95% that any given sequence will be present.[19] In view of possible modification of fungal DNA,[20] it may be advantageous to use *Mbo*I instead of *Sau*3A for fragmentation of genomic DNA. A library specifically of chromosomal DNA can be prepared from DNA from which mitochondrial and some of the ribosomal DNA have been excluded (see below). This may be useful for cloning strategies where a wild-type library is to be screened with labeled total DNA from wild-type and a deletion mutant.[21] If cloning strategies are to be used that employ differential screening of genomic libraries with labeled cDNA probes representing different mRNA populations, it should be noted that the number of genes in *Aspergillus nidulans* and *Schizophyllum commune* have been estimated as >7,000 and 10,000, respectively[22,23]; therefore, 15-kb insert fragments are expected to carry on average more than one gene.

Genome Size Estimation by Dot-Blot Hybridization

We have estimated the haploid genome size of *P. chrysosporium* as 44,000 kb (of which about 20–30% is mitochondrial and ribosomal DNA) in a dot-blot hybridization experiment which depends on the following concept. In a DNA mixture of two genomes (the genome in question and a cloning vector DNA of known size), where the mass ratio of the two DNAs corresponds to the size ratio of the two genomes, single-copy sequences of both genomes are present in equimolar amounts. Labeled DNA of a recombinant clone consisting of a fungal single-copy sequence and the vector sequence (with a known distribution of radioactivity in the two parts) is hybridized to denatured, filter-bound fungal DNA alone and fungal DNA mixed with various amounts of vector DNA, in a filter-driven hybridization. By measuring values of hybridized radioactivity and considering the label distribution in the fungal and the vector part of the probe DNA, one can determine the ratio of filter-bound fungal DNA : vector DNA at which equimolar amounts of fungal and vector DNAs would hybridize. This ratio is the ratio of the two genome sizes.[1]

[19] L. Clarke and J. Carbon, this series, Vol. 68, p. 396.

[20] F. Antequera, M. Tamame, J. R. Villanueva, and T. Santos, *Nucleic Acids Res.* **13,** 6545 (1985).

[21] M. X. Caddick, H. N. Arst, L. H. Taylor, R. I. Johnson, and A. G. Brownlee, *EMBO J.* **5,** 1087 (1986).

[22] W. E. Timberlake, *Dev. Biol.* **78,** 497 (1980).

[23] B. Zantinge, J. Springer, and J. G. H. Wessels, *Eur. J. Biochem.* **101,** 251 (1979).

Probe Preparation. Choose a clone that contains single copy fungal DNA (demonstrated by Southern hybridization, see below). It is important that vector and insert fragments can be separated by restriction and agarose gel electrophoresis. Restrict ~2 μg of recombinant clone DNA to separate vector and insert fragments and label by nick translation.[5] Run half of this DNA on a low-gelling-temperature agarose gel and cut out and combine agarose pieces with vector- and insert-derived DNA, respectively. Melt agarose at 70°, bring to equal volumes with H_2O, add each solution to 5 ml of Packard "Filter Count" (Packard Cat. No. 6013141), and measure radioactivity in a liquid scintillation counter. Remove unincorporated dNTP from the other aliquot on a Sephadex G-50-spun column[5] and use as a hybridization probe.

Preparation of DNA Mixtures. Determine the DNA concentration in solutions of total fungal DNA and vector DNA; the fungal DNA should be at a concentration of around 500 μg/ml. Add to 1 volume of fungal DNA 1 volume of 1 *M* NaOH and 2.5 *M* NaCl. Prepare dilutions of vector DNA in 0.5 *M* NaOH, 1.25 *M* NaCl at concentrations of, e.g., 1/500 and 1/5000 of that of the fungal DNA solution.[24] By combining equal volumes (e.g., 50 μl) of fungal DNA with 0.5 *M* NaOH and 1.25 *M* NaCl and with the different vector DNA solutions, prepare three solutions with fungal : vector DNA ratios of 1 : 0, 1 : 1/5000, and 1 : 1/500 (e.g., 3 × 100 μl with 25 μg fungal DNA, 25 μg fungal DNA plus 5 ng vector DNA, and 25 μg fungal DNA plus 50 ng vector DNA). As the DNAs are in 0.5 *M* NaOH and 1.25 *M* NaCl, they are denatured and ready to bind to nitrocellulose.

Dot-Blot Preparation and Hybridization. With a pencil, draw nine (3 × 3) 2 × 2-cm squares on a nitrocellulose filter (e.g., Schleicher and Schüll; handle the filter only with gloves). Fix the filter horizontally at its edges, so that the area with the squares does not touch any surface below (to avoid sample losses during application). Apply 20 μl of each sample (three replicates) to the different squares marked on the filter and leave to dry in this position for 20 min. Float the filter on 0.5 *M* Tris-HCl, pH 8, and 1.5 *M* NaCl until completely wet (this may take several minutes). Then submerge the filter in this solution for 5 min. Rinse in 6× SSC, remove excess liquid by blotting between filter paper, and bake for 1 hr under vacuum at 80°. Rewet filter by floating on 6× SSC.[5] Prehybridize the filter in a sealed plastic bag in 6× SSC, 10× Denhardt solution,[5] 0.5%

[24] In principle, it is sufficient to prepare the dot blot with fungal DNA and only one DNA mixture of fungal : vector DNA. In practice, it is useful to dot blot several different ratios of fungal : vector DNA, to serve as an internal control that the hybridization is filter driven. The mass ratio of fungal : vector DNA would be chosen around the expected size ratio of the two genomes, although this is not essential.

SDS, 1 mM EDTA ($\geq$200 μl/cm^2 filter) for 4 hr at 68° and hybridize with a prepared probe in the same solution as above (50 μl/cm^2 filter) overnight at 68°. Wash the filter 3× in 0.1× SSC and 0.5% SDS at 68° and take an autoradiograph to check for unspecific binding of radioactivity (which should not be present). Cut out the individual filter squares, dissolve overnight in a Packard Filter Count, and measure the radioactivity in the liquid scintillation counter. Possible homology between the fungal and vector DNA is tested in the same way in a separate hybridization of labeled vector DNA to filter-bound fungal DNA and, if present, is subtracted from the above measured values.

Evaluation. Draw a graph plotting the hybridized radioactivity (cpm) against the ratio of filter-bound fungal : vector DNA (where intercept corresponds to the fungal : fungal DNA hybridization and slope corresponds to additionally hybridized radioactivity due to the vector : vector DNA hybridization) and fit the line which should be straight. The fungal : vector DNA ratio at which equimolar hybridization would occur is then determined from the graph as the point where the ratio of intercept : slope value corresponds to the ratio of the radioactivity between the fungal and vector part of the hybridization probe.

In a particular experiment,[1] 10 μCi of [α-^{32}P]dATP were used to label 2 μg of DNA of a recombinant clone in λgtWES-T5 · 622[25] (with about 50% incorporation efficiency). Thirty percent of the incorporated radioactivity was in the insert fragment. (There may be differences in incorporation efficiency between vector and insert DNA, especially if they differ in GC content.) Each spot on the filter contained 6.7 μg of fungal DNA. To a spot with fungal DNA alone, 8000 cpm hybridized. The amount of hybridization increased linearly with the amount of vector DNA in the spots up to the maximum amount used [13.4 ng (i.e., 1/500 of the fungal DNA amount)], and the slope showed an extra 4700 μCi hybridized for each 1.34 ng (1/5000 of the fungal DNA amount) of vector DNA. The fungal and the vector genomes would be equimolar in a spot where the ratio of fungal : vector DNA hybridization is 30 : 70 (ratio of incorporation in the probe). This would occur if 6.7 μg of fungal DNA and 1.34(8000 × 70/4700 × 30) ng = 5.32 ng of vector DNA were present, i.e., the ratio of genome sizes is 6.7 μg : 5.32 ng = 1259. With the vector size taken as 35.1 kb, the haploid genome size of *P. chrysosporium* is calculated as 1259 × 35.1 kb = 44,000 kb. The genome size is similar to that of other basidiomycetes.[26,27]

[25] J. Davison, F. Brunel, and M. Merchez, *Gene* **8,** 69 (1979).

[26] S. K. Dutta, *Nucleic Acids Res.* **1,** 1411 (1974).

[27] J. M. M. Dons, O. H. M. DeVries, and J. H. G. Wessels, *Biochim. Biophys. Acta* **563,** 100 (1979).

Fractionation into Mitochondrial, Ribosomal, and Chromosomal DNA: Estimation of Percentage GC in DNA

Bisbenzimide (Hoechst dye #33258, Sigma Chemical Co., St. Louis, Missouri) binds to adenine residues in DNA thereby reducing its buoyant density in CsCl and increasing the percentage GC-dependent separation in CsCl gradients.[28] Total DNA from each of the four different strains of *P. chrysosporium*[9] separates in CsCl–bisbenzimide gradients into three bands. The uppermost (~15–20% of all DNA) contains mitochondrial DNA, the middle (~5%) contains ribosomal DNA (and in two of the four investigated strains, ME446 and Novobranova, a small amount of nonribosomal satellite DNA with apparently few restriction sites), and the lowest and main band contains bulk chromosomal DNA with some residual ribosomal DNA.[29] Restriction and agarose gel electrophoresis of chromosomal DNA of strain ME446 shows discrete bands, indicating the presence of repetitive DNA not of mitochondrial or ribosomal origin.

Experiments with marker DNAs of different percentage GC (*Staphylococcus aureus*, 33% GC; phage λ, 50% GC; and *Streptomyces coelicolor*, 75% GC) showed a linear relationship between GC contents and position in CsCl–bisbenzimide gradients. The initial concentration of CsCl and the rotor speed have to be carefully chosen for optimal separation. The CsCl concentration determines the position of the DNA of a particular GC content in the gradient, and the speed determines the steepness of the gradient. A low speed gives shallow gradients and thus better separation. We found the following conditions very suitable. Gradients prepared by mixing 8.3 ml of DNA solution with 10.2 g of CsCl and 1 ml of bisbenzimide (0.1 mg/ml H_2O) and by centrifugation in a Beckman Ti50 fixed angle rotor at 33,000 rpm for 65 hr at 17° separated marker DNAs by 0.78 mm/1% difference in the GC contents. By running three parallel gradients containing fungal DNA (0.5 mg), marker DNAs (20 μg each), and fungal plus marker DNAs, we estimated the fungal DNA percentage GC in relation to the marker DNAs as 59% for chromosomal DNA, 52% for ribosomal DNA, and 33% for mitochondrial DNA. The relatively high percentage GC for chromosomal DNA and low percentage GC for mitochondiral DNA are typical for basidiomycete DNA.[30,31] DNA in CsCl–

[28] W. Müller and F. Gautier, *Eur. J. Biochem.* **54,** 395 (1975).

[29] Ribosomal DNA in the satellite band and the bulk chromosomal DNA band show the same restriction patterns and should have the same percentage GC. The fact that only part of it bands separately is probably due to the high molecular weight of the total DNA, so that some ribosomal DNA is physically linked to and trapped in the more GC-rich bulk chromosomal DNA band.

[30] V. D. Villa and R. Storck, *J. Bacteriol.* **96,** 184 (1968).

[31] R. Storck and C. J. Alexopoulos, *Bacteriol. Rev.* **34,** 126 (1970).

bisbenzimide gradients is visualized and recovered as with CsCl–ethidium bromide gradients.[5]

Sequence Variation within the Species of *P. chrysosporium* and Use of RFLPs for Genetic Mapping

DNA Sequence Variation between Different Strains. DNA comparisons are a convenient way of studying the relatedness of strains. It is possible to use mitochondrial DNA, ribosomal DNA, and cloned chromosomal DNA fragments in hybridizations against Southern blots of restricted DNAs of various strains. It is expected that ribosomal DNA is more conserved and that mitochondrial DNA is less conserved than chromosomal DNA. In practice, the closer the relationship, the higher the degree of hybridization (hybrid stability), i.e., the stronger the hybridization signal intensity. The most stringent criterion in the comparison of similar strains is the conservation of restriction sites (also the spacing between them), which even within species are not highly conserved (see below). The stringency of washing conditions after hybridization in comparative experiments should be carefully chosen.[5] DNA duplex stability can be calculated as a function of GC content, temperature, Na^+ concentration,[32] and percentage of base pair mismatch.[33]

We compared four different strains of *P. chrysosporium*[1,9] using standard conditions of stringency (wash in 0.1% SSC and 0.5% SDS at 68°).[5] This showed high DNA homology between the two mainly investigated lignin degraders, ME446 and Novobranova (previously *Sporotrichum pulverulentum*), with high efficiency of cross-hybridization and conservation of most restriction sites. However, there are RFLPs between their chromosomal DNAs which allow one to distinguish between these two strains. In contrast, mitochondrial and chromosomal DNA of the other two strains hybridize under these conditions only weakly to cloned DNA of ME446 and Novobranova, but the hybridizing restriction patterns of these two other strains are sometimes similar.

Although we did not quantify sequence divergence, this study readily revealed the degree of relationships between these morphologically very similar strains.

RFLPs as a Genetic Tool. DNA from cultures of several individual conidiospores and basidiospores of ME446 were restricted with *Pst*I or *Sal*I and used in Southern blots with various genomic clones as the probes. In the majority of cases, there were RFLPs that segregated 1 : 1 in the basidio-

[32] R. Lathe, *J. Mol. Biol.* **183,** 1 (1985).

[33] T. M. Bonner, D. J. Brenner, B. R. Neufeld, and R. J. Britten, *J. Mol. Biol.* **81,** 121 (1973).

spores, i.e., each basidiospore DNA showed one or other of two types of simpler patterns, while the conidiospore DNAs showed complex patterns consisting of the combination of the two simpler patterns. This meant that the conidiospores contain two genome equivalents (and are thus presumably dikaryotic) while basidiospore formation had lead to a haploid form.

RFLPs can be used as genetic markers. We determined the linkage relationships between cloned sequences by analyzing whether RFLPs (i.e., the A and the B alleles of different cloned sequences) had segregated independently or nonindependently in a series of 53 single basidiospore-derived cultures.[3] If cloned sequences were derived from different chromosomes, their A and B alleles would segregate independently. If cloned sequences are neighbors on the same chromosome, the A alleles and the B alleles would cosegregate, except in cases where there has been a crossover between the two loci. Calculation using the binomial distribution[34] shows that the cosegregation values of $\geq 80\%$ in 53 meiotic events is a good indicator for genetic linkage (80% cosegregation corresponding to 20% crossover frequency, i.e., ~20 cM genetic distance).

It is important to use enzymes that are likely to reveal sufficient numbers of RFLPs in Southern hybridization. *Sal*I and *Pst*I are both enzymes that cut the relatively GC-rich chromosomal DNA frequently to give high numbers of DNA fragments between 1 and 10 kb for easy blotting. In combination with 15-kb probe sequences, *Sal*I and *Pst*I digests each revealed enough RFLPs to be used alone for efficient mapping. To prepare agarose gels, we used haircombs (Kent of London R9T) as slot formers (59 slots/20 cm). In parallel, up to 10 such agarose gels (70 ml of 0.8% agarose solution/200 cm^2 gel) were loaded (1–2 μg of restricted DNA in <5 μl/slot), run, and Southern blotted by the quick blot method using ammonium acetate.[35]

Experience with the use of classical genetical markers suggests that 50–100 markers are sufficient to give a very useful map for most organisms.

Acknowledgment

This work was part of a program jointly supported by British Petroleum's Venture Research Unit and the Agriculture and Food Research Council.

[34] The probability P that genetically unlinked loci cosegregate r times in n meiotic events is calculated by the binomial distribution. As the parental arrangement of alleles is not known (homokaryons among wild-type conidiospores were not detected), the assignment of As and Bs for the different allelic versions is arbitrary, and thus r can also be $(n - r)$. $P = 2[n!/(n - r)!r!](\frac{1}{2})^n$.

[35] G. E. Smith and M. D. Summers, *Anal. Biochem.* **109,** 123 (1980).

[21] Preparation and Characterization of mRNA from Ligninolytic Fungi

By RICHARD HAYLOCK and PAUL BRODA

The best characterized lignin-degrading fungus is *Phanerochaete chrysosporium (Sporotrichum pulverulentum)* (for a review, see Kirk and Fenn[1]). The methods described below have been optimized for this organism. They have, however, been shown to work well with an unrelated basidiomycete, *Coprinus cinereus*, and are probably applicable to other lignin-degrading basidiomycetes.

One characteristic of the *P. chrysosporium* lignin-degrading system is its repression by high nitrogen and/or high carbon levels in the medium.[2] This implies that mRNA encoded by genes of at least some components of the ligninolytic system is present in the mycelium only during nitrogen and/or carbon limitation, on the assumption that ligninolytic gene expression is regulated at the transcriptional or RNA processing stages.

Using an approach similar to that utilized by Timberlake's group[3] to identify genes implicated in *Aspergillus nidulans* sporulation, such differential gene expression can be exploited for the identification and isolation from a genomic library of clones that contain DNA sequences expressed only at the time of lignin degradation.[4] Once such genomic clones have been isolated the DNA from them can be labeled and used as probes in order to study the kinetics of ligninolytic gene expression. The same probes can be used to identify the corresponding cDNA clones in libraries also made from the mRNA populations. It is, therefore, important to have available a simple and reliable method of mRNA isolation.

Growth of the Fungus

Two sets of cultures should be grown in parallel, one set under conditions where lignin degradation is maximal and the other set under conditions where lignin degradation is minimal. As many culture parameters as possible should be kept constant between the two sets of cultures. For *P.*

[1] T. K. Kirk and P. Fenn, *Br. Mycol. Soc. Symp.* **4,** 67 (1982).

[2] T. K. Kirk, E. Schultz, W. J. Connors, L. F. Lorenz, and J. G. Zeikus, *Arch. Microbiol.* **117,** 277 (1978).

[3] C. R. Zimmermann, W. C. Orr, R. F. Leclerc, E. C. Barnard, and W. E. Timberlake, *Cell* **23,** 709 (1980).

[4] R. Haylock, U. Raeder, A. Paterson, and P. Broda, *Heredity* **54,** 420 (1985).

chrysosporium all cultures are grown in static culture under an oxygen atmosphere at 37°. The components kept constant in the medium for all cultures are, in grams per liter: KH_2PO_4, 0.2; $CaCl_2 \cdot 2H_2O$, 0.01; $MgSO_4 \cdot 7H_2O$, 0.05; thiamin, 0.0001; 2,2-dimethylsuccinic acid, 1.42; ball-milled barley straw, 1.0. In addition, 1 ml of a mineral solution described by Jefferies *et al.*[5] is added to each liter. The pH is adjusted to 4.5 by the addition of sodium hydroxide. To maximize lignin breakdown both the carbon and nitrogen sources are adjusted so that they would be limiting after 3 days growth. For this, 0.276 g/liter $NH_4H_2PO_4$ and 1.5 g/liter glucose are provided in the medium. To minimize lignin breakdown excess carbon and nitrogen are provided; such a medium contained 2.76 g/liter $NH_4H_2PO_4$ and 10 g/liter glucose. It is important to limit carbon as well as nitrogen supply, otherwise the large amounts of glucan produced make isolation of RNA impossible.

Harvesting of the Mycelium

Ligninolytic cultures of *P. chrysosporium* can be harvested after carbon and nitrogen become limiting. Nonligninolytic cultures are normally harvested 3½ days after inoculation. Mycelium is harvested by filtration through nylon mesh, washed in ice-cold 0.15 *M* NaCl, gently squeezed dry, and then frozen in small pieces in liquid nitrogen.

Breakage of the Cells

It is important that the breakage of the cell does not result in the shearing of the RNA. The frozen mycelium is ground in liquid nitrogen in a Waring blender. Three cycles of 20-sec high-speed grinding with intermediate addition of liquid nitrogen is found sufficient to achieve cell breakage. Much longer grinding has been used successfully by Garber and Yoder[6] as a method of isolating high-molecular-weight DNA from filamentous fungi. The method is therefore not prone to the shearing problems associated with most mechanical methods of nucleic acid isolation.

Extraction of RNA from Ground Mycelium

During the cell breakage, the low temperature prevents any nuclease activity. When the ground mycelium is melted in order to extract the RNA the primary objective must be to prevent the degradation of RNA by

[5] T. W. Jefferies, S. Choi, and T. K. Kirk, *Appl. Environ. Microbiol.* **42,** 290 (1981).

[6] R. C. Garber and O. C. Yoder, *Anal. Biochem.* **135,** 416 (1983).

nucleases. The other important objective of the extraction procedure is the separation of the RNA from other cell components, in particular the nucleoproteins. Many of the agents used for the inactivation of nucleases are also active in dissociating proteins from nucleic acids.

Most methods of RNA extraction used today fall into two general categories: (1) those using anionic detergent together with phenol or phenol/chloroform and (2) those using guanidinium hydrochloride or guanidinium isothiocyanate, examples of which are to be found in Maniatis *et al.*[7] A method utilizing the former approach is found to be most suited for the isolation of RNA from *P. chrysosporium.* The ground mycelial powder is melted into 3 vol of the following buffer, the composition of which is based on that of Parish and Kirby.[8]

1% Triisopropylnaphthalenesulfonic acid sodium salt (TNS)
6% 4-Aminosalicyclic acid sodium salt (PAS)
200 m*M* Tris–HCl + 25 m*M* EGTA, pH 7.8
250 m*M* NaCl

The two detergents, TNS and PAS, rapidly inactivate nucleases and dissociate the nucleoproteins from the RNA.[8] Their advantage over the more commonly used detergent, sodium dodecyl sulfate, is that they are soluble at low temperatures. Nucleases are therefore inactivated as soon as the mycelial powder melts. The temperature is maintained below 15° throughout the extraction procedure; this, together with the relatively high ionic strength of the extraction buffer, maintains the secondary structure of the RNA and renders it less liable to degradation by residual nucleases. Mycelial debris is removed by centrifugation at 10,000 rpm for 6 min in a Sorvall SS34 rotor at 4°. The pellet is washed with 1 vol of extraction buffer and the supernatants are pooled. To each milliliter of supernatant is added 0.5 g of solid phenol. At the low temperature of the extraction the phenol dissolves slowly when gently agitated. Such slow release is the equivalent to the drop-by-drop addition of aqueous phenol, in that it avoids the trapping of RNA by the denatured protein, but it is less time consuming. It has the additional advantage of keeping the extraction volume small. After all the phenol has dissolved, agitation is continued for a further minute. One half-volume of prechilled chloroform is added and the mixture shaken to homogeneity. The phases are separated by slow-speed centrifugation, the aqueous phase is collected, and the organic phase is discarded. The interphase (1 vol) is extracted with 4 vol of extraction buffer

[7] T. Maniatis, E. F. Fritsch, and J. Sambrook, "Molecular Cloning: A Laboratory Manual." Cold Spring Harbor Laboratory, Cold Spring Harbor, New York, 1982.
[8] K. S. Parish and J. H. Kirby, *Biochim. Biophys. Acta* **129,** 554 (1966).

followed by 4 vol of chloroform. After phase separation the aqueous phase is pooled with the previous one and they are reextracted twice with 1 vol of chloroform. If after the reextractions with chloroform the aqueous phase is cloudy it is cleared by centrifugation in a Sorvall SS34 rotor for 20 min at 18,000 rpm at 4°. The RNA in the aqueous phase is precipitated by the addition of 2 vol of ethanol, washed with 65% ethanol, and dried under vacuum.

Separation of mRNA from Other Nucleic Acid Species

The standard method of mRNA isolation depends on the presence of poly(A) tail at the 3′ end. Affinity columns with oligo(dT) or poly(U) attached to an inert support are used to bind the poly(A) under conditions of high ionic strength and low temperature. The poly(A)-containing mRNA is eluted from the column by lowering the ionic strength of the buffer, raising the temperature, adding a nucleic acid denaturant, or by a combination of these methods. For the isolation of *P. chrysosporium* mRNA poly(U) Sephadex, supplied by Bethesda Research Laboratories, is chosen because its strong poly(A)-binding characteristics make it more suitable than oligo(dT)-cellulose for the isolation of fungal mRNAs with their short poly(A) tails.

The RNA pellet is dissolved in about 10 ml of the poly(U)-binding buffer, previously described by Deeley *et al.*,[9] containing 250 m*M* NaCl, 10 m*M* Tris–HCl, pH 7.4, 0.5 m*M* EDTA, and 0.4% sodium dodecyl sulfate (SDS). The RNA solution is loaded onto poly(U) Sephadex pre-equilibrated with binding buffer, contained within a Bio-Rad Thermal Column (bed volume 5 ml) at room temperature. The flow rate is set at about 0.5 ml/min, to allow for complete binding. The column matrix is washed with wash buffer (50 m*M* NaCl, 10 m*M* Tris–HCl, pH 7.4, 0.5 m*M* EDTA, 0.4% SDS) at a flow rate of about 2.5 ml/min, until no more A_{260}-absorbing material is eluted. The bound poly(A)-containing RNA is recovered in a very small volume by eluting with a buffer containing 45% redistilled formamide, 10 m*M* Tris–HCl, pH 7.4, 0.5 m*M* EDTA, and 0.4% SDS and by raising the temperature of the column to 53°. Two volumes of ethanol is added to the eluate to precipitate the RNA, which is washed with 65% ethanol before drying under vacuum. The RNA is then taken up in sterile distilled water to a concentration of 1–5 μg/μl and stored at −70°.

[9] R. G. Deeley, J. I. Gordon, A. T. H. Burns, K. P. Mullinix, M. Bina-Stein, and R. F. Goldberger, *J. Biol. Chem.* **252,** 8310 (1977).

Characterization of the mRNA

The level of protein contamination of the mRNA can be rapidly assessed by monitoring the $A_{260}:A_{280}$ ratio of the solution. A rapid test for degradation of the RNA during isolation is electrophoresis of a sample on a standard 1.5% agarose gel in Tris-borate buffer (89 m*M* Tris base, 89 m*M* boric acid, 2 m*M* EDTA). When stained with ethidium bromide the two high-molecular-weight ribosomal RNA bands [rRNA contamination is unavoidable even after several cycles of poly(U)-Sephadex or oligo(dT) chromatography] should be well defined. The mRNA should be visible as a smear mainly above and between the two ribosomal bands. If the ribosomal bands are poorly defined and merge into a low-molecular-weight smear it can be assumed that degradation of the mRNA has occurred. It is also apparent from the lack of the characteristic high-molecular-weight band on the gel that the mRNA samples are not contaminated with DNA when the above procedures are used.

More rigorous characterizations of the mRNA can be made by assessing its ability to act as a template for reverse transcription and for protein synthesis in a cell-free system. Using one of the now readily available high-purity commercial reverse transcriptase preparations mRNA prepared as described above can be used to produce cDNA with an average size of 800–1000 nucleotides with a large proportion of full-length transcripts. It is found that oligo(dT) priming for reverse transcription is highly specific to the mRNA, with no apparent contamination with transcripts from the rRNA.

In vitro translation of mRNA is probably the best way to characterize an mRNA population. Single- or two-dimensional polyacrylamide gel electrophoresis (PAGE) of the products gives a good indication of the number and size distribution of the resulting polypeptides. The extent to which mRNA degradation has occurred can be judged by the amount of low-molecular-weight polypeptides and smearing that is seen. Specific polypeptide products can be identified by immunological techniques.

The *in vitro* rabbit reticulocyte translation system, adapted from that of Pelham and Jackson,[10] is found to give higher levels of ^{35}S incorporation than wheat germ systems. Moreover it offers the convenience that only four additions have to be made to the tube for each translation. Single-dimensional PAGE gives enough resolution to visualize differences in the mRNA populations isolated from *P. chrysosporium* mycelia grown under different culture conditions.[11]

[10] H. B. R. Pelham and R. J. Jackson, *Eur. J. Biochem.* **67**, 247 (1976).

[11] R. Haylock, R. Liwicki, and P. Broda, *J. Microbiol. Methods* **4**, 155 (1985).

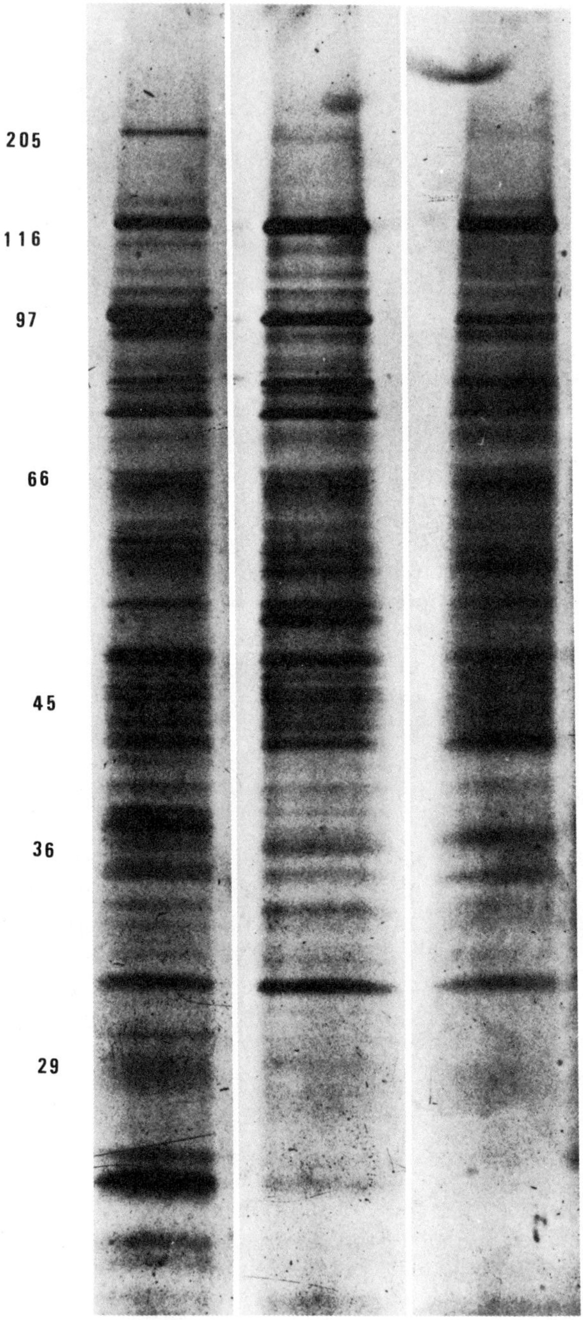

FIG. 1. An autoradiogram comparing the translation products of mRNA produced by strains mutant in their ligninolytic activities. Numbers indicate molecular weight (MW) ($\times 10^3$).

Three- to five-milligram aliquots of mRNA are translated in a cocktail containing 80% staphylococcal nuclease treated rabbit reticulocyte lysate, 100 m*M* KCl, 0.5 m*M* $MgCl_2$, 10 m*M* creatine phosphate, 0.05 m*M* of each of 19 amino acids (excluding methionine) and adjusted to pH 7, 1–5 μCi of [^{35}S]methionine (from Amersham), specific activity 800–1050 Ci/mmol, in a final reaction volume of 25 μl. The reaction is terminated by cooling on ice after 30 min at 34°. A small aliquot is taken and 0.5 ml of sterile water added. This solution is treated with 0.5 ml *M* NaOH, 5%, 100 vol hydrogen peroxide, and 1 mg/ml unlabeled methionine at 37° for 15 min. The polypeptides in the solution are precipitated by the addition of ice-cold 25% trichloroacetic acid. The number of TCA-precipitable counts is estimated after collection on glass fiber filters; the estimation of incorporated [^{35}S]methionine enables the loading of an equal number of counts onto each track of the gel, making comparisons between the tracks much easier. This is important as the incorporation varies between 25 and 250 times that of the background incorporation occurring in the absence of mRNA.

Translation samples are prepared for SDS–PAGE by the addition of 4 vol of a buffer containing 10 m*M* dithiothreitol, 2% filtered SDS, 80 m*M* Tris, pH 6.9, 10% glycerol, and 0.004% bromphenol blue,[12] followed by boiling for 5 min. After cooling it is important to remove any solid material from the sample by centrifugation before loading onto the gels. Failure to do this results in poor resolution of the polypeptide bands visualized on the autoradiogram. Samples are analyzed on gels containing 12.5% acrylamide, 0.17% *N,N'*-methylenebisacrylamide (bis), 375 m*M* Tris–HCl, pH 8.7, 0.1% SDS in the running gel and 5% acrylamide, 0.13% bis, 125 m*M* Tris–HCl, pH 6.9, and 0.1% SDS in the 2-cm depth stacking gel. Gels are set by the addition of 0.08% (0.2% in the stacking gel) *N,N,N',N'*-tetramethylethylenediamine (TEMED), and 0.03% ammonium persulfate (0.08% in the stacking gel), after oxygen is removed by deaeration under vacuum. The gels are run in a buffer containing 0.1% SDS, 7.2 g/liter glycine, and 1.5 g/liter Tris base until the bromphenol blue reaches the gel bottom. (The current should be set so that this takes about 5 hr on a 15-cm gel.) After electrophoresis, the gels are fixed overnight in 50% methanol and 7.5% acetic acid at 4° followed by two changes of 5% methanol and 0.75% acetic acid. Gels are then dried on a vacuum gel drier. The dried gels are exposed to preflashed X-ray film at −70° for 2 to 7 days depending on the number of counts loaded. A resulting autoradiograph is shown in Fig. 1.

[12] U. K. Laemmli, *Nature (London)* **227**, 680 (1970).

[22] Use of Synthetic Oligonucleotide Probes for Identifying Ligninase cDNA Clones

By YI-ZHENG ZHANG and C. ADINARAYANA REDDY

Ligninase, an extracellular, H_2O_2-dependent, glycosylated, heme protein, has recently been purified from a white rot basidiomycete, *Phanerochaete chrysosporium*.[1,2] This enzyme is synthesized under nitrogen-limited conditions during secondary metabolism. Ligninase has many potential practical applications, such as upgrading lignocellulosic materials via delignification for the efficient production of fuels, feeds, and chemicals; biobleaching of pulps; treatment of industrial wastes; hazardous environmental toxins such as PBBs, DDT, and benzopyrenes; controlled modification of lignins to produce aromatic chemicals; and cracking of petroleum. To better understand the nature, organization, expression, and regulation of the ligninase genes and to develop the full bioprocessing potential of this enzyme, we initiated studies to isolate and characterize the cDNA clones for ligninase.

Principle

Synthetic oligodeoxyribonucleotides are useful as specific probes for the detection and isolation of cloned cDNA or gene sequences of interest.[3-5] As a general approach, a chemically synthesized mixture of oligonucleotides whose sequences represent all possible codon combinations, predicted from a partial peptide sequence within a protein, is employed. Therefore, one of the oligonucleotides in the mixture must be complementary to a region of DNA coding for the protein. Since probes that form duplexes with a single base-pair mismatch have significantly less thermal stability than their perfectly matched counterpart, appropriate choice of hybridization temperature or filter wash temperature would virtually eliminate the formation of mismatched duplexes without affecting the formation of

[1] M. Tien and T. K. Kirk, *Proc. Natl. Acad. Sci. U.S.A.* **81,** 228 (1984).

[2] M. H. Gold, M. Kuwahara, A. A. Chiu, and J. K. Glenn, *Arch. Biochem. Biophys.* **234,** 353 (1984).

[3] R. B. Wallace, M. J. Johnson, T. Hirose, T. Miyake, E. Kawashima, and K. Itakura, *Nucleic Acids Res.* **9,** 879 (1981).

[4] J. W. Szostak, J. I. Stiles, T.-K. Tye, P. Chiu, F. Sherman, and R. Wu, this series, Vol. 68, p. 419.

[5] A. A. Reyes and R. B. Wallace, *Genet. Eng.* **6,** 157 (1984).

perfectly matched ones. Hence, the use of stringent hybridization criteria would allow the selection of the single correct sequence from the mixture.

The basic steps involved in cloning ligninase cDNA from *P. chrysosporium* are as follows[6]: (1) construction of cDNA library using poly(A) RNA from a 6-day-old lignin-degrading culture; (2) isolation of cDNA clones specific for 6-day culture using differential hybridization; (3) synthesis of oligonucleotide probes, deduced from partial amino acid sequences of ligninase; and (4) use of these probes to screen, isolate, and identify the ligninase clones from the 6-day-specific cDNA minilibrary.

Chemicals

Oligo(dT)-cellulose, X-Gal, dithiothreitol (DTT), and terminal deoxynucleotidyltransferase were purchased from Bethesda Research Laboratories (BRL), Gaithersburg, Maryland. Deoxyribonucleotides were purchased from Boehringer Mannheim Biochemicals, Indianapolis, Indiana. RNasin (RNase inhibitor) was purchased from Promega Biotech, Madison, Wisconsin. Carrier DNA (type III from salmon testes) was purchased from Sigma Chemical Company, St. Louis, Missouri. T4 polynucleotide kinase was purchased from International Biotechnologies, Inc., New Haven, Connecticut. HATF hybridization membranes were purchased from Millipore, Bedford, Massachusetts.

Isolation of Poly(A) RNA[7]

Phanerochaete chrysosporium strain BKM-F1767 (ATCC 24725) is grown in 50 ml of low nitrogen medium (modified to contain 20 m*M* NaOAc, pH 4.5, instead of 10 m*M* 2,2-dimethyl succinate) in 500-ml Erlenmeyer flasks.[6] Flasks are flushed with pure oxygen at the time of inoculation and reflushed every other day. A modified hot phenol extraction procedure is used for RNA isolation.[8] In this procedure, mycelia from 1 liter of a 6-day-old culture are harvested by centrifugation, washed twice with 50 m*M* NaOAc, (pH 5.2), and suspended in 20 ml of extraction buffer (0.15 *M* NaOAc, pH 5.2, 5% SDS, and 2 m*M* EDTA). The mycelial suspension is then mixed with 10 ml phenol and 25 g of glass beads (0.45-mm size) and blended in an Omni-Mixer (Sorvall) for 20 min. Ten

[6] Y. Z. Zhang, G. J. Zylstra, R. H. Olsen, and C. A. Reddy, *Biochem. Biophys. Res. Commun.* **137**, 649 (1986).

[7] Details of buffers, solutions, reagents, and procedures not described in detail in this chapter are adequately described by Maniatis *et al.*[8]

[8] T. Maniatis, E. F. Fritsch, and J. Sambrook, "Molecular Cloning: A Laboratory Manual." Cold Spring Harbor Laboratory, Cold Spring Harbor, New York, 1982.

milliliters of chloroform : isoamyl alcohol (24 : 1) is added to this mixture and the blending is carried out for an additional 10 min. The mixture is then heated at 60° for 15 min while gently shaking and is then chilled on ice. After centrifugation (10,400 *g* for 20 min at 4°), the upper phase is extracted with phenol–chloroform (1 : 1), RNA is precipitated with 2 vol of 100% ethanol, and the pellet is dissolved in 10 ml of 2 m*M* EDTA. After heating at 65° for 10 min, the RNA solution is chilled on ice for 10 min and 10 ml of 2× loading buffer (1 *M* NaCl, 20 m*M* Tris–HCl pH 7.5, 2 m*M* EDTA, and 1% SDS) is added. The RNA solution obtained is generally too viscous to pass through the oligo(dT)-cellulose column; hence, it is mixed with 0.1 g oligo(dT)-cellulose powder (BRL) and the mixture is gently shaken for 30 min at room temperature. The oligo(dT)-cellulose with the bound poly(A) RNA is then spun down, washed with loading buffer, and then packed in a small glass column (10 × 1 cm). The column is washed with loading buffer and the poly(A) RNA is eluted with TES buffer (10 m*M* Tris–HCl, pH 7.5, 1 m*M* EDTA, and 0.2% SDS). Elution of the RNA is monitored by measuring the absorbance of the elutant at 254 nm using UV monitor UA-5 (ISCO Instruments, Lincoln, NE). For isolating 2-day poly(A) RNA, the same procedure is used except that the RNA solution is directly loaded on the oligo(dT)-cellulose column. Using this procedure, 10 to 20 mg of RNA is obtained from 1 liter of culture and poly(A) RNA accounts for 1 to 2% of the total RNA.

Construction of cDNA Library

The double-stranded cDNA synthesis described here is based on the procedure described by Gubler and Hoffman.[9] The reaction mixture used for the first-strand cDNA synthesis contained in 40 μl: 50 m*M* Tris–HCl (pH 8.3), 10 m*M* $MgCl_2$, 10 m*M* DTT, 1.25 m*M* dATP, 1 m*M* dCTP, 1.25 m*M* dGTP, 1.25 m*M* dTTP, 20 μCi of [α-^{32}P]dCTP (≃ 3000 Ci/mmol, ICN, Irvine, CA), 0.75 μg oligo(dT) 12–18 bases long, 60 U RNasin, 10 μg poly(A) RNA, and 45 U avian myeloblastosis virus (AMV) reverse transcriptase (BRL, Gaithersberg, MD). The reaction mixture is incubated at 43° for 1 hr. The reaction is stopped by adding 2 μl of 0.5 *M* EDTA and 3 μl of 5 *M* NaCl. The reaction mixture is first extracted with an equal volume of phenol followed by extraction with an equal volume of chloroform. The product is then precipitated with 100% ethanol at −20° overnight, spun down in an Eppendorf centrifuge for 20 min at 4°, washed once with 70% and once with 100% ethanol, and dried in a vacuum desiccator. The amount of first-strand cDNA obtained is estimated by

[9] U. Gubler and B. J. Hoffman, *Gene* **25**, 263 (1983).

determining the TCA-insoluble radioactivity before and after the reaction. The single-stranded cDNA is then dissolved in double-distilled water at a concentration of 50 μg/ml. The reaction mixture for the second-strand cDNA synthesis contained in 100 μl: 2 m*M* Tris–HCl (pH 7.5), 5 m*M* $MgCl_2$, 10 m*M* $(NH_4)_2SO_4$, 100 m*M* KCl, 0.1 m*M* dATP, 0.1 m*M* dCTP, 0.1 m*M* dGTP, 0.1 m*M* dTTP, 20 μCi dCTP (same as above), 500 μg of single-stranded cDNA, 0.15 m*M* β-NAD, 5 μg of BSA, 1.8 U RNase H, 1 U T4 DNA ligase, and 25 U DNA polymerase I. The reaction is carried out at 14° for 60 min and then at 22° for an additional 60 min and is stopped by adding 4 μl of 0.5 *M* EDTA. After one extraction each with phenol and chloroform, the double-stranded cDNA is separated from unincorporated dNTP by passing the reaction mixture through a Sephadex G-50 column (10 × 1 cm) and eluting the cDNA with TE buffer (10 m*M* Tris–HCl, pH 8.0, 1 m*M* EDTA). The fractions showing the highest radioactivity are collected, pooled, and the cDNA is precipitated with ethanol. The efficiency of second-strand cDNA synthesis is estimated by determining the TCA-insoluble radioactivity before and after the reaction as described above.

The dC tailing of the double-stranded cDNA is performed in 150 μl of reaction system containing 13 μ*M* dCTP, 75 μg BSA, 500 ng of double-stranded cDNA, and 30 μl of 5 × tailing buffer provided by BRL (500 m*M* potassium cacodylate, pH 7.2, 10 m*M* $CoCl_2$, and 1.0 m*M* DTT). After incubating the reaction mixture for 5 min at 20° the tailing reaction is initiated by adding 90 U of terminal deoxynucleotidyltransferase (TdT) and the reaction is allowed to proceed for an additional 20 min at 20°.

The reaction is stopped by adding 6 μl of 0.5 *M* EDTA, chilled on ice for 10 min, 600 μl of sterile double-distilled water is added, extracted once with phenol and once with chloroform, and sodium acetate (3 *M*) is added to a final concentration of 0.3 *M*. After adding 10 μg of yeast tRNA, the tailed product is precipitated with 2 vol of ethanol, washed once with absolute ethanol, dried, and dissolved in 100 μl double-distilled water.

Vector pUC9, used for cloning the cDNA, contains an ampicillin resistance (Ap^r) marker, a multiple cloning site, and a segment of *lacZ* gene which encodes an α-donor peptide that complements an α-acceptor encoded by the host strain (*E. coli* JM83) to produce a functional β-galactosidase which hydrolyzes X-Gal to release 5-bromo-4-chloroindigo that colors the colonies deep blue.[10] On the other hand, in recombinant pUC9 containing cloned cDNA, the *lacZ* gene segment is inactivated; hence, colonies of *Escherichia coli* containing recombinant pUC9 will be white. For dG tailing of the vector, pUC9 is first completely digested with the

[10] J. Vieira and J. Messing, *Gene* **19**, 259 (1982).

restriction enzyme *Pst*I and the linear DNA is purified by electrophoresis through a 0.7% low-melting-temperature agarose gel.[8] The DNA is then extracted by melting the agarose gel strip containing the vector fragment.[8] This purification step is important because (1) it removes any RNA or low-molecular-weight DNA contaminating the plasmid or the restriction enzyme *Pst*I; and (2) it separates the linear plasmid DNA from any undigested circular plasmid molecules. The reaction system used for dG tailing of the plasmid contained 13 μM dGTP, 75 μg BSA, 2 μg pUC9, 30 μl of 5× tailing buffer (same as above), and 90 U TdT. The reaction mixture, except TdT, is preincubated for 5 min at 20°; the reaction is initiated by adding TdT, and is incubated at 37° for an additional 60 min. The stoppage of the reaction and the rest of the steps used in the isolation of dG-tailed products are the same as those described above in the isolation of dC-tailed product.

The dG-tailed vector is annealed with dC-tailed cDNA in 100 μl of mixture containing 0.1 *M* NaCl, 10 m*M* Tris–HCl (pH 7.6), 1 m*M* EDTA, 200 μg dG-tailed pUC9, and 50 ng dC-tailed double-stranded cDNA. The mixture is then incubated successively at 65° for 5 min, at 58° for 2 hr, and at 42° for 30 min, and chilled on ice for 20 min.

The annealed DNA molecules are transformed into *E. coli* JM83 (*ara* Δ*lacpro strA thi* ϕ80d*lacZ* ΔM15[10]) using the procedure of Hanahan.[11] The transformed cells are spread on 2 YT plates (1.6% Bacto-tryptone, 1% yeast extract, 0.5% NaCl, and 1.5% agar, pH 7.0)[12] supplemented with 100 μg/ml ampicillin and 40 μg/ml X-Gal[12] (5-bromo-4-chloro-3-indolyl-β-D-galactoside dissolved in dimethylformamide and stored at −20°). Ten thousand white *E. coli* colonies that are potential cDNA clones are picked. Each white colony is picked individually with a sterile toothpick and inoculated into 50 μl of sterile LB broth in one of the wells of a 96-well microtiter plate. The inoculated plates are incubated overnight at 37°. Fifty microliters of sterile glycerol is added to each well and the plates are then stored at −20° until further use.

Differential Hybridization

Rationale. Ligninase has been shown to be produced in 6-day-old idiophasic cultures of *P. chrysosporium* grown in low-nitrogen medium; the enzyme is not detectable in 1- or 2-day-old cultures in primary growth. Therefore, differential hybridization technique allows isolation of the cDNA clones specific for the idiophase. It is much easier to screen such a

[11] D. Hanahan, *J. Mol. Biol.* **166,** 557 (1983).

[12] J. H. Miller, "Experiments in Molecular Genetics." Cold Spring Harbor Laboratory, Cold Spring Harbor, New York, 1972.

minilibrary of idiophasic clones than to screen the total cDNA library for isolating cloned cDNA of interest.

Procedure. Nunc-TSP transferable solid-phase screening system (Vanguard International, Inc., Neptune, NJ) is used to transfer the cDNA clones on each microtiter plate onto a 137-mm HATF hybridization membrane placed on the surface of LB agar plates (1% Bacto-tryptone, 0.5% yeast extract, 1% NaCl, and 1.5% Bacto-agar, pH 7.2; 100 μg/ml ampicillin is added just before pouring plates), which are then incubated at 37° for 14 hr. The filter paper is peeled off and dried on 3MM Whatman chromatography paper for 20 min. One circular 3MM Whatman paper (about 137-mm diameter) is placed in each of four Petri dishes (150 × 10 mm) labeled 1, 2, 3, and 4. The papers are saturated with 1% SDS, denaturation solution (1.5 *M* NaCl and 0.5 *M* NaOH), neutralizing solution (1.5 *M* NaCl and 0.5 *M* Tris–HCl, pH 8.0), and 2× SSPE (0.36 *M* NaCl, 20 m*M* NaH_2PO_4, and 2 m*M* EDTA, pH 7.4)[8] in Petri dishes 1, 2, 3, and 4, respectively. The cDNA clones on the HATF paper are lysed, denatured, and neutralized in dishes 1, 2, and 3, respectively, by putting the paper in each of the plates for 5 min.[8] The filter is then placed in dish 4 for 5 min. The cDNA blots are then baked at 80° for 4 hr, wetted with 6× SSC (1× SCC, 0.15 *M* NaCl, and 0.015 *M* sodium citrate, pH 7.0),[8] and washed in prewashing solution (3× SSC and 0.1% SDS) at 65° for 10 hr with several changes of the same solution to completely remove all cell debris. After briefly blotting on a 3MM paper, every eight blots are put in one hybridization bag and 16 ml of hybridization solution [50% formamide, 5× Denhardt's solution (1% Ficoll, 1% polyvinylpyrrolidone, and 1% BSA),[8] 1 *M* NaCl, 10 m*M* Tris–HCl, pH 8.0, 1 m*M* EDTA, 50 m*M* NaH_2PO_4, pH 6.8, 10 μg/ml carrier DNA] is added. The blots are prehybridized at 42° overnight, the synthetic 2-day cDNA probe (see below) is added at a final concentration of 1×10^6 cpm/ml of hybridization solution, and the hybridization is carried out at 42° for 36 hr. The hybridized blots are washed in high stringent solution (10 m*M* Tris–HCl, pH 7.5, 1 m*M* EDTA, 0.1% SDS, 0.1% $Na_4P_2O_7$, and 50 m*M* NaCl) twice at room temperature and three times at 65° (each wash is for 15 min). The blots are then exposed to an X-ray film for a suitable length of time and the film is developed.[8] The 2-day cDNA probe is then washed off and the blots are hybridized with the 6-day cDNA probe (see below).

The 2- and 6-day cDNA probes are synthesized, respectively, from poly(A) RNA isolated from 2- and 6-day-old cultures using a modification of the procedure of Berlin and Yanofsky.[13] The reaction mixture used for probe synthesis contained in 50 μl: 50 m*M* Tris–HCl pH 8.3; 8 m*M*

[13] V. Berlin and C. Yanofsky, *Mol. Cell. Biol.* **5**, 849 (1985).

$MgCl_2$; 40 m*M* KCl; 2 m*M* DTT; 1 m*M* dATP, dTTP, and dGTP; 1 μg oligo(dT), 12–18 bases long; 1 μg poly(A) RNA; 50 U RNasin; and 200 μCi dCTP (~600 Ci/m*M*). The reaction is started by adding 300 U M-MLV (Moloney murine leukemia virus) reverse transcriptase (BRL). The reaction is carried out at 43° for 1 hr and then 2 μl of 0.5 *M* EDTA, 10 μl of 10 mg/ml carrier DNA, and 7 μl of 20% SDS are added. The unincorporated nucleotides are separated from the labeled cDNA by passing the reaction mixture through a Sephadex G-50 column. The fractions showing the highest radioactivity are combined and to this cDNA eluate 0.1 vol of 2 *M* NaOH is added and the mixture is heated at 65° for 5 min. After chilling on ice for 10 min, 2 *M* HCl (equal to 0.1 vol of the cDNA eluate) is added to neutralize the probe.

Of the 10,000 cDNA clones in our cDNA library, 850 clones are shown to be specific to the 6-day cDNA probe using the differential hybridization technique described above. A representative differential hybridization blot is shown in Fig. 1.

Identification of Ligninase cDNA Clones

Oligonucleotide Probes. Three oligonucleotide probes, deduced from the amino acid sequences of selected tryptic peptides of ligninase H8,[14] are used for screening the cDNA library. The sequences of these probes are shown below:

GTQ-TTQ-GGN-AAP-CA	PROBE 14.1
AAP-CAN-GTQ-TTQ	PROBE 14.2
CAN-AAP-GTP-CTP-CG	PROBE 25

N = AGCT/U, P = AG, Q = CT/U

Due to the redundancy of the genetic code all probes are a mixture of 32 different oligonucleotide sequences. (Note: Several commercial sources and university labs synthesize the desired oligonucleotide probes for a fee.) The oligonucleotides are end labeled with polynucleotide kinase in a 50 μl total reaction mixture containing kinase buffer (50 m*M* Tris–HCl, pH 8.0, 10 m*M* $MgCl_2$, and 15 m*M* DTT), 500 ng synthetic oligonucleotide (dissolved in water), 100 μCi [γ-^{32}P]ATP (4500 Ci/μmol; ICN, Irvine, CA) and 10 U T4 polynucleotide kinase (IBI, New Haven, CT). The mixture is incubated at 37° for 30 min and then cooled on ice until use.

Preparation of cDNA Blots. The cDNA blots are prepared as described above using 6-day-specific cDNA clones.

[14] T. K. Kirk, S. Croan, M. Tien, K. E. Murtagh, and R. L. Farrell, *Enzyme Microb. Technol.* **8,** 27 (1986).

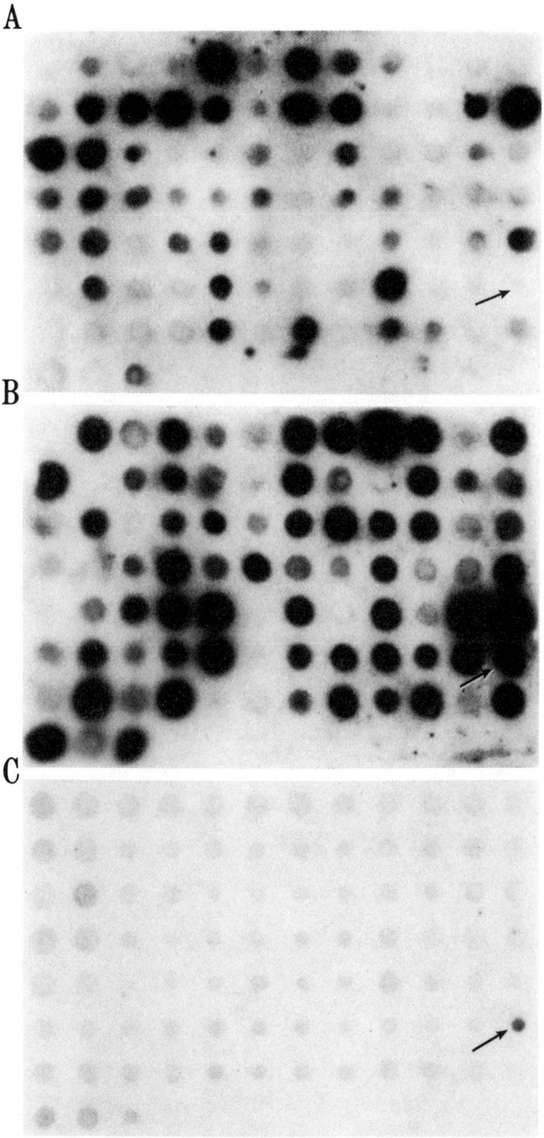

FIG. 1. Differential hybridization and identification of a ligninase cDNA clone. The cDNA clones in the library are hybridized with the 2-day cDNA probe (A), 6-day cDNA probe (B), or synthetic oligonucleotide probe 14.1 (C). Note that the ligninase cDNA clone indicated by the arrow in each panel shows hybridization with the 6-day cDNA probe, but not with the 2-day cDNA probe. From Zhang *et al.*[6]

Hybridization of cDNA Blots with Synthetic Probes. The cDNA blots are prehybridized at 37° for 4 hr in a solution containing 6× SSC, 1× Denhardt's solution, 0.5% SDS, 0.05% $Na_4P_2O_7$, and 10 μg/ml carrier DNA. The prehybridization solution is drained out and hybridization solution (6×SSC, 1× Denhardt's solution, 0.05% $Na_4P_2O_7$, and 20 μg/ml tRNA), along with probe 14.1, is then added. The hybridization is carried out at room temperature for 1 hr. The hybridized blots are washed once at room temperature in a solution containing 6× SSC and 0.05% $Na_4P_2O_7$ for 30 min and once at 42° in the same solution for 10 min.

Four cDNA clones showed strong hybridization with probe 14.1. A representative clone is shown in Fig. 1.

To further identify these clones, the plasmid DNA isolated from these clones is digested with different restriction enzymes in such a way that each clone gives three unequal fragments, one from the vector and two from the

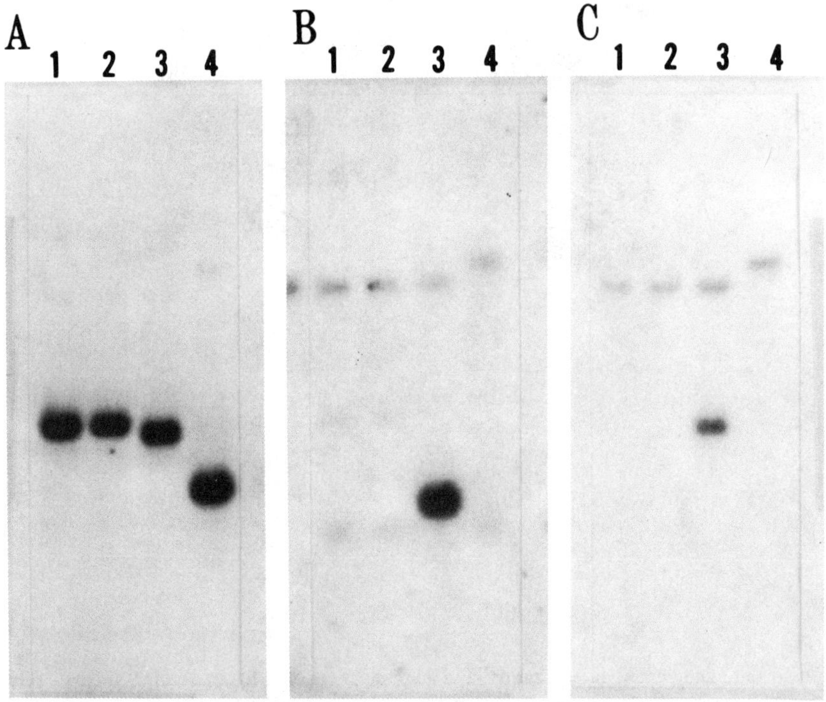

FIG. 2. Hybridization of ligninase cDNA clones with three oligonucleotide probes. Clones pCLG3 and pCLG4 are digested with *Bam*HI and *Hin*dIII; pCLG5 is digested with *Bam*HI, *Hin*dIII, and *Pst*I; and pCLG6 is digested with *Hin*dIII and *Pst*I. The DNA blots are hybridized with probe 14.1 (A), 14.2 (B), and 25 (C). Different lanes contain pCLG3 (lane 1), pCLG4 (lane 2), pCLG5 (lane 3), or pCLG6 (lane 4).

cDNA insert. The fragments are transferred onto nitrocellulose paper using Southern blotting[8] and the DNA blots are hybridized with the above three synthetic probes as described above. The temperature for the second washing for probe 14.2 is 32°. The results show that only the cDNA insert in clone pCLG5 hybridizes with all the three probes, whereas the other three clones (pCLG3, pCLG4, and pCLG6) show detectable hybridization with probe 14.1 only (Fig. 2). Furthermore, each of the three probes hybridizes with only one cDNA fragment from a given clone, indicating that the probes are specific for specific sequences in the cDNA. The restriction and Southern hybridization data showed that the above four cDNA clones represent two types of ligninase cDNA, one consisting of pCLG5 and the other consisting of pCLG3, pCLG4, and pCLG6. Additional studies showed that the product expressed by the cDNA insert of pCLG5 is immunoreactive with the ligninase (H8) antibody. Our nucleotide sequence data for the cDNA inserts in pCLG4 and pCLG5 showed that both are full-length cDNA clones, have a high degree of sequence homology (71.5%), and that the sequences of the synthetic probes are present in the cDNA sequences. The experimentally determined N-terminal sequence of mature ligninase H8 is found in the deduced amino acid sequence of pCLG5 cDNA. These collective data lead us to the conclusion that the pCLG5 and pCLG4 cDNA represent two of the ligninase genes in *P. chrysosporium* (Y. Z. Zhang and C. A. Reddy, unpublished data).

The procedure described above for isolating the ligninase cDNA of *P. chrysosporium* is reliable and potentially applicable for isolating the ligninase genes from other organisms. Besides, the procedure described can be used to isolate genes for other secondary metabolic enzymes such as glucose oxidase, which has recently been purified from *P. chrysosporium.*[15]

Acknowledgments

This research was supported, in part, by Grant DE-FG02-85#13369 from the U.S. Department of Energy, Division of Basic Biological Sciences, NSF Grant DMB-844271, and a grant from the Michigan Agriculture Experiment Station. This is publication No. 12063 from the Michigan Agricultural Experiment Station.

[15] R. L. Kelley and C. A. Reddy, *J. Bacteriol.* **166,** 269 (1986).

[23] Lignin Peroxidase of *Phanerochaete chrysosporium*

By MING TIEN and T. KENT KIRK

Introduction

Ligninase is a generic name for a group of isozymes that catalyze the oxidative depolymerization of lignin. Although undoubtedly produced by other lignin-degrading fungi, these isozymes to data have been isolated only from the basidiomycete *Phanerochaete chrysosporium* Burds.[1,2] These ligninases are extracellular and are produced during secondary metabolism, brought about by nutrient starvation. Nitrogen limitation is usually employed, as described here, but carbon-limited cultures have also been used for ligninase production.[3] The ligninases exhibit a high degree of homology. They are all heme-containing glycoproteins and all cross react with a polyclonal antibody raised to the predominant ligninases.[4] Since they all have overlapping substrate specificities, the exact role of this multiplicity is not yet understood. The number of genes encoding for ligninases is not yet known.

The major isozyme, ligninase H8, has been extensively characterized and is the protein initially isolated by Tien and Kirk.[5] Based on kinetic[6] and spectroscopic data,[7] this ligninase has been characterized as a peroxidase containing one high-spin ferric heme per enzyme molecule.[8] Like horseradish peroxidase, the ligninases are capable of catalyzing a wide range of one- and two-electron oxidations. The substrates of ligninase, however, exhibit much higher reduction potentials. This property, along with its low pH optimum,[6] imparts ligninase with the unique ability to catalyze the oxidative depolymerization of lignin and the oxidation of methoxybenzene-containing lignin-like substrates.[9,10]

[1] M. Tien and T. K. Kirk, *Science* **221,** 661 (1983).

[2] J. K. Glenn, M. A. Morgan, M. B. Mayfield, M. Kuwahara, and M. H. Gold, *Biochem. Biophys. Res. Commun.* **114,** 1077 (1983).

[3] B. D. Faison and T. K. Kirk, *Appl. Environ. Microbiol.* **49,** 299 (1985).

[4] T. K. Kirk, S. C. Croan, M. Tien, K. E. Murtagh, and R. Farrell, *Enzyme Microb. Technol.* **8,** 27 (1985).

[5] M. Tien and T. K. Kirk, *Proc. Natl. Acad. Sci. U.S.A.* **81,** 2280 (1984).

[6] M. Tien, T. K. Kirk, C. Bull, and J. A. Fee, *J. Biol. Chem.* **261,** 1687 (1986).

[7] D. Kuila, M. Tien, J. A. Fee, and M. R. Ondrias, *Biochemistry* **24,** 3394 (1985).

[8] L. A. Anderson, V. Renganathan, A. A. Chiu, T. M. Loehr, and M. H. Gold, *J. Biol. Chem.* **260,** 6080 (1985).

[9] P. Kersten, M. Tien, B. Kalyanaraman, and T. K. Kirk, *J. Biol. Chem.* **260,** 2609 (1985).

Several procedures have been described for growing *P. chrysosporium* for ligninase production. These procedures differ somewhat in the medium formulation and types of growth vessels: (1) shallow stationary cultures, (2) agitated liquid cultures, and (3) rotating biological contactors (RBCs; disk fermenters). Because the RBCs employ a mutant strain that adheres to the plastic disk[4] and equipment that has to be constructed, their use is not described here. The more recently developed use of agitated culture for production of ligninase permits easier "scale up."[11] Although ligninase can be produced in agitated flask cultures, the reliable use of stirred tank fermenters awaits further development, which is ongoing in several laboratories. In the following we describe the production of ligninase in shallow stationary cultures and in agitated cultures. The stationary cultures give somewhat more reliable and reproducible results than the agitated cultures.

Maintenance of Fungus and Preparation of Spore Inoculum

Cultures of *P. chrysosporium* (strain BKM-F-1767; ATCC 24725) are maintained on supplemented malt agar slants; the medium is described below. Of the strains that have been studied, strain BKM-F-1767 produces highest ligninase activity, although activity is produced by all examined wild-type strains.[12]

Composition of agar for maintenance and spore production (per liter):

Glucose, 10 g
Malt extract, 10 g
Peptone, 2 g
Yeast extract, 2 g
Asparagine, 1 g
KH_2PO_4, 2 g
$MgSO_4 \cdot 7H_2O$, 1 g
Thiamin-HCl, 1 mg
Agar, 20 g

Spore production in the slants usually requires 2 to 5 days of growth at 39°. Spores (conidia) are prepared by suspension in sterile water followed by passage through sterile glass wool to free it of contaminating mycelia. Spore concentration is determined by measuring absorbance at 650 nm (an absorbance of 1.0 cm^{-1} is approximately 5×10^6 spores/ml).

[10] K. E. Hammel, M. Tien, B. Kalyanaraman, and T. K. Kirk, *J. Biol. Chem.* **260,** 8348 (1985).

[11] A. Jäger, S. Croan, and T. K. Kirk, *Enzyme Microb. Technol. Appl. Environ. Microbiol.* **50,** 1274 (1985).

[12] R. K. Kirk, M. Tien, S. C. Johnsrud, and K.-E. Eriksson, *Enzyme Microb. Technol.* **8,** 75 (1986).

Culture Media

Stock Reagents

1. Basal III medium (per liter):

   KH_2PO_4, 20 g
   $MgSO_4$, 5 g
   $CaCl_2$, 1 g
   Trace elements solution (see below), 100 ml

2. Trace element solution (per liter):

   $MgSO_4$, 3 g
   $MnSO_4$, 0.5 g
   NaCl, 1.0 g
   $FeSO_4 \cdot 7H_2O$, 0.1 g
   $CoCl_2$, 0.1 g
   $ZnSO_4 \cdot 7H_2O$, 0.1 g
   $CuSO_4$, 0.1 g
   $AlK(SO_4)_2 \cdot 12H_2O$, 10 mg
   H_3BO_3, 10 mg
   $Na_2MoO_4 \cdot 2H_2O$, 10 mg
   Nitrilotriacetate,[13] 1.5 g

Culture Composition (Shallow Stationary Cultures)

The following items are added per liter of shallow stationary cultures:

Basal III medium (filter sterilized), 100 ml
10% glucose (autoclaved), 100 ml
0.1 *M* 2,2-dimethylsuccinate, pH 4.2 (autoclaved), 100 ml
Thiamin (100 mg/liter stock, filter sterilized), 10 ml
Ammonium tartrate (8 g/liter stock, autoclaved), 25 ml
Spores (absorbance at 650 nm = 0.5), 100 ml
Veratryl alcohol (0.4 *M* stock, filter sterilized), 100 ml
Trace elements (filter sterilized), 60 ml

Culture Composition (Agitated Cultures)

The medium for agitated cultures has the same composition as that for stationary cultures except that 0.05% Tween 20 or Tween 80 is added, and the fungus is introduced as a mycelial suspension instead of a spore suspen-

[13] Dissolve nitrilotriacetate in 800 ml H_2O, adjust pH to ~ 6.5 with 1 *N* KOH, add each component, and then bring the volume to 1 liter.

sion. The detergent is solubilized and sterilized by autoclaving a 1% solution in distilled water; 50 ml of this solution is added to the above medium. The mycelial inoculum is prepared by growing the fungus from spore suspension in stationary 2.8-liter Fernbach flasks containing 50 ml of the above medium (without detergent). After 48 hr at 39°, the mycelium plus medium is blended for 1 min in a blender (100 ml; 45 mg dry wt). The resulting suspension is substituted for the spores in the above culture formulation.

Growth and Harvest

Shallow stationary cultures (10 ml) are grown in rubber-stoppered, 125-ml Erlenmyer flasks at 39° under 100% oxygen. They are flushed with oxygen at the time of inoculation and again on day 3. Preparations typically utilize 400 flask cultures yielding about 3.8 liters of ligninase-containing culture supernatant. Care is taken not to perturb the cultures after the mycelial mats have formed, which takes about 24 hr. Attempts to scale up production via a proportional increase in both culture volume and flask size or with the use of shallow pans resulted in lower activity.

Agitated cultures, 45 or 750 ml, are grown in either 125-ml Erlenmeyer flasks, or 2-liter Erlenmeyer flasks, respectively. The cultures are grown at 39° on a rotary shaker with a 2.5-cm-diameter cycle, the small flasks at about 200 rpm, and the larger ones at about 125 rpm. The rubber-stoppered culture flasks are flushed with 100% O_2 at the time of inoculation, and daily thereafter. Enough cultures are grown to yield approximately 3.8 liters of culture supernatant (about 4.2 liters of cultures).

Mycelial growth under the nitrogen-limited conditions stops by day 2 and ligninase activity appears in the extracellular fluid on day 4, coinciding with development of a brown coloration on the mycelia (which is only observed with the excess trace elements solution). Under both stationary and shaken incubation, activity reaches a maximum on days 5 and 6. When the maximum is reached, the supernatant is obtained by centrifugation at 10,000 *g* for 5 min at 4°. The yellow supernatant (3.3 liters, Table I), which contains all of the activity (0.075 U/ml), is then concentrated by ultrafiltration (Millipore Minitan unit) using a 10-kDa cut-off membrane. After concentration to approximately 40 ml, the preparation is filtered (0.45-μm pore size), which removes precipitated mycelial slime, then further concentrated (Amicon, 10-kDa cut-off) to a final volume of 13.5 ml. The sample is then dialyzed overnight against 4 liters of either 10 m*M* sodium acetate, pH 6, for Mono-Q chromatography, or 5 m*M* sodium succinate, pH 5.5, for chromatography on DEAE-BioGel A (see below). As shown in Table I, the concentration and dialysis step results in very little

TABLE I
PURIFICATION OF LIGNINASE ISOZYMES[a]

Sample	Volume (ml)	Activity (U/ml)	Protein (mg/ml)	Total activity (U)	Specific activity (U/mg)	Recovery (%)
ECF[b]	3300	0.076	0.013	251	5.71	100
ECF						
Concentrated (Minitan/Amicon)	13.5	16.9	1.23	229	13.8	89
Pre-FPLC (dialyzed/filtered)	17	12.95	0.7	220	18.5	88
FPLC purified[c]						
H1	8.7	0.52	0.07	4.5	7.24	1.8
H2	17.7	2.1	0.13	36.3	16.4	14
H6	13.1	0.33	0.06	4.36	5.5	1.7
H7	4.3	0.33	0.1	1.4	3.28	0.5
H8	31.8	1.56	0.21	49.6	7.6	20
H10	25.7	0.25	0.09	6.35	2.7	2.5

[a] Ligninolytic cultures of BKM were grown and harvested as described by Kirk *et al.*[4]

[b] ECF, Extracellular fluid. Please note that the specific activity increased after concentration due to loss of low-molecular-weight components which contributed background in the protein assay.

[c] Peaks from repeated injections were pooled, dialyzed against 5 m*M* sodium tartrate, pH 4.5, and assayed for protein activity.[4]

loss in total activity. Total percentage recovery is usually in the high eighties (Table I). We typically concentrate on the day of harvest and then dialyze overnight.

Assay Method

Principle

Ligninase catalyzes the oxidation of veratryl alcohol by H_2O_2[6] to veratraldehyde. The alcohol exhibits no absorbance at 310 nm whereas the aldehyde absorbs strongly (molar extinction coefficient = 9300 M^{-1} cm^{-1}). Use is made of this property in a continuous spectrophotometric assay.

Reagents

10 m*M* veratryl alcohol
0.25 *M* *d*-tartaric acid, pH 2.5
5 m*M* H_2O_2 (prepared daily)

Procedure

Reaction mixtures contain 2 m*M* veratryl alcohol ($K_m = 60\ \mu M$), 0.4 m*M* H_2O_2 ($K_m = 80\ \mu M$), 50 m*M* tartaric acid, and enough ligninase to give an absorbance change of 0.2/min.

Comments on Assay

Although the ligninase is most active at pH below 3, it is not very stable; thus reaction rates are linear only for about 2 min. The ligninase is also inactivated by H_2O_2 in the absence of a reducing substrate, such as veratryl alcohol. Consequently, care should be taken to minimize the preincubation of ligninase with buffers of low pH (pH < 3.0) or with H_2O_2 in the absence of veratryl alcohol. For reproducible results, the temperature should be held constant because the ligninase shows a high temperature dependence; the rate approximately doubles with every 7° increase.[6]

Reagents for enzyme activity are commercially available and, except for veratryl alcohol, do not require further purification. Prior to use, veratryl alcohol is vacuum distilled to free it of the trace contaminant methyl-3-methoxy-4-hydroxybenozate, which is a better ligninase substrate than veratryl alcohol.[6] This contaminant probably can also be removed by extracting a solution in ether or dichloromethane with aqueous alkali. Unless removed, this phenolic contaminant causes distinct lag periods in the initial rate, most noticeable at low enzyme activity.

Purification of Ligninase(s)

Multiple ligninases of *P. chrysosporium* can be separated by either ordinary column chromatography using DEAE-BioGel A or by FPLC (or HPLC) with the Mono-Q anion-exchange column of Pharamacia. For chromatography on DEAE-BioGel A, the column (1 × 16 cm) is equilibrated with 5 m*M* sodium succinate buffer, pH 5.5; the sample is loaded and then eluted with a NaCl gradient (0–0.14 *M*, total volume of 600 ml). All steps are performed at 4°.

Mono-Q is the method of choice due to its superior resolution. The results of the Mono-Q separation are summarized in Table I. Following dialysis, the sample is again filtered through a 0.45-μm filter (Gilson Co., Madison, WI, low protein-binding filter). The Mono-Q column capacity is 25 mg total protein or 5 mg/peak. The 10-ml preparation described above can be purified by five injections of 2 ml. The sample is loaded with 10 m*M* sodium acetate, pH 6.0, and eluted with a gradient from 10 m*M* to 1 *M* sodium acetate, pH 6.0, over a 40-min period at 2 ml/min. We use

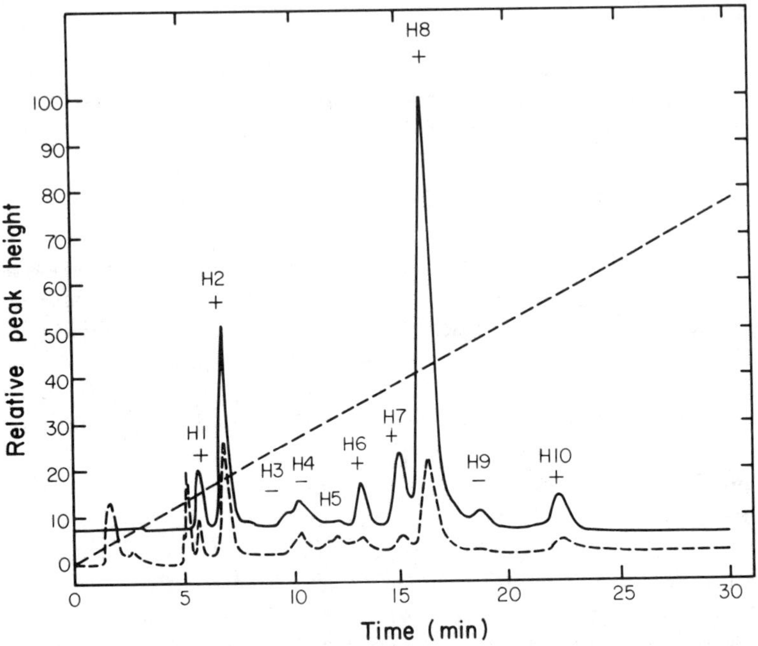

FIG. 1. FPLC profile of extracellular fluid from 5-day flask cultures. Full and dashed lines show absorbance at 409 and 280 nm, respectively. Veratryl alcohol-oxidizing activity is indicated qualitatively as positive (+) or negative (−). The sloping line shows the acetate gradient. Reproduced from Kirk *et al.*[4]

this method at room temperature, but return the protein to 4° after elution.

Figure 1 shows the profile from the FPLC column showing the absorbance at 409 nm (heme absorbance) and 280 nm (total protein). Figure 2 shows the profile from the DEAE-BioGel A column (showing only the 409-nm absorbance). As clearly demonstrated by the elution profiles, the resolution is much better on the Mono-Q than the DEAE-BioGel A column. Both profiles indicate the presence of numerous proteins. Over 13 proteins can be detected from the Mono-Q column; most of them are baseline resolved. The peaks designated H1, H2, H6, H7, H8, and H10 all have veratryl alcohol-oxidizing activity in addition to activity toward various dimeric models of lignin.[4] These enzymes are the ligninases. The other peaks (H3, H4, H5, and H9) are the Mn-dependent peroxidases characterized by Glenn and Gold[14] and Paszczyński *et al.*[15] The recovery of each of the ligninase isozymes from the Mono-Q column is given in Table I. The

[14] J. K. Glenn and M. H. Gold, *Arch. Biochem. Biophys.* **242,** 329 (1985).

[15] A. Paszczyński, V.-B. Huynh, and R. Crawford, *Arch. Biochem. Biophys.* **244,** 750 (1986).

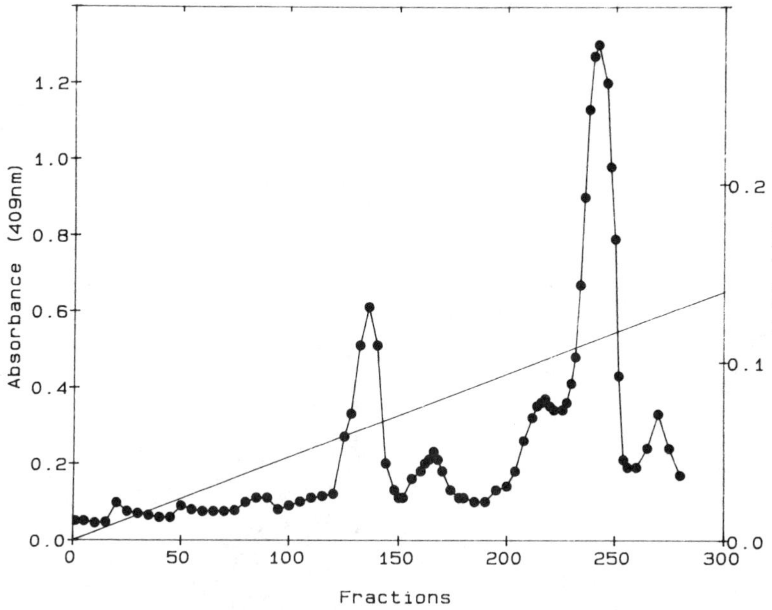

FIG. 2. DEAE-BioGel A profile of sample similar to that shown in Fig. 1. Only absorbance at 409 nm is shown. Sloping line shows NaCl gradient. Fractions of 2 ml were collected.

total activity (accounted for by the six isozymes listed above) recovered from the Mono-Q is usually 50% of the original activity (Table I). The relative amount of each isozyme varies depending on culture additives such as veratryl alcohol and the trace elements solution.[4] Differences are also seen between the stationary and agitated cultures at different harvest times. Consequently, a range of 20 to 40% of the total veratryl alcohol-oxidizing activity can be recovered in the purified H8 fraction.

Under the growth conditions described above, H8 is the predominant ligninase. This isozyme is the enzyme previously purified by Tien and Kirk[5] and most likely the same as that characterized by Gold *et al.*[16] This isozyme is the most extensively studied and characterized. Collecting the H8 peak manually from the Mono-Q column provides a highly purified H8 fraction. Reinjecting the purified peak into the Mono-Q indicates that it is over 98% pure (Fig. 3).[17] Subjecting the purified H8 to SDS-polyacrylamide gel electrophoresis also indicates that the preparation is homogeneous (Fig. 3).

[16] M. H. Gold, M. Kuwahara, A. A. Chiu, and J. K. Glenn, *Arch. Biochem. Biophys.* **234,** 353 (1984).

[17] U. K. Laemmli, *Nature (London)* **227,** 680 (1970).

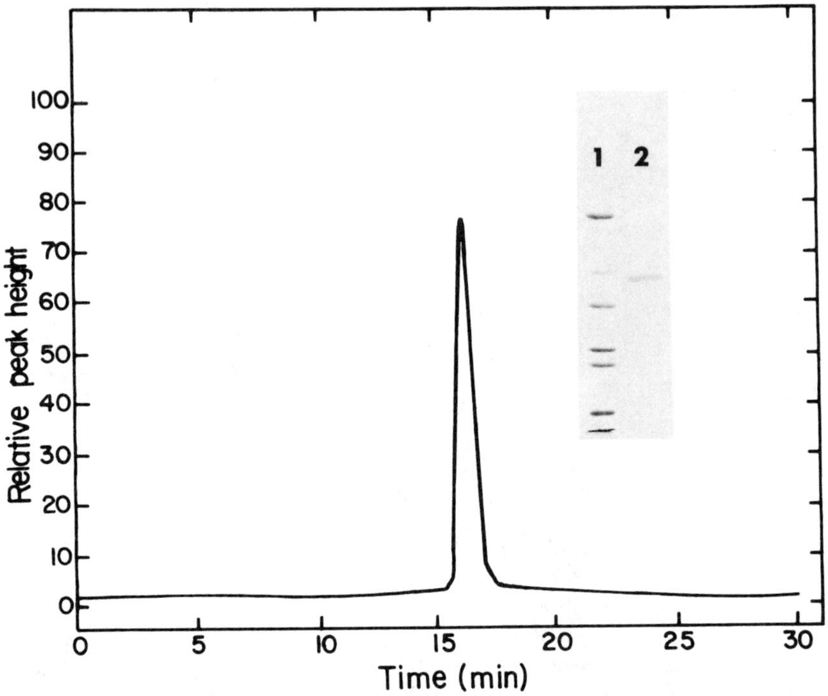

FIG. 3. Purity of H8 as determined by two techniques. A sample of H8, collected from an elution shown in Fig. 1, was rechromatographed on Mono-Q and also subjected to SDS–polyacrylamide gel electrophoresis. Chromatography on Mono-Q was as described in the legend of Fig. 1 except that absorbance at 280 nm was monitored. Electrophoresis was performed with 10% acrylamide by the method of Laemmli.[17] Lane 2 contains 5 μg of ligninase H8. Lane 1 contains molecular weight markers (from top): 66K, 45K, 36K, 29K, 24K, and 14.2K.

Storage

The purified ligninase (H8) from either DEAE-BioGel A or Mono-Q chromatography is then concentrated and dialyzed against 5 m*M* potassium phosphate buffer, pH 6.5. Rapid freezing with liquid nitrogen and storage at −20° or below yields a preparation stable for months.

Properties of the Ligninase Isozymes

The ligninase isozymes are similar in structure and function. Some of their physical properties are summarized in Table II. The molecular weight of the isozymes, as determined by SDS–polyacrylamide gel electrophoresis, vary from 38,000 for H1 and H2 to 46,000 for H10 (Table II). The molecular weight must be considered an upper estimation because the ligninases are all glycoproteins, as demonstrated by their ability to bind to

TABLE II
PHYSICAL PROPERTIES OF LIGNINASE ISOZYMES

Isozyme	Molecular weight[a]	Carbohydrate[b]	ϵ_{409} ($mM^{-1}cm^{-1}$)[b]	Peptide homology[c]
H1	38,000	+	169	H2(++);H7,H8(+)
H2	38,000	+	165	H1(++);H7,H8(+)
H6	43,000	+	162	H10(+);H7,H8(+)
H7	42,000	+	177	H8(++);H1,H2(+)
H8	42,000	+	168	H7(++);H1,H2(+)
H10	46,000	+	182	H6(+);H7,H8(+)

[a] M. Tien, unpublished.
[b] From Farrell *et al.*[18]
[c] From Kirk *et al.*[4]

conconavalin A-Sepharose.[18] The absorption spectrum of the ligninases [exhibiting 409 nm (Soret) absorbance] suggests that they are all heme proteins. This was verified with formation of a diagnostic pyridine hemochromogen complex.[18] The extinction coefficients of the various ligninases, as determined by quantitation of the heme content with the pyridine hemochromogen method, range from 162 $mM^{-1}cm^{-1}$ for H6 to 182 $mM^{-1}cm^{-1}$ for H10.[18]

The isoenzymes are fairly homologous in primary and tertiary structure. Polyclonal antibodies prepared against H8 cross react with ligninase H2, H10, and H8 (to itself), indicating homology between these different enzymes.[4] Analysis of the peptides produced after protease (V8) digestion by electrophoresis on SDS–PAGE gels indicated that H1 and H2 are almost identical.[4] Peptides from H8 are similar to H1 and H2, but lack at least two major peptides.[4] The peptides produced from H6 and H10 are most similar to H7 and H8.[4]

The ligninase isozymes are also similar in their catalytic properties. Results from Farrell *et al.*[18] indicate that the K_m and V_{max} exhibited by the isozymes for H_2O_2 with various aromatic substrates are not significantly different. Based on the lack of significant kinetic properties, it is difficult to ascertain the physiological significance of the multiple ligninases.

Physical and Kinetic Properties of Ligninase H8

Ligninase H8 contains one protophorphyrin IX-derived heme per enzyme molecule and is composed of 15% by weight carbohydrate.[6] The electron absorption spectrum of ligninase H8 is typical of most heme

[18] R. A. Farrell, K. E. Murtagh, M. Tien, M. D. Mozuch, and T. K. Kirk, *J. Biol. Chem.*, submitted.

proteins, showing a Soret peak at 409 nm and visible absorption bands at 498 and 630 nm.[8] The ferric (resting) ligninase forms complexes with cyanide and azide.[8] The ligninase can be reduced with dithionite or deazaflavin; the artificially reduced enzyme complexes with CO, NO and O_2.[8] The reduced enzyme is not involved in catalysis since CO is not an inhibitor.[6]

The ESR spectrum of the ferriligninase shows *g* values at 5.83 and 1.99, indictive of a high-spin ferric heme.[8] Resonance Raman results by Kuila *et al.*[7] describes the heme as most similar to those of peroxidases. Kuila and co-workers based their results on resonance Raman spectrum at the low-frequency range, which showed striking similarities between the ferroligninase and ferro-horseradish peroxidase. Kinetic results[6] are in close accord with the results of Kuila *et al.*,[7] indicating a mechanism similar to other peroxidase.

The kinetics of ligninase catalysis have been studied by both steady state and transient state techniques.[6] The steady state studies of veratryl alcohol oxidation indicate that the mechanism of catalysis is Ping-Pong.[6] Ping-Pong kinetics are consistent with the mechanism of other peroxidases.[19] The initial step in catalysis is the reaction of ligninase with H_2O_2, resulting in formation of an oxidized enzyme intermediate. This intermediate returns to the resting state by oxidizing its aromatic substrates. The productive binding rate for ligninase with H_2O_2 (V/K) is $1.0 \times 10^5\ M^{-1}sec^{-1}$.[6] This rate constant is approximately 100 times lower than that observed with most other peroxidases.

Transient state kinetic studies of the ligninase show that two intermediate states of the enzyme are formed during catalysis. These two states are similar to those formed by other peroxidases; they are the classical intermediates compounds I and II characterized by Chance.[20] Formation of ligninase compounds I and II were detected by stopped flow rapid-scan spectral analysis of the reaction between ligninase and H_2O_2.[6] The initial step in catalysis is the reaction of ligninase with H_2O_2 to form compound I, which is two-electron oxidized. This reaction procedes with a second-order rate constant of $5.8 \times 10^5\ M^{-1}sec^{-1}$, which is in close agreement with the productive binding rate obtained from steady state kinetics. Compound I then reacts with a substrate molecule to form product and the compound II intermediate of ligninase, which is one-electron oxidized. Compound II returns to resting enzyme by reacting with another molecule of substrate.

The catalytic cycle described above, where two molecules of free radical products are formed per turnover, is common for all peroxidases. Formation of free radical products during ligninase catalysis has been demon-

[19] G. L. Kedderis and P. F. Hollenberg, *J. Biol. Chem.* **258,** 12413 (1983).
[20] B. Chance, *Arch. Biochem. Biophys.* **41,** 416 (1952).

strated by ESR spectroscopy. Kersten *et al.*[9] detected the cation radicals of methoxybenzenes by ESR spectroscopy; Hammel *et al.*[10] detected radicals from dimeric model compounds of lignin through ESR spin-trapping techniques. Largely through the results of these two studies, a generalized mechanism for lignin degradation can be formulated. This generalized mechanism involves a central role for substrate aryl cation free radicals. Cation radicals can undergo a wide range of reactions; the type of reactions can be affected by the ring substituents. Substrates with α-hydroxy-containing propyl side chains (prominent in lignin) preferentially undergo carbon–carbon bond cleavage.[10] Methoxybenzenes cation radicals tend to hydrate and demethylate for form formaldehyde and benzoquinones.[9] Because the cation radicals are stable enough to diffuse away from the active site into the "bulk phase," their fate can also be dependent on the components of the bulk phase. Thus the pH, concentration of dioxygen, and concentration of other radicals can affect the addition of H_2O, addition of dioxygen, and dimerization with other radicals (reactions all observed with lignin models).

Much of our understanding of the chemistry of cation radicals has been provided by Snook and Hamilton.[21] These workers studied the formation and degradation of aryl cation radicals in chemical systems. These studies have provided a model for ligninase catalysis; they have indicated that mechanistically, the chemistry of cation radicals accounts for most if not all of the prominent reactions observed in lignin biodegradation. Carbon–carbon bond cleavage of propyl side chains, loss of methoxyls, oxidation of benzylic hydroxyls, and ring opening are mechanically consistent with a free radical mechanism. It is thus apparent that the future utilization of ligninase will require not only an understanding of ligninase catalysis, but also the chemistry of cation radicals.

[21] M. E. Snook and G. A. Hamilton, *J. Am. Chem. Soc.* **96,** 860 (1974).

[24] Lignin-Depolymerizing Activity of *Streptomyces*

By Don L. Crawford and Anthony L. Pometto III

Lignin is a complex phenylpropane polymer consisting of coumaryl, guaiacyl, and syringyl moieties linked together by numerous linkages, but primarily by the β-aryl ether bond.[1] Biodegradation of this recalcitrant

[1] R. L. Crawford, "Lignin Biodegradation and Transformation," Wiley (Interscience), New York, 1981.

polymer by *Streptomyces* is by an oxidative process in which the ether bonds are cleaved and the phenylpropane side chains are modified or removed.[2] Some low-molecular-weight aromatic intermediates are released from lignin as a result of these reactions, but once produced they may or may not be catabolized by specific *Streptomyces*.[2] A principal result of β-aryl ether bond cleavage is the substantial depolymerization of lignin. While *Streptomyces*-mediated depolymerization results in the production of relatively small amounts of single-ring aromatic intermediates, it results in the release of large amounts of a microbially modified water-soluble lignin polymer that is precipitable from aqueous solutions when they are acidified. This polymer has been named acid-precipitable polymeric lignin (APPL).[3] APPLs are the principal initial product of lignin degradation by the ligninolytic *Streptomyces* that have been studied thus far.[4] In this chapter, we describe a simple spectrophotometric assay for monitoring *Streptomyces*-mediated solubilization of lignin when cultures are degrading lignocellulosic substrates in submerged cultures or in solid-state fermentations. We also describe a simplified lignin acidolysis procedure which is important for chemically characterizing APPLs in order to establish their lignin origin, to confirm that β-aryl ether bonds in the original lignin were cleaved, and to demonstrate that lignin depolymerization occurred. Both the spectrophotometric and chemical assays are valuable for use in screening for new lignocellulose-decomposing microorganisms having the ability to depolymerize and simultaneously solubilize lignin.

Cultivation of *Streptomyces* on Lignocellulose

The preparation of the lignocellulose substrate and of log phase cells to be used as inoculum in solid-state fermentation and submerged culture systems is described in Chapter [5] in this volume. In this discussion, solid-state fermentations are described for *Streptomyces viridosporus* T7A (ATCC No. 39115) and a submerged culture system is described for *Streptomyces badius* 252 (ATCC No. 39117). These *Streptomyces* are examples of the two known types of lignin-depolymerizing actinomycetes,[4] either of which might be encountered when using the lignin-depolymerization assay.

Solid-State Fermentation

By following the procedures outlined in Chapter [5], 5 ml of late log phase cells of *S. viridosporus* is inoculated into 0.5 g of sterile ground and

[2] D. L. Crawford and R. L. Crawford, *Enzyme Microb. Technol.* **2,** 11 (1980).
[3] D. L. Crawford, A. L. Pometto III, and R. L. Crawford, *Appl. Environ. Microbiol.* **45,** 898 (1983).
[4] J. R. Borgmeyer and D. L. Crawford, *Appl. Environ. Micrbiol.* **49,** 273 (1985).

extracted corn *(Zea mays)* lignocellulose, and incubations are carried out in cotton-plugged 250-ml flasks. Incubation is at 37° in a humid incubator. Sufficient replicates are incubated so that at least three flasks can be harvested at each time interval to be examined, and a similar number of controls containing sterile lignocellulose moistened with only sterile medium are incubated as well. The number of samplings per experiment will vary with the metabolic rate of the organism being studied; however, in initial experiments flasks should probably be harvested in replicates of three at 4-day intervals. After inoculation, time zero inoculated and control flasks are retained for immediate assay, and the remaining flasks are placed into the incubator. To prevent moisture loss flasks can be placed in a plastic bag along with a beaker of water to ensure that a humid environment is maintained. At periodic time intervals, three inoculated and three control flasks are removed for assay. Each flask is assayed for its content of polymeric water-soluble lignin (APPL), using the turbidometric procedure described below. Results are reported as averages of three replicate values with standard deviations.

Submerged Culture System

Following the procedures described in Chapter [5] in this volume, 50 ml of late log phase cells of *S. badius* is inoculated into 5 g of sterile ground and extracted corn lignocellulose in a 2-liter flask, and the culture is preincubated as a solid-state fermentation at 37° in a humid incubator for 3 to 4 days. At the end of this period, 1.0 liter of additional liquid medium [0.6% (w/v) yeast extract in mineral salts] is added. This medium consists of 6.0 g of yeast extract (Difco Laboratories, Detroit, MI), 5.3 g Na_2HPO_4, 1.98 g KH_2PO_4, 0.2 g $MgSO_4 \cdot 7H_2O$, 0.2 g NaCl, 0.05 g $CaCl_2 \cdot 2H_2O$, plus 1.0 ml of trace element solution (6.4 g $CuSO_4 \cdot 5H_2O$, 1.1 g $FeSO_4 \cdot 7H_2O$, 7.9 g $MnCl_2 \cdot 4H_2O$, 1.5 g $ZnSO_4 \cdot 7H_2O$/liter of water)[5] per liter of deionized H_2O, adjusted to pH 7.1–7.2. After addition of the liquid medium, the culture is shifted to shaking incubation at 100 rpm and 37°, and cultures are incubated in replicates of three. Three uninoculated controls prepared using only sterile medium in place of cell suspension are also incubated, and these controls are carried through the same sequence of manipulations as were the inoculated cultures. A zero time 10-ml sample of culture medium is taken aseptically at the time that the liter of medium is added to each of the solid-state fermentation flasks. Additional 10-ml samples are taken at 2-day intervals. Each sample is then assayed for soluble lignin (APPL) by the turbidometric procedure described below, and results are reported as averages of three replicate values with standard deviations.

[5] T. G. Pridham and G. Gottlieb, *J. Bacteriol.* **56**, 107 (1948).

Turbidometric Assay for Water-Soluble Polymeric Lignin Fragments

Solid-State Fermentation

To each solid-state fermentation flask to be assayed, 125 ml of distilled water is added. The flask is then placed in a boiling water bath or steamer for 1 hr, during which time it is shaken periodically. Next, the suspension is filtered through preweighed filter paper (Whatman No. 54) to recover the residual lignocellulose/biomass for substrate weight loss determinations and later chemical analysis of the lignocellulosic residue (see below). For spectrophotometric assay of APPL, a 2.0-ml portion of the filtrate is pipetted into a cuvette, 0.1 ml of concentrated HCl is added to acidify the sample, and the solution is thoroughly mixed. This acidified sample is allowed to stand at least 1 hr at 4°, over which time any APPL or other acid-precipitable material present will precipitate. The sample is then remixed, and its optical density (OD) is determined at 600 nm against a water blank. Several repeat readings are taken to minimize variation of data due to the settling of the precipitate during measurement. An average OD reading is then calculated. When a reading is taken, it is also advisable to calculate average values based upon three to four separate 2.0-ml samples. By periodically harvesting flasks and performing turbidometric assays as described, a turbidity versus time plot can be drawn, and production of APPL by the culture over time can be monitored.

Submerged Culture System

The assay for lignin solubilization by cells growing as submerged cultures is simpler than for those growing as solid-state fermentations since the cultures do not have to be subjected to the hot water extraction procedure. Ten-milliliter samples of culture supernatant are aseptically withdrawn from the incubating culture at 2-day intervals, and they are filtered to remove any insoluble material. The filtrate is then assayed for solubilized lignin turbidometrically at 600 nm as above.

Correction for Protein in the Precipitates

The only disadvantage of relying solely on the turbidometric assay for following APPL production by *Streptomyces* is that extracellular proteins may also precipitate when samples are acidified, and thus inflate the APPL values.[6] This problem seems to be more evident in the submerged culture

[6] A. L. Pometto III and D. L. Crawford, *Appl. Environ. Microbiol.* **51,** 171 (1986).

system than in the solid-state fermentation system. It is relatively easy to correct for protein contamination by assaying air-dried APPL precipitates for crude protein content by the Kjeldahl procedure (see below and Chapter [5] in this volume).

Confirmation of Lignin-Solubilizing Activity

If a specific microbial strain grows well on lignocellulose in solid-state or submerged culture and produces increasing amounts of APPL-like material over time as shown by the turbidometric assay, then it can be considered tentatively as a lignin-depolymerizing microorganism. The acid-precipitable material measured by the turbidometric assay must, however, be examined to confirm that it is lignin in origin and that a correlation exists such that the appearance of solubilized lignin corresponds to a similar pattern of loss of insoluble lignin from the lignocellulose substrate (see Chapter [5] in this volume).[4]

Confirmation for Solid-State Fermentations

It is desirable first to quantify the exact amount of APPL present in filtrates taken from the different-aged cultures discussed above. When each flask is harvested, the total volume of filtrate is determined, and then the filtrate is acidified to pH 1–2 with concentrated HCl. The precipitate that forms is collected by centrifugation (16,000 *g*) in preweighed centrifuge bottles. The supernatants are discarded. Each precipitate is washed once with acidic water, and then it is air dried at 50° for 24 to 48 hr. The weight of the total recovered APPL is next determined to ± 0.1 mg accuracy, and with the total volume removed for turbidometric assay taken into consideration, the total milligram weight of APPL recovered from the culture is calculated. After the dry weight of APPL is determined, the Klason lignin, carbohydrate, protein, and ash contents of the APPL are determined (see Chapter [5]). From these data it is possible to correlate the actual weight of lignin-derived APPL with the OD_{600} readings, and thereby generate a standard APPL curve for the organism under study. The standard curve is presented as a graph of OD_{600} against mg/ml APPL values. One should also calculate the percentage weight loss of lignocellulosic substrate at each time interval by weighing the air-dried residues recovered at each filtration step. The crude protein, lignin, and carbohydrate contents of the insoluble residues can also be determined so that cell mass increases and lignin and carbohydrate depletion from the lignocellulosic substrate can be calculated and compared with the rate and extent of APPL production by specific ligninolytic *Streptomyces* (see Chapter [5] in this volume).

Submerged Culture System

Once the time course of apparent lignin solubilization has been determined, tentatively identified lignin-solubilizing cultures are again grown on lignocellulose as described above. However, this time the cultures should be grown in 250-ml flasks containing 0.5 g of lignocellulose, and three replicate flasks should be harvested at each time interval. Sterile uninoculated controls should be incubated in the same fashion. Preincubation as solid-state fermentations is used as before, which in this case requires an inoculum of 5 ml of late log cells (for *S. badius*), followed by 3–4 days of preincubation at 37°. Then 100 ml of the 0.6% (w/v) yeast extract–mineral salts medium is added prior to aerobic shaking incubation at 100 rpm and 37°. At the time of initiating the submerged culture portion of the incubation and at 2- to 3-day intervals thereafter, three flasks each from the inoculated cultures and the uninoculated sterile controls are harvested. The flasks are placed in a boiling water bath or steamer for 1 hr, and the residual lignocelluloses are recovered by filtration through preweighed filter paper. The residues are air dried and weighed as described above so that the rate of lignocellulose weight loss over time can be calculated. Each residue is also chemically analyzed for its crude protein, lignin, and carbohydrate contents (see Chapter [5] in this volume) so that increases in actinomycete cell mass and lignin and carbohydrate depletion rates can be calculated. The filtrates are transferred into preweighed centrifuge bottles, acidified to pH 1–2 with concentrated HCl, and the resulting precipitates are collected by centrifugation (16,000 *g*). After discarding the supernatants, the pellets are washed once with acidic water and air dried at 50° for 24 to 48 hr. The weight of each precipitate is determined to ±0.1 mg accuracy, and then the Klason lignin, carbohydrate, protein, and ash contents of each APPL are determined (see Chapter [5] in this volume). Using the data obtained from all of these analyses, a mg APPL versus 600 nm standard curve (corrected for protein contamination) can be constructed, and an APPL versus time production curve can be drawn and compared with curves showing the rates of lignin and carbohydrate depletion and the rate of actinomycete cell mass increase over time.

Additional Comments

In all cases, APPLs derived from lignin should contain a high percentage (60–80%) of acid-insoluble component as measured by the Klason lignin assay, and they should also contain some carbohydrate (5–10%), mostly in the form of lignin-associated hemicelluloses still complexed with the solubilized lignin.[4] In addition, any APPL recovered should be offset by a similar or greater loss of lignin from the lignocellulosic substrate.[3,4] However, there is one problem that may be encountered with certain ligninolytic *Streptomyces*. If an organism produces a highly modified lignin-derived APPL, that APPL may not assay as high in lignin content as expected because it may contain considerable amounts of acid-soluble lignin.[4] To conclusively prove the lignin origin of the APPL, chemical characterization of the lignin by acidolysis, permanganate oxidation, and ester hydrolysis is required (see Chapter [5] in this volume).

Confirmation of β-Ether Linkage Cleavage Activity

Another procedure for confirming lignin depolymerization is acidolysis, a chemical degradative procedure that shows whether or not there is a lower number of β-aryl ether bonds in degraded lignins as compared to the native lignin from which they were derived.[3,4] Conclusions are drawn from quantitative data on the amounts of two key acidolysis products. The amount of one product, phenylpropane Ketol I, should be lower in depolymerized lignins as compared to native lignin, while the amount of a second product, vanillic acid, should be greater. This procedure requires a sample of purified native lignin and degraded lignin from the lignocellulose substrate being used. Minimally, 5 g corn lignocellulose (solid-state or submerged culture fermentations) is incubated as described above for a sufficient length of time to ensure that substantial lignin degradation has occurred (4–6 weeks). The partially degraded lignocellulose residues and APPLs are then recovered as described above. Purified milled corn lignins (MCL) are prepared from the degraded lignocellulose and from an undegraded control lignocellulose using the neutral solvent extraction procedure of Björkman[7] (see Chapter [3] in this volume). Then the APPL and the purified lignins are chemically degraded using the simplified acidolysis procedure of Pometto and Crawford[8] and the yields of Ketol I and vanillic acid obtained from each lignin and APPL are then compared. The results for the control MCL are considered as native lignin baseline values, and the yield of Ketol I should be considerably higher than the yield of vanillic acid. The ratio of Ketol I to vanillic acid should, however, decrease for degraded lignins and APPLs. For example, values reported previously for control MCLs and for APPLs resulting from lignin depolymerization by *S. viridosporus* are 5.4/0.7% and 2.9/2.2%, respectively (Ketol I/vanillic acid yields, as a percentage of the lignin or APPL subjected to acidolysis).[3]

[7] A. Björkman, *Sven. Papperstidn.* **59**, 477 (1956).
[8] A. L. Pometto III and D. L. Crawford, *Appl. Environ. Microbiol.* **49**, 879 (1985).

Acidolysis

In this procedure, APPLs and Björkman lignins from decayed and control lignocelluloses are subjected to a 6-hr hydrolysis in acidic dioxane.[8] Hydrolyzed samples are then solvent extracted to recover single-ring acidolysis products which are in turn identified and quantified by gas–liquid chromatography (GLC) using procedures described in Chapter [16] of this volume. The work-up procedure is tedious and requires attention to detail. A solution of 15 mg of Björkman lignin or APPL/ml of 0.2 *M* HCl in dioxane–water (9 : 1) is placed in a glass ampoule, which is then cooled in an ice bath, flushed with nitrogen, and flame sealed. The sealed ampoule is placed into a heating block at 87° for 6 hr. Glycerol is added to the block wells to promote more even heat transfer. After cooling, the ampoule is opened, diluted with water to give a final dioxane/water ratio of 1 : 1, and then quantitatively transferred to a separatory funnel using acidic dioxane–water (1 : 1) washes to ensure complete transfer. The acidolysis mixture is extracted four times with chloroform–acetone (1 : 1) and then once with chloroform only. Combined extracts are dried under a vacuum at 45–50°, then the dry extracts are again extracted, this time with dioxane–chloroform (1 : 1), until extracts are colorless. The combined extracts are transferred into a preweighed beaker and evaporated to dryness in a hood at room temperature. Then the weight of recovered solids is determined. These residues are dissolved in ethyl acetate–ethanol 95% (1 : 1) to a final concentration of 10 mg/ml. A 0.3-ml (=3 mg) sample is pipetted into a dry (predesiccated), preweighed vial, and the solution is evaporated to dryness in a hood. Next the vial is desiccated for 72 hr under nitrogen to avoid oxidation of any acidolysis products. The weight of the residue is then determined to ±0.1 mg accuracy. The products and known standards are next converted to their trimethylsilyl derivatives and each is then quantified by GLC (see Chapter [16] in this volume).

Other Comments

Twelve important single-ring aromatic acidolysis products are typically recovered from softwood lignins after acidolysis.[8,9] Similar products, some with different ring substitution patterns, are recovered from hardwood and grass lignins.[1] The most important and most dominant product found in nondegraded Björkman lignin is Ketol I (1-hydroxy-3-[4-hydroxy-3-methoxyphenyl]-2-propanone), which is produced only when the β-aryl ether bond of the polymer is intact at the time of acidolysis. When extensive depolymerization has occurred, vanillic acid (4-hydroxy-3-methoxyben-

[9] K. Lundquist and T. K. Kirk, *Acta Chem. Scand.* **25**, 889 (1971).

zoic acid) typically becomes a more dominant product, and the yield of Ketol I decreases. Lignins and APPLs from degraded lignocelluloses will often have a Ketol I/vanillic acid ratio of less than 1 while the corresponding native lignin will have a ratio greater than 1.[3,4] For example, with softwood lignins from spruce *(Picea pungens)* the ratio from undegraded Björkman lignin was reported to be 1.3 (4.2% Ketol I/3.2% vanillic acid), whereas the ratio from a lignin derived from *S. viridosporus*-degraded spruce lignocellulose was 0.5 (2.4% Ketol I/5.1% vanillic acid).[10] Similarly, the ratio of Ketol I to vanillic acid for an *S. viridosporus* APPL derived from corn lignocellulose was 1.3 (2.9% Ketol I/2.2% vanillic acid) as compared to a ratio of 3.7 (5.2% Ketol I/1.4% vanillic acid) for the control MCL.[3] These results demonstrated that the APPL, though modified, was definitely derived from a true lignin polymer containing β-aryl ether bonds. On the other hand, acidolysis can sometimes give unexpected results. For example, acidolysis of an *S. badius* APPL produced an acidolysis product mixture that was distinctly nonligninlike (yielding no Ketol I or vanillic acid),[4] an indication that this APPL was either not a lignin-derived product, or else it was so extensively modified by this actinomycete that it no longer resembled lignin.

Assaying Previously Unstudied Organisms for Lignin-Depolymerizing/Solubilizing Activity

The turbidometric assay for determination of solubilized polymeric lignin fragments can be utilized to screen a wide variety of microorganisms for lignin-depolymerizing ability. In general the assay is employed in the screening of cultures isolated previously by selection on lignocellulose-containing media, or selected based upon the ability to mineralize [^{14}C]lignin-labeled lignocelluloses[1] (see Chapter [3] on [^{14}C]lignin degradation). In particular, the turbidometric assay will be useful for examining filamentous actinomycetes and fungi because these microbes do not produce turbidity since they grow filamentously. However, it is possible to study nonfilamentous bacteria if assay samples are centrifuged to remove the bacterial cells prior to the acidification step that precipitates the APPLs. Whenever a new organism is found to solubilize lignin, it will also be necessary to chemically characterize the APPL-like product to confirm its lignin origin, and to confirm a corresponding loss of lignin from the lignocellulose substrate.

With the *Streptomyces* thus far studied, APPLs appear to be essentially

[10] D. L. Crawford, M. J. Barder, A. L. Pometto III, and R. L. Crawford, *Arch. Microbiol.* **131,** 140 (1982).

a terminal product of lignin metabolism, or at most an intermediate that is only slowly metabolized further.[4,6] However, some as yet undiscovered organisms may produce APPLs and then metabolize them rapidly. With these organisms, APPLs would be a transitory intermediate and would likely not accumulate in amounts equivalent to the lignin lost from the lignocellulose. For example, the white rot fungus *Phanerochaete chrysosporium* is capable of completely degrading lignin to CO_2 and H_2O, but it produces only small amounts of an APPL-like intermediate when it is growing on lignocellulose.[3,11] The chemistry of this product has not been extensively studied. Another factor to be considered is that the enzymatic mechanism for APPL production may be different for different microorganisms. For example, while the extracellular enzymes involved in the initial oxidation of lignin by *P. chrysosporium* have now been identified,[11] the enzymes responsible for APPL release by *Streptomyces* remain to be discovered and are probably quite different from the ligninases of *P. chrysosporium*.[12] This complexity of variable must always be considered when new cultures are being examined for ligninolytic activities.

[11] R. L. Crawford and D. L. Crawford, *Enzyme Microb. Technol.* **6,** 434 (1984).

[12] D. L. Crawford, A. L. Pometto III, and L. A. Deobald, "Recent Advances in Lignin Biodegradation Research," pp. 78–95. Uni Publ., Tokyo, 1983.

[25] Manganese Peroxidase from *Phanerochaete chrysosporium*

By MICHAEL H. GOLD and JEFFREY K. GLENN

$$2Mn(II) + H_2O_2 + 2H^+ \rightarrow 2Mn(III) + 2H_2O$$

Principle. Mn(II) peroxidase is an extracellular enzyme expressed during secondary metabolism as part of the lignin-degradative system of *Phanerochaete chrysosporium.* Mn(II) peroxidase oxidizes Mn(II) to Mn(III). Mn(III) is a nonspecific oxidant which in turn oxidizes a variety of organic compounds.[1–3]

[1] M. Kuwahara, J. K. Glenn, M. A. Morgan, and M. H. Gold, *FEBS Lett.* **169,** 247 (1984).

[2] J. K. Glenn and M. H. Gold, *Arch. Biochem. Biophys.* **242,** 329 (1985).

[3] A. Paszczyński, V.-B. Huynh, and R. Crawford, *Arch. Biochem. Biophys.* **244,** 750 (1986).

Assay Method

1. Oxidation of ABTS [Diammonium 2,2′-Azinobis(3-ethyl-6-benzothiazoline Sulfonate)]

Reagents

A. Sodium succinate, sodium lactate buffer, each at 100 mM, pH 4.5, containing 6 mg/ml egg albumin, $MnSO_4$, 200 μM, and ABTS, 80 μg/ml

B. H_2O_2, 100 μM, in H_2O

Reaction mixture, 1 ml, consists of 500 μl of reagent A and 500 μl of B. The reaction is initiated at room temperature by the addition of 75 ng of enzyme (1 – 10 μl) and the initial rate of ABTS oxidation is determined spectrophotometrically by following the increase in absorbance at 415 nm.

2. Oxidation of Mn(II)

Reagents

A. Sodium lactate buffer, 100 mM, pH 4.5, containing $MnSO_4$, 100 μM

B. H_2O_2, 100 μM, in H_2O

Reaction mixture, 1 ml, consists of 500 μl of reagent A and 500 μl of B. The reaction is initiated at room temperature by the addition of 100 ng of enzyme (1 – 15 μl) and the initial rate of Mn(III) lactate formation is determined spectrophotometrically by following the initial increase in absorbance at 240 nm.

Purification Procedure

The method given here is essentially that described previously.[2,4]

Culture Conditions

Mn(II) peroxidase can be isolated from any wild-type strain of *P. chrysosporium*. However, we used a strain, OGC101, which was isolated in our laboratory as previously described.[4] Vegetative cultures are maintained on slants of Vogel medium N supplemented with 3% malt extract and 0.25% yeast extract.[5] Conidia are washed from slants, filtered, and diluted in distilled water. Stationary cultures (250 ml) inoculated with conidia are

[4] M. H. Gold, M. Kuwahara, A. A. Chiu, and J. K. Glenn, *Arch. Biochem. Biophys.* **234,** 353 (1984).

[5] M. H. Gold and T. M. Cheng, *Appl. Environ. Microbiol.* **35,** 1223 (1978).

as previously described[4,6] and incubated for 2 days. The cultures are homogenized (20 sec) and used to inoculate a 2-liter flask containing 1 liter of medium (cf. Ref. 6) except that 20 m*M* acetate replaces the dimethyl succinate buffer and the $MnSO_4$ concentration is increased to 180 μM. The flasks are incubated on a shaker at 28° under air for 3 days, after which they are purged with 100% O_2 at 24-hr intervals.[4] After 5 days, cultures are filtered through glass wool, and the filtrate is stored at 4°.

Purification

Step 1. Acetone (−10°) is added to the culture filtrate to 25% (v/v), after which an insoluble slime material is removed. The acetone concentration is raised to 66% (v/v), and the precipitated protein is centrifuged and resuspended in 20 m*M* sodium succinate, pH 4.5 (buffer A).

Step 2. The resuspended protein is applied to a DEAE-Sepharose column (1.5 × 15 cm) equilibrated with buffer A and the column is washed with buffer A. The Mn(II) peroxidase activity does not bind to the DEAE-Sepharose column under these conditions. A linear salt gradient is then used to elute the lignin peroxidase activity from the column.[4]

Step 3. The breakthrough fractions are pooled, concentrated, and washed by membrane ultrafiltration using 50 m*M* sodium succinate, pH 4.5 (buffer B). All of the manganese peroxidase is adsorbed to a column of Reactive Blue 2–crosslinked agarose (0.8 × 18 cm) equilibrated with buffer B. The column is washed with buffer B and the enzyme is eluted at pH 4.5 with a linear gradient consisting of buffer B in the mixing chamber and buffer B + 400 m*M* NaCl in the reservoir. One major peak of activity (Fig. 1) is pooled and concentrated.

Step 4. This concentrated Blue Agarose eluate is applied to a Sephadex G-100 column (1.7 × 50 cm) previously equilibrated with buffer B + 100 m*M* NaCl. The results in Fig. 2 show the Sephadex G-100 elution profile for the Mn(II) peroxidase. The enzyme elutes as a single symmetrical peak. The yield of the purified enzyme is 25% with respect to the crude filtrate, with a 12-fold increase in specific activity. However, the amount of enzyme in the culture filtrate of individual batches varies considerably. The manganese peroxidase retains full activity for at least 3 months when stored at −20° as a concentrated solution (1 mg/ml).

Properties of the Enzyme

Homogeneity and Molecular Mass of the Extracellular Manganese Peroxidase. When the purified enzyme is subjected to SDS-PAGE only

[6] T. K. Kirk, E. Schultz, W. J. Connors, L. F. Lorenz, and J. G. Zeikus, *Arch. Microbiol.* **117,** 227 (1978).

one band of protein is detected (Fig. 2, inset). A plot of the mobilities of standard proteins versus log of molecular mass gives an M_r for the Mn(II) peroxidase of ~46,000.

Spectral Properties of the Manganese Peroxidase. From the pyridine hemochrome spectrum the iron protoporphyrin IX content of manganese peroxidase is calculated to be 0.94 mol/mol of enzyme. The absorption spectra of the native enzyme and manganese peroxidase compound II are shown in Fig. 3.

Mn(II) Dependence. In its oxidation of all organic substrates, the Mn(II) peroxidase is totally dependent on Mn(II). No other metal will

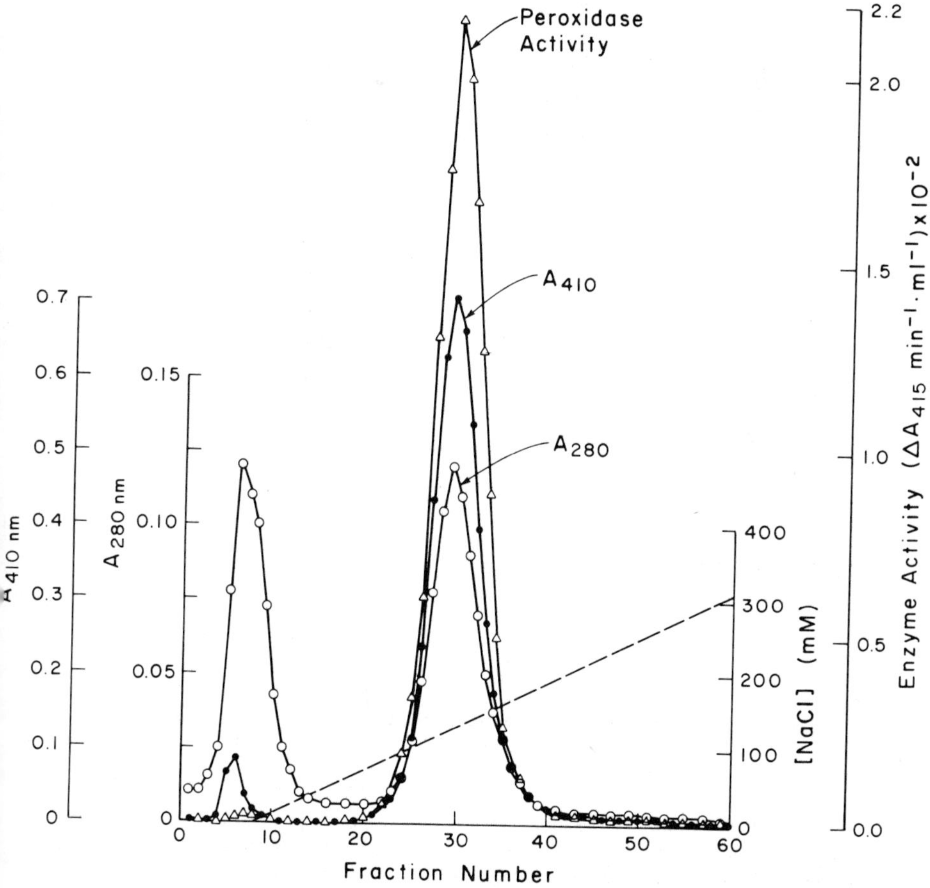

FIG. 1. Chromatography of the manganese peroxidase on Blue Agarose. Procedures are described in the text and in Ref. 2. Absorbance at 410 nm (●); absorbance at 280 (○); peroxidase activity as measured by ABTS oxidation (△).

substitute. Furthermore, the manganese peroxidase oxidizes Mn(II) to form Mn(III).[2] The product is measured by the characteristic spectrum of its complex with pyrophosphate[2] or lactate. The K_m for Mn(II) is approximately 80 μM. At equivalent concentrations only Mn(II) will reduce the

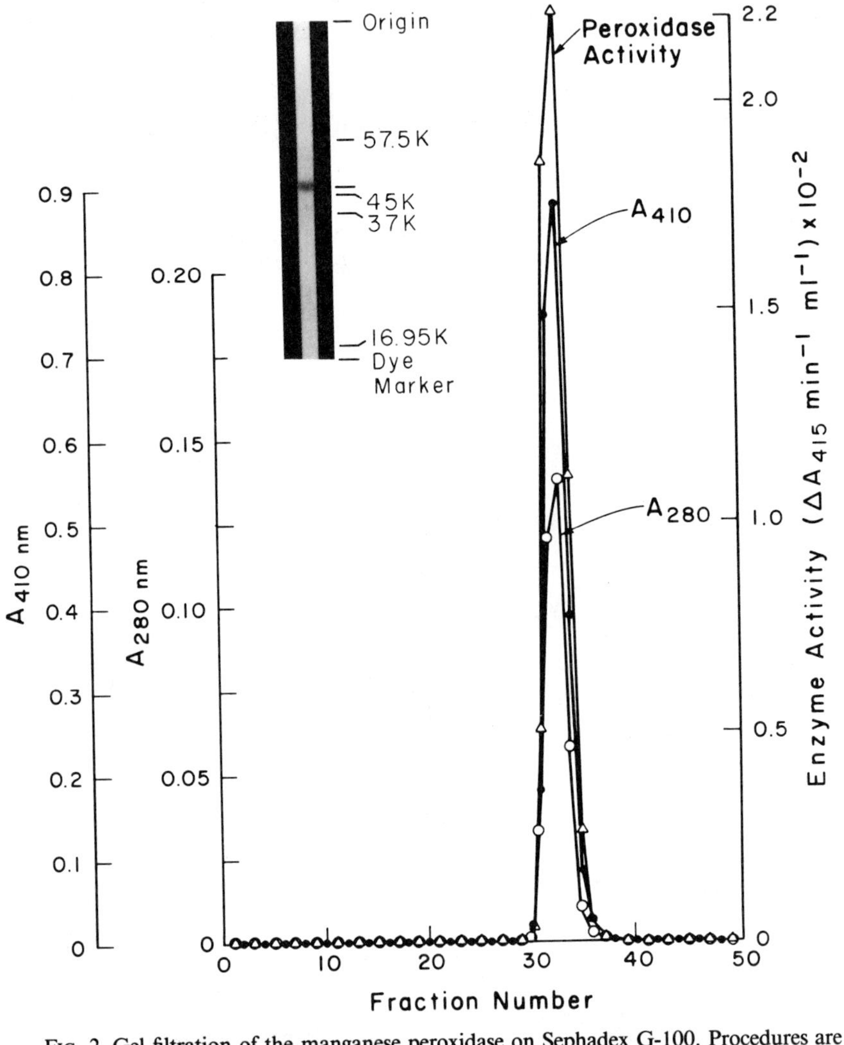

FIG. 2. Gel filtration of the manganese peroxidase on Sephadex G-100. Procedures are described in the text and in Ref. 2. Absorbance at 410 nm (●), absorbance at 280 nm (○), and peroxidase activity (△). *Inset:* SDS–PAGE of the purified enzyme. The gel was stained with Coomassie blue R-250. The following marker proteins were electrophoresed on a separate gel: catalase, 57.5K; ovalbumin, 45K; glyceraldehyde-3-phosphate dehydrogenase, 37K; and myoglobin, 16.95K.

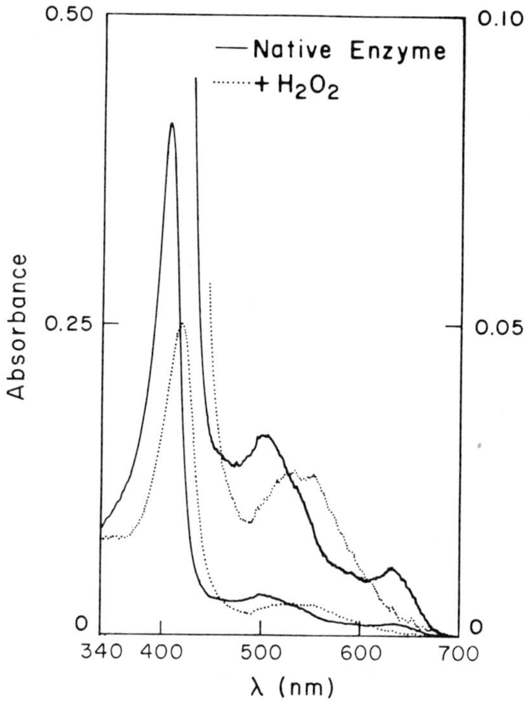

FIG. 3. Electronic absorption spectra of native manganese peroxidase (——) and manganese peroxidase compound II[2] (· · ·).

manganese peroxidase compound II back to the native enzyme. Finally, Mn(III) lactate prepared by the addition of Mn(III) acetate to lactate solutions will oxidize all of the organic substrates oxidized by the enzyme. As shown in the scheme below, all of these results indicate that the H_2O_2-oxidized enzyme oxidizes Mn(II) to Mn(III), which in turn oxidizes the substrates.

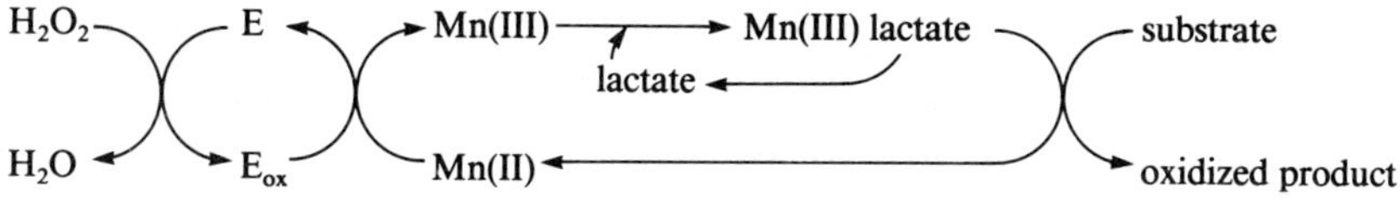

SCHEME 1.

Dependence on α-Hydroxy Acids and Protein. ABTS oxidation by the peroxidase is dependent on α-hydroxy acids such as lactate, malate, tartrate, and citrate. These compounds probably chelate the Mn(III), leading to its stabilization without reducing its reduction potential.[7] Bulk protein

[7] W. A. Waters and J. S. Littler, *in* "Oxidation in Organic Chemistry" (K. B. Wiberg, ed.). Academic Press, New York, 1985.

also stimulates the manganese peroxidase to varying degrees, depending on the substrate used.[2]

Substrate Specificity. In the presence of H_2O_2 the enzyme oxidizes polymeric and other dyes, lignin model compounds, and various phenols.[2,3] Mn(III) is capable of oxidizing all of the organic substrates which are oxidized by the enzyme system.

In the absence of exogenous H_2O_2 the enzyme also acts as an NAD(P)H oxidase[2,3] generating H_2O_2. This reaction can be coupled to the oxidation of ABTS in the absence of exogenous H_2O_2, suggesting that the manganese peroxidase may play a role in H_2O_2 production by the fungus under ligninolytic conditions.[2,3]

[26] Manganese Peroxidase of *Phanerochaete chrysosporium*: Purification[1]

By ANDRZEJ PASZCZYŃSKI, RONALD L. CRAWFORD, and VAN-BA HUYNH

Two types of extracellular peroxidases have been discovered in the growth medium of ligninolytic cultures of white rot fungus *Phanerochaete chrysosporium*.[2,3] One type has been termed "ligninase"[4] or "diarylpropane oxygenase,"[5] and appears to oxidize lignin and many lignin model compounds by extracting an electron from an aromatic nucleus, creating an unstable cation radical species which undergoes numerous degradative transformations leading to substrate decomposition.[6] A second type of peroxidase excreted by *P. chrysosporium* has been called manganese peroxidase, as it oxidizes Mn(II) to Mn(III).[7,8] This enzyme also shows oxidase activity, producing hydrogen peroxide by oxidation of reduced substrates like NAD(P)H, glutathione (GSH), dithiothreitol (DTE), and

[1] This work was performed at the University of Minnesota, Gray Freshwater Biological Institute, Navarre, Minnesota 55392 under National Science Foundation Grant PCM-8318151.

[2] M. Tien and T. K. Kirk, *Science* **221,** 661 (1983).

[3] M. Kuwahara, J. K. Glenn, M. A. Morgan, and M. H. Gold, *FEBS Lett.* **169,** 247 (1984).

[4] M. Tien and T. K. Kirk, *Proc. Natl. Acad. Sci. U.S.A.* **81,** 2280 (1984).

[5] J. K. Glenn and M. H. Gold, *Arch. Biochem. Biophys.* **242,** 329 (1985).

[6] P. J. Kersten, M. Tien, B. Kalyanaraman, and T. K. Kirk, *J. Biol. Chem.* **260,** 2609 (1985).

[7] A. Paszczyński, V. B. Huynh, and R. Crawford, *FEMS Microbiol. Lett.* **29,** 37 (1985).

[8] J. K. Glenn and M. H. Gold, *Arch. Biochem. Biophys.* **242,** 329 (1985).

dihydroxymaleic acid.[7,9] Both Mn(III) and hydrogen peroxide produced by the enzyme may be involved in lignin biodegradation.[9] Here we outline a method for purification of the manganese peroxidase of *P. chrysosporium*.

Enzyme Production

Phanerochaete chrysosporium (we employed strain BKM-1767, ATCC 24725) was maintained and grown as described in previous work.[7] Growth was in a defined medium at pH 4.5 on glucose in the presence of growth-limiting amounts of nitrogen to ensure that the fungus produced its ligninolytic system.[10] One liter of medium was prepared by mixing 100 ml of 10× concentrated basal medium (L-asparagine, 1 g; NH_4NO_3, 0.5 g; KH_2PO_4, 2 g; $MgSO_4 \cdot 7H_2O$, 0.5 g; $CaCl_2 \cdot 2H_2O$, 0.1 g; thiamin, 0.001 g; mineral elixer, 10 ml; distilled water to 1 liter; filter sterilized), 100 ml of 0.1 *M* sodium dimethylsuccinate (pH 4.5, filter sterilized), 100 ml of a *P. chrysosporium* conidiospore suspension ($A_{650} \approx 0.5$), and 600 ml of autoclaved distilled water. The mineral elixer used in the basal medium contained (in g/liter distilled water): nitrilotriacetic acid, 1.5; $MgSO_4$, 3.0; $MnSO_4$, 0.5; NaCl, 1.0; $FeSO_4$, 0.1; $CaCl_2$, 0.1; $CoCl_2$, 0.1; $ZnSO_4$, 0.1; $CuSO_4$, 0.01; $AlK(SO_4)_2$, 0.01; H_3BO_3, 0.01; and $NaMoO_4$, 0.01. Fungal mycelium was grown attached to the roughened interior walls of a 20-liter polyethylene carboy containing 1 liter of medium and rotated at about 0.2 rpm. Every second day the carboy was filled with pure oxygen by passing O_2 (filter sterilized) through a tube passed through the carboy's screw-top cap. Two additional tubes passed through the cap: an oxygen outlet (equipped with a bacteriological filter) and a sampling port. All three ports were clamped closed during incubations, which were done at 40°. One day prior to harvesting the culture fluid for isolation of enzymes, veratryl alcohol (0.4 m*M*) and Tween 80 (20 mg/liter) were added to the medium to increase enzyme production.[11]

Alternatively, investigators may use the above medium in agitated submerged cultures if Tween 80 (0.1%), Tween 20 (0.05%), or 3-[(3-cholamidopropyl)dimethylammonio]-1-propane sulfonate (0.05%) is added to overcome the previously known necessity to grow ligninolytic *P. chrysosporium* in stationary culture.[11] Medium (600–1000 ml) is placed in a 2-liter Erlenmeyer flask, inoculated with conidia, and shaken at 150–200

[9] A. Paszczyński, V. B. Huynh, and R. Crawford, *Arch. Biochem. Biophys.* **244,** 750 (1986).

[10] K. Kirk, F. Shultz, W. J. Connors, L. F. Lorenz, and J. G. Zeikus, *Arch. Microbiol.* **117,** 277 (1978).

[11] A. Jäger, S. Croan, and T. K. Kirk, *Appl. Environ. Microbiol.* **50,** 1274 (1985).

rpm at 40°. Development of ligninolytic activity is idiophasic,[10,11] as in stationary cultures. Addition of veratryl alcohol (0.4 mM) still enhances enzyme yields.

Assay Methods for Manganese

Manganese peroxidase may be assayed using a variety of aromatic substrates, particularly those that are employed for assays of common peroxidases such as horseradish peroxidase. Reaction mixtures, however, must be supplemented with Mn(II) ions. Table I lists some useable substrates along with their extinction coefficients at wavelengths to be monitored during the peroxidatic reaction.[7]

Reagents

A. 0.5 M sodium tartrate buffer, pH 5.0
B. 1 mM substrate in water (water-soluble compounds) or 50% aqueous N',N-dimethylformamide (water-insoluble compounds)
C. 1 mM $MnSO_4$
D. 1 mM H_2O_2
E. Enzyme solution containing about 0.2 U ml^{-1}

Assay Procedure

One milliliter of assay solution contains about 0.02 U of peroxidase, 0.1 M sodium tartrate (pH 5.0), 0.1 mM substrate (Table I), 0.1 mM

TABLE I
ASSAY SUBSTRATES FOR MANGANESE PEROXIDASE

Substrate	Wavelength (nm)	E (M^{-1} cm^{-1})
1. TMPD[a]	610	11,600
2. Vanillylacetone[b]	336	18,300
3. 2,6-Dimethoxyphenol	568	[10,000][c]
4. Syringic acid	260	8,050
5. Guaiacol	465	12,100
6. Curcumin	430	23,100
7. Syringaldazine	525	65,000
8. Coniferyl alcohol	263	13,400
9. *o*-Dianisidine (-2HCl)	460	29,400

[a] *N,N,N,N*-Tetramethyl-1,4-phenylenediamine (-2HCl).
[b] 4-(4-Hydroxy-3-methoxyphenyl)-3-buten-2-one.
[c] Estimated for a nonhomogeneous, polymeric product; disappearance of absorbance (removal of substrate) is monitored for substrates 2, 4, and 6 while increases in absorbance (product formation) are monitored for substrates 1, 3, 5, 7, 8, and 9.

H_2O_2, and 0.1 mM $MnSO_4$. The solution is contained in a 1.5-ml quartz cuvette of 1-cm path length. The spectrophotometer is set at the wavelength appropriate to the assay substrate chosen (Table I). As a standard assay, vanillylacetone was used as substrate during purification of the manganese peroxidase. Decrease in absorbance at 336 nm was observed ($\epsilon = 18{,}300$) and used to calculate enzyme activity. Assays are performed at room temperature (about 22°). One unit (U) of peroxidase oxidizes 1 μmol of substrate/min, and units may be calculated based upon U mg^{-1} of protein ml^{-1} of enzyme solution. Assays should be carried out in duplicate, and are initiated by addition of H_2O_2. The reference cuvette does not receive H_2O_2.

Assay of Manganese Peroxidase Using Mn(II) as Substrate

A very convenient assay of manganese peroxidase activity involves monitoring the enzyme's oxidation of Mn(II) to Mn(III). This assay is best used with purified preparations of the peroxidase, as contaminating metals such as iron and copper inhibit the reaction. The reaction catalyzed is as follows: $2Mn^{2+} + H_2O_2 \rightarrow 2Mn^{3+} + H_2O$. The reaction mixture contains enzyme, 0.1 M sodium tartrate (pH 5.0), 0.1 mM H_2O_2, and 0.1 mM $MnSO_4$. The product, Mn(III), forms a transiently stable complex with tartaric acid, showing a characteristic absorbance at 238 nm ($\epsilon = 6500$). Reactions are initiated by addition of H_2O_2, and the reference cuvette contains (1-cm path length) all components except Mn(II). Increase in $A_{238\ nm}$ is monitored during the first 5–30 sec of reaction. One unit of peroxidase oxidizes 1 μmol of Mn(II)/min.

Assay for Ligninase

The enzyme purification employed here will allow purification of the principal ligninase of *P. chrysosporium* simultaneously with the manganese peroxidase. The most convenient assay for ligninase is a spectrophotometric assay that monitors the oxidation of veratryl alcohol to veratryl aldehyde.[4]

Purification Procedures

All procedures are carried out at 4°. Culture fluids of the fungus are separated from the mycelium by filtration through glass wool. The fresh filtrate is concentrated (each 1 liter to 30 ml) by ultrafiltration (e.g., using a stirred cell ultrafiltration device equipped with an M_r 10,000 cut-off mem-

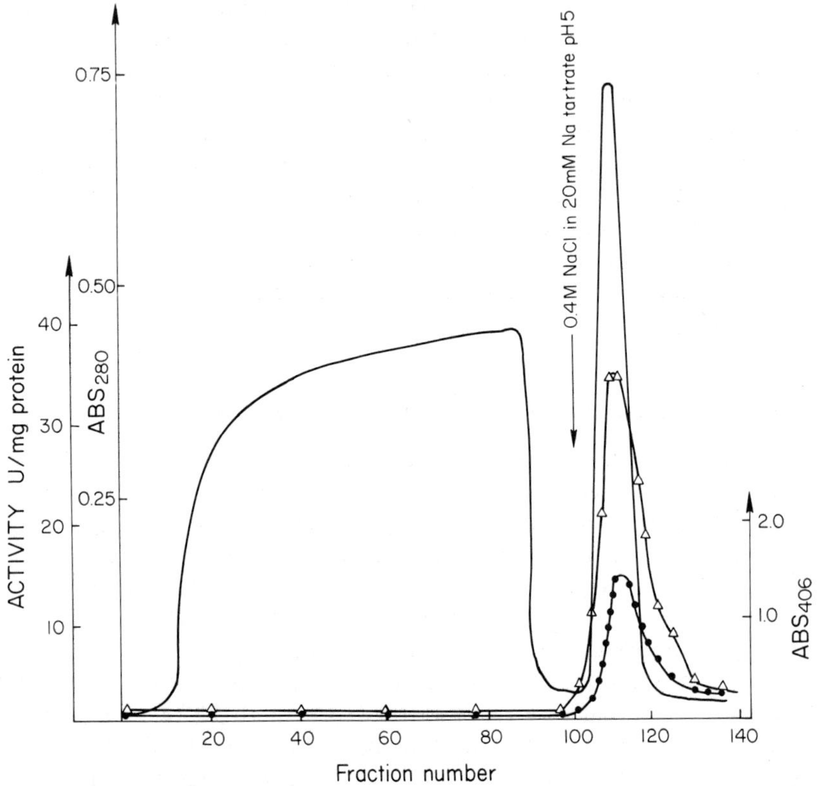

FIG. 1. Chromatography of the crude extracellular proteins of *Phanerochaete chrysosporium* on fast-flow DEAE-Sepharose. We employed a Pharmacia K15 column packed to a bed height of 10 cm. Enzyme solution (200 ml; concentrated by ultrafiltration on an Amicon PM10 membrane and diluted 1 : 1 with water) was applied to the column and then the column was washed with 100 ml H_2O. Elution conditions: flow rate, 70 ml hr^{-1}; fraction volume, 2.2 ml; the column first was washed with 50 ml of H_2O and then proteins desorbed with 20 m*M* sodium tartrate (pH 5.0) containing 0.4 *M* NaCl; (△–△) manganese peroxidase activity; (●–●) absorbance at 406 nm; (—) absorbance at 280 nm.

brane). Crude culture filtrate usually contains 6–8 $\mu g\ ml^{-1}$ of protein, as determined by the Coomassie blue method.[12] After ultrafiltration, the protein concentration is increased to about 150 $\mu g\ ml^{-1}$. Ultrafiltered enzyme (100 ml) is diluted 1 : 1 with deionized water and applied to a column (1.6 × 10 cm) containing DEAE(fast flow)-Sepharose (prewashed with 200 ml of distilled water). Proteins absorb to the column as a dark

[12] M. M. Bradford, *Anal. Chem.* **72,** 248 (1976).

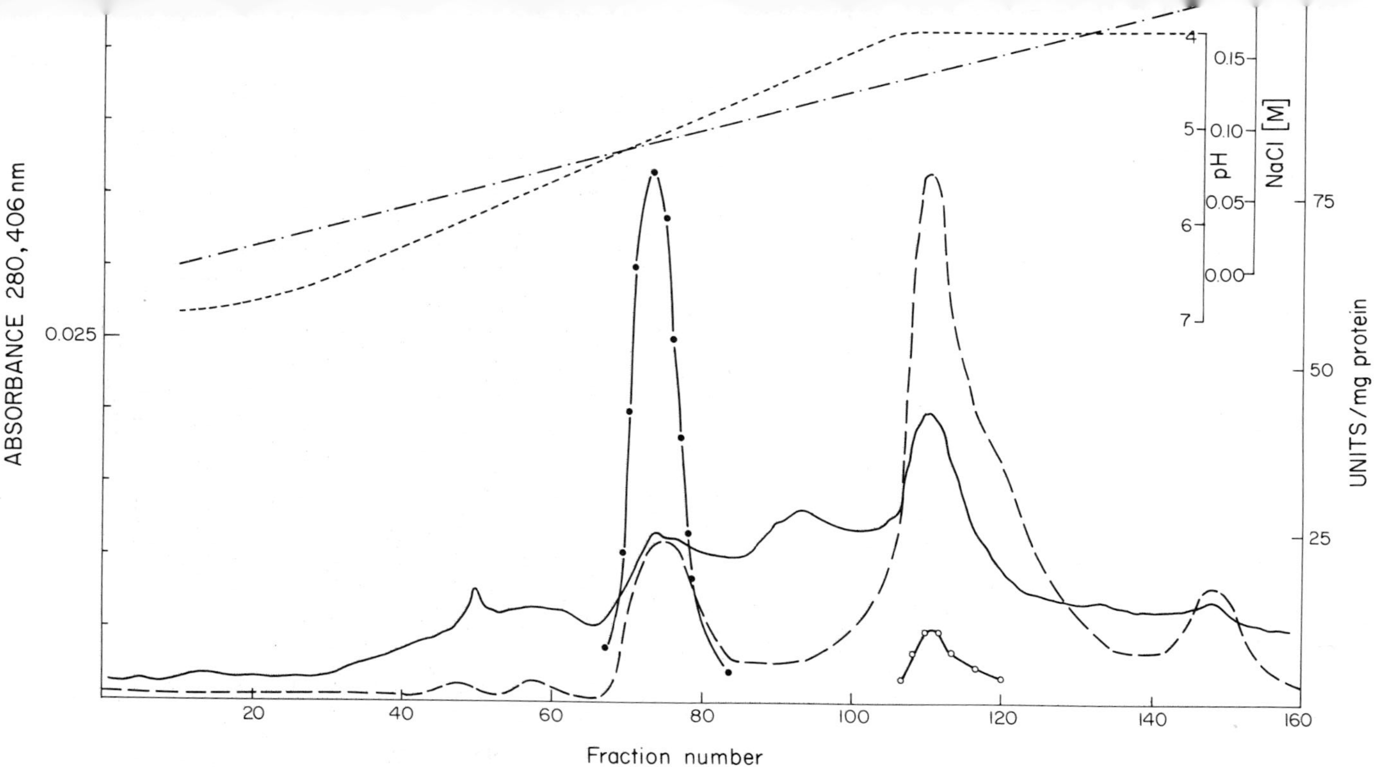

FIG. 2. Chromatofocusing of partially purified manganese peroxidase and lignin peroxidase of PBE 94 ion-exchange resin. Here we used a Pharmacia K9 column packed to a bed height of 50 cm. Sample (5 ml; peak activity of manganese peroxidase from the DEAE-Sepharose column, concentrated by ultrafiltration on an Amicon PM10 membrane) was dissolved in 25 m*M* imidazole–HCl (pH 7.4) and applied to the column (previously equilibrated with the same buffer). Elution conditions: flow rate, 20 ml hr^{-1}; fraction volume, 2.7 ml; elution buffer, Pharmacia Polybuffer 74-HCl at pH 4.0 and diluted 1 : 8 before use and containing a linear gradient of 0 to 0.2 *M* NaCl; (●–●) Mn-peroxidase activity; (O–O) lignin peroxidase activity; (—) absorbance at 280 nm; (---) absorbance at 406 nm.

TABLE II
PURIFICATION OF MANGANESE PEROXIDASE (AND LIGNINASE) FROM THE GROWTH MEDIUM OF *Phanerochaete chrysosporium*

Step[a]	Volume (ml)	U ml^{-1}		Total units		U mg^{-1} protein		Yield (%)		RZ (P)[c]	Purification	
		P[b]	L	P	L	P	L	P	L		P	
1	3000	0.06	0.04	189	126	9	6	—	—	0.02	—	-
2	100	1.5	1.2	150	120	16	8	79	95	0.24	1.7	1
3	5	16	12	80	60	26	12	42	47	1.5	2.8	2
4	5 (P) 5 (L)	12	1.6	60	8	100	15	31.7	6.3	3.6	11.1	2

[a] 1, Crude filtrate; 2, ultrafiltrate (10,000-Da cut-off); 3, DEAE-Sepharose; 4, PBE-94 chromat cusing column chromatography and ultrafiltration.
[b] P, Manganese peroxidase; L, ligninase; both enzymes will store (freeze dried in sealed ampoule −20° with a half-life of about 1 year. Both enzymes contain appreciable amounts of carbohydr
[c] $RZ = A_{406\ nm}/A_{280\ nm}$.

yellow band. The loaded column is washed with 50 ml of water, and the protein eluted using 20 m*M* sodium tartrate containing 0.4 *M* NaCl. The manganese peroxidase peak from the DEAE column also contains ligninase activity (Fig. 1). Peak fractions are combined and desalted/concentrated by ultrafiltration. This preparation may be freeze dried and stored at −20° up to 1 year without appreciable loss of activity. Freeze-dried enzyme is dissolved in 10 ml of 0.025 *M* imidazole buffer (pH 7.4) and applied to a column (35 × 0.9 cm) of Polybuffer Exchanger (PBE-94, Pharmacia) that had been preequilibrated with the same buffer (see Fig. 2). The column is then eluted with Polybuffer 74-HCl (Pharmacia) diluted 1 : 8, at pH 4.0, and containing a linear gradient of NaCl (0–0.2 *M*). About 10 fractions containing the highest activities of manganese peroxidase and ligninase are pooled separately and concentrated by ultrafiltration to a final volume of about 5 ml. The concentrated solutions are diluted to 30 ml with deionized water and reconcentrated, repeating this operation three times to remove dissolved salts and buffer. Both enzyme preparations should show a single band of protein upon polyacrylamide gel electrophoresis in the presence of SDS. After the final ultrafiltration, both enzyme preparations may be freeze dried for long-term storage in sealed ampoules. A summary of the above purification scheme is shown in Table II.

[27] NAD(P)H Dehydrogenase (Quinone) from *Sporotrichum pulverulentum*

By JOHN A. BUSWELL and KARL-ERIK ERIKSSON

The intracellular quinone oxidoreductase system [NAD(P)H dehydrogenase (quinone), EC 1.6.99.2; NAD(P)H: quinone oxidoreductase] from *Sporotrichum pulverulentum*[1] catalyzes the reduction of a quinone to the corresponding phenol using reduced pyridine nucleotides as electron donors according to the equation:

$$\text{Quinone} + \text{NAD(P)H} + \text{H}^+ \rightarrow \text{phenol} + \text{NAD(P)}$$

Assay Method

The enzyme is assayed by measuring the rate of decrease in absorbance at 340 nm, due to the oxidation of reduced pyridine nucleotide, using 2-methoxyquinone as electron acceptor.

Reagents

Citrate buffer, 100 m*M*, pH 5.6
NAD(P)H, 10 m*M*, freshly prepared, 0.05 ml
2-Methoxyquinone, 10 m*M* in methanol, 0.03 ml
Cell extract protein, 14–18 μg, 0.1 ml

Procedure

Citrate buffer, 250 μmol, NAD(P)H, 0.5 μmol, and 2-methoxyquinone, 0.3 μmol, are added in a total volume of 2.9 ml to a cuvette of 1-cm light path. After equilibration at 22°, the reaction is initiated by addition of 0.1 ml cell extract protein. The control cuvette is without NAD(P)H. The rate of decrease in absorbance at 340 nm is nonlinear and enzyme activity is calculated from the absorbance decrease observed during the initial 90 sec of the assay period. Considerable dilution of the crude extract is necessary to obtain a measurable rate. However, since quinone:oxidoreductase activity is labile in dilute solution, small volumes of concentrated extract should be diluted immediately prior to use and discarded after 1 hr. Enzyme activity values require correction for the decrease in A_{340} due to reduction of 2-methoxyquinone by NAD(P)H in the absence of fungal

[1] J. A. Buswell, S. Hamp, and K.-E. Eriksson, *FEBS Lett.* **108,** 229 (1979).

extract. Also, observed rates of NAD(P)H oxidation will be slight underestimates due to the relatively small increase in A_{340} resulting from methoxyhydroquinone formation. It should be noted, however, that increases in A_{340} due to hydroquinone formation may be considerably more significant when quinones other than methoxyquinone serve as substrates.

Spectral changes associated with the formation of methoxyhydroquinone may be observed during the course of NAD(P)H oxidation: i.e., the strong methoxyquinone absorption peak at 255 nm decreases and a new peak at 285 nm appears. Identification of methoxyhydroquinone as the reaction product is confirmed by gas chromatography of the trimethylsilyloxy derivative.[2]

Definition of Enzyme Activity

Enzyme-specific activity is expressed as micromoles NAD(P)H oxidized $min^{-1}mg^{-1}$ protein. Protein is determined by the method of Bradford[3] using bovine serum albumin as standard.

Growth of the Fungus and Preparation of Mycelial Extract

Sporotrichum pulverulentum (ATCC 32629, anamorph of *Phanerochaete chrysosporium*) is grown in shake culture at 28° in 1-liter conical flasks containing 300 ml of a modified Norkrans' medium[4] to which 3 m*M* vanillate is added. A spore suspension[5] ($\sim 2.5 \times 10^6$ spores) serves as inoculum and mycelial pellets are harvested on muslin after 50 hr and may be used either immediately or stored frozen at −20° until required.

Fungal extracts are prepared by suspending fresh or frozen mycelium in 2–3 vol of 0.1 *M* KH_2PO_4–NaOH buffer (pH 7.4) and breaking in a 50-ml homogenizer (Thomas Co., Philadelphia, PA) for 3 min at 0–5°. The homogenate is clarified by centrifugation at 30,000 *g* for 30 min and extracts, containing 8–12 mg/ml protein, are passed through a Sephadex PD-10 column (Pharmacia, Uppsala, Sweden) to remove endogenous metabolites which interfere with the enzyme assay.

Properties

General. In crude mycelial extracts, the enzyme is active over a broad pH range (pH 4–8) with an optimum near pH 5.6. Enzyme activity is

[2] J. A. Buswell and K.-E. Eriksson, this volume [28].
[3] M. M. Bradford, *Anal. Biochem.* **72,** 248 (1976).
[4] J. A. Buswell, P. Ander, B. Pettersson, and K.-E. Eriksson, *FEBS Lett.* **103,** 98 (1979).
[5] K.-E. Eriksson and S. C. Johnsrud, *Enzyme Microbiol. Technol.* **5,** 425 (1983).

reduced by approximately 50% when citrate buffer is replaced by phosphate.

Cofactor Requirements. Both NADH or NADPH are about as equally effective in serving as electron donors for the reduction of methoxyquinone to methoxyhydroquinone by crude extracts of *S. pulverulentum*. On the other hand, NAD(P)H : quinone oxidoreductases in other fungi appear to exhibit a preference for either one of the reduced pyridine nucleotide cofactors.[6] However, until purification procedures are established, it remains unclear if crude extracts contain just one enzyme able to use both electron donors, or two enzymes specific for either NADH or NADPH.

Substrate Specificity. Several quinones are reduced by the NAD(P)H: quinone oxidoreductase activity in crude mycelial extracts of *S. pulverulentum*. Highest enzyme activity is observed with the *p*-quinones methoxyquinone, toluquinone, and 1,4-benzoquinone, although the *o*-quinone 4,5-dimethoxy-1,2-benzoquinone is also rapidly reduced. Naphthoquinones are less effective substrates. However, since crude fungal extracts are again involved, similar qualifications will apply with respect to a single enzyme of low specificity versus several enzymes, each one specific for an individual quinone substrate.

Stoichiometry. The stoichiometry of the NAD(P)H : 2-methoxyquinone oxidoreductase reaction is difficult to establish unequivocally due to low rates of nonenzymatic reduction of quinone by NAD(P)H, small increases in A_{340} due to product formation, and to spontaneous reoxidation of methoxyhydroquinone. However, under conditions where the rate of product autooxidation is low (i.e., citrate buffer, pH 5.6; 100–150 μg cell extract protein; 100–300 nmol methoxyquinone), ~1 mol NADH is consumed for every mole methoxyquinone reduced.

Role and Distribution of the Enzyme. Mycelial extracts of several soft rot, brown rot, and white rot fungi grown on glucose, cellobiose, or cellulose contain appreciable levels of NAD(P)H : quinone oxidoreductase activity.[6] In many cases, including *S. pulverulentum* where a 5-fold increase is observed, enzyme levels are higher when vanillic acid is included in the culture medium.[1,6]

Quinones are readily formed through the action of phenol oxidases induced during the growth of white rot fungi on both low-molecular-weight phenolic compounds (e.g., vanillate) and lignin, and also by autooxidation of polyhydroxylated catabolic intermediates. They are generally highly reactive and are known to inhibit a wide range of metabolic processes, including key enzymes in the fungal metabolism of aromatic

[6] J. A. Buswell, K.-E. Eriksson, J. K. Gupta, S. G. Hamp, and I. Nordh, *Arch. Microbiol.* **131**, 366 (1982).

compounds.[7,8] Thus, NAD(P)H : quinone oxidoreductases may play a role in reversing any enzymatic or nonenzymatic conversion of phenols within the fungal cytoplasm, thereby avoiding inhibitory effects and ensuring further conversion of the benzenoid compounds to intermediates of central metabolism. Pyridine nucleotide : quinone reductase systems may also participate in electron transfer between respiratory substrates and polyphenol oxidases.[9]

[7] R. F. Bilton and R. B. Cain, *Biochem. J.* **108,** 829 (1968).
[8] J. M. Varga and H. Y. Neujahr, *Acta Chem. Scand.* **26,** 509 (1972).
[9] W. D. Wosilait, N. Nason, and A. J. Terrell, *J. Biol. Chem.* **206,** 271 (1954).

[28] Vanillate Hydroxylase from *Sporotrichum pulverulentum*

By JOHN A. BUSWELL and KARL-ERIK ERIKSSON

Vanillate hydroxylase[1-3] catalyzes the oxidative decarboxylation of vanillic acid to 2-methoxyhydroquinone (Scheme 1). Activity of the enzyme may be determined by measuring (1) $^{14}CO_2$ evolution from [^{14}C-*carboxyl*]vanillic acid, (2) O_2 consumption using the oxygen electrode, (3) the decrease in absorbance at 340 nm resulting from the oxidation of the NADH or NADPH cofactor, or (4) methoxyhydroquinone production by gas–liquid chromatography.

Assay Methods

1. Enzyme Assay Based on $^{14}CO_2$ Evolution from [^{14}C]Carboxyl-labeled Vanillic Acid

Reagents

Potassium phosphate buffer, 100 m*M*, pH 7.4
NAD(P)H, 10 m*M*, prepared fresh daily, 0.3 ml
[^{14}C-carboxyl]Vanillic acid (6.2×10^6 dpm/mg dissolved in absolute ethanol, 29 nmol in 5 μl (26,000 dpm)
Enzyme protein, 0.1–0.3 ml

[1] J. A. Buswell, P. Ander, B. Pettersson, and K.-E. Eriksson, *FEBS Lett.* **103,** 98 (1979).
[2] J. A. Buswell, K.-E. Eriksson, and B. Pettersson, *J. Chromatogr.* **215,** 99 (1981).
[3] Y. Yajima, A. Enoki, M. B. Mayfield, and M. H. Gold, *Arch. Microbiol.* **123,** 319 (1979).

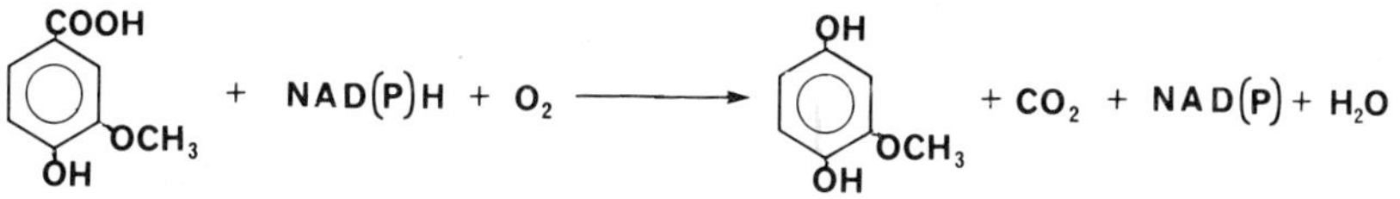

SCHEME 1. Reaction scheme for vanillate hydroxylase.

Procedure. Potassium phosphate buffer, 200 μmol, enzyme protein (crude extract or purified enzyme), and NAD(P)H, 3 μmol, in a total volume of 3.0 ml in a 125-ml conical flask are equilibriated at 30° in a shaker water bath. In cases where crude extract is used, reduced pyridine nucleotide is added to the reaction vessel just prior to initiation of the reaction by addition of radiolabeled vanillic acid substrate. This is to avoid extensive oxidation of cofactor which may occur due to NAD(P)H oxidase activity in the extracts. After addition of substrate, flasks are tightly sealed with rubber stoppers from which small glass tubes, containing 1 ml 1 *N* NaOH to absorb $^{14}CO_2$, are suspended.[4]

At appropriate intervals, the NaOH is transferred into scintillation vials. The glass tubes are rinsed twice with 0.5 ml H_2O and the washings also added to the vial. To this 2 ml is added 10 ml Picofluor 30, containing 1% Carbosorb (Packard), and the vials allowed to stand for at least 1 hr at 4° before measuring radioactivity in a scintillation counter.[4] To improve the efficiency of this technique in terms of quantitative rate determinations of enzyme activity, rubber stoppers can be fitted with a flushing device and $^{14}CO_2$ air-flushed directly into vials containing the scintillation cocktail.

2. Enzyme Assay Based on Oxygen Uptake Measurements

Reagents

Potassium phosphate buffer, 100 m*M*, pH 7.4
NAD(P)H, 10 m*M*, 0.1 ml
Vanillic acid in distilled water adjusted to pH 6.8 with 1 *N* NaOH, 0.1 ml
Enzyme protein, 0.1–0.3 ml

Procedure. Potassium phosphate buffer, 250 μmol, enzyme protein, and NAD(P)H, 1.0 μmol, are added in a total volume of 2.9 ml to the reaction vessel of a Clark oxygen electrode (Rank, Bottisham, England). After equilibration at 30°, oxygen consumption is measured following addition of 1 μmol (in 0.1 ml) vanillate and corrected for oxygen uptake in the absence of substrate.

[4] P. Ander, A. Hatakka, and K.-E. Eriksson, *Arch. Microbiol.* **125,** 189 (1980).

3. Enzyme Assay Based on Spectrophotometric Measurement

Reagents. As for assay (2) above.

Procedure. Potassium phosphate buffer, 250 μmol, enzyme protein, and NAD(P)H, 1.0 μmol, are added in a total volume of 2.9 ml to a cuvette of 1.0-cm light path. After equilibration at 30°, 1.0 μmol (in 0.1 ml) vanillate is added and the rate of NAD(P)H oxidation measured from the decrease in absorbance at 340 nm. Values are corrected for NAD(P)H oxidation in the absence of substrate.

4. Enzyme Assay Based on Methoxyhydroquinone Production

Reagents

Potassium phosphate buffer, 100 m*M*, pH 7.4
Vanillic acid, 10 m*M*
NAD(P)H, 10 m*M*
Enzyme protein

Procedure. Potassium phosphate buffer (7 ml), enzyme protein (0.1–1.0 ml), and vanillic acid (1.0 ml), in a total volume of 10 ml in a 125-ml conical flask, are equilibrated at 30° in a shaker water bath. The reaction is initiated by addition of 20 μmol NAD(P)H (in 2.0 ml) and allowed to proceed for 1 hr before acidifying with 0.5 ml concentrated HCl. Precipitated protein is removed by centrifugation; syringol is added as an internal standard and the supernatant is extracted three times with 10 ml diethyl ether. Combined ethereal extracts are dried over anhydrous Na_2SO_4 and, after evaporating off the ether in a stream of nitrogen, the residue taken up in 0.4 ml dry pyridine. An aliquot (100 μl) is silylated with bistrimethylsilyltrifluoroacetamide (BSTFA) for 1 hr at room temperature and methoxyhydroquinone determined by gas–liquid chromatography using a Packard model 427 with flame-ionization detector. Identification is based on comparison with the retention time of authentic methoxyhydroquinone. Separation is achieved using a glass capillary column SE-30 (25 m × 0.36 mm) and the following operating conditions: injection 220°, detection 250°, program 8 min, 150°, rise 6°/min to 170°, final time 7 min.

Definition of Enzyme Unit and Specific Activity

One unit of vanillate hydroxylase is the amount which converts 1 μmol of vanillic acid into methoxyhydroquinone and carbon dioxide per minute at 30°. Specific activity is expressed as U/mg protein, as determined by the method of Bradford,[5] with bovine serum albumin as the protein standard.

[5] M. M. Bradford, *Anal. Biochem.* **72,** 248 (1976).

Purification Procedure

Growth of the Fungus

Sporotrichum pulverulentum (ATCC 32629 anamorph of *Phanerochaete chrysosporium*) is grown in shake culture at 28° in 1-liter conical flasks containing 300 ml of a modified Norkrans' medium[1] to which 3 m*M* vanillate is added. A spore suspension[6] (2.5×10^6 spores) serves as inoculum and mycelial pellets are harvested after 50 hr and may be used either immediately or stored frozen at −20° until required. The purification procedure, carried out at 0–5° unless otherwise stated, involves four steps starting with the fungal mycelium: (1) preparation of crude extract and precipitation with potassium phosphate, (2) fractionation on a phenyl-Sepharose bed, (3) chromatofocusing, and (4) affinity chromatography on phenyl-Sepharose.

Step 1: Preparation of Crude Extract. A total of approximately 120 g wet weight of fresh or frozen mycelium is suspended in 4 vol 0.1 *M* KH_2PO_4–NaOH buffer (pH 7.4) and broken in separate batches in a 50-ml homogenizer (Thomas Co., Philadelphia, PA) for 3 min. Combined homogenates are clarified by centrifugation at 30,000 *g* for 30 min. Following dialysis against 0.001 *M* potassium phosphate buffer (pH 7.4), the enzyme solution is concentrated by freeze drying and, if necessary, may be stored at −20° in this form. The freeze-dried material is dissolved in 0.1 *M* potassium phosphate buffer (pH 7.0) so that a total absorbance at 280 nm of ~9500 is obtained. An equal volume of 2.0 *M* potassium phosphate buffer (pH 7.0) is then added and the solution slowly stirred at room temperature for 30 min. The precipitate is removed by centrifugation and the supernatant retained.

Step 2: Phenyl-Sepharose Chromatography. Supernatant material from step 1 is passed through a phenyl-Sepharose bed, 40 × 50 mm (Pharmacia, Uppsala, Sweden), equilibriated with 1.0 *M* potassium phosphate buffer (pH 7.0) at a rate of 8 ml/min. Under these conditions, vanillate hydroxylase is quantitatively retained on the phenyl-Sepharose. The bed is washed with 1.0 *M* potassium phosphate buffer (pH 7.0) and then successively with 500 ml each of 0.5 and 0.25 *M* potassium phosphate buffer until the absorbance reading of the washing solution at 280 nm is zero in both cases. The enzyme is finally eluted from the phenyl-Sepharose bed with a mixture of 0.2 *M* potassium phosphate buffer (pH 7.0) and an equal amount of ethylene glycol. The pooled active fractions are dialyzed against distilled water for 2 hr to remove the ethylene glycol and then concentrated by

[6] K.-E. Eriksson and S. C. Johnsrud, *Enzyme Microbiol. Technol.* **5**, 425 (1983).

ultrafiltration using an Immersible Molecular Separator (Millipore, Bedford, MA) to a total volume of 15 ml.

Step 3: Chromatofocusing. After desalting on a PD-10 column (Pharmacia), the concentrated enzyme solution is applied to a PBE 94 (Pharmacia) column (200 × 9 mm) previously equilibriated with 200 ml of 20 m*M* Tris buffer (pH 8.0). The column is then eluated with Polybuffer 96 until adjusted to pH 6.0 with glacial acetic acid.

Step 4: Affinity Chromatography on Phenyl-Sepharose-Vanillic Acid Gel. Phenyl-Sepharose-vanillic acid gel is prepared by adding a solution of vanillic acid (200 mg) in acetone (20 ml) to 13 g phenyl-Sepharose gel suspended in 50 ml of acetone. The gel is kept in suspension by end-over-end rotation and diluted stepwise with water, allowing 1 hr between each dilution step for equilibration. The acetone concentration is thereby reduced sequentially from 100% to 66, 33, 17, and 8%. After the last step, the acetone is removed by washing with water and the gel finally equilibriated with 1 *M* potassium phosphate buffer (pH 7.0).

Active fractions from step 3 are pooled and concentrated to 1.5 ml by ultrafiltration using Immersible Molecular Separators. The concentrated enzyme is applied to a phenyl-Sepharose-vanillic acid column (120 × 6 mm) and eluted with a linear gradient (total volume 200 ml), simultaneously decreasing from 1.0 to 0.05 *M* potassium phosphate (pH 7.0) and increasing from 0 to 50% ethylene glycol. Purified enzyme protein gives a single band when examined using analytical isoelectric focusing and sodium dodecyl sulfate gel electrophoresis. The chemicals, apparatus, preparation of gels, and technique for isoelectric focusing are as described by Vesterberg[7] and Ayers *et al.*[8] Data from a typical preparation of vanillate hydroxylase, resulting in an approximately 240-fold purification and an overall yield of 13.3%, are summarized in Table I.

Properties

General. Based on protein molecular weight standards, the molecular weight of vanillate hydroxylase is estimated to be 65,000. In crude mycelial extracts the enzyme is active in potassium phosphate buffer over a wide pH range (5.8–8.0) and activity peaks are observed at pH 6.6 and 7.8. In common with other aromatic hydroxylases, vanillate hydroxylase activity is markedly reduced in Tris–HCl buffer (40% inhibition compared with activity in potassium phosphate at pH 7.2). Addition of 0.1 *M* KCl or NaCl to assay mixtures using potassium phosphate buffer reduces enzyme

[7] O. Vesterberg, *Biochim. Biophys. Acta* **257,** 11 (1972).

[8] A. R. Ayers, S. B. Ayers, and K.-E. Eriksson, *Eur. J. Biochem.* **90,** 171 (1978).

TABLE I

PURIFICATION OF VANILLATE HYDROXYLASE[a]

Step	Volume (ml)	Total absorbance, A_{280} (nm)	Total amount of enzyme (mU)	Specific activity (mU/A_{280})	Purification factor	Yield (%)
Crude mycelial extract	250	9495.5	63.1	6.64×10^{-3}	1.0	100
After precipitation with phosphate buffer	500	6086.0	45.2	7.43×10^{-3}	1.0	71.6
Phenyl-Sepharose chromatography	52	145.6	32.2	2.21×15^{-1}	33.5	51.1
Chromatofocusing	15	8.0	11.6	1.45	218.4	18.3
Affinity chromatography	10	5.2	8.4	1.61	243.1	13.3

[a] From Buswell *et al.*[2]

activity by approximately 70%, indicating that inhibition is probably due to chloride ions.

Cofactor Supplementation. Both NADH and NADPH serve as electron donors for vanillate hydroxylase although enzyme activity with NADH is only about 85% of that observed with NADPH using the $^{14}CO_2$ evolution assay. Activity is stimulated still further when FAD is used in combination with NADPH but not NADH, although there is no evidence of a flavin component being associated with the enzyme.

Substrate Specificity. Several substrate analogs promote NADH oxidation by partially purified vanillate hydroxylase, indicating the enzyme to be specific for phenolic compounds with a hydroxyl group located para to a carboxyl substituent attached directly to the aromatic ring. NADH oxidation proceeds at about the same rate when vanillate is replaced by either protocatechuate, *p*-hydroxybenzoate, or 2,4-dihydroxybenzoate. Gallate, 3-*O*-methylgallate, 2,4,6-trihydroxybenzoate, and 2,3,4-trihydroxybenzoate also promote high rates of NADH oxidation although activity with syringate is $< 10\%$ compared to vanillate.

Inhibitors. Tiron and heavy metals (Cu^{2+}, Ag^{2+}, Hg^{2+}) at 1 m*M* concentrations and 0.1 m*M* *p*-chloromercuribenzoate completely inhibit vanillate hydroxylase. Inactivation by the latter is partially reversible by addition of stoichiometric amounts of reduced glutathione or dithiothreitol. Cyanide (1 m*M*) and α,α'-dipyridyl (1 m*M*) also depress enzyme activity by about 15 and 39%, respectively, but arsenite, azide, EDTA, and diethyl dithiocarbamate at 1 m*M* concentrations have no significant inhibitory effect.

Stoichiometry. Vanillate hydroxylase is presumed to catalyze a typical monooxygenase reaction although the exact stoichiometry is difficult to establish since the reaction product, methoxyhydroquinone, undergoes nonenzymatic oxidation to the corresponding quinone, which in turn is reduced by any excess of NADPH present in the reaction mixture. Thus, observed oxygen uptake is usually slightly more than ascribed to a monooxygenase reaction although the rate of quinone formation is relatively slow at slightly acidic pH values and the presence of crude or partially purified fungal extract retards the rate of nonenzymatic oxidation even further. However, vanillate hydroxylase oxidatively decarboxylates protocatechuate and 2,4-dihydroxybenzoate to hydroxyquinol which, in turn, undergoes intradiol ring cleavage to maleylacetate. Ring fission is catalyzed by a dioxygenase present in crude mycelial extracts. In reaction mixtures containing 100–200 nmol of protocatechuate or 2,4-dihydroxybenzoate and crude extract, consumption of 2 nmol oxygen/nmol of substrate is observed. By analogy, oxidative decarboxylation of vanillate consumes 1 nmol of oxygen/nmol vanillate converted to methoxyhydroquinone. The stoichiometry of NADPH consumption is not available due to the asymptotic nature of the decrease in absorbance at 340 nm.

Role and Distribution of Enzyme. Vanillic acid is found in relatively high yield in extracts of wood following fungal decay.[9] It is a breakdown product of the lignin component and a catabolic intermediate in the degradation of lignin-related compounds by white rot fungi and other microorganisms.[10,11] Vanillate hydroxylase is found in many brown rot and white rot fungi and oxidative decarboxylation via methoxyhydroquinone may serve as the major route for vanillic acid catabolism in these two groups of wood-decaying fungi.[12,13]

[9] E. Adler, *Wood Sci. Technol.* **11,** 169 (1977).
[10] J. K. Gupta, S. G. Hamp, J. A. Buswell, and K.-E. Eriksson, *Arch. Microbiol.* **128,** 349 (1981).
[11] M. Ohta, T. Higuchi, and S. Iwahara, *Arch. Microbiol.* **121,** 23 (1979).
[12] T. K. Kirk and L. F. Lorenz, *Appl. Microbiol.* **26,** 173 (1973).
[13] J. A. Buswell, K.-E. Eriksson, J. K. Gupta, S. G. Hamp, and I. Nordh, *Arch. Microbiol.* **131,** 366 (1982).

[29] 4-Methoxybenzoate Monooxygenase from *Pseudomonas putida*: Isolation, Biochemical Properties, Substrate Specificity, and Reaction Mechanisms of the Enzyme Components

By FRITHJOF-HANS BERNHARDT, ECKHARD BILL, ALFRED XAVER TRAUTWEIN, and HANS TWILFER

Importance of 4-Methoxybenzoate Monooxygenase in Bacterial Metabolism

Various authors who have investigated the biological degradation of lignanes as model substances for lignin have pointed out that certain soil fungi belonging to the Basidiomycetes and Ascomycetes are able to use the plant structural substance lignin as a carbon source.[1-4]

Lignin, which is highly polymerized and water insoluble, is a major component of plant residues which are degraded in the soil by microorganisms. In the degradation of lignin by soil microorganisms, the cleavage of intramolecular aryl–alkyl ethers plays a central role. Studies on the mecha-

[1] W. F. Van Vliet, *Biochim. Biophys. Acta* **15,** 211 (1954).
[2] H. Ishikawa, W. J. Schubert, and F. F. Nord, *Arch. Biochem. Biophys.* **100,** 131 (1963).
[3] T. K. Kirk, W. J. Connors, R. D. Bleam, W. F. Hackett, and J. G. Zeikus, *Proc. Natl. Acad. Sci. U.S.A.* **72,** 2515 (1975).
[4] T. K. Kirk, W. J. Connors, and J. G. Zeikus, *Appl. Environ. Microbiol.* **32,** 192 (1976).

nism of biological degradation of lignin or lignin model substances by fungi showed that degradation of lignin down to vanillic acid followed the pathway involving successively the intermediates: α-guaiacyl glycerol-coniferyl ether, 4-hydroxy-3-methoxyphenylpyruvic acid, or 4-hydroxy-3-methoxycinnamic acid and vanillin.[5,6]

Also some soil bacteria, e.g., species of the genera *Pseudomonas, Flavobacteria, Achromobacter,* and *Agrobacteria,*[7-9] can degrade lignin model substances first into vanillic acid or isovanillic acid. However, no genus of bacteria has so far been reported to be able to degrade lignin itself. This indicates that soil bacteria can use lignin as a source of carbon only in connection with fungi, which first—probably by means of extracellular enzymes—convert lignin into water-soluble substances accessible to the bacteria.

Further degradation of the dissociated salts of vanillic acid and isovanillic acid by fungi and bacteria is initiated by O-demethylating monooxygenases which O-demethylate the two isomers of methoxybenzoic acid in the presence of molecular oxygen and of NAD(P)H as electron donor, yielding protocatechuic acid and formaldehyde.[10-14] Protocatechuic acid is then further degraded into acetyl-CoA and succinyl-CoA.[15,16]

The results of investigations performed on intact cells,[17] on cell-free extracts from bacteria induced with salts of vanillic acid or 4-methoxybenzoic acid, or on enriched or purified preparations of 3- and 4-methoxybenzoate monooxygenases from bacteria[18,19] showed that these enzymes behave differently in their substrate specificity. Depending on the microorganism and on the alkoxybenzoic acid used for induction, these monooxygenases demethylate only an alkoxy group para[14,20] or meta[11,13] to the permanent carboxyl group. Alternatively, they may demethylate alkoxy groups in both positions, but with different activities.[18,19,21]

[5] T. Fukuzumi, H. Takatuka, and K. Minami, *Arch. Biochem. Biophys.* **129,** 396 (1969).
[6] H. Ishikawa, W. J. Schubert, and F. F. Nord, *Arch. Biochem. Biophys.* **100,** 140 (1963).
[7] H. H. Tabak, C. W. Chambers, and P. W. Kabler, *J. Bacteriol.* **78,** 469 (1959).
[8] H. Sörensen, *J. Gen. Microbiol.* **27,** 21 (1962).
[9] V. Sundman, *J. Gen. Microbiol.* **36,** 171 (1964).
[10] M. E. K. Henderson, *J. Gen. Microbiol.* **26,** 155 (1961).
[11] N. J. Cartwright and A. R. W. Smith, *Biochem. J.* **102,** 826 (1967).
[12] N. J. Cartwright and J. A. Buswell, *Biochem. J.* **105,** 767 (1967).
[13] D. W. Ribbons, *FEBS Lett.* **8,** 101 (1970).
[14] J. A. Buswell and A. Mahmood, *Arch, Mikrobiol.* **84,** 275 (1972).
[15] L. N. Ornston and R. Y. Stanier, *J. Biol. Chem.* **241,** 3776 (1966).
[16] L. N. Ornston, *J. Biol. Chem.* **241,** 3787 and 3800 (1966).
[17] M. E. K. Henderson, *J. Gen. Microbiol.* **16,** 686 (1957).
[18] D. W. Ribbons, *FEBS Lett.* **12,** 161 (1970).
[19] F.-H. Bernhardt, N. Erdin, and H. Staudinger, *Eur. J. Biochem.* **35,** 126 (1973).
[20] N. J. Cartwright, K. S. Holdom, and D. A. Broadbent, *Microbios* **3,** 113 (1971).
[21] N. J. Cartwright and J. A. Buswell, *Microbios* **1A,** 31 (1969).

The low substrate specificity of the various inducible alkoxybenzoate monooxygenases (O-demethylating) of the lignin degradation pathway suggests that the fairly nonspecific inducible 4-methoxybenzoate monooxygenase (EC 1.14.99.15) from *Pseudomonas putida* (DSM-No. 1868) described here is also an enzyme that belongs in this biologic metabolic pathway. In the context of this degradative pathway, the enzyme could well have the physiological role of O-demethylating the various mono- and dimethoxybenzoic acids produced.

Bacterial Culture Conditions

The strain of *Pseudomonas putida* used throughout our studies and available from DSM (Göttingen, FRG; DSM-No. 1868), is kept on slants of Merck standard I nutrient agar.

In our initial studies[22] the inoculum for a 10-liter culture is grown in 500 ml of a slightly modified liquid medium as described in Ref. 23 and later on in a culture medium, pH 6.9, containing in 1000 ml: 2.0 g KH_2PO_4, 1.0 g $(NH_4)_2SO_4$, 0.2 g $MgSO_4 \cdot 7H_2O$, 0.25 g NaCl, 20 mg $CaCl_2 \cdot 2H_2O$, 2.9 mg $FeSO_4 \cdot 7H_2O$, 50 μg KI, 20 μg $Na_2MoO_4 \cdot H_2O$, 40 μg $MnSO_4 \cdot 4H_2O$, 40 μg H_3BO_4, 100 μg $CuSO_4 \cdot 5H_2O$, 100 μg $ZnSO_4 \cdot 7H_2O$, and 4.5 g 4-methoxybenzoic acid. After 1 to 3 days of aerobic growth at room temperature on a shaker the 500-ml culture reaches an optical density of about 1 at 436 nm and is transferred into a culture flask containing 10 liters of medium of the same composition. Aerobic growth is continued at room temperature by bubbling sterile air through the culture medium. Silicone polymer was used as antifoam agent. The cells are harvested by centrifugation within the logarithmic growth phase between 48 to 72 hr of growth. One liter of culture medium yielded 1.5 to 2.0 g wet weight bacterial paste which could be kept in 20-g batches at $-20°$ for 12 months without showing significant loss of their O-demethylating activity.

Interestingly, bacterial cells cultured by the described procedure contain a much higher O-demethylating activity than those bacterial cells which are cultured in a commercially available 10-liter fermenter under optimal growth conditions.

For Mössbauer measurements the 4-methoxybenzoate monooxygenase is isolated from bacterial cells enriched *in vivo* with ^{57}Fe to about 80% by replacing the 2.9 mg $FeSO_4 \cdot 7H_2O$ in 1 liter of culture medium by 0.6 mg ^{57}Fe dissolved in 0.12 ml 25% HCl.

[22] F.-H. Bernhardt, H. Staudinger, and V. Ullrich, *Hoppe-Seyler's Z. Physiol. Chem.* **351,** 467 (1970).

[23] D. W. Ribbons and W. C. Evans, *Biochem. J.* **83,** 482 (1962).

Purification of the Components of 4-Methoxybenzoate Monooxygenase

The enzyme activity of the cell-free crude extract is not proportional to protein concentration in the assay mixture, especially at low protein concentrations.[22] From this behavior we conclude that 4-methoxybenzoate monooxygenase is a dissociable enzyme system, i.e., it consists of several components. This is confirmed by the isolation of two components, a reductase and a dioxygen-activating protein. The reconstitution of these two components reveals full enzymatic activity.[24,25]

Under aerobic conditions cell-free crude extracts lose more than 50% of their activity within 24 hr at 0–4°, because of the enzyme's extreme sensitivity to atmospheric oxygen and to variations in pH. Under anaerobic conditions, however, it is possible to stabilize the enzyme activity over more than 24 days in the crude extract. Additional stabilizing factors are (1) supplementing the crude extract and the buffer solutions with 5–15% ethanol and (2) adding 4-methoxybenzoate (10^{-4} *M*). Under these conditions isolation of the enzyme can be achieved by standard purification techniques.[22,24,25]

Isolation of Putidamonooxin (PMO)

Isolation of the dioxygen-activating component (PMO) of 4-methoxybenzoate monoxygenase[25] from a cell-free crude extract after sonication of a bacterial suspension followed by 100,000 *g* centrifugation is performed in the presence of 10% ethanol at pH 8.0 and 4°.

All buffer solutions contain 10% ethanol and are deoxygenated by repeated evacuation and nitrogen flushing and finally by the addition of 0.3 m*M* sodium dithionite and 1 m*M* dithioerythritol. All chromatographic steps are carried out under strictly anaerobic conditions in a nitrogen atmosphere.[26]

Otherwise, PMO is purified by using standard purification techniques in eight steps[22,24,25]:

Step 1. About 1000 g wet weight bacterial paste is suspended in 1250 ml 50 m*M* potassium phosphate buffer. Portions (45 ml) of the suspension supplemented with 5 ml of ethanol are sonicated at 7 A for 5–6 min using a Branson sonifier S-125 fitted with the standard tip; they are kept at 0–8° and centrifugated at 100,000 *g* for 1 hr.

[24] F.-H. Bernhardt, H. H. Ruf, H. Staudinger, and V. Ullrich, *Hoppe-Seyler's Z. Physiol. Chem.* **352,** 1091 (1971).
[25] F.-H. Bernhardt, H. Pachowsky, and H. Staudinger, *Eur. J. Biochem.* **57,** 241 (1975).
[26] W. Sakami, *Anal. Biochem.* **3,** 358 (1962).

Step 2. The nucleic acids of the 100,000 *g* supernatant are precipitated by addition of a solution of 100% (w/v) streptomycin sulfate in 50 m*M* potassium phosphate buffer, pH 8.0, yielding a concentration of 6–7%. The solution is stirred for 15 min and then kept overnight in an N_2 atmosphere after being supplemented with a solution of 0.3 *M* dithioerythritol to give a final concentration of 1.25 m*M*. The precipitate is sedimented by centrifugation at 15,000 *g* for 15 min and discarded.

Step 3. The protein of the supernatant is fractionated by ammonium sulfate precipitation using a saturated ammonium sulfate solution containing 10% (v/v) ethanol, adjusted to pH 8.0 with 25% ammonia. Just before use a 0.1 *M* solution of sodium dithionite in 1 *M* Tris–HCl buffer, pH 8.0, is added to give a final concentration of 0.3 m*M*. The protein fraction, which precipitates between 60 and 80% ammonium sulfate saturation (100% = 2.82 *M* ammonium sulfate determined by titration with 25 m*M* $BaCl_2$ in presence of alizarin red S as indicator according to Bergmeyer *et al.*[27]), is collected by centrifugation and dissolved in a minimum amount of 50 m*M* potassium phosphate buffer, yielding a volume of about 500 ml of a dark reddish brown protein solution.

Step 4. The protein solution, protected against oxygen by addition of solutions of 120 mg dithioerythritol in 2.5 ml of 50 m*M* potassium phosphate buffer, pH 8.0, and 60 mg sodium dithionite in 2.5 ml of 1 *M* Tris–HCl buffer, pH 8.0, is applied on a Sephadex G-100 column (90 × 8 cm) equilibrated and eluted with 5 m*M* potassium phosphate buffer. About 700 ml of a reddish-brown protein fraction is eluted and concentrated to a volume of 200 ml by ultrafiltration in an Amicon cell equipped with a PM 30 membrane.

Step 5. Subsequently the concentrated protein solution is adsorbed on a DEAE-Sephadex A-50 column (68 × 5.5 cm) equilibrated with 5 m*M* potassium phosphate buffer saturated with $FeSO_4$. Elution is performed using 2 liters of a linear concentration gradient of potassium chloride from 50 m*M* to 0.5 *M* in the equilibration buffer, and then continued with 1 liter of 0.5 *M* KCl in the same buffer. About 300 ml of the reddish-brown enzyme fraction is collected and again concentrated by ultrafiltration to about 100 ml.

Step 6. Step 4 is repeated using a Sephadex G-100 column (100 × 5 cm) equilibrated and eluted with 50 m*M* potassium phosphate buffer containing 0.3 *M* KCl. About 200 ml of PMO fraction is eluted, concentrated to about 60 ml, and desalted by filtration on a Sephadex G-25 column (40 × 4.5 cm) equilibrated and eluted with 5 m*M* potassium phosphate buffer.

Step 7. The KCl-free enzyme fraction is concentrated up to 80 ml by ultrafiltration and rechromatographed on a DEAE-Sephadex A-50 column

(60 × 2.7 cm) equilibrated with 5 m*M* potassium phosphate buffer and eluted with 1.2 liter of the linear concentration gradient of KCl as used in step 5.

Step 8. For final purification the concentrated PMO fraction (60 ml) from step 7 is rechromatographed as described in step 6.

The purified, reddish-brown PMO fractions from step 8 containing between 10 and 15 mg of protein (per milliliter) can be stored in absence of oxygen for several months at 1–4° in the presence of 0.3 m*M* sodium dithionite and 1 m*M* dithioerythritol without significant loss of activity.

Isolation of NADH-PMO Oxidoreductase

The purification of the reductase does not require anaerobic working conditions, despite the fact that the purified enzyme is even more sensitive to atmospheric oxygen than PMO. The reductase is isolated[25] as follows:

Step 1. About 400 g wet weight bacterial paste is suspended in 700 ml of 50 m*M* potassium phosphate buffer, pH 6.8. Then 45-ml portions of the suspension are sonicated as described under the purification procedure of PMO (however, in the absence of ethanol), and centrifugated at 100,000 *g* for 1 hr.

Step 2. The nucleic acids of the 100,000 *g* supernatant are precipitated with streptomycin sulfate as reported in the purification procedure for PMO, with the exception that a 50–60% (w/v) solution of streptomycin sulfate, dissolved in 50 m*M* potassium phosphate buffer and adjusted to pH 6.8, is employed.

Step 3. The protein of the supernatant from step 2 is fractionated by ammonium sulfate precipitation using a saturated ammonium sulfate solution, pH 6.8. The portion precipitating between 30 and 60% (100% = 3.75 *M* ammonium sulfate as titrated according to Bergmeyer *et al.*[27]) is collected by centrifugation and dissolved in a minimum amount of 50 m*M* potassium phosphate buffer, pH 6.8, containing 1 m*M* mercaptoethanol and 0.1 m*M* EDTA. The protein solution with a final volume of about 300 ml is dialyzed for 24 hr against 5 liters of the same buffer solution used as solvent.

Step 4. The dialyzed protein solution is loaded onto a Sephadex G-100 column (100 × 8 cm). Equilibration and elution are performed with the same buffer solution used as solvent in step 3. Under these conditions the fraction which contains the reductase travelled as a coffee-colored band. This fraction changes its color to maize yellow when eluted and freeze-dried.

[27] H. U. Bergmeyer, G. Holz, E. M. Kauder, H. Möllering, and O. Wieland, *Biochem. Z.* **333**, 471 (1961).

Step 5. For further purification a vertical preparative polyacrylamide gel electrophoresis apparatus is employed. Four hundred and fifty to 500 mg of the freeze-dried reductase fraction is dissolved in 2 ml of distilled water and applied on the long and narrow side of a slice of 6% polyacrylamide gel (4.5 × 3.5 × 0.8 cm) in 0.3 *M* Tris–HCl buffer, pH 8.9. The electrode buffer is Tris (10 m*M*)–glycine (80 m*M*), pH 8.2, and the eluting buffer is the same as the electrode buffer, but contains 20% (w/v) glucose. Under these conditions and at a constant current of 40 mA the reductase migrates as a coffee-colored band just behind the buffer front. The combined reductase fractions of this step have a protein content of about 8 mg/ml and are normally used for further experiments.

Step 6. For specific purposes only (i.e., determination of the nature of the flavin chromophore), a final purification step is followed. The combined maize yellow reductase fractions (about 10 ml) are adsorbed on a DEAE-Sephadex A-50 column (20 × 1.3 cm) equilibrated with 5 m*M* potassium phosphate buffer, pH 7.5. Elution is carried out using 200 ml of a linear concentration gradient of KCl from 50 m*M* to 0.3 *M* in the equilibration buffer.

The reductase preparation obtained after step 4 can be freeze-dried and stored at −20° for several months without any significant loss of activity. On the other hand, the solution of purified reductase obtained after step 5 or 6 is extremely unstable under aerobic conditions. However, the enzyme can be stabilized by the addition of NADH and then can be kept for 3 to 4 weeks at −20° under nitrogen atmosphere.

Activity Tests

Although 4-methoxybenzoate is O-demethylated under concomitant equimolar consumption of dioxygen and NADH, the polarographic and spectrophotometric measurement of the rate of dioxygen uptake and NADH oxidation, respectively, as a criterion for enzyme activity, is not sensitive enough for testing low-grade enriched preparations routinely. These two methods, however, are well recommended when testing the purified, reconstituted 4-methoxybenzoate monooxygenase in the presence of 4-methoxybenzoate.

The O-demethylation of 4-methoxybenzoate by the 4-methoxybenzoate monooxygenase cannot be monitored directly; but two substrates, 3-nitro-4-methoxybenzoate and 3-phenyl-4-methoxybenzoate, can be used to determine spectrophotometrically the activity of the enzyme system directly.

The O-demethylation of 3-nitro-4-methoxybenzoate may be followed by monitoring at 405 nm the appearance of the yellow color of the reaction product (3-nitro-4-hydroxybenzoate). The extinction coefficient at 405 nm of this product at pH 8.0 is 4.04 mM^{-1} cm^{-1}.[28]

A typical assay system of final volume 2 ml contains 50 m*M* potassium phosphate buffer, pH 8.0, 0.25 m*M* 3-nitro-4-methoxybenzoate, 0.5 m*M* NADH, and about 80 μg of putidamonooxin. The reaction is started by the addition of about 45 μg reductase, and the rate of increase in absorbance at 405 nm is recorded at 30° in a 1-cm quartz cuvette.

This test can also be used to determine 4-methoxybenzoate monooxygenase activity in cell-free extracts, because the reaction product is not metabolized further by the NADPH-dependent, highly substrate-specific 4-hydroxybenzoate monooxygenase, which is always present in the crude extracts.[29] For this purpose the same assay system as described above, but without putidamonooxin, was started by the addition of 2 mg of protein from crude cell-free extracts.

The conversion of 3-phenyl-4-methoxybenzoate into 3-phenyl-4-hydroxybenzoate can be measured fluorometrically. The excitation spectrum of 3-phenyl-4-hydroxybenzoate has a peak at 295 nm, and emission occurs at 405 nm. Alternatively an Eppendorf photometer equipped with the fluorescence accessory fitted with the 313-nm excitation filter and the 404-nm secondary filter can be used. Since the NADH oxidation concomitant with the O-demethylation interferes slightly with the fluorescence measurements, a NADH-regenerating system should be used. A standard assay system of final volume of 2 ml contains 50 m*M* potassium phosphate buffer, pH 8.0, 0.25 m*M* 3-phenyl-4-methoxybenzoate, 0.1 m*M* NADH, about 90 μg of putidamonooxin, and as NADH-regenerating system 0.75 U galactose dehydrogenase (EC 1.1.1.48) and 5 m*M* D-galactose. The reactions are initiated by addition of about 180 μg of reductase, and the increase of fluorescence is monitored at 30° in a 1-cm quartz cuvette. For calibration, a known amount (20 nmol) of 3-phenyl-4-hydroxybenzoate is added to the reaction mixtures at the end of each experiment.

Properties of the NADH-PMO Oxidoreductase

The molecular weight of the NADH-PMO oxidoreductase is 42,000.[25] The enzyme is a conjugated iron–sulfur protein and contains a [2Fe-2S] cluster[30,31] and a FMN-chromophore.[25] In the presence of NADH or in

[28] F.-H. Bernhardt, W. Nastainczyk, and V. Seydewitz, *Eur. J. Biochem.* **72,** 107 (1977).

[29] V. Ullrich, F.-H. Bernhardt, H. Diehl, N. Erdin, and H. H. Ruf, *Z. Naturforsch.* **27b,** 1067 (1972).

[30] H. H. Ruf, F.-H. Bernhardt, V. Ullrich, and H. Staudinger, *Fed. Eur. Biochem. Soc. Symp., 8th,* Abstr. No. 447 (1972).

[31] H. Twilfer, F.-H. Bernhardt, and K. Gersonde, *Eur. J. Biochem,* **119,** 595 (1981).

the sodium dithionite half-reduced state a blue semiquinone form of FMN appears, established by its optical absorbance at about 600 nm and by the characteristic linewidth of its isotropic EPR signal at $g = 2.003$.[24,25,30,31] By further reduction of the reductase by sodium dithionite the isotropic EPR line diminishes and finally disappears completely at the fully reduced state. However, an anisotropic EPR signal with rhombic symmetry ($g_1 = 2.032$, $g_2 = 1.942$, $g_3 = 1.893$) remains, which is assigned to the [2Fe-2S] cluster. In the oxidized state both the [2Fe-2S] cluster and the FMN chromophore are EPR silent.[31]

The NADH-PMO oxidoreductase acts as an electron-transport chain and furnishes the electron flow from reduced pyridine nucleotides onto the PMO–substrate complex. NADH and NADPH can serve as electron donors. However, the affinity of the reductase within the reconstituted 4-methoxybenzoate monooxygenase is more than 200-fold higher for NADH (K_m: 0.63 μM) than that toward NADPH (K_m: 140 μM); and with respect to the O-demethylation activity of the reconstituted enzyme system NADPH is 40% less effective than NADH.[28] This suggests that NADH is the physiological electron donor for the reductase.

The NADH-PMO oxidoreductase also exhibits a lipoamide dehydrogenase (diaphorase) activity in the presence of both one-electron and two-electron acceptors such as ferricyanide ($E_0' = +420$ mV), cytochrome *c* ($E_0' = +255$ mV), 2,6-dichlorophenolindophenol ($E_0' = +217$ mV), cytochrome b_5 ($E_0' = +20$ mV), methylene blue ($E_0' = +11$ mV), indigo tetrasulfonate ($E_0' = -46$ mV), and indigo disulfonate ($E_0' = -125$ mV), while phenosafranin ($E_0' = -240$ mV) is reduced only up to 70%. These findings indicate that the redox potential (E_0') of the reductase is limited to the range between -200 and -240 mV.[25,28,32] The pH optimum of the reductase is at pH 8.0.[25]

Properties of Putidamonooxin (PMO)

PMO, the dioxygen-activating component of the 4-methoxybenzoate monooxygenase, has a molecular weight of 126,000, as derived from ultracentrifugation and gel filtration.[25,33] It is an oligomeric protein composed of either three or four identical subunits as suggested by the amino acid composition which yields a minimal molecular weight of 33,000, or by sodium dodecyl sulfate polyacrylamide gel electrophoresis of monomeric or cross-linked subunits, which leads to a molecular weight of 41,500.[25,33] The sequence of the last four amino acids at the C-terminus of each

[32] H. H. Ruf, Ph.D. thesis. Universität des Saarlandes, Federal Republic of Germany, 1974.
[33] F.-H. Bernhardt, E. Heymann, and P. S. Traylor, *Eur. J. Biochem.* **92,** 209 (1978).

subunit was found to be -Val-Ala-Leu-Thr.[34] Taking the average molecular weight to be 126,000 the millimolar absorption coefficients at 280 and 455 nm (λ_{max} in the visible range of light) were calculated from oxidized, highly purified PMO preparations to be 161 and 14.7 mM^{-1} cm^{-1}, respectively.[33]

The isoelectric point of PMO is at pH 4.73, and only traces of carbohydrates (about 0.07%) were detected in the protein.[33] PMO is strongly inactivated by sulfhydryl-modifying reagents. This inhibition can be reversed by several thiol compounds, indicating that a sulfhydryl group of the protein is necessary for catalytic activity.[22,28,35]

Chemical analyses[33] and spectrophotometric,[24,25] kinetic,[35] EPR,[24,31,36] and Mössbauer studies[37,38] revealed PMO as a conjugated iron–sulfur protein containing in its active sites an iron–sulfur cluster and as cofactor a mononuclear nonheme iron, in a 1 : 1 ratio. Chemical determination of iron and acid-labile sulfur contents as well as Mössbauer measurements on oxidized PMO substrate complexes show the existence of at least two or more likely three [2Fe–2S] clusters per molecule of cofactor iron-depleated PMO.[33,37]

The iron–sulfur cluster was identified by EPR and Mössbauer studies as a Rieske-type [2Fe–2S] cluster with a pronounced g-anisotropy of $g_1 = 2.008$, $g_2 = 1.913$, and $g_3 = 1.72$.[24,31,37] The mononuclear nonheme iron is in its ferric state weakly bound to PMO and therefore escapes easily out of the active sites, leading to a fast enzyme inactivation. However, in the presence of substrate the ferric mononuclear nonheme iron remains strongly bound.[35] In its ferrous state it is strongly bound in the presence as well as absence of substrate. EPR and Mössbauer studies indicate that it is penta- or hexacoordinated in its ferric and ferrous state and most likely pentacoordinated in its NO-ligated form.[36,38] Both together, the [2Fe–2S] cluster and the mononuclear nonheme iron, form in their reduced state the dioxygen-activating unit within the active sites of PMO. The cofactor iron acts as dioxygen-binding site, and the [2Fe–2S] cluster functions as a one-electron storage center. This structure of the dioxygen-activating unit provides the rapid transfer (<5 msec) of two electrons to dioxygen, as indicated by freeze-quench measurements.[39]

The electronic structure of the mononuclear nonheme iron depends on

[34] W. Adrian and F.-H. Bernhardt, unpublished results.

[35] F.-H. Bernhardt and H.-U. Meisch, *Biochem. Biophys. Res. Commun.* **93,** 1247 (1980).

[36] H. Twilfer, F.-H. Bernhardt, and K. Gersonde, *Eur. J. Biochem.* **147,** 171 (1985).

[37] E. Bill, F.-H. Bernhardt, and A. X. Trautwein, *Eur. J. Biochem.* **121,** 39 (1981).

[38] E. Bill, F.-H. Bernhardt, A. X. Trautwein, and H. Winkler, *Eur. J. Biochem.* **147,** 177 (1985).

[39] E. Bill, Ph.D. thesis. Universität des Saarlandes, Federal Republic of Germany, 1985.

the nature of the substrate bound to the active sites of PMO,[31,40] accounting probably for the varying affinities of this iron to dioxygen with K_m values ranging from 1.9 μM in the presence of 4-methoxybenzoate to 55 μM in the presence of 4-methylbenzoate.[19]

The electronic structure of the oxidized [2Fe-2S] cluster depends on the nature of the substrate and on the binding of cofactor iron to PMO.[37,41] This structure is optimized with respect to the physiological reduction of PMO by its reductase by binding either of 4-methoxybenzoate or cofactor iron plus any substrate analog.[25,37]

The redox potential of the [2Fe-2S] cluster and of the mononuclear nonheme iron of putidamonooxin was determined to be about $E_0' = 0 \pm 5$ mV at pH 7.8 and $E_0' = +220$ to $+360$ mV at pH 8.0, respectively.[32,36]

Protein-Protein Interactions between PMO and Its Reductase

By ultracentrifugation no interaction between PMO and its native reductase was detected.[25] However, the optical absorption of PMO and/or its reductase changes when mixed in the presence or absence of substrate. This behavior probably can be attributed to protein-protein interactions between both components.[42] The affinity of NADH-PMO oxidoreductase to NADPH and NADH decreases by a factor of about 3 and 10, respectively, when cytochrome *c* or 2,6-dichlorophenolindophenol were used as electron acceptors instead of the PMO-substrate complex itself. This may be the result of conformational changes of reductase caused by protein-protein interactions.[28] Furthermore, the binding of the mononuclear nonheme iron to the active sites of PMO, or the interactions between PMO and its native reductase, or both effects together lead to an increase of the affinity of PMO to its substrates as seen by comparing the Michaelis-Menten constants (K_m) with the dissociation constants (K_s). The K_m values are two orders of magnitude smaller than the K_s values, which had been derived from mononuclear nonheme iron-depleted PMO in the absence of the native reductase.[28,33]

The activity of the reconstituted 4-methoxybenzoate monooxygenase is controlled by the properties of the reductase. This is concluded from the influence of temperature, ionic strength, and pH upon the activity of (1) the reconstituted 4-methoxybenzoate monooxygenase and (2) the NADH-PMO oxidoreductase.[25,42]

[40] F.-H. Bernhardt, K. Gersonde, H. Twilfer, P. Wende, E. Bill, A. X. Trautwein, and K. Pfleger, *in* "Oxygenases and Oxygen Metabolism" (M. Nozaki, S. Yamamato, Y. Ishimura, M. J. Coon, L. Ernster, and R. W. Estabrook, eds.), pp. 63-77. Academic Press, New York, 1982.

[41] F.-H. Bernhardt, H. H. Ruf, and H. Ehrig, *FEBS Lett.* **43,** 53 (1974).

[42] F. Eich, P. J. Geary, and F.-H. Bernhardt, *Eur. J. Biochem.* **153,** 407 (1985).

Kinetic studies proved that the semiquinone form of the NADH-PMO oxidoreductase rather than the fully reduced reductase is the most powerful reductant of PMO–substrate complex.[43] Under saturation condition of reductase the maximum turnover rate of PMO was found to be 80 sec^{-1}.[28] The K_m value of the reductase for the PMO–4-methoxybenzoate complex is about 30 μM.[42]

Substrate Specificity and Reaction Mechanism

The 4-methoxybenzoate monooxygenase demethylates, under stoichiometric consumption of NADH and O_2, its physiological substrate by inserting one oxygen atom of dioxygen into a CH bond of the 4-methoxy group, while the second oxygen atom is reduced to water.[44] The formed semiacetal intermediate decays spontaneously into 4-hydroxybenzoate and formaldehyde.[45]

The enzyme was found to be fairly nonspecific. It catalyzes by a monooxygenation reaction: (1) O-, S-, and N-demethylation, (2) dealkylation of methoxy and of higher alkoxy groups in para as well as in meta position of the carboxy group of benzoate, (3) oxygenation of 4-methylbenzoate under formation of 4-carboxybenzyl alcohol, and (4) attack of the aromatic ring of 4- and 3-hydroxybenzoate and 4-aminobenzoate, yielding 3,4-dihydroxybenzoate and 4-amino-3-hydroxybenzoate, respectively.[19,40,46]

The enzyme metabolizes only substrates which fulfill the following two requirements: a dissociable carboxy group directly at the aromatic ring and the absolute planarity of the substrate molecule. Nonaromatic ring systems such as cyclohexanecarboxylic acid and 3-cyclohexene-1-carboxylic acid will not be oxygenated. Because of the lack of the coplanarity they probably do not interact with the active sites of PMO.[46] The negatively charged carboxy group cannot be substituted by the acetic acid or sulfo group.[19,28,33]

With the turnover rate for 4-methoxybenzoate being 100% the turnover rates for substrate analogs which are also oxygenated by the enzyme are as follows: 3-nitro-4-methoxybenzoate (110%), 3,4-methylenedioxybenzoate (100%), 3,4-dimethoxybenzoate (96%), 3-phenyl-4-methoxybenzoate (89%), 4-ethoxybenzoate (87%), 4-methylmercaptobenzoate (83%), 4-methylbenzoate (77%), *N,N*-dimethyl-4-aminobenzoate (66%), *N*-methyl-4-aminobenzoate (49%), 3,5-dimethoxybenoate (37%), 3-methoxybenzoate (29%), 4-hydroxybenzoate (16%), 4-hydroxy-3-methoxybenzoate

[43] F.-H. Bernhardt and H. Kuthan, *Eur. J. Biochem.* **130,** 99 (1983).
[44] F.-H. Bernhardt and H. H. Ruf, *Biochem. Soc. Trans.* **3,** 878 (1975).
[45] J. Axelrod, *Biochem. J.* **63,** 634 (1956).
[46] P. Wende, K. Pfleger, and F.-H. Bernhardt, unpublished results.

(6%), 3-hydroxybenzoate (5%), and 4-aminobenzoate (4%).[19,44,46] However, some benzoic acid derivatives like 3- and 4-chlorobenzoate, 4-bromobenzoate, 2-hydroxybenzoate, 2- and 3-aminobenzoate, 4-trifluoromethylbenzoate, 4-*tert*-butylbenzoate, and benzoate itself are bound by the active sites of PMO without being oxygenated.[19,46] Although these derivatives are not oxygenated, oxygen consumption and NADH oxidation occur under stoichiometric formation of H_2O_2.[19,44,46] Obviously in this case the formation of the active oxygen species occurs uncoupled from an oxygenation reaction and leads to the formation of hydrogen peroxide by protonation.[19,40,44,46] On the basis of these results the substrate analogs can be classified with respect to the stoichiometry of O_2 uptake, NADH oxidation, and product formation as follows: "tight couplers" (oxygenated under stoichiometric consumption of O_2 and NADH), "uncouplers" (cannot be affected by the active oxygen species with the result that the activated oxygen is released under formation of H_2O_2), and "partial uncouplers" (oxygenated, but account for only a fraction of O_2 consumed while the rest is reduced to H_2O_2).[40,44]

Furthermore, by a *substrate-modulated reaction,* the 4-methoxybenzoate monooxygenase can act as an external dioxygenase, i.e., incorporation of both atoms of activated dioxygen into a substrate molecule. This was demonstrated with 4-vinylbenzoate as substrate analog in the presence of NADH and $^{18}O_2$. Both atoms of activated $^{18}O_2$ were incorporated into 4-vinylbenzoate under formation of 4-[$^{18}O_2$](1,2-dihydroxy ethyl)benzoate[47] with a turnover rate of 60% compared to that for 4-methoxybenzoate. This finding indicates that the iron peroxo complex $[FeO_2]^+$ is the active oxygenating species in the active sites of PMO. This is supported by the fact that under uncoupling conditions H_2O_2 is formed directly and not indirectly by a rapid bimolecular disproportionation of superoxide anion radicals (O_2^-).[48]

With deuterated and nondeuterated 3-phenyl-4-methoxybenzoate, and 3-[$^1H_3$], 5-[$^2H_3$]dimethoxybenzoate as substrates the intermolecular and the intramolecular isotope effect was determined to be $k_{^1H}/k_{^2H} = 1.47$ and 0.74, respectively.[34,44]

From the influence of both the nature of substituents and their position at the aromatic ring of benzoic acid on the product pattern and on the magnitude of uncoupling we conclude the following.

1. The attack of the active oxygen species of 4-methoxybenzoate monooxygenase occurs on the aromatic ring according to the empirical

[47] P. Wende, K. Pfleger, and F.-H. Bernhardt, *Biochem. Biophys. Res. Commun.* **104,** 527 (1982).

[48] F.-H. Bernhardt and H. Kuthan, *Eur. J. Biochem.* **120,** 547 (1981).

rules of electrophilic substitutions and by steric effects on the aliphatic substituents.[46]

2. There exists a competition between the oxygenation reaction at a substrate molecule and the detoxification of the active oxygen species by protonation.[44,49]

3. The 4-methoxybenzoate monooxygenase acts by a *substrate-modulated reaction* either as a monooxygenase, a peroxotransferase (external dioxygenase), an oxidase, or a peroxidase.[40,47]

4. The rate of formation of the active oxygen species $[FeO_2]^+$ appears to be independent of the nature of the substrate. The lifetime of this species is determined (a) by the rate of the oxygenation reaction in the presence of tight couplers; (b) by the rate of its detoxification by protonation in the presence of 4-aminobenzoate, a nearly complete uncoupler; (c) by the rates of detoxification as well as oxygenation in the presence of partial uncouplers; and (d) by the rate of its reduction to water in presence of uncouplers and D_2O as solvent.[40,44,48,49]

Acknowledgments

Dedicated to Professor Robert Ammon for his eighty-fifth birthday. This work was supported by the Deutsche Forschungsgemeinschaft.

[49] P. Wende and F.-H. Bernhardt, *Biophys, Struct. Mech. (Suppl.)* **6**, 62 (1980).

[30] Vanillate *O*-Demethylase from *Pseudomonas* Species

By JOHN A. BUSWELL and DOUGLAS W. RIBBONS

Vanillate *O*-demethylase[1-3] is an inducible enzyme which catalyzes the O-demethylation of vanillic acid (4-hydroxy-3-methoxybenzoic acid) to protocatechuic acid and formaldehyde according to the scheme:

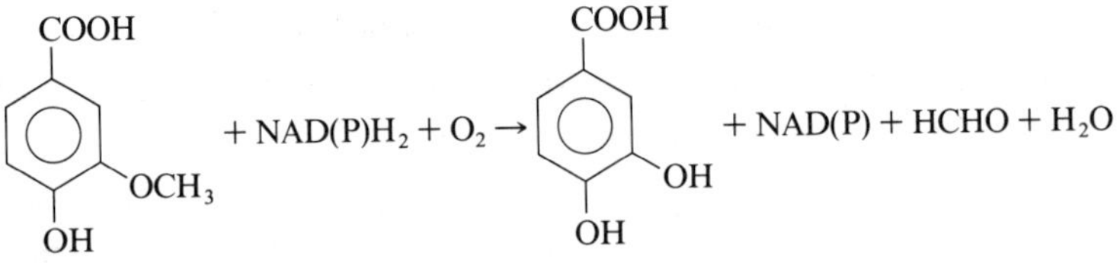

[1] N. J. Cartwright and A. R. W. Smith, *Biochem. J.* **102**, 826 (1967).
[2] D. W. Ribbons, *FEBS Lett.* **8**, 101 (1970).
[3] A. Toms and J. M. Wood, *Biochemistry* **9**, 337 (1970).

Activity of the enzyme may be determined by measuring (1) O_2 consumption with the Clark oxygen electrode or in the Warburg respirometer using standard manometric techniques, or (2) the decrease in absorbance at 340 nm resulting from the oxidation of the NADH or NADPH cofactor.

Assay Methods

1. Enzyme Assay Based on Oxygen Uptake Measurements

a. Manometric Assay
Reagents

Potassium phosphate buffer, 67 m*M*, pH 7.8
NAD(P)H, 10m*M*
Vanillic acid, 10 m*M*, pH 7
Cell extract, ~20 mg protein (from *Pseudomonas fluorescens*, strain T)

Procedure. Warburg flasks contain potassium phosphate buffer, 2.0 ml; NAD(P)H, 1 μmol; and cell extract in a total volume of 2.5 ml in the main compartment; the center well contains 0.2 ml of 20% (w/v) KOH. After equilibration at 30°, the reaction is initiated by mixing 3 μmol of vanillate (0.3 ml) from the side arm. Controls contain water in place of vanillate and oxygen uptake values are corrected for NAD(P)H oxidation.

b. Polarographic Assay
Reagents

Tris–HCl buffer, 50 m*M*, pH 7.2, equilibrated and air saturated at 30°
NAD(P)H, 25 m*M*
Vanillic acid, 25 m*M*, pH 7
Cell extract, ~12 mg protein (from *Pseudomonas testosteroni*)

Procedure. Tris–HCl buffer (2.0 ml), cell extract (0.2–0.3 ml), NAD(P)H (40 μl), and water to a total volume of 3 ml are added to the reaction vessel of a Clark oxygen electrode (Rank, Bottisham, England). Oxygen consumption due to NADH oxidation is monitored for 1–2 min before addition of vanillate (20 μl), and again for 2 min (or until a linear reaction rate is established). Oxygen consumption is corrected for uptake in the absence of substrate.

2. Spectrophotometric Assay

Reagents. As for Part 1,b above.

Procedure. Tris–HCl buffer (2.0 ml), cell extract (0.2–3 ml), NAD(P)H (20 μl), and water to a total volume of 3 ml are added to a cuvette of 1-cm light path in a thermostatted compartment. The reaction is initiated by addition of vanillate (7 μl). The rate of NAD(P)H oxidation is determined by measuring the decrease in absorbance at 340 nm before (1 min) and after addition of vanillate. Values are corrected for NAD(P)H oxidation prior to the addition of substrate.

A simultaneous polarographic and spectrophotometric assay is available using a cuvette fitted with an electrode.[4]

Definition of Enzyme Unit and Specific Activity

One unit of vanillate *O*-demethylase is the amount of enzyme which converts 1 μmol of vanillic acid into protocatechuate and formaldehyde per minute at 30°. Specific activity is expressed as units per milligram of protein, as determined by the method of Bradford[5] with bovine serum albumin as the protein standard.

Growth of Bacteria

Pseudomonas fluorescens strain T (NCIB[6] 12374)[1,7] is grown on a semidefined medium consisting of (g/liter): KH_2PO_4, 0.4; $(NH_4)_2SO_4$, 1.0; $MgSO_4 \cdot 7H_2O$, 0.01; yeast extract (Oxoid), 0.1; and trace element solution,[8] 10 ml/liter, adjusted to pH 7.2–7.5 with 2 *N* NaOH. Final concentrations (mg/liter) of the trace elements are as follow: H_3BO_3, 0.5; $CuSO_4 \cdot 5H_2O$, 0.04; KI, 0.1; $FeCl_3$, 0.2; $MnSO_4 \cdot 4H_2O$, 0.4; $(NH_4)_6Mo_7O_{24} \cdot 4H_2O$, 0.2; and $ZnSO_4 \cdot 7H_2O$, 0.4. A solution of *trans*-ferulic acid as its sodium salt (to give a final concentration of 1.5 g/liter) is prepared separately and sterilized by filtration before adding to the autoclaved mineral salt–yeast extract medium. Bacteria are grown aerobically at 24–26° and harvested by centrifugation during late exponential phase (~ 16 hr). Cells are washed three times with 0.9% NaCl, 67 m*M* potassium phosphate buffer (pH 7.8), or 10 m*M* Tris–HCl buffer (pH 7.8) prior to breakage.

Cultures of *P. testosteroni* (NCIB 8893)[2] and *P. aeruginosa T1*[2] (strain available from DWR) are grown at 30° on a medium[9] made by mixing sterile solutions A and B [3:2 (v/v)]. Solution A, adjusted to pH 7.0 with NaOH, contains (g/liter): KH_2PO_4, 9.0; $(NH_4)_2SO_4$, 2.0; yeast extract, 1.0; and Bacto-peptone, 1.0. Solution B contains 0.15% vanillic acid (neutral-

[4] D. W. Ribbons, F. A. Smith, and A. J. W. Hewitt, *Biotechnol. Bioeng.* **10,** 238 (1968).
[5] M. M. Bradford, *Anal. Biochem.* **72,** 248 (1976).
[6] NCIB, National Collection of Industrial Bacteria.
[7] N. J. Cartwright and J. A. Buswell, *Biochem. J.* **105,** 767 (1967).
[8] J. A. Barnett and M. Ingram, *J. Appl. Bacteriol.* **28,** 131 (1955).
[9] D. W. Ribbons, *J. Gen. Microbiol.* **44,** 221 (1966).

ized to pH 7.0 with NaOH) and 2.5 ml of a trace element mixture consisting of (g/liter): MgO, 11.12; $CaCO_3$, 2.0; ZnO, 0.41; $FeCl_3 \cdot 6H_2O$, 5.4; $MnCl_2 \cdot 4H_2O$, 0.99; $CuCl_2 \cdot 2H_2O$, 0.17; $CoCl_2 \cdot 6H_2O$, 0.24; H_3BO_3, 0.062; and concentrated HCl (58 ml/liter). Bacteria are harvested as described above and pellets washed twice in 50 m*M* Tris–HCl buffer, pH 7.6, prior to disintegration.

Pseudomonas acidovorans is grown at 30° on a medium consisting of (g/liter): *trans*-ferulic acid, 0.7; KH_2PO_4, 2.0; $(NH_4)_2SO_4$, 1.0; $MgSO_4$, 0.05; and $FeSO_4$, 0.01, adjusted to pH 7.2 with 5 *N* NaOH. Cells are harvested as described above and washed with 50 m*M* Tris–HCl buffer (pH 7.2) prior to disruption.

Preparation of Cell Extracts

a. Ultrasonic Treatment. A cell-free extract of *P. fluorescens* is prepared by ultrasonic disruption (3 × 2 min) of a thick slurry of cells (1 vol) in 67 m*M* potassium phosphate buffer, pH 7.8 (1 vol), maintained below 10° with an ice–salt water bath. Coarse particulate material is removed by centrifugation at 72,000 g_{max} for 30 min. Protein concentration of the supernatant fraction (F1) is normally about 45 mg/ml.

b. Press Treatment. In order to avoid inactivation of vanillate *O*-demethylase from *P. acidovorans*,[3] and to obtain more concentrated bacterial extracts, cell breakage can be achieved by press treatment using either the French press,[2] the Hughes press,[3,10] or a modified Hughes press, the X-press (A.B. Biox, Nacka, Sweden).[1] Good breakage of *P. fluorescens* cells is obtained by three or four passages of a cell slurry (see above) through the orifice of the X-press precooled to −20°. Viscosity of the cell break is sufficiently reduced by treatment (15 min at room temperature) with a few crystals of DNase to allow unbroken cells and cell debris to be removed by centrifugation at 72,000 *g* for 45 min. This procedure consistently yields an active supernatant fraction containing 45–60 mg protein/ml or higher.

Enzyme Stability

Vanillate *O*-demethylase activity in cell extracts of *Pseudomonas* spp. is proportional to protein concentration above a minimum threshold and is rapidly and irreversibly deactivated by dilution and oxidation in air.[1,2] The enzyme from *P. fluorescens* is also sensitive to dialysis even under nitrogen atmosphere.[1] Thus, normal procedures available for enzyme purification (e.g., ammonium sulfate fractionation, column chromatography on ion exchangers and gels) are unsuitable for the demethylase. However, dilution

[10] D. E. Hughes, *Br. J. Exp. Pathol.* **32,** 97 (1951).

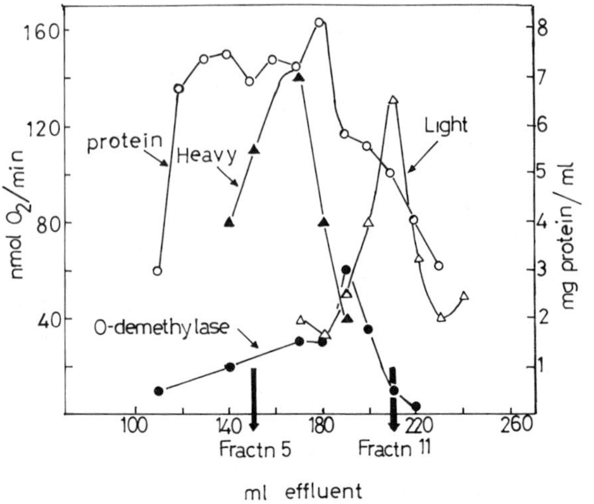

FIG. 1. Separation of vanillate *O*-demethylase fractions on Sephadex G-150. Enzyme activity is determined using 0.5 ml of each fraction as specified under the conditions described for the polarographic assay. (●—●) Vanillate *O*-demethylase in single fractions. (▲—▲) Heavy components assayed in combination with fraction No. 11. (△—△) Light components assayed in combination with fraction No. 5. (○—○) Protein.

of cell extracts of *P. testosteroni* under nitrogen (10- or 20-fold for 24 hr) does not result in loss of demethylase activity provided a minimum protein concentration (3–4 mg) is supplied in the assay mixture. By using anaerobic techniques, vanillate *O*-demethylase in *P. testosteroni* is resolvable into two protein fractions, both of which are required for activity and appear to be oxygen labile.[11] The procedure for separating the two protein bands is as follows.

Pseudomonas testosteroni is grown in a mineral salt medium[9] supplemented with 0.15% vanillic acid, 0.1% yeast extract, and 0.1% Bacto-peptone at 30°. Cells are harvested during late exponential phase (16–17 hr), suspended in 2 vol 50 m*M* Tris–HCl buffer, pH 7.6, and broken in a French press. Extracts are treated with DNase and RNase and centrifuged at 17,000 *g* for 20 min and 100,000 *g* for 2 hr. Supernatant material (100,000 *g* sedimentation), 28 ml, is loaded onto a Sephadex G-150 column (2.5 × 82 cm) previously equilibrated with nitrogen-saturated 20 m*M* Tris–HCl buffer, pH 7.6. Fractions (9–12 ml) are collected manually into rubber-capped bottles with continuous flushing of N_2. Fractions are stoppered under N_2 and stored at 0° prior to assay polarographically.

[11] D. W. Ribbons, *FEBS Lett.* **12,** 161 (1971).

Vanillate *O*-demethylase activity in single and combined fractions is shown in Fig. 1.

The two protein fractions have yet to be characterized although one fraction catalyzes electron transport from $NADH_2$ to 2,6-dichlorophenolindophenol.[11]

Dealkylases from other *Pseudomonas* and *Nocardia* spp. catalyzing both meta- and para-O-demethylation have also been purified and characterized as multicomponent systems.[12,13]

Identification of Protocatechuic Acid and Formaldehyde as Products of Demethylase Activity

Extracts of bacterial cells induced to oxidize vanillate also contain high levels of protocatechuate 3,4- or 4,5-dioxygenase activity.[1-3,7] Since O-demethylation is the rate-limiting step, and vanillate *O*-demethylase is sensitive to traditional enzyme purification techniques, protocatechuate is not readily isolated as the reaction product. However, bacterial extracts which rapidly oxidize vanillate when supplemented with reduced pyridine nucleotide but which exhibit reduced ring-cleavage activity toward protocatchuate may be obtained using an ultracentrifugation procedure.[7]

A cell-free fraction of *P. fluorescens* (fraction F1) is prepared by ultrasonic disruption as described above and diluted with 67 m*M* potassium phosphate buffer to give a concentration of 35 mg protein /ml. Further centrifugation of diluted fraction F1 for 3 hr at 350,000 g_{max} results in a supernatant fraction (F3), arbitrarily taken as the upper 3.5 ml of 10 ml of material contained in the centrifuge tube, which contains high levels of vanillate *O*-demethylase but is virtually devoid of protocatechuate 3,4-dioxygenase. Separation is almost completely reproducible with careful standardization of conditions.

Isolation of protocatechuate as the catabolic product of vanillate *O*-demethylase in fraction F3 is achieved as follows. A pilot Warburg flask containing 5 μmol of vanillate is set up in conjunction with five flasks, each containing 20 μmol of vanillate and the corresponding quantities of buffer, NADH, and enzyme under the conditions described for the manometric assay. After completion of the initial rapid oxygen consumption registered with the pilot flask, the contents of all flasks are combined, 2 ml of 10% (w/v) trichloroacetic acid added, and the precipitated protein removed by centrifugation. The supernatant is continuously extracted with ether for 8 hr, residual solvent removed by evaporation, and the solid residue recrystallized from 2 ml of hot water (yield = ~ 10 mg). Formalde-

[12] F.-H. Bernhardt, H. H. Ruf, and H. Staudinger, *Hoppe-Seyler's Z. Physiol. Chem.* **352,** 1091 (1971).

[13] N. J. Cartwright and D. A. Broadbent, *Microbios* **10,** 87 (1974).

hyde can be detected qualitatively using chromotropic acid[14] or acetylacetone[15] reagents.

Cofactor Requirements. Both NADH and NADPH serve as electron donors for the vanillate *O*-demethylase in extracts of *P. fluorescens,*[1] *P. aeruginosa,*[2] and *P. acidovorans,*[3] although a preference for one or the other cofactor is shown depending on the species. However, transhydrogenase activity in extracts could account for this substitution. Demethylation by the cell-free system from *P. fluorescens* is also reported to require reduced glutathione (GSH).[1]

Substrate Specificity. Activity of crude cell extracts of *P. fluorescens* toward several methoxybenzoate and ethoxybenzoate derivatives indicate that the *O*-demethylase activity is specific for a methoxyl group meta to a carboxyl group.[1] Both substituents but not the hydroxyl group are necessary for oxidation. Thus, 3-ethoxy-4-hydroxybenzoate and guaiacol are not attacked while *m*-methoxybenzoate and veratrate (3,4-dimethoxybenzoate) are converted to *m*-hydroxybenzoate and isovanillate, respectively, at a rate about 60% that for vanillate. Piperonylate (3,4-methylenedioxybenzoate) is not oxidized by vanillate *O*-demethylase from this bacterium.

A similar specificity for 3-O-methyl groups is shown by cell extracts of *P. aeruginosa* grown with vanillate.[2,16] The *O*-demethylase from *P. testosteroni* apparently exhibits a broader spectrum of demethylase activity since extracts of cells induced with vanillate demethylate both 3- and 4-methoxybenzoates and also hydroxylate methyl substituents.[16] However, since crude extracts are involved, it is not clear yet if activity is due to a nonspecific *O*-demethylase or to the presence of different enzymes induced during growth of *P. testosteroni* on vanillate.

Inhibitors. EDTA and 8-hydroxyquinoline at 1 m*M* concentrations completely inhibit *O*-demethylase activity in *P. fluorescens,* which is not restored by addition of Mg^{2+}, Fe^{2+}, Co^{2+}, or Mn^{2+}. Vanillate oxidation is also totally inhibited by 10 m*M* cyanide and by 20% at 1 m*M* concentration.[1]

Stoichiometry. Vanillate *O*-demethylase catalyzes a typical monooxygenase reaction in which vanillate is converted to protocatechuate and formaldehyde with the consumption of 1 mol of O_2 and 1 mol of NADH for each mole of vanillate oxidized.[2] Protocatechuate is further metabolized via aromatic ring cleavage, which consumes a second mole of oxygen per mole of substrate. However, 3-O-demethylation of the substrate ana-

[14] D. A. MacFayden, *J. Biol. Chem.* **158,** 107 (1945).

[15] T. Nash, *Biochem. J.* **55,** 416 (1953).

[16] D. W. Ribbons and J. E. Harrison, *in* "Degradation of Synthetic Organic Molecules in the Biosphere," p. 98. Natl. Acad. Sci., Washington, D.C., 1972.

logs *m*-methoxybenzoate and veratrate is also catalyzed by these extracts with concomitant consumption of 1 mol of O_2 and oxidation of 1 mol NADH/mol of substrate and the quantitative accumulation of the respective phenolic products *m*-hydroxybenzoate and isovanillate.[2]

Although there is little direct evidence, O-demethylation may proceed via hydroxylation of the methoxyl carbon, forming an unstable hemiacetal intermediate which then undergoes hydrolysis to give the reaction products, protocatechuate and formaldehyde. The observation that *m*-toluate and *p*-toluate, which serve as substrates for vanillate *O*-demethylase and a 4-*O*-demethylase, respectively, are converted to the corresponding hydroxymethylbenzoate provides indirect support for this mechanism.[11,17]

[17] F.-H. Bernhardt, H. Staudinger, and V. Ullrich, *Hoppe-Seyler's Z. Physiol. Chem.* **351,** 467 (1970).

[31] Purification of Coniferyl Alcohol Dehydrogenase from *Rhodococcus erythropolis*

By E. JAEGER

Coniferyl alcohol + $NAD^+ \rightarrow$ coniferyl aldehyde + NADH + H^+

The coniferyl alcohol dehydrogenase from *Rhodococcus erythropolis* catalyzes the pyridine nucleotide-dependent oxidation of coniferyl alcohol with the formation of NADH and the corresponding aldehyde. In the organism, the coniferyl aldehyde is further catabolized via ferulic acid and vanillic acid to protocatechuic acid.[1,2] The coniferyl alcohol dehydrogenase also accepts other aromatic alcohols bearing the α,β-unsaturated side chain of coniferyl alcohol, including dilignols. The enzyme is intracellular and may be induced by coniferyl alcohol, ferulic acid, or vanillic acid as a growth substrate.

Recently an intracellular NAD-dependent coniferyl alcohol dehydrogenase of the lignolytic bacteria *Xanthomonas* sp. 99 was described.[3] This enzyme shows nearly the same migration rate as the coniferyl alcohol dehydrogenase from Rhodococcus erythropolis.[4]

[1] L. Eggeling and H. Sahm, *Arch. Microbiol.* **126,** 141 (1980).

[2] E. Jaeger, Thesis. University of Düsseldorf, Düsseldorf, Federal Republic of Germany, 1980.

[3] H. W. Kern, *Arch. Microbiol.* **138,** 18 (1984).

[4] H. W. Kern, L. E. Webb, and L. Eggeling, *Syst. Appl. Microbiol.* **5,** 433 (1984).

Assay Method

Principle

This assay is based on the spectrophotometric determination of coniferyl aldehyde formed with coniferyl alcohol as substrate. For assays with different substrates the equivalent reduction of NAD^+ at 340 nm is measured.

Procedure

In a total volume of 1 ml, the reaction mixture contains the following:

0.2 mmol Tris–HCl buffer, pH 9.0
2.0 μmol NAD^+
4.0 μmol coniferyl alcohol (or other substrates)
0.1 mmol semicarbazide

The reaction is carried out at 24° in a spectrophotometer cell with a 1.0-cm light path and started by the addition of the enzyme solution. The absorbance at 400 or 340 nm is measured, due to the formation of coniferyl aldehyde ($\epsilon = 22{,}400\ M^{-1}\ cm^{-1}$) or NADH ($\epsilon = 6230\ M^{-1}\ cm^{-1}$), respectively.

Definition of Unit and Specific Activity

One unit of enzyme is defined as the amount that catalyzes the oxidation of 1 μmol of coniferyl alcohol or the reduction of 1 μmol of NAD^+ respectively, per minute under the above conditions. Specific activity is expressed as units per milligram protein.

Organism and Cultivation

Rhodococcus erythropolis has been previously described as a *Nocardia* species[5] and is kept at the German Culture Collection, Göttingen (DSM 1069).

Growth Conditions

In preculture, the organism is grown in a complete medium containing, per liter, 10 g yeast extract, 10 g glucose, 1 g KH_2PO_4, and 0.5 g $MgSO_4$. The incubation is performed at pH 7.0 and 27° for 22 hr in 500-ml flasks

[5] J. Trojanowski, K. Haider, and V. Sundman, *Arch. Microbiol.* **114,** 149 (1977).

containing 100 ml medium on a reciprocal shaker. After growth the cells are harvested by centrifugation, washed twice with 10 m*M* potassium phosphate buffer, pH 7.5, and transferred into a 12-liter Biostat V fermenter (Braun, Melsungen, FRG) with a working volume of 8 liters. Fermentation is carried out at 27° with aeration (1 vol/vol/min) and a stirring speed of 1300 rpm. The minimal medium contains, per liter, 0.4 g KH_2PO_4, 1.6 g K_2HPO_4, 0.5 g NH_4NO_3, 0.2 g $MgSO_4 \cdot 7H_2O$, 1 ml of Fe-EDTA solution, and 1 ml of a trace element solution.

The Fe-EDTA solution is prepared as described by Jacobson[6]: 6.53 g EDTA-II is dissolved in 27 ml 1 *N* KOH and after adding 6.25 g $FeSO_4 \cdot 7H_2O$ diluted to 250 ml and aerated for 16 hr to oxidize Fe^{2+} to Fe^{3+}. The trace element solution consists of 50 mg boric acid, 50 mg $CaCl_2$, 100 mg $CuSO_4 \cdot 5H_2O$, 50 mg $MnSO_4 \cdot H_2O$, and 100 mg $ZnSO_4 \cdot 7H_2O$ in 100 ml distilled water. The pH value is maintained at 7.0 with 1 *N* NaOH. Filter-sterilized ferulic acid (0.05%) is used as a carbon and energy source, which gives rise to about 70% of coniferyl alcohol dehydrogenase activity, as compared to growth on coniferyl alcohol. Ferulic acid is added 24 and 39 hr after incubation, 0.05% each time, to give a final yield of 25 g wet weight of cells.

Purification Procedure

All purification steps are carried out at 4° in buffers containing 0.01% sodium azide.

Step 1: Preparation of Cell-Free Extract. The washed cells of one fermentation (25 g) are suspended in 10 m*M* potassium phosphate buffer, pH 7.5, and disrupted with an X-press (AB-Biox, Nacka, Sweden). The homogenate thus obtained is centrifuged at 42,000 *g* and 4° for 30 min. The resulting clear supernatant is used as a cell-free extract.

Step 2: DEAE-Cellulose Chromatography. The cell-free extract is dialyzed overnight against 10 m*M* potassisum phosphate buffer, pH 7.5, containing 0.175 m*M* NaCl. This dialyzed extract is applied to a DEAE-cellulose column (2.6 × 40 cm) equilibrated with 10 m*M* potassium phosphate buffer, pH 7.5, containing 0.175 *M* NaCl (equilibration buffer). After loading the column is washed with 1500 ml of equilibration buffer.

Subsequently a linear gradient is applied. The starting buffer is 10 m*M* potassium phosphate buffer, pH 7.5, containing 0.175 *M* NaCl; the limiting buffer is 10 m*M* potassium phosphate buffer, pH 7.5, containing 0.35 *M* NaCl. The total gradient volume is 500 ml. The flow rate is 50 ml/hr. Fractions exhibiting enzyme activity are combined (about 140 ml)

[6] L. Jacobson, *Plant Physiol.* **25,** 411 (1951).

and concentrated under negative pressure with collodion bags to give a final volume of 6.4 ml.

Step 3: BioGel A 1.5 Filtration. The enzyme solution is applied to a 2.6 × 85 cm column of BioGel A 1.5 (200–400 mesh). The coniferyl alcohol dehydrogenase activity is eluted with 10 m*M* potassium phosphate buffer at a flow rate of 10 ml/hr between 280 and 340 ml. The most active fractions are combined (about 30 ml) and concentrated with collodion bags. The resulting 3.5 ml is dialyzed for 16 hr against 5 m*M* potassium phosphate buffer, pH 7.5.

Step 4: Hydroxyapatite Treatment. The dialyzed enzyme solution is placed on a column (1.6 × 14 cm) packed with hydroxyapatite, which is equilibrated with 5 m*M* potassium phosphate buffer, pH 7.5. The enzyme is not absorbed under these conditions and can be recovered in the void volume. The pooled fractions (45 ml) are concentrated with collodion bags to 5 ml.

Step 5: Sephadex G-200 Filtration. The concentrated enzyme solution derived from the hydroxyapatite chromatography is dialyzed for 16 hr against 10 m*M* Tris–HCl buffer, pH 7.2, containing 0.1 *M* NaCl and 10 m*M* 2-mercaptoethanol. The dialyzed solution is placed on a Sephadex G-200 column (2.6 × 80 cm) equilibrated with the same buffer. Fractions of 4.6 ml are collected at a flow rate of 3.9 ml/hr. The six most active fractions are combined, concentrated to a volume of 5.0 ml, and stored at −20° for use in the experiments. A typical purification is summarized in Table I.

Properties

Purity

When the enzyme fraction obtained from the above purification procedure is subjected to polyacrylamide gel electrophoresis, three major protein bands are visible after staining with Coomassie blue. One of these bands coincides with the coniferyl alcohol dehydrogenase as judged from gels stained for enzyme activity. Attempts to further purify the enzyme by affinity chromatography (Blue Sepharose CL-6B, 5-AMP-Sepharose 4B) are without success. For this reason characterization studies are performed on the 1200-fold enriched preparation of the coniferyl alcohol dehydrogenase.

Stability

The enzyme is stable for 6 weeks in 10 m*M* potassium phosphate buffer, pH 7.5, containing 10 m*M* dithiothreitol or 10 m*M* 2-mercap-

TABLE I

PURIFICATION OF CONIFERYL ALCOHOL DEHYDROGENASE FROM *Rhodococcus erythropolis*[a]

Step	Volume (ml)	Activity (U)	Protein (mg)	Specific activity (U/mg)	Yield (%)	Purification (-fold)
Cell-free extract	160	928	1824.0	0.54	100	1
DEAE-cellulose	6.4	724	85.8	9.3	80.0	18
BioGel A 1.5	3.4	683	40.8	16.75	73.6	31
Hydroxyapatite	5.0	239	6.9	34.64	24.0	64
Sephadex G-200	5.0	132	0.21	644.2	13.3	1193

[a] Typical data obtained with 25 g wet weight of cells.

toethanol. Incubation for 10 min at 40° results in a 40% loss of activity. The enzyme is stored at −20° in 10 m*M* potassium phosphate buffer, pH 7.5, for at least 2 months.

pH Optimum

The optimum pH for the oxidation of coniferyl alcohol is at pH 9.0 in 0.1 *M* Tris–HCl buffer.

Substrate Specificity

The coniferyl alcohol dehydrogenase of *Rhodococcus erythropolis* catalyzes the oxidation of coniferyl alcohol to coniferyl aldehyde. The K_m value calculated from Michaelis–Menten plots by nonlinear regression is found to be 0.645 m*M*. At higher concentrations the substrate inhibits the enzyme activity with a K_i value of 18.3 m*M*.[2,7] A positive reaction is also obtained with the dimeric compounds dehydrodiconiferyl alcohol and guaiacylglycerol β-coniferyl ether. Both monomeric free compounds are dissolved in 10 μl dimethylformamide and given to the test system. The reaction is started by adding the enzyme solution. Determination of the reaction rate is not possible because of the interference in extinction at 340 nm with NADH and the unknown extinction coefficients of the aldehydes formed. Cinnamyl alcohol, which contains the α,β-unsaturated side chain like coniferyl alcohol, and vanillyl alcohol are also attacked with low activity compared to coniferyl alcohol (Table II). Very weak activity is obtained with 4-(4-methoxyphenyl)-1-butanol and 3-(3,4-dimethoxyphenyl)-1-propanol. Other aromatic alcohols (Table II), aliphatic primary

[7] E. Jaeger, L. Eggeling, and H. Sahm, *Curr. Microbiol.* **6**, 333 (1981).

TABLE II
RELATIVE ACTIVITY OF CONIFERYL ALCOHOL DEHYDROGENASE FROM *Rhodococcus erythropolis* WITH DIFFERENT AROMATIC COMPOUNDS (4 m*M*)

Substrate	Activity (%)
4-(4-Methoxyphenyl)-1-butanol	2
Coniferyl alcohol	100
Cinnamyl alcohol	36
3-(3,4-Dimethoxyphenyl)-1-propanol	2
3-(4-Hydroxyphenyl)-1-propanol	0
1-Phenyl-1-propanol	0
Homovanillyl alcohol	0
Veratryl alcohol	0
Vanillyl alcohol	14
4-Hydroxybenzyl alcohol	0
Benzyl alcohol	0

alcohols, and the dimeric compound dehydrodivanillyl alcohol are not accepted as substrates.

As electron acceptor the enzyme needs NAD. The K_m value for NAD^+ is determined to be 0.220 m*M*, whereas the affinity to $NADP^+$ as cosubstrate is characterized by a relatively high K_m value of 12.05 m*M*.

Inhibitors

The enzyme is inhibited completely after a 10-min incubation at 24° with *p*-chloromercuribenzoate at a final concentration of 0.1 μM. A 50% loss of activity results by incubating the enzyme with 0.11 m*M* iodoacetamide or 2.95 m*M* *N*-ethylmaleimide. These results indicate that a cysteine residue is required for activity or for maintaining the proper structure of the protein. No loss of activity was observed with the metal chelators EDTA and α,α'-bipyridyl.

Molecular Weight

The molecular weight of the enzyme is estimated to be 200,000 from its elution profile on a column of Sephadex G-200.

[32] Glucose Oxidase of *Phanerochaete chrysosporium*

By ROBERT L. KELLEY and C. ADINARAYANA REDDY

Glucose oxidase (β-D-glucose: oxygen oxidoreductase, EC 1.1.3.4) has been identified as the predominant source of hydrogen peroxide (H_2O_2) in ligninolytic cultures of *Phanerochaete chrysosporium*,[1-4] a white rot basidiomycete that is being extensively used worldwide in studies on the physiology, biochemistry, and genetics of lignin biodegradation. H_2O_2 plays a central role in lignin biodegradation by this fungus.[3] For example, H_2O_2 is obligately required for the activity of ligninases, a family of lignin peroxidases that is important in the oxidative depolymerization of lignin.[5-8] Both glucose oxidase and ligninase activities are triggered after the cessation of primary growth in response to nitrogen or carbohydrate starvation.[1,9] Both enzyme activities are repressed on addition of exogenous nitrogen sources such as glutamate to nitrogen-starved idiophasic cultures.[1,9] Glucose oxidase-negative (*gox*$^-$) mutants appear to lack both glucose oxidase and ligninase activities, in addition to a few other secondary metabolic activities.[2,4] *gox*$^+$ revertants regain glucose oxidase activity as well as the other activities that are lacking in *gox*$^-$ mutants.

Glucose oxidate catalyzes the oxidation of D-glucose to δ-D-gluconolactone and H_2O_2 in the presence of molecular oxygen[10] (see reaction sequence below). In a subsequent step δ-D-gluconolactone is nonenzymatically hydrolyzed to D-gluconic acid. This enzyme has been demonstrated in various *Aspergillus* and *Penicillium* species[10-12] and has recently been purified from *P. chrysosporium*.[13]

[1] R. L. Kelley and C. A. Reddy, *Arch. Microbiol.* **144,** 248 (1986).
[2] R. L. Kelley, K. Ramasamy, and C. A. Reddy, *Arch. Microbiol.* **144,** 254 (1986).
[3] C. A. Reddy and R. L. Kelley, *in* "Biodeterioration 6" (S. Barry, D. Houghton, C. O'Rear, and G. C. Llewellyn, eds.). Commonw. Agric. Bur., London, 1986.
[4] K. Ramasamy, R. L. Kelley, and C. A. Reddy, *Biochem. Biophys. Res. Commun.* **131,** 436 (1985).
[5] M. Tien and T. K. Kirk, *Proc. Natl. Acad. Sci. U.S.A.* **81,** 2280 (1984).
[6] M. H. Gold, M. Kuwahara, A. A. Chiu, and J. K. Glenn, *Arch. Biochem. Biophys.* **234,** 353 (1984).
[7] T. K. Kirk, S. C. Croan, M. Tien, K. E. Murtagh, and R. L. Farrell, *Enzyme Microbiol. Technol.* **8,** 27 (1986).
[8] V. Renganathan, K. Miki, and M. H. Gold, *Arch. Biochem. Biophys.* **241,** 304 (1985).
[9] B. Faison and T. K. Kirk, *Appl. Environ. Microbiol.* **49,** 299 (1985).
[10] R. Bentley, this series, Vol. 6, p. 37.
[11] B. E. P. Swoboda and V. Massey, *J. Biol. Chem.* **240,** 2209 (1965).

$$\beta\text{-D-Glucose} + \text{enzyme-FAD} \rightleftharpoons \text{enzyme-FADH}_2 + \delta\text{-D-gluconolactone} \quad (1)$$

$$\text{Enzyme-FADH}_2 + O_2 \rightarrow \text{enzyme-FAD} + H_2O_2 \quad (2)$$

Certain enzymes in animal tissues also catalyze oxidation of D-glucose (or derivatives) to δ-D-gluconolactone, but these are readily differentiated from glucose oxidase because they do not require molecular oxygen and H_2O_2 is not a product.[12]

Assay Method

Principle

Glucose oxidase activity can be measured conveniently by a spectrophotometric assay that makes use of the following coupled enzyme reactions:

$$\text{D-Glucose} + O_2 + H_2O_2 \xrightarrow{\text{glucose oxidase}} \text{D-gluconic acid} + H_2O_2 \quad (3)$$

$$H_2O_2 + \text{reduced chromogen} \xrightarrow{\text{peroxidase}} \text{oxidized chromogen} + H_2O \quad (4)$$

H_2O_2 generated by the action of glucose oxidase in reaction 1 is utilized by horseradish peroxidase (EC 1.11.1.7) to oxidize *o*-dianisidine (reduced chromogen) to yield a brown-colored product (oxidized chromogen) and H_2O. The change in absorbance at 460 nm due to the peroxidative oxidation of *o*-dianisidine is a measure of the glucose oxidase activity.

Reagents

- 0.1 *M* citrate–sodium phosphate buffer (pH 4.5)
- 0.31 m*M* *o*-dianisidine in glass-distilled water
- 1.0 *M* D-glucose in glass-distilled water
- 60 U horseradish peroxidase (Sigma Chemical Co., St. Louis, MO) in 1 ml glass-distilled water
- 0.15 U glucose oxidase, type VII (Sigma Chemical Co.) in 1 ml glass-distilled water

Procedure. Reaction mixtures contain 1.5 ml citrate–sodium phosphate buffer, 1.0 ml *o*-dianisidine, 0.3 ml D-glucose, 0.1 ml horseradish peroxidase (HRP; 60 U/ml), and 0.1 ml glucose oxidase solution or *P. chrysosporium* enzyme preparation. The reaction mixture is saturated with O_2 (by bubbling with 100% O_2 for 10 min) prior to the addition of the

---

[12] J. R. Whitaker, "Principles of Enzymology for the Food Sciences," p. 561. Dekker, New York, 1972.

[13] R. L. Kelley and C. A. Reddy, *J. Bacteriol.* **166,** 269 (1986).

glucose oxidase preparation, and incubated in the spectrophotometer for 3 to 5 min to achieve temperature equilibration and to establish blank rate.

The glucose oxidase enzyme preparation is then added and the increase in absorbance at 460 nm is recorded for 3 to 5 min. The change in absorbance (ΔA_{460}) is calculated from the initial linear portion of the curve. Specific activity (U/mg protein) is measured by using a molar extinction coefficient of 8.3.

$$\text{U/mg} = (\Delta A_{460}/\text{min})/(8.3 \times \text{mg enzyme/ml reaction})$$

One unit of activity represents the oxidation of 1 μmol *o*-dianisidine/min at 37° and pH 4.5.

Purification Procedure

Organism and Growth Conditions. P. chrysosporium (ATCC 34541) is maintained on slants of malt extract (ME) agar. This medium contained, per liter: malt extract, 20 g; glucose, 20 g; Bacto-peptone, 1.0 g; and Bacto-agar, 20 g. The pH of the medium is 4.5. For preparing conidial inoculum, the organism is grown on plates of ME medium for 4 to 6 days. Ten milliliters of sterile glass-distilled water is added to each plate to suspend the conidia. This conidial suspension in water is passed through sterile glass wool to remove contaminating mycelia. This filtrate, which contains approximately 5×10^7 conidia/ml, is diluted to give 1.25×10^6 conidia/ml. Foam-stoppered 500-ml Erlenmeyer flasks containing 50 ml of low N medium[14] are inoculated with the above conidial suspension (0.5 ml inoculum/10 ml medium) and are incubated at 37° without agitation for 6 days.

Preparation of Cell Extracts. Mycelia from 10 liters of culture are collected by filtration through six layers of gauze in a Büchner funnel. The mycelial mat (~15 g dry wt) is washed three times with 200 ml of 0.1 *M* sodium phosphate buffer (PO_4 buffer; pH 6.8), resuspended in 100 ml of the same buffer, mixed with glass beads (0.1 mm) in a 1:1 ratio (glass beads:mycelial wet wt), and blended at 4° using an Omni-mixer (Sorvall, Inc., Newton, CT) for 15 min. The glass beads and unbroken mycelia are removed by centrifugation (4080 *g* for 10 min at 4°). The supernatant is collected and the pellet is resuspended in 100 ml of PO_4 buffer and blended for an additional 15 min at 4°. The supernatants are combined and frozen until used. There is negligible loss of glucose oxidase activity even after several weeks of storage in the freezer.

[14] T. K. Kirk, E. Schultz, W. J. Connors, L. F. Lorenz, and J. G. Zeikus, *Arch. Microbiol.* **117**, 277 (1978).

TABLE I
PURIFICATION OF GLUCOSE OXIDASE FROM LIGNINOLYTIC CULTURES OF *Phanerochaete chrysosporium*[a]

Fraction	Specific activity (U/mg)	Total protein (mg)	Total activity (U)	Yield (%)	Purification (-fold)
Crude cell extract	0.17	285	48.4	100	1.0
DEAE-Sephadex	0.91	32	29.1	60	5.4
Sephacryl (S-300)	6.39	3.1	19.8	41	37.8
DEAE-Sepharose	15.1	0.28	4.2	9	89.3

[a] From Kelley and Reddy.[13]

General. Unless mentioned otherwise, all the purification steps are carried out at 4° and the pH of the phosphate buffer used is 6.8. Protein in the enzyme preparations is determined by the method of Lowry *et al.*,[15] crystalline bovine serum albumin (IV; Sigma) being used as the standard.

Step 1: DEAE-Sephadex Chromatography. The frozen cell extracts are thawed, clarified by centrifugation at 27,000 *g* for 15 min, and diluted 5-fold with distilled water. Both previous steps are necessary to remove suspended matter and reduce the viscosity of the cell extracts. The extract is then applied to a DEAE-Sephadex (A50, Pharmacia) column (50 ml of gel, 2.4 × 16 cm gel bed) preequilibrated with 0.01 *M* PO_4 buffer. The column is washed with 50 ml of 10 m*M* PO_4 buffer followed by batch elution with 40-ml vol of PO_4 buffer containing 0.05, 0.10, and 0.25 *M* NaCl. Fractions (2.25 ml) are collected and tested for glucose oxidase activity as described above. In this step about 60% of the total activity present in the crude cell extracts was recovered in the 0.25 *M* NaCl eluate, giving about a 5-fold increase in specific activity (see Table I). Fractions of 0.25 *M* NaCl eluate are desalted on a Sephadex G-25 column and those with the highest activity are pooled and concentrated by ultrafiltration.

Step 2: Sephacryl Chromatography. Concentrated pooled fraction from step 1 is loaded onto a Sephacryl S-300 column (Pharmacia, 170 ml of gel; 2.4 × 46 cm gel bed) equilibrated with 0.1 *M* PO_4 buffer. The column is eluted with the same buffer at a flow rate of 0.25 ml/min. Fractions (1.5 ml) are collected and those with the highest activity are pooled (see Fig. 1A). This step gives a 38-fold enrichment in specific activity and a 41% recovery of total activity (see Table I).

[15] O. H. Lowry, N. J. Rosebrough, A. L. Farr, and R. J. Randall, *J. Biol. Chem.* **193,** 265 (1951).

Step 3: DEAE-Sepharose Chromatography. The pooled fractions from step 2 are applied to a DEAE-Sepharose CL-6B column (Pharmacia; 20 ml of gel, 1.6 × 20 cm gel bed) previously equilibrated with 0.01 *M* PO_4 buffer. Protein is eluted from the column with a linear salt gradient (440 ml total volume, 0 to 100 m*M* NaCl gradient) in 0.01 *M* PO_4 buffer. The NaCl concentration in each fraction is calculated by using conductivity values as compared to those of NaCl standards. The flow rate is 0.25 ml/min and 1.5-ml fractions are collected and tested for activity. The elution profile from the DEAE-Sepharose column (Fig. 1B) shows that a single protein peak has all the glucose oxidase activity with approximately 90-fold enrichment in specific activity and an enzyme recovery of 9% (Table I). This DEAE-Sepharose protein fraction is found to be homogeneous based on SDS-PAGE analysis (Fig. 2). The purification procedure described here is consistently reproducible in our hands. The estimated purity is 99% and the purified protein is stable in the frozen state for at least 1 week.

Properties

Molecular Weight. Based on gel filtration chromatography on a Sephacryl S-300 column, the apparent molecular weight of purified glucose oxidase is estimated to be 180,000. The denatured molecular weight, determined by SDS-PAGE, is estimated to be 80,000 (Fig. 2A). Presumably this enzyme, similar to other glucose oxidases, consists of two identical polypeptides (M_R 80,000 each). The overestimation of the native molecular weight may perhaps be due to hydrodynamic properties of this enzyme being different from those of other glucose oxidases.

Carbohydrate and Flavin Content. Staining of the purified glucose oxidase from *P. chrysosporium* for protein-bound carbohydrate by the dansyl hydrazine method[16] shows no detectable carbohydrate, whereas an equal amount of commercially prepared glucose oxidase from *Aspergillus niger*, which is known to be a glycoprotein,[10,11] stains positive (Fig. 2B). This finding is of interest in light of the previous observations that a majority of peroxisomal proteins appear not to be glycosylated[17] and that H_2O_2 production in ligninolytic cultures of *P. chrysosporium*, presumed to be due to glucose oxidase activity, has been shown to be localized in periplasmic, peroxisome-like structures.[18]

Flavin analysis indicated that the purified enzyme contains 1.5 mol of

[16] A. E. Eckhardt, C. E. Hayes, and I. J. Goldstein, *Anal. Biochem.* **75,** 192 (1976).
[17] A. Volkl and P. B. Lazarow, *Ann. N.Y. Acad. Sci.* **386,** 504 (1982).
[18] L. J. Forney, C. A. Reddy, and H. S. Pankratz, *Appl. Environ. Microbiol.* **44,** 732 (1982).

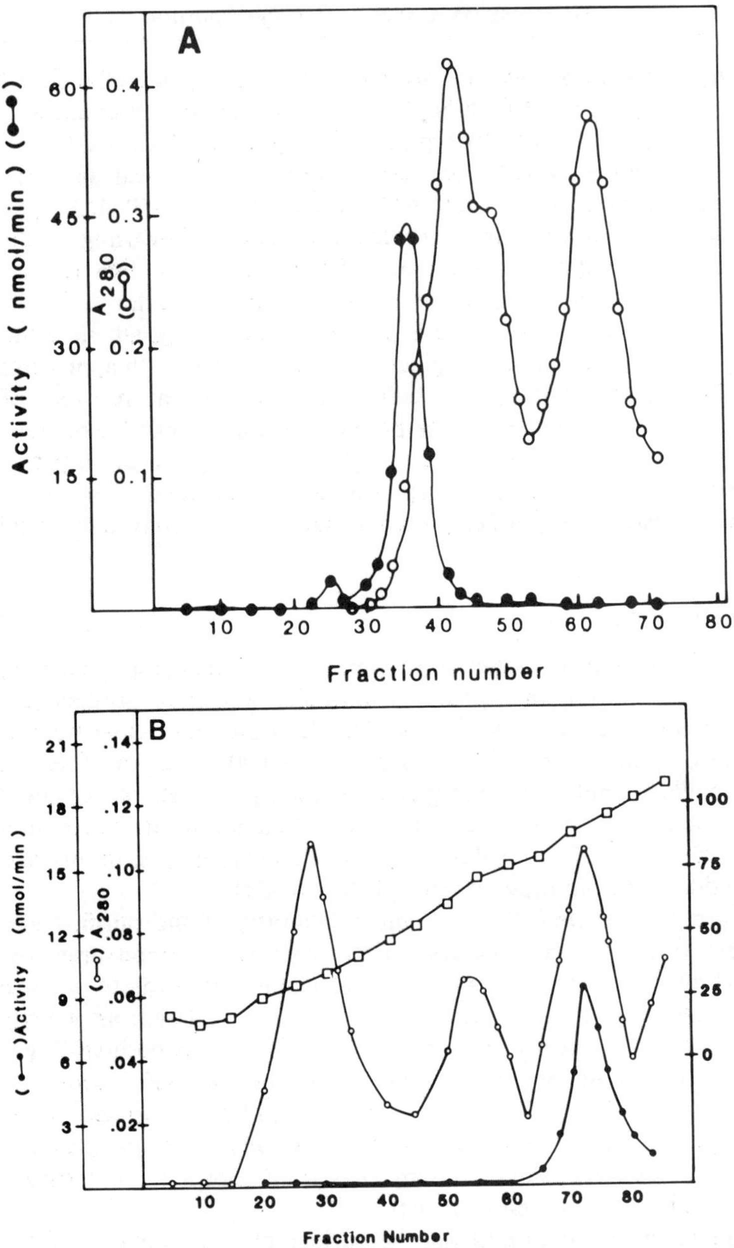

FIG. 1. (A) Elution profile of glucose oxidase activity (●—●: nmol/min/mg protein) and A_{280} (○—○) on a Sephacryl S-300 column. (B) Elution profile of protein (○—○) and glucose oxidase activity (●—●) from a column of DEAE-Sepharose. Protein is eluted from the column with a linear salt gradient (see Methods). Each data point for salt concentration (□—□) is determined by conductivity measurements. Reproduced from Kelley and Reddy.[13]

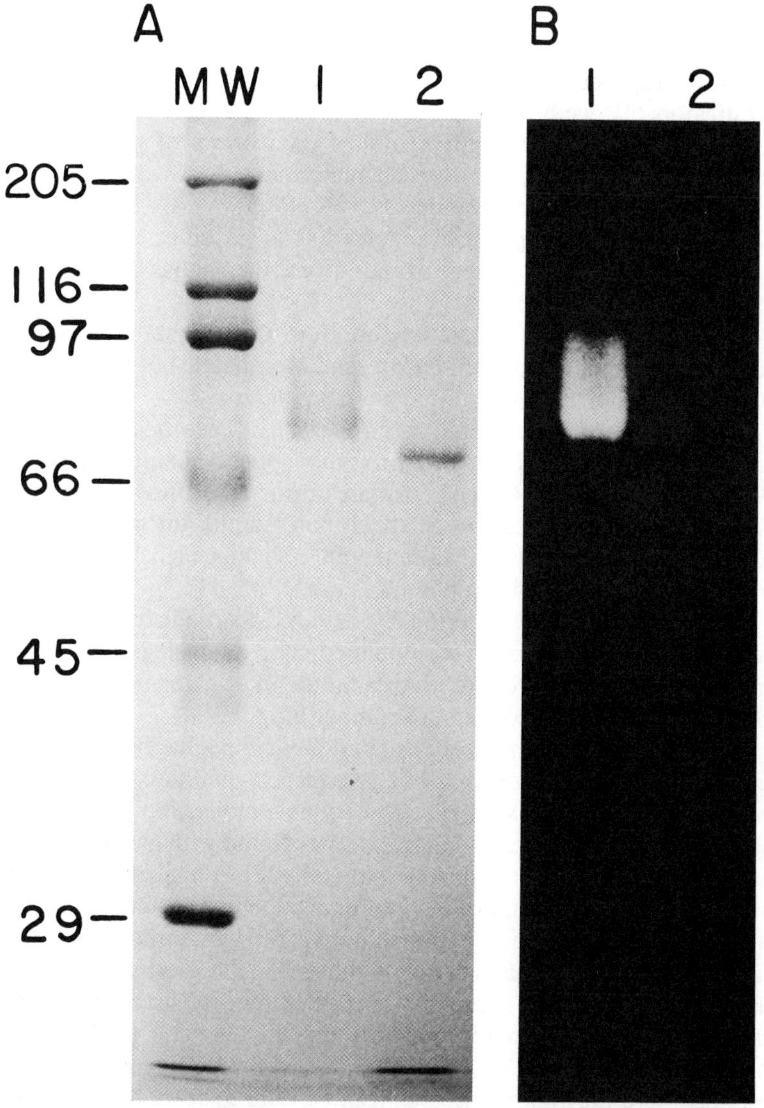

FIG. 2. Staining for carbohydrate and protein after SDS-PAGE of commercial glucose oxidase from *A. niger* and purified glucose oxidase from *P. chrysosporium*. (A) Molecular weight standards (lane MW), *A. niger* glucose oxidase (lane 1), and glucose oxidase from *P. chrysosporium* (lane 2) are stained with Coomassie blue. (B) *Aspergillus niger* glucose oxidase (lane 1) and glucose oxidase from *P. chrysosporium* (lane 2) are stained for carbohydrate by the dansylhydrazine method.[16] In A and B, 1 μg of the respective enzyme is used. Reproduced from Kelley and Reddy.[13]

flavin/mol of protein. Using identical procedures, we showed that *A. niger* glucose oxidase contains 1.6 flavins/mol of protein. Since *A. niger* enzyme has previously been shown to have 2 flavins/mol of protein by earlier investigators,[10,11] we believe that both *P. chrysosporium* and *A. niger* glucose oxidases actually contain 2 mol of flavins/mol of protein and the lower values of 1.5 to 1.6 we obtained experimentally are apparently due to a limitation of the analytical procedure employed by us.[19]

pH Optimum. The purified enzyme has a pH optimum between 4.6 and 5.0. In comparison, glucose oxidase from *A. niger* has been reported to have a pH optimum of 5.6.[10]

Enzyme Inhibition. Glucose oxidase from *P. chrysosporium*, similar to that from *A. niger*, is inhibited by Ag^+ and *o*-phthalate but is not inhibited by Cu^{2+}, KCN, or NaF.

Kinetic Properties and Substrate Specificity. The apparent K_m values for glucose and O_2 for *P. chrysosporium* glucose oxidase are 38 and 0.95 m*M*, respectively. Glucose oxidases from other fungal sources have also been shown to possess a relatively low affinity for glucose with K_m values ranging from 0.11 to 33 m*M* and a slightly higher affinity for O_2 with K_m values ranging from 0.2 to 0.83 m*M*.[10,11]

D-Glucose was stoichiometrically oxidized to D-gluconate; from 28.6 mol of D-glucose oxidized, we obtained 26.1 mol of gluconate, which amounts to 91.2% recovery. These results are in agreement with the results obtained with glucose oxidases from other fungi.[10,11]

Glucose oxidase from *Aspergillus* and *Penicillium* has been shown to be highly specific for β-D-glucose.[11,20] Although D-mannose, D-galactose, 2-deoxy-D-glucose, and D-xylose have been shown to exhibit low activities as substrates, no greater than 2% of the activity found with glucose was found with these or 50 other carbohydrates tested.[11,20] The enzyme from *P. chrysosporium*, on the other hand, appears to be less specific in that it gave 33, 13, and 7% specific activity, respectively, with sorbose, xylose, and maltose compared to that seen with glucose as the substrate. However, a comparison of the V_{max}/K_m ratios for the different substrates clearly shows that glucose is the primary substrate for this enzyme (Table II). The apparent low substrate specificity of *P. chrysosporium* glucose oxidase described here may allow the organism to utilize sugars derived not only from cellulose but also from hemicellulose found in wood material, its natural habitat, to produce H_2O_2 which is known to be important to the ligninolytic system.

[19] P. Cerletti, R. Strom, and M. Giordano, *Arch. Biochem. Biophys.* **101,** 423 (1963).
[20] E. Adams, R. Mast, and A. Free, *Arch. Biochem. Biophys.* **91,** 320 (1960).

TABLE II
SUBSTRATE SPECIFICITY OF GLUCOSE OXIDASE[a]

Substrate	V_{max} (μmol) min^{-1} ml^{-1}	K_m (m*M*)	V_{max}/K_m	Percentage
D-Glucose	15	38.0	0.395	100
L-Sorbose	5	217.4	0.023	5.8
D-Xylose	2	105.2	0.019	4.8
D-Maltose	1	55.5	0.018	4.5

[a] From Kelley and Reddy.[13]

Addendum

Recently, a number of H_2O_2-producing enzymes have been isolated from ligninolytic cultures of *P. chrysosporium.* These enzymes include a number of extracellular peroxidases which require extracellular coenzyme(s),[21-23] a glucose-2-oxidase,[24] and an extracellular, glyoxal oxidase.[25] It is doubtful that the extracellular proxidases are major sources of H_2O_2 in lignin-degrading cultures of *P. chrysosporium* because these are present in low amounts and it has not adequately been demonstrated that this organism can generate the reduced coenzyme(s) or other substrates in amounts needed for extracellular H_2O_2 production. The extracellular, glucose-2-oxidase isolated by Eriksson *et al.*,[24] from the K-3 isolate of *P. chrysosporium*, is similar to the enzyme that we have isolated, except that GLC analysis of the products of the latter enzyme indicated that it was a glucose-1-oxidase.[13] It is possible that glucose-2-oxidase[24] represents a different source of H_2O_2 in idiophasic cells of *P. chrysporium.* Kersten and Kirk[25] described a glyoxal oxidase enzyme which may be important as a source of H_2O_2 in lignin degradation. These authors indicated that the specific activity of this enzyme is roughly the same as that of the glucose oxidase that we have isolated[13]; however, no evidence has been presented to show that there is sufficient concentration of glyoxal or methylglyoxal in the culture fluid of *P. chrysosporium* to support this activity. Kersten and Kirk[25] also claim that an extracellular enzyme is more likely the source of

[21] Y. Asada, M. Miyabe, M. Kikkawa, and M. Kuwahara, *Agric. Biol. Chem.* **50,** 525 (1986).
[22] J. Glenn and M. Gold, *Arch. Biochem. Biophys.* **242,** 329 (1985).
[23] A. Paszczyński, V.-B. Huynh, and R. Crawford, *Arch. Biochem. Biophys.* **244,** 750 (1986).
[24] K.-E. Eriksson, B. Pettersson, J. Volc, and V. Musilek, *Appl. Microbiol. Biotechnol.* **23,** 257 (1986).
[25] P. Kersten and T. K. Kirk, *J. Bacteriol.* **169,** 2195 (1987).

H_2O_2 involved in ligninolytic activity. However, glucose oxidase activity is not necessarily intracellular. Forney *et al.*[18] reported that unique periplasmic microbodies are the predominant sites of H_2O_2 production in ligninolytic cultures of *P. chrysosporium* and H_2O_2 may leak out from these periplasmic sites into the extracellular environment. In conclusion, although several different enzymes of *P. chrysosporium* have been implicated in H_2O_2 production, we believe that current evidence still favors glucose-1-oxidase (and perhaps glucose-2-oxidase) as the primary physiological source of H_2O_2 in ligninolytic cultures of *P. chrysosporium*.

Acknowledgments

This research was supported, in part, by Grant DE-FG02-85#13369 from the U.S. Department of Energy, Division of Basic Biological Sciences. This is publication No. 12051 from the Michigan Agricultural Experiment Station.

[33] Pyranose 2-Oxidase from *Phanerochaete chrysosporium*

By J. Volc and Karl-Erik Eriksson

$$\text{D-Glucose} + O_2 \rightarrow \text{D-}\textit{arabino}\text{-2-hexosulose} + H_2O_2$$

Introduction

Enzymes with glucose 2-oxidase activity, tentatively assigned as pyranose 2-oxidase (EC 1.1.3.10), have so far been isolated from mycelial cultures of two basidiomycetes, *Polyporus obtusus*[1] and *Trametes versicolor*.[2] The enzyme is also present in extracts of a number of other higher fungi[2,3] and appears to play an important metabolic role.[4,5] In contrast to glucose oxidase (EC 1.1.3.4), which is specific for C-1 of glucose, pyranose

[1] H. W. Ruelius, R. M. Kerwin, and F. W. Janssen, *Biochim. Biophys. Acta* **167**, 493 (1968); F. W. Janssen and H. W. Ruelius, *Biochim. Biophys. Acta* **167**, 501 (1968).

[2] Y. Machida and T. Nakanishi, *Agric. Biol. Chem.* **48**, 2463 (1984).

[3] J. Volc, N. P. Denisova, F. Nerud, and V. Musilek, *Folia Microbiol.* **30**, 141 (1985).

[4] J. Volc, P. Sedmera, and V. Musilek, *Folia Microbiol.* **23**, 292 (1978).

[5] K.-E. Eriksson, B. Pettersson, J. Volc, and V. Musilek, *Appl. Microbiol. Biotechnol.* **23**, 257 (1986).

2-oxidase catalyzes the oxidation of several carbohydrates at C-2 to produce 2-keto derivatives and hydrogen peroxide.[1,2,4] The preferred substrate for the enzyme is D-glucose.

Assay Method

Principle. Hydrogen peroxide produced during C-2 oxidation of D-glucose is determined spectrophotometrically by a peroxidase–chromogen assay based on the oxidative coupling of 3-methyl-2-benzothiazolinone hydrazone (MBTH) and 3-(dimethylamoino)benzoic acid (DMAB) to produce an intense blue, probably indamine, dye. Oxidase activity is calculated from the increase in absorbance at 590 nm using a molar absorption coefficient of 3.2×10^4.

Reagents

Sodium phosphate buffer, 100 m*M*, pH 6.5
Glucose–peroxidase reagent: 100 m*M* D-glucose and horseradish peroxidase (5 U/ml) in the above buffer
DMAB reagent: 25 m*M* 3-(dimethylamino)benzoic acid in the above buffer
MBTH reagent: 1 m*M* 3-methyl-2-benzothiazolinone hydrazone hydrochloride monohydrate (Fluka) in distilled H_2O
Potassium cyanide, 100 m*M* in H_2O

Procedure. Glucose 2-oxidase activity is measured using a modification of the chromogenic assay for peroxidase.[6] A standard assay mixture (2 ml total volume) contains 1 ml glucose–peroxidase reagent, 200 μl DMAB reagent, 100 μl MBTH reagent, phosphate buffer, pH 6.5, and the test enzyme solution (≤ 100 μl) which is added to initiate the reaction. In the fixed time method, the reaction is terminated after 5 min incubation at 25° by the addition of 1 ml 100 m*M* KCN and the absorbance at 590 nm read against a blank containing every component except the enzyme. The color produced is stable for at least 20 min. In kinetic studies, the enzyme (0.03 U) is incubated with variable substrate concentrations substituting for 50 m*M* glucose in the standard assay mixture and the change in absorbance is monitored continuously. A calibration graph for the determination of H_2O_2 in the above reaction mixture is linear over the range $3-40 \times 10^{-6}$ *M* H_2O_2. Appropriate dilutions of commercially available H_2O_2 solutions are prepared using a molar absorption coefficient at 240 nm of 43.6.[7]

[6] T. T. Ngo and H. M. Lenhoff, *Anal. Biochem.* **105,** 389 (1980).
[7] A. G. Hildebrand and I. Roots, *Arch. Biochem. Biophys.* **171,** 385 (1975).

Definition of Enzyme Unit and Specific Activity. One unit (U) of glucose 2-oxidase activity catalyzes the formation of 1 μmol H_2O_2/min when D-glucose is oxidized to D-*arabino*-2-hexosulose (D-glucosone) under the assay conditions. Specific activity is expressed as units per milligram of protein, as determined by the Folin reagent[8] with bovine serum albumin as the protein standard.

Purification Procedure

Growth of the Organism. The basidiomycete *Phanerochaete chrysosporium* Burds, strain K-3,[9] is cultivated at 25° in 500-ml shake flasks (reciprocal frequency 2 Hz) containing 80 ml of a complex glucose–corn steep medium[5] and vegetative inoculum (4%). The composition of the glucose–corn steep medium is the following: 20 g D-glucose, 7 g corn steep (Novo Industry A/S, Denmark), and 1.5 g $MgSO_4 \cdot 7H_2O$ made up to 1000 ml with tap water. The pH is adjusted to 5.5 with 6 *M* NaOH prior to autoclaving. The inoculum is prepared by mild homogenization (Ultra-Turrax homogenizer) under aseptic conditions of 8-day culture grown up on the same medium from an agar slant inoculum. After 13 days of cultivation the mycelium is harvested by filtration, washed with water, and stored at −20°. Cell mass from one flask (0.6 g dry weight) contains approximately 15 U of glucose 2-oxidase.

Crude Enzyme Extract. Frozen mycelium (60 g wet weight) is thawed in 2 vol of 50 m*M* phosphate buffer, pH 6.5, supplemented with 12 g powdered microcrystalline cellulose and disrupted in an Ultra-Turrax homogenizer (Shaft 18KG, full speed, 3 × 1 min under cooling). The homogenate is centrifuged (10,000 *g*, 20 min) and the supernatant (136 ml) retained as the crude extract. Solid ammonium sulfate is then added, with gentle stirring, to 25% saturation (1.1 *M*) at 4°. Precipitated protein is removed by centrifugation and the supernatant purified at room temperature using the following three chromatographic steps.

Step I: Hydrophobic interaction chromatography on phenyl-Sepharose CL-4B (Pharmacia): The pH of the ammonium sulfate supernatant is adjusted to 6.5 with 1 *M* Tris and the entire volume of supernatant applied to a phenyl-Sepharose bed ($V_t = 44$ ml, K 26/40 column, Pharmacia) previously equilibrated with eluent A [1 *M* $(NH_4)_2SO_4$ in 50 m*M* sodium phosphate, pH 6.5]. After washing the column with eluant A, the adsorbed material is eluted with a linear gradient (from 20 to 100%) of eluent B (30% ethylene glycol in 30 m*M* sodium phosphate, pH 6.5) in 440 ml and frac-

[8] E. F. Hartree, *Anal. Biochem.* **48,** 422 (1972).

[9] S. C. Johnsrud and K.-E. Eriksson, *Appl. Microbiol. Biotechnol.* **21,** 320 (1985).

tions (8 ml) collected at a flow rate of 3 ml/min. Pooled active fractions (94 ml) show a 12% increase in the original glucose oxidase activity applied to the column, which is probably due to removal of a glucose 2-oxidase inhibitor.

Step II: Anion-exchange chromatography on a Mono-Q column: Pooled active fractions from the phenyl-Sepharose column are concentrated (to 12 ml) and transferred to buffer A (20 m*M* Tris–HCl, pH 6.5) using the Amicon model 52 UF cell, XM100 membrane. Aliquots of sample (4 ml/run; 13 mg protein each) are applied via 10 ml Superloop to a Mono-Q HR 5/5 column equipped with fast protein liquid chromatography (FPLC) attachment (Pharmacia) and washed with three column volumes of eluant A. Elution is then continued with a linear gradient from 0 to 40% B (1 *M* NaCl in buffer A) in 22 ml and effluent monitored at 280 nm. A peak of glucose 2-oxidase activity appears at 16% B. Fractions of 0.5 ml are collected at a flow rate of 1 ml/min.

Step III: FPLC chromatofocusing on a Mono-P column: Pooled active fractions from Step II (9 ml, three runs) are concentrated and transferred to buffer A (25 m*M* piperazine–HCl buffer, pH 6.3) by passage through a PD-10 column (Pharmacia) (6 ml). After equilibration of the column with A (45 ml), 2-ml aliquots of the sample per run (2.5 mg protein each) are applied to the Mono-P HR 5/20 column. Elution is initiated with 3 ml of buffer A before introducing the pH gradient with eluent B [10% Polybuffer 74 (Pharmacia), pH 4]. The gradient volume is 40 ml and the flow rate 1 ml/min; fraction size is 0.3 ml. After each run, the column is regenerated with 1 ml 2 *M* NaCl and reequilibrated as above. The elution profile (280 nm, 2.0 absorbance units full scale) shows incomplete separation of the glucose 2-oxidase peak from step II into four main components of very similar elution pH values: 4.95, 4.92, 4.88, and 4.84. Fractions corresponding to the double peak at pH 4.95 and 4.92 are combined (2 ml containing 45% activity) and the Polybuffer removed using a Sephadex G-75 column. This material is adopted for the partial characterization of the enzyme described below. Peaks at pH 4.88 and 4.84 have lower specific activities and have not been further examined.

Data from a typical purification of glucose 2-oxidase are summarized in Table I. The procedure provides high yields (96% recovery) of substantially pure enzyme in only two steps; the third step (chromatofocusing) appears to cause a partial separation of active isoenzyme components.

Properties of the Isolated Glucose 2-Oxidase

General. The purified enzyme has a specific activity of 15.5 U/mg protein. It is stable at 4° in 30 m*M* sodium phosphate buffer, pH 6.5 (less

TABLE I
PURIFICATION OF GLUCOSE 2-OXIDASE FROM MYCELIUM OF THE BASIDIOMYCETE
Phanerochaete chrysosporium

Step	Total protein (mg)	Total enzyme (U)	Specific activity (U/mg protein)	Purification factor (-fold)	Yield (%)
Crude extract	267.9	93	0.35	—	100
Phenyl-Sepharose	38.5	104	2.7	8	112
Mono-Q	7.4	96	13.5	39	96
Mono-P[a]	2.9	45	15.5	45	45

[a] Fraction of isoforms eluted at pH 4.95 and 4.92.

than 1% loss in activity after 3 weeks), and has a broad pH optimum with maximum activity at pH 7.5.[5] The enzyme is completely inactivated at 80° for 15 min. Electrophoresis on SDS-polyacrylamide reveals a close double band of protein corresponding to a subunit molecular weight of about 70,000. This double band probably consists of almost identical subunits, each originating from one of the two coisolated enzyme isoforms of p*I* 4.95 and 4.92, respectively, based on their chromatofocusing elution pH values.

Spectral Properties. The oxidized form of the enzyme exhibits absorption maxima at 360 and 457 nm. Reduction of the enzyme by the substrate glucose under nitrogen atmosphere, or by sodium dithionite, results in elimination of the absorption maximum at 457 nm. These spectral properties are characteristic of a flavoprotein. The flavin moiety has not yet been identified.

Substrate Specificity. Of the sugars investigated (Table II), D-glucose is the preferred substrate. Activities with D-xylose and L-sorbose are almost the same as recorded for glucose 2-oxidase from *Polyporus obtusus*.[1] Sugars which contain a glucosidic bond are not substrates for the enzyme. K_m values determined from Lineweaver–Burk plots for D-glucose and D-xylose are 1.03 and 20 m*M*, respectively, at pH 6.5.

Inhibitors. Glucose 2-oxidase is inhibited by Ag^+, Hg^{2+}, and PCMB at 1 m*M* concentration. Sodium azide, Cu^{2+}, and α,α'-dipyridyl are only slightly inhibitory at this concentration (>85% activity is retained).

Analysis of the D-Glucose Oxidation Product. The glucose oxidation product exhibits the R_f value on TLC as authentic D-*arabino*-2-hexosulose (D-glucosone) and yields a characteristic blue color with diphenylamine–aniline–phosphoric acid detection reagent which differs from the color

TABLE II
SUBSTRATE SPECIFICITY OF GLUCOSE 2-OXIDASE FROM *P. chrysosporium*[a]

Substrate	Relative activity	Substrate	Relative activity
D-Glucose	100	L-Arabinose	<5
6-Deoxy-D-glucose	73	L-Fucose	<5
β-D-Gluconolactone[b]	60	D-Lyxose	<5
L-Sorbose	52	*N*-Acetyl-D-glucosamine	<5
D-Xylose	37	Lactose	<5
2-Deoxy-D-glucose	18	Maltose	<5
D-Galactose	5	Cellobiose	<5
D-Mannose	<5	Methyl-β-D-glucoside	<5
D-Fructose	<5	Methyl-α-D-glucoside	<5
D-Arabinose	<5		

[a] Values expressed as a percentage of the activity toward glucose. The activity is determined by incubating 100 μmol of each compound with 0.03 U of the oxidase in the standard reaction mixture. From Eriksson *et al.*[5]

[b] A freshly prepared solution was used for the measurement.

reactions of monocarboxyl sugars. TLC analysis is performed on cellulose-coated sheets (Lucefol, Kavalier, Czechoslovakia) using the solvent system butyl acetate–acetic acid–acetone–water (140:100:33:80). The analysis further demonstrated that D-*arabino*-2-hexosulose is not accumulated in the reaction mixture at higher yields but appears as a transient metabolite, which is continuously transformed to another product. The same end product is detected (dephenylamine reagent, yellow color) when authentic D-*arabino*-2-hexosulose is incubated with the enzyme preparation. Recently we have demonstrated that the transient primary product of D-glucose transformation by the purified enzyme, D-*arabino*-2-hexosulose, is further converted to the pyrone cortalcerone, first discovered as an antibiotic metabolite of the fungus *Corticium caeruleum*.[10] TLC on silica gel sheets (DC-Plastikfolien Kieselgel $60F_{254}$, Merck; solvent system *n*-butanol–acetic acid–water, 4 : 1 : 1) and staining for α-diols[11] fails to show the presence of either δ-gluconolactone or gluconic acid. Production of chromogen in the assay system with peroxidase implies that the other reaction product is hydrogen peroxide.

All the characteristics presented suggest that the enzyme described here can be assigned the EC number 1.1.3.10 originally ascribed to pyranose oxidase from the fungus *Polyporus obtusus*.[1]

[10] M.-A. Baute, R. Baute, J. Deffieux, and M.-J. Filleau, *Phytochemistry* **16,** 1895 (1977).

[11] J. B. Weiss and I. Smith, *Nature (London)* **215,** 638 (1967).

Metabolic Role. Recent studies have provided evidence for the central role of hydrogen peroxide as a cosubstrate in ligninolytic systems (H_2O_2 dependence of lignin-depolymerizing peroxidases, first demonstrated in 1983,[12,13] and of lignin-demethoxylating activities[14]). Thus, the finding that glucose 2-oxidase gives rise to H_2O_2 production during idiophase,[5] i.e., the ligninolytic phase, allows the suggestion that the enzyme plays an important role in lignin degradation by *P. chrysosporium*. Ramasamy *et al.*[15] suggested recently that the main physiological source of H_2O_2 in ligninolytic cultures of the fungus is glucose oxidase (EC 1.1.3.4) and described the isolation of glucose oxidase-negative mutants which are deficient in lignin-degrading capacity. However, since the product of glucose oxidase was not identified it is not possible to evaluate which enzyme, glucose 1-oxidase and/or glucose 2-oxidase, is involved.

[12] M. Tien and T. K. Kirk, *Science* **221,** 661 (1983).
[13] J. K. Glenn, M. A. Morgan, M. B. Mayfield, M. Kuwahara, and M. H. Gold, *Biochem. Biophys. Res. Commun.* **114,** 1077 (1983).
[14] P. Ander and K.-E. Eriksson, *Appl. Microbiol. Biotechnol.* **21,** 96 (1985).
[15] K. Ramasamy, R. L. Kelley, and C. A. Reddy, *Biochem. Biophys. Res. Commun.* **131,** 436 (1985).

[34] Methanol Oxidase of *Phanerochaete chrysosporium*

By KARL-ERIK ERIKSSON and ATSUMI NISHIDA

$$CH_3OH + O_2 \rightarrow HCHO + H_2O_2$$

Methanol Oxidase (EC 1.1.3.13) catalyzes the oxidation of methanol and other low-molecular-weight primary alcohols to the corresponding aldehydes and H_2O_2. We report here the isolation and characterization of this enzyme from the mycelium of *Phanerochaete chrysosporium*, a basidiomycete.

In addtion to the organism studied here, several other basidiomycetes, including *Polyporus obtusus*, *Polyporus versicolor*, and *Lenzites trabea*, have also been demonstrated to produce this enzyme.[1-3] However, the

[1] F. W. Janssen, R. M. Kerwin, and H. W. Ruelius, *Biochem. Biophys. Res. Commun.* **20,** 630 (1965).
[2] F. W. Janssen and H. W. Ruelius, *Biochim. Biophys. Acta* **151,** 330 (1968).
[3] R. M. Kerwin and H. W. Ruelius, *Appl. Microbiol.* **17,** 347 (1969).

enzyme methanol oxidase is produced not only in basidiomycetes, but also in yeasts such as *Kloeckera* sp. No. 2201[4] and *Hansenula polymorpha*.[5]

Assay Method

Principle. The enzyme activity is measured by estimation of oxygen consumption using an oxygen electrode.

Reagents. Sodium phosphate buffer (0.1 *M*, pH 7.0), methanol, and crude mycelial extract or partially purified enzyme solution are used.

Procedure. A reaction mixture containing 250 μmol phosphate buffer, 300 μmol methanol, and crude mycelial extract (1.0 ml) or partially purified enzyme solution (0.1 to 0.5 ml) is mixed and adjusted with water to a total volume of 3.0 ml. The activity is measured at 30° by estimation of oxygen consumption using an oxygen electrode (Yellow Springs Instrument Co., Inc., Yellow Springs, USA).

Definition of Enzyme Unit and Specific Activity. One unit (U) of methanol oxidase is defined as the amount of enzyme consuming 1 μmol of oxygen per minute under the assay conditions. Specific activity is expressed as units per milligram of protein as determined by the method of Lowry *et al.*[6] with bovine serum albumin (Sigma Chemical Co., St. Louis, MO) as the protein standard.

Production of Methanol Oxidase

Growth of the Fungus. The basidiomycete *Phanerochaete chrysosporium* Burds, strain K-3[7] (anamorph *Sporotrichum pulverulentum* ATCC 32629), was used. The fungus has been cultivated in either 6- (4 liters of medium, 30°) or 30-liter (20 liters of medium, 39°) jar fermenters. The medium used contains the following components per liter: 2.0 g $(NH_4)H_2PO_4$, 2.5 g glucose, 0.6 g KH_2PO_4, 0.4 g K_2HPO_4, 0.5 g $MgSO_4 \cdot 7H_2O$, 0.1 g yeast extract, 74 mg $CaCl_2 \cdot 2H_2O$, 12 mg ferricitrate$\cdot 3H_2O$, 6.6 mg $ZnSO_4 \cdot 7H_2O$, 5 mg $MnSO_4 \cdot 8H_2O$, 1 mg $CoCl_2 \cdot 6H_2O$, 0.1 g thiamin-HCl. The pH was 5.8.

A conidiospore suspension serves as inoculum (final concentration, 1×10^6 spores/ml). After 1 day of cultivation, vanillate solution (sterilized by a 0.45 μm Millipore filter) is added to the medium (final vanillate concentration, 1 m*M*). After another 48 hr of incubation (total 72 hr),

[4] Y. Tani, T. Miya, H. Nishikawa, and K. Ogata, *Agric. Biol. Chem.* **36,** 68 (1972).

[5] B. Sherry and R. H. Abeles, *Biochemistry* **24,** 2594 (1985).

[6] O. H. Lowry, N. J. Rosebrough, A. L. Farr, and R. J. Randall, *J. Biol. Chem.* **193,** 265 (1951).

[7] S. C. Johnsrud and K.-E. Eriksson, *Appl. Microbiol. Biotechnol.* **21,** 320 (1985).

mycelial pellets are harvested by filtration on a Büchner funnel and washed with small amounts of water. The washed mycelium is either used immediately or stored frozen at −25° until required.

The jar fermenter conditions are as follows: rotation of the impeller at 300 rpm, aeration by 0.5 vol of air per min/volume of medium and pressure at 0.15 kg/cm^2 (6-liter jar) or 0.3 kg/cm^2 (30-liter jar). Under these conditions, approximately 180 g mycelium (wet weight) is obtained from 20 liters of medium.

Purification Procedure

Step 1. Fermenter-grown mycelium is harvested after 72 hr of growth. A crude enzyme extract is prepared by suspending 350 g (wet weight) of mycelium in 800 ml of 0.1 *M* phosphate buffer, pH 7.2. The mycelium is dispersed by a Physcotron (Polytron type, Niti-On, Ltd., Japan) disperser 3× for 2 min, 20,000 rpm, at 4°. It is then homogenized under N_2 at 4° in a Vibrogen cell mill (Edmund Bühler, Tübingen, West Germany) using a continuous vessel with a flow rate of 300 ml/hr. The crude extract is centrifuged at 14,000 *g* for 60 min, and the supernatant (1050 ml) is used as the cell-free extract (cf. Table I).

Step 2: Protamine Sulfate Treatment. Three hundred and seventy milligrams Na_2SO_4 (ca. 1 mg protamine sulfate/10 mg protein) is neutralized in 30 ml dilute ammonium hydroxide. The solution is added to the crude extract and the mixture agitated for 30 min.

The precipitate is removed by centrifugation at 14,000 *g* for 60 min. The supernatant obtained (1050 ml) contains the enzyme and is further treated as described in Table I.

Step 3: Fractionation with Ammonium Sulfate. Crystalline ammonium sulfate is gradually added to the supernatant of step 2 to 40% saturation.

TABLE I
PURIFICATION OF METHANOL OXIDASE FROM *Phanerochaete chrysosporium*

Step	Volume (ml)	Protein (mg)	Total activity (U)	Specific activity (U/mg)	Yield (%)	Purifica (-fol
1. Crude extract	1050	5874	739	0.126	100	1
2. Protamine treatment	1050	5174	739	0.143	100	1.
3. $(NH_4)_2SO_4$ fractionation (40–55%)	63	2163	779	0.360	106	2.
4. Sephadex G-50 chromatography	100	1942	860	0.443	116	3.
5. DEAE-Sepharose chromatography	140	274	776	2.837	105	22.
6. Sephacryl S-300 chromatography	77	71.2	1232	17.30	167	137

The pH value is adjusted to pH 7.7 with dilute ammonium hydroxide. After 1 hr the precipitate is removed by centrifugation at 14,000 *g* for 60 min. Ammonium sulfate is added to the supernatant to 55% saturation. The mixture is stored overnight at 4°. The precipitate containing the enzyme is collected by centrifugation at 14,000 *g* for 60 min and dissolved in 10 m*M* phosphate buffer, pH 7.2; the volume is 63 ml.

Step 4: Desalting by Sephadex G-50 Column Chromatography. The resulting solution from step 3 is desalted on a Sephadex G-50 column (2.6 × 93 cm) equilibrated with 10 m*M* phosphate buffer, pH 7.2. The eluate (100 ml) is used in step 5.

Step 5: Chromatography on a DEAE-Sepharose CL-6B Column. The fractions containing enzyme activity are applied to a DEAE-Sepharose column (5.0 × 27 cm) and equilibrated with 10 m*M* phosphate buffer, pH 7.2. The elution is carried out with 750 ml of the same buffer and subsequently with a gradient of phosphate buffer (10 to 300 m*M*), total volume 2 liters. The elution rate is 50 ml/hr. The most active fractions are combined and the enzyme precipitated by addition of solid ammonium sulfate to 70% saturation. The precipitate is collected after standing overnight by centrifugation at 20,000 *g* for 30 min. The precipitate is dissolved in 30 ml of 50 m*M* phosphate buffer, pH 7.2.

Step 6: Chromatography on a Sephacryl S-300 Column. The solution obtained in the preceding step is applied to a Sephacryl S-300 column (2.6 × 95 cm) and equilibrated with 50 m*M* phosphate buffer, pH 7.9. The elution is carried out with the same buffer. The most active fractions are combined (77 ml) and used for enzyme characterization.

Properties

General. Based on protein molecular weight standards, the molecular weight of methanol oxidase is estimated to be 310,000. The purity of the enzyme and the molecular weight investigation on SDS-polyacrylamide slab gels give one single band with a calculated molecular weight of approximately 75,000, which suggests that the enzyme is composed of four subunits of the same molecular weight.

The isoelectric point estimated on an isoelectric focusing column was found to be at pH 5.4. The enzyme exhibits a broad pH optimum with no activity differences in the range pH 6.0–10.5. However, there is a rapid drop in activity below pH 6.

The relation between oxygen uptake and formaldehyde and H_2O_2 formation, studied at various enzyme concentrations, shows that a given oxygen consumption corresponds to somewhat higher than stoichiometric yields of both aldehyde and H_2O_2.

Methanol oxidase of *Phanerochaete chrysosporium* is a flavoprotein where FAD is the flavin component. The FAD content, calculated from spectrophotometric data, is shown to be 5.62 mol/mol of the enzyme.

Substrate Specificity. Methanol oxidase of *Phanerochaete chrysosporium* catalyzes the oxidation of several other simple primary alcohols but at varying rates: ethanol (91.9), *n*-propanol (34.4), and *n*-butanol (10.6). The figures in parentheses give relative activity (methanol, 100) toward primary alcohols. Toward higher alcohols, the activity is negligible.

Role of the Enzyme. Methanol oxidase is found in relatively large amounts in extracts of wood-degrading basidiomycetes.[3] The enzyme gives rise to hydrogen peroxide when the fungus exhibits secondary metabolism, the only phase during which lignin is degraded.[8] It has recently been shown that hydrogen peroxide, normally produced by the oxidation of glucose and other sugars,[9] is absolutely necessary for lignin degradation.[10] We have recently shown that the methoxyl groups in lignin are converted to methanol,[11,12] and the enzyme methanol oxidase, through its involvement in the production of hydrogen peroxide from methanol, can facilitate the ability of white rot fungi to degrade lignin.

[8] P. Keyser, T. K. Kirk, and J. G. Zeikus, *J. Bacteriol.* **73,** 294 (1978).
[9] K.-E. Eriksson, B. Pettersson, J. Volc, and V. Musilek, *Appl. Microbiol. Biotechnol.* **23,** 257 (1986).
[10] R. L. Crawford and D. L. Crawford, *Enzyme Microb. Technol.* **6,** 434 (1984).
[11] P. Ander and K.-E. Eriksson, *Appl. Microbiol. Biotechnol.* **21,** 96 (1985).
[12] P. Ander, K.-E. Eriksson, and M. E. R. Eriksson, *Physiol. Plant.* **65,** 317 (1985).

Section II

Pectin

A. Assays for Pectin-Degrading Enzymes
Article 35

B. Purification of Pectin-Degrading Enzymes
Articles 36 through 43

[35] Assay Methods for Pectic Enzymes

By ALAN COLLMER, JEFFREY L. RIED, and MARK S. MOUNT

Pectic enzymes are produced in large quantities by many plant-associated microorganisms.[1,2] Pectic polymers (chains of predominantly 1,4-linked α-D-galacturonic acid and methoxylated derivatives) are major constituents of the middle lamellae and primary cell walls of higher plants.[3] Because of the structural importance and enzymatic vulnerability of these polymers, pectic enzymes can cause plant tissue maceration, cell lysis, and modification of cell wall structure, allowing other depolymerases to act on their respective substrates.[3-5]

As typified by the well-characterized extracellular enzyme complexes of the plant pathogens *Erwinia chrysanthemi* and *Erwinia carotovora*, pectolytic microorganisms produce a battery of pectic enzymes differing in substrate preference, reaction mechanism, and action pattern.[6,7] Pectate lyase[8,9] (EC 4.2.2.2), pectin lyase[10,11] (EC 4.2.2.10), and exopolygalacturonate lyase[12] (EC 4.2.2.9) cleave by β-elimination and generate products with a 4,5-unsaturated residue at the nonreducing end. Polygalacturonase[13] (EC 3.2.1.15) and exo-poly-α-D-galacturonosidase[14,15] (3.2.1.82) cleave by hydrolysis. The lyases have pH optima of ca. 8.5 and (except for pectin lyase) require divalent cations. The hydrolases have pH optima at pH 6.0 or lower and do not require divalent cations. Pectate lyase, pectin lyase, and polygalacturonase cleave internal glycosidic bonds and generate a series of oligomeric products. Exopolygalacturonate lyase and exo-poly-

[1] D. F. Bateman and R. L. Millar, *Annu. Rev. Phytopathol.* **4,** 119 (1966).
[2] F. M. Rombouts and W. Pilnik, *Econ. Microbiol.* **5,** 227 (1980).
[3] M. McNeil, A. G. Darvill, S. C. Fry, and P. Albersheim, *Annu. Rev. Biochem.* **53,** 625 (1984).
[4] D. F. Bateman and H. G. Basham, *Encycl. Plant Physiol., New Ser.* **4,** 316 (1976).
[5] A. Collmer and N. T. Keen, *Annu. Rev. Phythopathol.* **24,** 383 (1986).
[6] J. P. Stack, M. S. Mount, P. M. Berman, and J. P. Hubbard, *Phytopathology* **70,** 267 (1980).
[7] J. L. Ried and A. Collmer, *Appl. Environ. Microbiol.* **52,** 305 (1986).
[8] A. Garibaldi and D. F. Bateman, *Physiol. Plant Pathol.* **1,** 25 (1971).
[9] F. Moran, S. Nasuno, and M. P. Starr, *Arch. Biochem. Biophys.* **123,** 298 (1968).
[10] S. Kamimiya, T. Nishiya, K. Izaki, and H. Takahashi, *Agric. Biol. Chem.* **38,** 1071 (1974).
[11] S. Tsuyumu, T. Funakubo, and A. K. Chatterjee, *Physiol. Plant Pathol.* **27,** 119 (1985).
[12] C. Hatanaka and J. Ozawa, *Agric. Biol. Chem.* **37,** 593 (1973).
[13] S. Nasuno and M. P. Starr, *J. Biol. Chem.* **241,** 5298 (1966).
[14] A. Collmer, C. H. Whalen, S. V. Beer, and D. F. Bateman, *J. Bacteriol.* **149,** 626 (1982).
[15] C. Hatanaka and J. Ozawa, *Agric. Biol. Chem.* **33,** 116 (1969).

α-D-galacturonosidase attack chain termini, release only digalacturonates, and reduce the viscosity of solutions containing pectic polymers more slowly than do the endo-attacking enzymes. Pectin lyase attacks pectin (polymethoxygalacturonide); the other enzymes attack pectate (polygalacturonate). Assays for pectinesterase (EC 3.1.1.11), which deesterifies pectin to pectate and methanol, are comprehensively discussed by Rexová-Benková and Markovič.[16]

The characterization and purification of a pectic enzyme is often complicated by the presence of other pectic enzymes produced by the source microorganism. Three assay procedures that address this problem will be described; pectic enzymes from *Erwinia* spp. will be used as examples. The assays are readily adapted to the analysis of pectic enzyme complexes of other organisms and can be used with crude preparations. Results from plant tissue extracts should be interpreted cautiously, however, because of the prevalence of pectic enzyme inhibitors.[5] The activity stains in the first procedure enable rapid approximation of the number and type of pectic enzymes in the sample and can guide purification strategies, particularly when used in conjunction with titration curves.[17] The second and third procedures permit quantitative assay of hydrolytic and β-eliminative pectic enzymes, respectively. The action patterns of purified pectic enzymes are determined by viscometric[18] and reaction product analyses; oligogalacturonides are resolved by paper chromatography[19] or thin-layer chromatography[20] and detected with bromphenol blue[21] or thiobarbituric acid spray reagent[22] (which specifically detects 4,5-unsaturated oligogalacturonides).

Activity Stain for Rapid Characterization of Pectic Enzymes Resolved by Electrophoretic Procedures

Principle. Pectic enzymes diffuse from electrophoresis gels into ultrathin polygalacturonate-agarose overlays. After a period of incubation, the overlays are placed in a solution of cetyltrimethylammonium bromide or ruthenium red, which precipitates and stains the undegraded substrate, leaving a clear zone in the overlay that corresponds to the location of the enzymes in the electrophoresis gel.[17,23] Pectic hydrolases and lyases present

[16] L. Rexová-Benková and O. Markovič, *Adv. Carbohydr. Chem. Biochem.* **33,** 323 (1976).

[17] Y. Bertheau, E. Madgidi-Hervan, A. Kotoujansky, C. Nguyen-The, T. Andro, and A. Coleno, *Anal. Biochem.* **139,** 383 (1984).

[18] D. F. Bateman, *Phytopathology* **53,** 1178 (1963).

[19] C. W. Nagel and M. M. Anderson, *Arch. Biochem. Biophys.* **112,** 322 (1965).

[20] R. T. Zink and A. K. Chatterjee, *Appl. Environ. Microbiol.* **49,** 714 (1985).

[21] A. L. Demain and H. J. Phaff, *Arch. Biochem. Biophys.* **51,** 114 (1954).

[22] L. Warren, *Nature (London)* **186,** 237 (1960).

[23] J. L. Ried and A. Collmer, *Appl. Environ. Microbiol.* **50,** 615 (1985).

in the same gel are differentially assayed by manipulating reaction conditions in the overlays.

Reagents

GelBond support film for agarose, 110 × 205 mm (FMC Bioproducts, Rockland, ME)
Glass plates, two, 110 × 205 mm
Teflon spacers, 0.35 × 5 mm
Agarose (Bethesda Research Laboratories, Rockville, MD; electrophoresis grade)
Polygalacturonic acid (Pfaltz and Bauer, Inc., Waterbury, CT; product P21750)
Sodium hydroxide, 1 *N*
Pectate lyase substrate: 0.1 *M* tris(hydroxymethyl)aminomethane (Tris)–HCl buffer, pH 8.5, 1.5 m*M* calcium chloride, 0.1% (w/v) polygalacturonic acid
Exo-poly-α-D-galacturonosidase substrate: 0.1 *M* potassium phosphate buffer, pH 6.0, 10 m*M* ethylenediaminetetraacetic acid (EDTA), 0.1% (w/v) polygalacturonic acid
Polygalacturonase substrate: 0.1 *M* sodium acetate buffer, pH 5.3, 10 m*M* EDTA, 0.1% (w/v) polygalacturonic acid
Ruthenium red (Sigma Chemical Co., St. Louis, MO), 0.05% (w/v) in water
Cetyltrimethylammonium bromide (Sigma Chemical Co., St. Louis, MO), 1.0% (w/v) in water

Procedure. The substrate solutions are prepared in advance and stored frozen. To prevent formation of an insoluble calcium-polygalacturonate complex during preparation of the pectate lyase substrate, 0.2 g of polygalacturonic acid is first dissolved in 100 ml of 0.2 *M* Tris–HCl, pH 8.7, and then 100 ml of 3.0 m*M* calcium chloride is added. Each of the substrate solutions must be titrated back to the desired pH with 1 *N* sodium hydroxide after addition of polygalacturonic acid. Agarose (0.25 g) is then added to 25 ml of the appropriate substrate solution and heated in a boiling water bath until it dissolves. The hot solution is then immediately pipetted into the gel mold with a preheated pipet. The gel mold is assembled by the following steps. Several milliliters of water are spread on a 110 × 205 mm glass plate. A sheet of GelBond support film with the hydrophilic side up is lowered onto the plate. Paper towels are placed on the support film and excess water is pressed out from beneath it with a rubber roller. A 0.35 × 5 × 205 mm spacer is placed along one edge of the plate, and then three 0.35 × 5 × 105 mm spacers are placed perpendicularly at the middle and outer edges. A second glass plate of the same size is laid on the spacers, offset so that 5 mm of the support film is exposed at the open end of the mold. The plates are clamped together and heated to 50° in an oven. The hot polygalacturonate-agarose solution (a little less than 5 ml) is then carefully delivered to each offset opening in the mold, allowing none to flow between the support film and the bottom plate. After 30 min at room temperature, the plates are separated and the support film is cut in half with scissors along the gap left by the middle spacer. The two overlays can be used immediately or covered with Saran Wrap (Dow Chemical Corp., Indianapolis, IN) and stored at 4° in a plastic box containing a wet paper towel for at least 3 weeks. Assays are initiated by placing the overlay on the surface of an electrophoresis gel (typically an ultrathin layer isoelectric focusing gel cast on support film of the same dimensions). Bubbles are pressed out with a glass rod. After a period of incubation at 30° varying from a few minutes to several hours, depending on the desired sensitivity, the overlays are placed in solutions of either ruthenium red or cetyltrimethylammonium bromide for at least 20 min. Overlays stained with ruthenium red are then rinsed with water and can be photographed through a green filter (Wratten No. 58) or dried and stored between plastic sheets. Overlays precipitated with cetyltrimethylammonium bromide have a white background and must be stored in the staining solution. The staining reagents can be reused for several gels. Because the toxicity of these reagents is unknown, both types of gels are handled with gloves.

Comments. The technique is sufficiently sensitive to detect an isoelectrically focused band containing 6.4×10^{-6} U of pectate lyase activity.[23] The sensitivity and high resolution results from the assay mechanism of detecting zones of enzymatic degradation in a background of limited and relatively slowly diffusing substrate. The same electrophoresis gel can be assayed sequentially with substrate overlays buffered to detect pectate lyases and pectate hydrolases. Overlays buffered for exo-poly-α-D-galacturonosidase or polygalacturonase detection can be interchanged without substantial loss in sensitivity. Pectic enzymes with an exo-cleaving action pattern will produce partially cleared bands in the overlay. In principle, pectin lyases should be detectable with overlays containing pectin. When the p*I* of the enzyme is substantially different than the pH optimum, the sensitivity of the assay is enhanced by incubating the electrophoresis gel in the appropriate buffer before applying the overlay (5 min is optimum for 0.35-mm 5% polyacrylamide isoelectric focusing gels). Pectic enzymes in 0.75-mm 10% polyacrylamide, sodium dodecyl sulfate gels can be renatured by incubating the gels for 2 hr with shaking in three 100-ml changes of the appropriate buffer (10 m*M*) before assaying with substrate overlays. Not all enzymes renature equally well, and some are inactivated by treatment with dithiothreitol.[23]

Assay for Polygalacturonase

Principle. The increase in reducing groups resulting from release of oligogalacturonates from polygalacturonate is measured with the copper-arsenomolybdate reagents of Nelson,[24] as modified by Somogyi.[25]

Reagents

Potassium sodium tartrate
Anhydrous sodium carbonate
Cupric sulfate pentahydrate
Sodium hydrogen carbonate
Anhydrous sodium sulfate
Ammonium molybdate
Sulfuric acid
Disodium hydrogen arsenate heptahydrate
Substrate stock solution: 60 m*M* sodium acetate buffer, pH 5.3, 0.12 *M* sodium chloride, 6.0 m*M* EDTA, 0.24% (w/v) polygalacturonic acid
D-Galacturonic acid (Sigma Chemical Co., St. Louis, MO)

Procedure. The copper reagent is prepared by dissolving 12 g of sodium potassium tartrate and 24 g of anhydrous sodium carbonate in 250 ml of water. The following reagents are then added in sequence while stirring: (1) a solution of 4.0 g of cupric sulfate pentahydrate in 100 ml of water; (2) 16 g of sodium hydrogen carbonate; and (3) a solution of 180 g of anhydrous sodium sulfate in 500 ml of water, boiled before addition to expel air. The volume of the final solution is brought to 1 liter with water. A precipitate that forms during the first week is removed by filtration or sedimentation and the clear blue supernatant is then used in assays. The arsenomolybdate reagent is prepared by dissolving 25 g of ammonium molybdate in 450 ml of water and adding 21 ml of 96% sulfuric acid and a solution of 3.0 g of disodium hydrogen arsenate heptahydrate in 25 ml of water. The final solution is incubated at 37° for 24 hr before use and is stored in a brown bottle. The substrate stock solution is prepared by adding 20 ml of 0.6 *M* sodium chloride with rapid mixing to 80 ml of a solution containing 75 m*M* sodium acetate, pH 5.3, 7.5 m*M* EDTA, and 0.3% (w/v) polygalacturonic acid (final pH adjusted to 5.3 with 1 *N* sodium hydroxide). The substrate stock solution can be preserved with 0.02% sodium azide and stored at 4°.

Enzyme assays are initiated by mixing 2.5 ml of substrate stock solution and 0.5 ml of enzyme sample plus water. The reactions are incubated

[24] N. Nelson, *J. Biol. Chem.* **153,** 375 (1944).
[25] M. Somogyi, *J. Biol. Chem.* **195,** 19 (1952).

at 30°. Samples (0.5 ml) are removed from the reaction mixture at timed intervals and immediately mixed with 0.5 ml of copper reagent (which stops the reaction). Sample tubes are subsequently incubated in a vigorously boiling water bath (with a marble over each tube) for 10 min. After the tubes have cooled to room temperature, 1.0 ml of the arsenomolybdate reagent is added. After a 15- to 40-min incubation at room temperature, the assay mixture is centrifuged in a clinical centrifuge to remove precipitated residual substrate, and the absorbance of the supernatant is read at 500 nm in a spectrophotometer. The increase in absorbance over time in reaction samples, relative to a substrate blank or relative to samples taken at several intervals, is then calculated. These values are used in reference to a D-galacturonic acid standard curve to determine the rate of oligogalacturonic acid formation. One unit of enzyme forms 1 μmol of oligogalacturonate in 1 min under the conditions of the assay. Oligogalacturonates as large as the pentamer yield the same absorbance at 500 nm as galacturonic acid in the Somogyi–Nelson assay.[26]

Comments. Purification of polygalacturonase from *E. carotovora* has been described.[13] EDTA is present in the reaction mixture only to inhibit the activity of contaminating pectate lyases. Citrate buffer should be avoided because it interferes with the Somogyi–Nelson assay.[27] The *E. chrysanthemi* exo-poly-α-D-galacturonosidase is assayed by a similar procedure.[14] A reducing sugar assay based on the formation of UV-absorbing complexes with 2-cyanoacetamide has been reported to be simpler and more sensitive than the Somogyi–Nelson assay for pectic enzymes but has not yet met wide use.[28,29]

Assay for Pectate Lyase

Principle. The increasing absorbance at 232 nm of 4,5-unsaturated reaction products is monitored with a recording spectrophotometer.

Reagents

Substrate stock solution: 60 mM Tris–HCl, pH 8.5, 0.6 mM calcium chloride, 0.24% (w/v) polygalacturonic acid

Procedure. The substrate stock solution is prepared in the manner described for the pectate lyase substrate in the activity stain procedure. To initiate the assay, 2.5 ml of the substrate stock and 0.5 ml of enzyme

[26] L. Rexová-Benková, *Eur. J. Biochem.* **39,** 109 (1973).
[27] L. G. Paleg, *Anal. Chem.* **31,** 1902 (1959).
[28] K. C. Gross, *HortScience* **17,** 933 (1982).
[29] S. Honda, Y. Matsuda, M. Takahashi, and K. Kakehl, *Anal. Chem.* **52,** 1079 (1980).

sample plus water (both previously equilibrated to 30°) are rapidly mixed in a 3-ml cuvette with a 1-cm light path. The subsequent increase in absorbance at 232 nm is monitored as a function of time with a recording spectrophotometer. The amounts of enzyme and water that are added must be empirically adjusted to yield a linear rate of reaction for at least 30 sec. One unit of enzyme forms 1 μmol of 4,5-unsaturated product in 1 min under the conditions of the assay. The molar extinction coefficient for the unsaturated product at 232 nm is 4600 M^{-1} cm^{-1}.[19]

Comments. Purification of pectate lyase from *E. chrysanthemi* and *Escherichia coli* clones carrying individual pectate lyase isozyme genes has been described.[8,30,31] The same assay procedure can be used with exopolygalacturonate lyase.[32] The assay for *Erwinia chrysanthemi* pectin lyase is altered by the deletion of calcium chloride from the substrate solution and by the substitution of highly methoxylated Link pectin for polygalacturonate.[11] Tsuyumu *et al.*[11] describe a recent modification of the method of Morell *et al.*[33] for the preparation of this substrate. The molar extinction coefficient of the unsaturated product is 5550 M^{-1} cm^{-1}.[34,35] The thiobarbituric acid procedure of Waravdekar and Saslaw[36] modified by Weissbach and Hurwitz,[37] provides an alternative procedure for assaying pectic lyases; 4,5-unsaturated products, following treatment with periodic acid, form a complex with thiobarbituric acid that absorbs maximally at 545–550 nm.

[30] C. Schoedel and A. Collmer, *J. Bacteriol.* **167,** 117 (1986).
[31] N. T. Keen, D. Dahlbeck, B. Staskawicz, and W. Belser, *J. Bacteriol.* **159,** 825 (1984).
[32] J. D. Macmillan and H. J. Phaff, this series, Vol. 8, p. 632.
[33] S. Morell, L. Bauer, and K. P. Link, *J. Biol. Chem.* **105,** 1 (1934).
[34] P. Albersheim, this series, Vol. 8, p. 628.
[35] R. D. Edstrom and H. J. Phaff, *J. Biol. Chem.* **239,** 2403 (1964).
[36] V. S. Waravdekar and L. D. Saslaw, *Biochim. Biophys. Acta* **24,** 439 (1957).
[37] A. Weissbach and J. Hurwitz, *J. Biol. Chem.* **234,** 705 (1959).

[36] Protopectinase from Yeasts and a Yeastlike Fungus

By TAKUO SAKAI

Protopectin is a water-insoluble pectic substance found in plants. The decomposition of protopectin was originally attributed to one specific enzyme, "protopectinase,"[1] but further research on pectolytic enzymes

[1] C. S. Brinton, W. H. Dore, H. J. Wichmann, J. J. Willaman, and C. P. Wilson, *J. Am. Chem. Soc.* **49,** 38 (1927).

showed that such decomposition is due to the action of a group of enzymes, including pectinesterase (pectin pectylhydrolase, EC 3.1.1.11), polygalacturonase [poly(1,4-α-D-galacturonide) glycanohydrolase, EC 3.2.1.15], and pectin lyase [poly(methoxygalacturonide) lyase, EC 4.2.2.10]. However, in most studies, the protopectin-solubilizing enzymes have been taken to be enzymes that macerate plant tissues. Recently, Sakai *et al.*[2-7] found microorganisms producing a protopectin-solubilizing enzyme (and names it protopectinase) that liberates highly polymerized water-soluble pectic substances from protopectin without marceration of plant tissue. Here, the isolation and properties of the protopectinase are discussed.

Assay Methods

Assay of Protopectinase Activity

Principles. Protopectinase activity is assayed by measuring the amount of pectic substance liberated from protopectin by the carbazole–sulfuric acid method.[8]

Reagents. The reagents used for the assay of enzyme activity are shown in Table I.

Protopectin is prepared by the following procedure. The albedo layer of lemon (*Citrus limon* Burm) peel is scooped out, pooled, washed with distilled water until the water-soluble substances that react with carbazole–sulfuric acid are washed off, and then lyophilized. The dried protopectin preparation is powdered to a 100–200 mesh powder. The protopectin is stocked in the refrigerator.

Procedure. The reaction mixture contains 10 mg of protopectin, 40 μmol of acetate buffer containing 50 μg/ml of bovine serum albumin, pH 5.0, and 0.5 ml of enzyme solution, in a total volume of 2.5 ml. The reaction mixture is incubated beforehand for 10 min at 37°. The reaction is started by the introduction of 0.5 ml of enzyme solution, and the mixture is kept for 30 min at 37°. The reaction is stopped by the cooling of the reaction mixture in an ice bath. The control blank is run using heat-denatured enzyme solution. After the reaction, the mixture is filtered on a Toyo

[2] T. Sakai and M. Okushima, *Agric. Biol. Chem.* **42,** 2427 (1978).

[3] T. Sakai and M. Okushima, *Appl. Environ. Microbiol.* **39,** 908 (1980).

[4] T. Sakai and M. Okushima, *Agric. Biol. Chem.* **46,** 667 (1982).

[5] T. Sakai, M. Okushima, and M. Sawada, *Agric. Biol. Chem.* **46,** 2233 (1982).

[6] T. Sakai and S. Yoshitake, *Agric. Biol. Chem.* **48,** 1941 (1984).

[7] T. Sakai and S. Yoshitake, *Agric. Biol. Chem.* **48,** 1951 (1984).

[8] S. Furutani and Y. Osajima, *Sci. Bull. Fac. Agric., Kyushu Univ.* **22,** 35 (1965).

TABLE I
REAGENTS USED FOR ENZYME ASSAY

Reagents	Amounts for one sample
For protopectinase activity	
Protopectin	10 mg
Acetate buffer (20 m*M*), containing 50 μg/ml of bovine serum albumin, pH 5.0	2 ml
Enzyme solution	0.5 ml
Carbazole solution (0.2%, in ethanol)	0.5 ml
Sulfuric acid (32 *N*)	6 ml
For polygalacturonase activity	
Pectic acid solution (1%) (adjusted to pH 5.0 with 0.1 *N* NaOH)	3 ml
McIlvaine's buffer, pH 5.0	3 ml
Enzyme solution	1 ml

No. 2 filter paper (Toyo Roshi Co., Ltd., Tokyo, Japan). To a test tube containing 0.5 ml of filtrate of the reaction mixture is introduced 6 ml of chilled 32 *N* H_2SO_4 solution, and then 0.5 ml of 2% carbazole solution in ethanol. This step is done in an ice bath. Then the assay mixture is heated at 80° for 20 min and cooled to room temperature, after which the optical density at 525 nm is measured. The pectin concentration is measured as D-galacturonic acid from the standard assay curve with D-galacturonic acid.

Definition of Units and Specific Activity. One unit of protopectinase activity is defined as the activity that liberates pectic substance corresponding to 1 μmol of D-galacturonic acid/ml of reaction mixture at 37° in 30 min, and specific activity is expressed as units per milligram of protein.

Assay of Polygalacturonase Activity

Polygalacturonase activity is assayed by the viscosity reduction of pectic acid solution using Ostwald's viscometer, as follows. First, 6 ml of 0.5% pectic acid solution in McIlvaine's buffer, pH 5.0, in Ostwald's viscometer, is incubated at 37° for 3 min, and to this is added 1 ml of enzyme solution, and the mixture is incubated at 37° for 5 min. The rate of viscosity reduction *(A)* is calculated using the equation:

$$A = [(T_a - T)/(T_a - T_0)]100$$

where T is the flow time (sec) of the reaction mixture, T_a is the flow time (sec) of pectic acid solution added to the heat-inactivated enzyme, and T_0 is the flow time (sec) of water added to the heat-inactivated enzyme. One unit of enzyme activity is defined as the activity reducing the viscosity by 50%.

Purification of Protopectinases

Protopectinase is an extracellular protein produced by *Kluyveromyces fragilis, Galactomyces reessii,* and *Trichosporon penicillatum.* The protein is obtained by purification of culture filtrates.

Purification Procedure for Protopectinase from Kluyveromyces fragilis (Protopectinase F)[7]

Step 1: Production of Enzyme. Kluyveromyces fragilis IFO 0288 is used for the production of protopectinase F. The microorganism is maintained on agar slants of a medium containing 2% glucose, 0.6% peptone, and 0.5% yeast extract, pH 5.0. For enzyme production, the microorganism is aerobically cultured in a medium containing 3% glucose, 0.6% peptone, 0.2% yeast extract, and 0.08% Silicone KM-70 (an antifoaming agent from Sin-Etsu Chemical Co., Ltd., Tokyo, Japan), pH 5.0, at 30°. The enzyme is first produced after 5 hr of cultivation and production reaches a maximum at 15 hr of cultivation. Thirty-seven liters of culture filtrate is concentrated by evaporation *in vacuo* at 30° to 1.5 liters and used for enzyme purification.

Step 2: CM-Sephadex C-50 Column Chromatography. The concentrated culture filtrate is dialyzed thoroughly against 20 m*M* acetate buffer, pH 5.0, and then put on a CM-Sephadex C-50 column (3 × 50 cm) equilibrated with 20 m*M* acetate buffer, pH 5.0. The column is washed thoroughly with acetate buffer, pH 5.0, and then the enzyme is eluted with 350 ml of linear gradient of 0 to 400 m*M* NaCl in the same buffer, at a flow rate of 20 ml/hr. The fractions containing enzyme activity are pooled and concentrated to about 5 ml by evaporation *in vacuo* at 30°.

Step 3: Sephadex G-75 Column Chromatography. The concentrated enzyme solution is chromatographed using a Sephadex G-75 column (2.2 × 80 cm) equilibrated with 20 m*M* acetate buffer, pH 5.0, containing 200 m*M* NaCl, and elution is done at a flow rate of 6.7 ml/hr. Protopectinase activities are recovered in two peaks (Fig. 1a). Most of the activity is recovered in F II, which is concentrated to 2 ml by evaporation *in vacuo* at 30°. The chromatography is repeated once more and the enzyme solution obtained is concentrated to 10 ml.

Step 4: Crystallization. To this fraction is added solid ammonium sulfate until faint turbidity is observed. After being left for 1 week in a

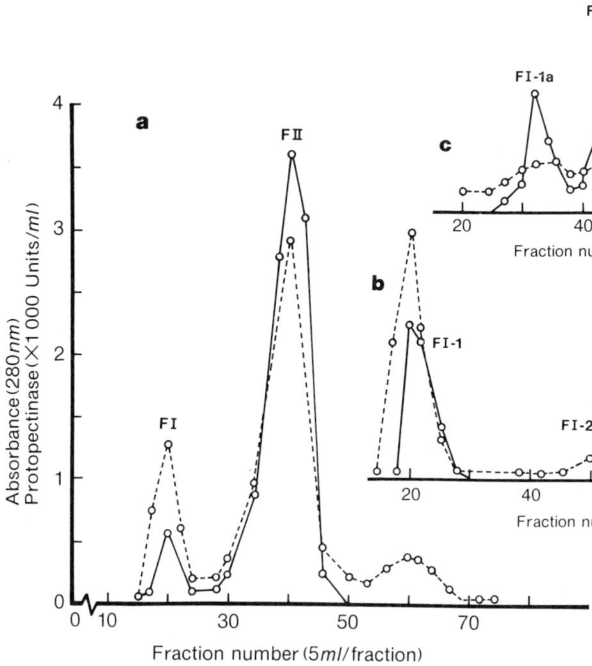

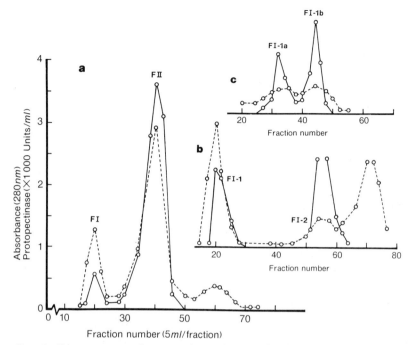

FIG. 1. Chromatograms of protopectinases from *K. fragilis* IFO 0288 on Sephadex columns. Chromatographies were performed on Sephadex (a) G-75, (b) G-100, and (c) G-200. —, Enzyme activity, ---, protein.

TABLE II
PURIFICATION OF PROTOPECTINASE F FROM CULTURE FILTRATE OF *K. fragilis* IFO 0288

Step	Total protein (mg)	Total activity (×10⁴ U)	Specific activity (U/mg)	Relative purification (-fold)	Recovery (%)
1. Culture filtrate	7844	75	96	1	100
2. CM-Sephadex chromatography	204	68.4	3353	35	91
3. First Sephadex G-75 chromatography	98	48.5	4949	52	65
4. Second Sephadex G-75 chromatography	65	38.5	5923	62	51
5. Crystallization	50	29.5	5900	61	39

TABLE III
PURIFICATION OF PROTOPECTINASE L FROM CULTURE FILTRATE OF *G. reessii* L.

Step	Total protein (mg)	Total activity ($\times 10^4$ U)	Specific activity (U/mg)	Relative purification (-fold)	Recovery (%)
1. Culture filtrate	6450	73.5	114	1	100
2. First CM-Sephadex chromatography	69	18.3	2652	23	25
3. Second CM-Sephadex chromatography	30	11.9	3957	35	16

refrigerator, the enzyme forms needle-like crystals. The crystallization is repeated two more times.

Table II is a summary of the enzyme purification. From 37 liters of culture filtrate, about 50 mg of crystalline enzyme is obtained, with a recovery of about 40%.

Protopectinase of Kluyveromyces fragilis IFO 0288

F I, which is eluted at around the void volume in Sephadex G-75 column chromatography, is put on a Sephadex G-100 column (2.5 × 60 cm) and chromatography is done using 20 m*M* acetate buffer, pH 5.0, containing 200 m*M* NaCl at a flow rate of 2.7 ml/hr. The activities are recovered in two peaks (Fig. 1b). The molecular weight of F I-1 is high, so rather than Sephadex G-100, Sephadex G-200 is used, the column (1.6 × 80 cm) being equilibrated with 20 m*M* acetate buffer, pH 5.0, containing 200 m*M* NaCl. By elution with the same buffer at a flow rate of 0.7 ml/hr, activity is recovered in two peaks, as shown in Fig. 1c. Thus, this strain seems to produce at least four protopectinases of different molecular weights.

Purification Procedure for Protopectinase from Galactomyces reessii (Protopectinase L)[6]

Step 1: Production of Enzyme. Galactomyces reessii L. is used for the production of protopectinase L. The strain was isolated from infected lemon (*Citrus limon* Burm) peel and is kept in the Institute of Applied Microbiology of the University of Tokyo (collection number IAM 1247.[9] The microorganism is maintained on agar slants of a medium containing 1% glucose, 0.6% peptone, and 0.2% yeast extract, pH 5.0. Enzyme pro-

[9] J. Sugiyama, J. D. Lee, and T. Sakai, *Trans. Mycol. Soc. Jpn.* **24,** 53 (1983).

TABLE IV
PURIFICATION OF PROTOPECTINASE S FROM CULTURE FILTRATE OF *T. penicillatum* SNO 3

Step	Total protein (mg)	Total activity ($\times 10^4$ U)	Specific activity (U/mg protein)	Relative purification (-fold)	Yield (%)
Culture filtrate	—	60.0	—	—	100
Ammonium sulfate fractionation	4614	59.5	129	1	99
CM-Sephadex chromatography	80	38.7	4838	38	65
First Sephadex G-75 chromatography	78	44.9	5756	45	75
Second Sephadex G-75 chromatography	75	43.3	5773	45	72
Crystallization	52	30.0	5769	45	50

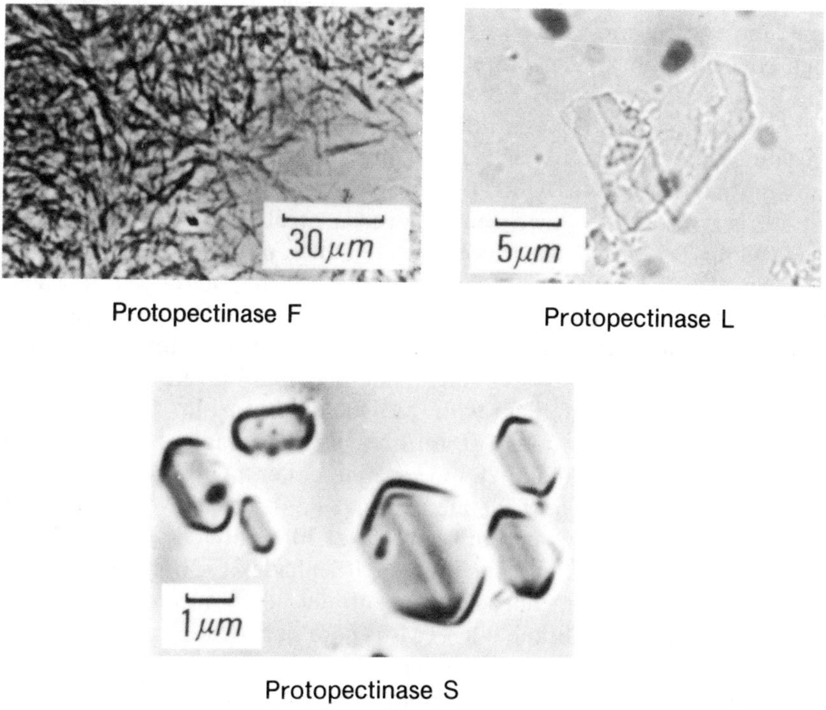

FIG. 2. Microphotographs of crystals of protopectinases.

TABLE V
PHYSICAL AND CHEMICAL PROPERTIES OF PROTOPECTINASES

Property	Protopectinase F	Protopectinase L	Protopectinase S
Molecular weight			
By electrophoresis	40,000	40,000	40,000
By gel filtration	33,000	30,000	30,000
By sedimentation	32,800	29,300	29,300
$s_{20,w}$	2.99S	3.77S	3.66S
$E^{1\%}_{280\ nm}$	10.0	11.9	9.20
Isoelectric point	5.0	8.4–8.5	7.6–7.8
N-Terminal amino acid	—	Glycine	Glycine
Sugar content (%)	5.9[a]	5.1[b]	1.7[a]

[a] Determined as mannose.
[b] Determined as rhamnose.

duction is influenced by the carbon and nitrogen sources in the culture medium. As a nitrogen source, peptone is suitable, and glucose and galactose are good carbon sources although they repress enzyme formation at high concentrations. A medium containing 0.5% glucose, 0.6% peptone, and 0.2% yeast extract, pH 5.0, is an appropriate medium. Cultivation of the microorganism is done in this medium under aerobic conditions. The enzyme is first produced after 5 hr of cultivation and production reaches a maximum at 20 hr, after which the activity decreases. Fifty liters of culture filtrate is concentrated to 5 liters *in vacuo* at 30°, and the solution is dialyzed for 2 days against 20 m*M* acetate buffer, pH 5.0, at 5°.

Step 2: CM-Sephadex C-50 Column Chromatography. The dialyzed enzyme solution is put on a CM-Sephadex C-50 column (5 × 45 cm) equilibrated with 20 m*M* acetate buffer, pH 5.0. The column is washed thoroughly with the dialysis buffer and then the enzyme is chromatographed with a linear gradient with 0 to 400 m*M* NaCl in 20 m*M* acetate buffer, pH 5.0, at a flow rate of 20 ml/hr. The linear gradient is made up of two column volumes, one of 20 m*M* acetate buffer, pH 5.0, and the other of 400 m*M* NaCl in 20 m*M* acetate buffer, pH 5.0. The active fractions are pooled and concentrated to 10 ml *in vacuo* at 30°, dialyzed against 20 m*M* acetate buffer, pH 5.0, and put on a CM-Sephadex C-50 column again. The active fractions are collected, concentrated to 10 ml *in vacuo* at 30°, filtered through a membrane filter TM-3 (pore size 3 μm, Toyo Roshi Co., Ltd., Tokyo, Japan), and used as the purified enzyme.

The purification procedure is summarized in Table III. By this proce-

FIG. 3. Double-diffusion analysis of rabbit antiserum to protopectinase S. Undiluted antiserum was added to the center well, and each well was filled about 60 μl of enzyme solution (100 μg/ml). F, Protopectinase F; L, protopectinase L; S, protopectinase S.

dure, the enzyme is purified 35-fold from the culture filtrate, with about 16% yield.

Step 3: Crystallization of Purified Enzyme. Purified enzyme solution is placed in a glass tube (1.5-cm diameter), which is then sealed with a dialysis cellulose sheet, and the enzyme solution is dialyzed against 2.5 *M* (60% saturated) ammonium sulfate solution at 5° for 2 weeks. The enzyme forms fragile plate crystals.

Purification Procedure for Protopectinase from Trichosporon penicillatum[2,4] *(Protopectinase S)*

Step 1: Production of Enzyme. Trichosporon penicillatum SNO 3, which was isolated as a protopectinase-producing strain,[3] is used for the production of protopectinase S. The strain is kept in the American Type Culture Collection with the collection number of ATCC 42397. The organism is maintained on agar slants containing 1% glucose, 0.2% peptone, 0.2% pectin, and 0.1% yeast extract, pH 5.0. For the production of the enzyme, the microorganism is grown aerobically in a medium containing 2% glucose, 0.4% peptone, 0.2% yeast extract, and 0.08% Silicone KM-70, pH 5.0, at 30°. Enzyme production begins after 5 hr of cultivation and reaches a maximum (80 U/ml) at 25 hr, after which the activity disappears rapidly.

Step 2: Ammonium Sulfate Precipitation. Forty liters of culture filtrate is concentrated to 4 liters by evaporation *in vacuo* at 30°. Solid ammonium sulfate is gradually added to 4 liters of concentrated culture filtrate, with mechanical stirring, to 2.9 *M* (70% saturation), and the solution is left overnight at 5°. The precipitate is collected by centrifugation and dissolved in 100 ml of 20 m*M* acetate buffer, pH 5.0. The enzyme solution obtained is dialyzed thoroughly against 20 m*M* acetate buffer.

Step 3: CM-Sephadex C-50 Column Chromatography. The enzyme solution obtained in step 2 is put on a CM-Sephadex C-50 column (3 × 70 cm) equilibrated with the dialysis buffer. The column is washed thoroughly with dialysis buffer, and the enzyme is eluted with a linear NaCl concentration gradient at a flow rate of 20 ml/hr. The linear gradient is made up of two column volumes, one of 20 m*M* acetate buffer, pH 5.0, and the other of 400 m*M* NaCl in 20 m*M* acetate buffer, pH 5.0. The active fractions are pooled and concentrated to 5 ml by evaporation *in vacuo* at 30°.

Step 4: Sephadex G-75 Column Chromatography. The concentrated enzyme solution (5 ml) is put on a column of Sephadex G-75 (2 × 70 cm) equilibrated with 20 m*M* acetate buffer, pH 5.0, containing 200 m*M* NaCl, and chromatography is done at a flow rate of 7 ml/hr. This chromatography is repeated once, and the active fractions obtained are pooled and concentrated *in vacuo* at 30° to 20 ml. The second Sephadex G-75 column chromatography yields a single symmetrical protein peak, and the enzyme activity is entirely associated with the peak.

Step 5: Crystallization. To the concentrated enzyme solution is added solid ammonium sulfate until faint turbidity is observed. After being left overnight in a refrigerator, the enzyme has crystallized. The crystallization is repeated three times. The yield and activity during purification of the enzyme are summarized in Table IV.

```
                  1               5                   10                  15
Protopectinase L  Gly-Gly-Ala- ? -Val-Phe-Lys-Asp-Ala-Gln-Ser-Ala-Ile-Ala-Gly-

Protopectinase S  Gly-Gly-Ala- ? -Val-Phe-Lys-Asp-Ala-Gln-Ser-Ala-Ile-Ala-Gly


                  16              20                  25                  30
Protopectinase L  Lys-Ala-Ser- ? - ? -Ser-Ile-Thr-Leu-Gln-Asn-Phe-Ala-Val-Pro

Protopectinase S  Lys-Ala-Ser-Ser-Ser-Ser-Ile- ? -Leu-Glu-Asn-Phe-
```

FIG. 4. Amino acid sequence of protopectinase L and S from the N-terminus of the intact enzyme protein.

TABLE VI
AMINO ACID COMPOSITION OF PROTOPECTINASES

	Moles of amino acid per 100 mol of protopectinase			Amino acid residues[a] per molecule of protopectinase		
Amino acid	F	L	S	F	L	S
Lysine	6.4	5.9	4.6	15	16	13
Histidine	1.6	2.2	1.7	4	6	5
Arginine	1.7	1.4	1.3	4	4	4
Tryptophan[b]	1.9	3.4	1.9	5	9	5
Aspartic acid	16.7	13.4	13.4	40	35	37
Threonine[c]	13.1	7.1	11.0	31	19	30
Serine[c]	13.6	10.9	13.9	32	29	39
Glutamic acid	4.3	6.3	7.4	10	17	20
Proline	1.9	2.4	2.0	5	6	6
Glycine	12.6	15.7	12.4	30	42	34
Alanine	4.0	6.3	5.8	10	17	16
Half-cystine[d]	1.6	0.4	2.2	4	1	6
Valine	5.5	6.7	5.6	13	18	16
Methionine	0.4	Not detected	0.4	1	0	1
Isoleucine	5.4	7.7	7.5	13	20	21
Leucine	4.7	4.2	3.9	11	11	11
Tyrosine	1.7	2.0	1.4	4	5	4
Phenylalanine	2.6	2.9	3.5	7	8	10

[a] Calculations were based on a molecular weight of 30,000.
[b] Determined spectrophotometrically by the method of Edelhoch.[11]
[c] The enzymes were hydrolyzed for 24, 48, and 72 hr and determined by extrapolation to zero time of hydrolysis.
[d] Determined by the method of Crestfield *et al.*[12]

Homogeneity of the Enzyme Preparations

Photomicrographs of the crystals of protopectinases are shown in Fig. 2. The enzyme preparations are homogeneous as judged by polyacrylamide gel electrophoresis both in the presence and absence of SDS and by sedimentation analysis.

Physical and Chemical Properties of Protopectinases

Physical properties of three protopectinases (protopectinase F, L, and S), including molecular weight, sedimentation coefficient, extinction coefficient, and isoelectric point, are shown in Table V. The molecular weight of three of the protopectinases are similar, around 30,000, both calculated by the sedimentation equilibrium method or by gel filtration on a column of Sephadex G-75 in 200 m*M* acetate buffer, pH 5.0, containing 200 m*M* NaCl. SDS-polyacrylamide gel electrophoresis gave 40,000 as the molecular weight. These protopectinases are the glycoprotein, and the molecular weight by SDS-polyacrylamide gel electrophoresis is probably in error.[10] Thus, the molecular weights of these enzymes are around 30,000.

Protopectinases L and S are basic proteins whereas protopectinase F is an acidic protein.

Table VI[11,12] shows the amino acid compositions of protopectinases. The amino acid compositions of these enzymes are different, although all are poor in methionine.

Immunological Properties

To confirm the homology of protopectinase, immunological analysis has been done. Antibody to protopectinase S was raised in a rabbit. The rabbit was immunized with 2 mg of protopectinase S emulsified in Freund's complete adjuvant. The emulsion was injected subcutaneously into a Japanese white rabbit. Seven injections were given at intervals of 7 days, and a titer of 1:32 was obtained. Double-diffusion analysis of the antiserum is shown in Fig. 3. The antiserum to protopectinase S gave precipitin lines with protopectinases L and S, but was unreactive with protopectinase F. Therefore, protopectinases L and S are homologous in their protein structures but protopectinase F is different. Homology of protopectinases L and S has been observed in their N-terminal amino acid sequences. When the N-terminal amino acid sequences of the intact protein were identified, 27 residues of protopectinases L and S were identical (Fig. 4).[13]

Biological Properties

Some biological properties of protopectinases, including optimum pH, optimum temperature, stability, and enzyme activities, are listed in Table VII. The biological properties of the enzymes are similar, except that

[10] J. P. Segrest, R. L. Jackson, E. P. Andrews, and V. T. Marchesi, *Biochem. Biophys. Res. Commun.* **44,** 390 (1971).

[11] H. Edelhoch, *Biochemistry* **6,** 1948 (1967).

[12] A. M. Crestfield, S. Moore, and W. H. Stein, *J. Biol. Chem.* **238,** 622 (1963).

[13] T. Sakai, I. Svendsen, and M. Ottesen, *Carlsberg Res. Commun.*, manuscript in preparation.

TABLE VII
BIOCHEMICAL PROPERTIES OF PROTOPECTINASE

Properties	Protopectinase F	Protopectinase L	Protopectinase S
Optimum pH	5.0	5.0	5.0
Optimum temperature (°C)	60	55	50
Inhibitor	Hg^{2+}, Hg^{+}, Ag^{+}, Ba^{2+}, Ca^{2+}, Pb^{2+}	Hg^{2+}, Hg^{+}, Ca^{2+}, Ba^{2+}, Co^{2+}	Hg^{2+}, Hg^{+}, Ca^{2+}, Ba^{2+}, Co^{2+}
Thermostability (°C)	Up to 40	Up to 50	Up to 55
pH stability	2–8	3–7	3–7
Activity (U/mg)			
Protopectinase	556	3,945	5,770
Polygalacturonase	2,053	16,219	21,107
K_m value (mg/ml)			
For protopectin	90	50	30
For polygalacturonic acid[a]	6.6	7.7	9.0

[a] Polygalacturonic acid, having a mean polymerization degree of 130, was used for the determination of the K_m value.

TABLE VIII
RELEASE OF PECTIC SUBSTANCE FROM VARIOUS PROTOPECTINS BY PROTOPECTINASE[a]

Origin of protopectin	Pectin (g) per 10 g protopectin — Protopectinase F	L	S	Yield (%) of pectin[b] — Protopectinase F	L	S
Mandarin orange (*Citrus unshiu*, peel)	0.63	0.97	0.78	63	97	78
Burdock	1.36	1.47	1.16	91	98	78
Radish	1.43	1.45	1.18	99	100	82
Watermelon (peel)	2.16	1.63	1.78	98	74	81
Carrot	1.46	1.90	1.63	73	95	82

[a] The reactions were performed in a reaction mixture containing 10 g protopectin and 100 U enzyme (as polygalacturonase activity) in 20 mM acetate buffer, pH 5.0, in a total volume of 20 ml at 37° for 2 hr.

[b] Percentage of pectin released to whole pectic substance in the sample. Whole pectic substance was determined by the method described by Stoddart *et al.*[14]

TABLE IX
ACTION MODE OF PROTOPECTINASES ON GALACTURONIC ACID OLIGOMERS[a]

Enzyme	Substrates	Reaction products	K_m (mM)	V_{max} (μM/U/min)
	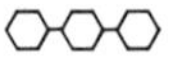	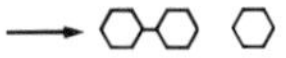	$>10^2$	$<10^{-10}$
Protopectinase F	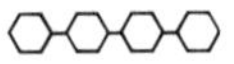	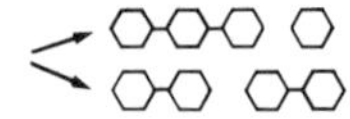	1.82	1.90×10^{-2}
	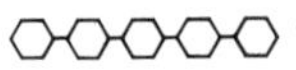	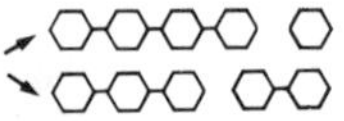	5.95×10^{-1}	2.95×10^{-1}
	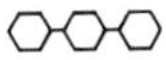	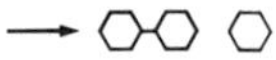	3.98	1.96×10^{-3}
Protopectinase L	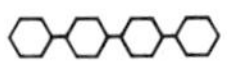	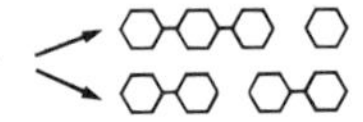	2.77	2.37×10^{-2}
	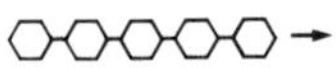	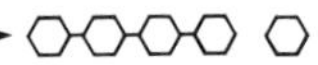	7.09×10^{-1}	9.79×10^{-2}
	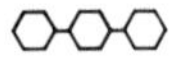	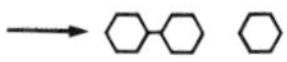	4.26	4.79×10^{-4}
Protopectinase S	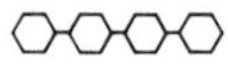	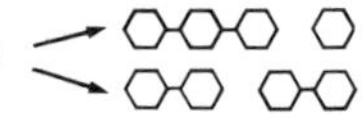	2.20	5.20×10^{-2}
	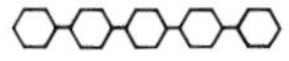	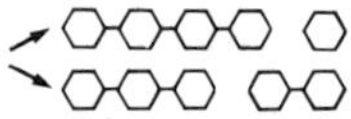	8.69×10^{-1}	2.85×10^{-1}

[a] Reactions were done under the optimum conditions for each enzyme. Hexagons express D-galacturonic acid molecules.

specific activity is different for each enzyme; the activity of protopectinase F is one order lower than those of L and S.

Catalytic Properties. The enzymes have pectin-releasing activity on protopectins from various origins (Table VIII).[14] The enzymes also catalyze the hydrolysis of polygalacturonic acid and markedly decrease the viscosity of the polygalacturonic acid-containing reaction medium, while slightly increasing the reducing value. The findings suggest that the enzymes are polygalacturonase [poly(1,4-α-D-galacturonide) glycanohydrolyase, EC 3.2.1.15).

Action of the Enzymes on Galacturonic Acid Oligomers. The enzymes catalyze the hydrolysis of galacturonic acid oligomers. The enzyme activities in hydrolyzing galacturonic acid oligomers are different for each enzyme. Table IX shows the mode of action of the hydrolysis of galacturonic acid oligomers and the kinetic constants K_m and V_{max} for the reaction. Three different patterns of action toward galacturonic acid oligomers are known for polygalacturonases.[15] Protopectinase S is novel in its action

[14] R. W. Stoddart, A. J. Barrett, and D. H. Northcote, *Biochem. J.* **102,** 194 (1967).
[15] L. Rexová-Benková and A. Slezárik, *Collect. Czech. Chem. Commun.* **31,** 122 (1965).

pattern toward oligogalacturonic acids. The kinetic constants K_m and V_{max} change with the substrate chain length; the K_m values tend to decrease, whereas the V_{max} values tend to increase with increasing chain length of the substrate. The greatest difference in V_{max} is found between trigalacturonic acid and tetragalacturonic acid. On the other hand, the number of methoxyl groups of the substrate affects the molecular weight of the reaction products; this molecular weight increases as the number of methoxyl groups of the substrate increase (Fig. 5).

Postulated Mechanism of Protopectinase Activity

On the basis of the kinetic properties of the enzymes, the mechanism of release of pectin from protopectin seems to be as follows. The enzymes react with the pectin molecule in protopectin at sites having three or more nonmethoxylated galacturonic acid chains (actually, four or more nonmethoxylated galacturonic acid chains, considering the reaction velocities for galacturonic acid oligomers) and cleave glycoside bonds in pectin

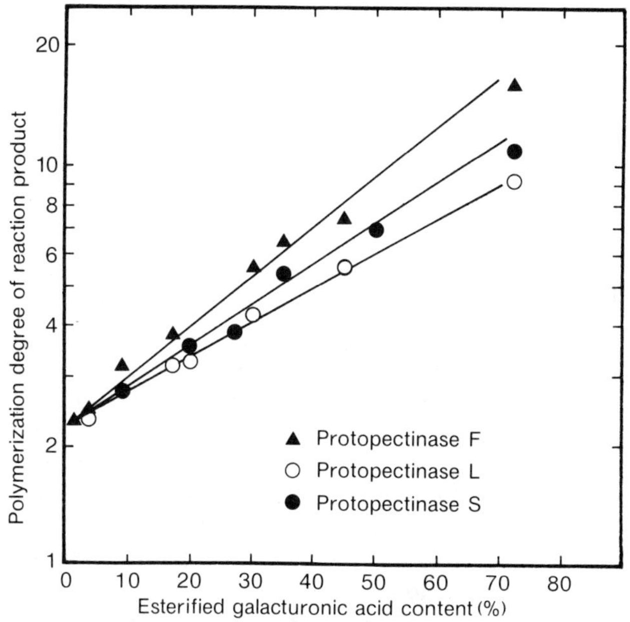

FIG. 5. Relationship between methoxyl groups of the substrate and the molecular weight of its degradation product with the enzyme reaction. Reactions were performed using polygalacturonic acid (mean polymerization degree, 33) containing the various numbers of methoxyl groups indicated. Polymerization degrees were determined from the ratio of total reducing groups to total galacturonic acid molecules.

molecules. Generally, 50% of the galacturonic acid of pectin in protopectin is methoxylated at random. Therefore, the pectin molecule in protopectin may be cleaved at restricted sites so as to form highly polymerized pectin.

These three enzymes degrade polygalacturonic acid as well as having protopectinase activities, and they are regarded as polygalacturonases.

Later, Sakai and Takaoka isolated a polygalacturonase from the culture filtrate of *Aureobasidium pullulans*, which has potent polygalacturonic acid-degrading activity but less protopectinase activity.[16] The enzyme shows a lower affinity for protopectin than do protopectinases F, L, or S; the K_m for protopectin is one order of magnitude larger than those of the values for the other three protopectinases, while the four enzymes have almost the same affinity for polygalacturonic acid.[16] The difference in affinity for protopectin seems to be one reason for polygalacturonases having the different protopectinase activity.

Thus, protopectinase activity seems not to be a common property of polygalacturonase, and protopectinases F, L, and S are unusual in having this property.

[16] T. Sakai and A. Takaoka, *Agric. Biol. Chem.* **49,** 449 (1985).

[37] Pectin Lyase from *Phoma medicaginis* var. *pinodella*

By D. Pitt

Nomenclature and Terminology

Pectin lyase [poly(methoxygalacturonide) lyase, EC 4.2.2.10][1] eliminates 6-methyl-Δ-4,5-D-galacturonate residues from pectin and thereby promotes depolymerization.[2,3] It was originally described as pectin *trans*-eliminase, which released Δ-4,5-unsaturated galacturonic acid groups specifically from the methyl ester of pectic acid (pectin) with the reaction proceeding by a trans elimination.[2] Recent reclassification as pectin lyase distinguishes this enzyme from pectate lyase (EC 4.2.2.2), which liberates Δ-4,5-D-galacturonate residues from pectate and other polygalacturonides but not from pectin.[3] Lyases have been historically subdivided into endoenzymes, which attack internal regions of α-1,4-linked polygalacturon-

[1] R. A. Plumbley and D. Pitt, *Physiol. Plant Pathol.* **14,** 313 (1979).
[2] P. Albersheim, H. Neukom, and H. Deuel, *Helv. Chim. Acta* **43,** 1422 (1960).
[3] M. Dixon and E. C. Webb, "Enzymes," 3rd Ed. Longmans, London, 1979.

ide chains, and exo forms, which act terminally, but this view may require revision.[4] It is considered that where tissue maceration in plant pathogenesis is extensive, enzymes of the endo-type are primarily responsible. It is generally believed that among plant pathogens promoting pectolysis, fungi usually secrete pectin lyases while bacteria produce predominantly pectate lyases. The role of pectolytic enzymes in degradation of plant cell wall polymers has been reviewed recently.[4]

Assay Method

Principle. In the present method enzyme activity was measured by the reaction between unsaturated end product of pectin degradation and thiobarbituric acid.[5] One unit of activity is that amount of enzyme causing a change in absorbance of 0.01 under the conditions of the assay.

Protein concentration was determined by the method of Lowry *et al.*[6]

An alternative enzyme assay method is suitable whereby end product is measured directly by spectrophotometry at 240 nm using a molar absorption of 4600 for the unsaturated uronide product.[7,8]

Reagents

Tris–HCl buffer, 0.05 *M*, pH 8.5
Pectin, 1% (w/v) in Tris buffer
$CaCl_2$, 0.01 *M* in distilled water
$ZnSO_4 \cdot 7H_2O$, 9% (w/v) in distilled water
NaOH, 0.5 *M* in distilled water
Thiobarbituric acid, 0.04 *M* in distilled water
Hydrochloric acid, 0.1 *M*
Enzyme preparation

Procedure. The reaction mixture contains 5.0 ml of pectin solution (Sigma Chemical Co., from citrus fruit), 1.0 ml of calcium chloride solution, and an aliquot of enzyme solution (1.0 ml) in a final reaction volume made up to 10 ml with distilled water. After incubation for 2 hours at 30°, 0.6 ml of $ZnSO_4$ solution is added followed by NaOH (0.6 ml). Precipitated protein and unused substrate are removed by centrifugation at 3000 *g* for 10 min and 5.0 ml of the supernatant is added to a mixture of

[4] R. M. Cooper, *in* "Biochemical Plant Pathology" (J. A. Callow, ed.), p. 101. Wiley, Chichester, England, 1983.
[5] W. A. Ayers, G. C. Papavizas, and A. F. Diem, *Phytopathology* **56,** 1006 (1966).
[6] O. H. Lowry, N. J. Rosebrough, A. L. Farr, and R. J. Randall, *J. Biol. Chem.* **193,** 265 (1951).
[7] R. L. C. Wijesundera, J. A. Bailey, and R. J. W. Byrde, *J. Gen. Microbiol.* **130,** 285 (1984).
[8] E. C. Hislop, J. P. R. Keon, and A. H. Fielding, *Physiol. Plant Pathol.* **14,** 371 (1979).

thiobarbituric acid (3.0 ml), hydrochloric acid (1.5 ml), and distilled water (0.5 ml). The mixture is heated on a boiling water bath for 30 min before cooling and the absorbance is read at 550 nm against a reference cuvette which contains the same reagents as the experimental cuvette but which is incubated for zero time before the addition of the $ZnSO_4$ and NaOH.

Preparation of Enzyme

Growth of Organism. Phoma medicaginis var. *pinodella* isolated from pea seeds is grown and induced to spore on pea decoction agar in Petri dishes for 4 days at 25° under a 12-hr alternating black light photoperiod.[1] Pea decoction agar is made by adding 2% (w/v) agar powder to a filtrate obtained by straining boiled pea seeds through cheesecloth (100 g dried seeds soaked in water for 24 hr and boiled for 15 min with 800 ml water and the filtrate made to 1 liter). For enzyme production the fungus is grown in 1-liter conical flasks containing 400 ml of pea decoction liquid. Medium is inoculated with a spore suspension and shaken for 7 days in a rotary incubator at 140 cpm at 25°.

Purification of Enzyme

All of the following manipulations are done at 0 to 4°.

Step 1: Preparation of Extract. Shaken cultures (3 liters) are harvested by separating mycelium from the medium by suction filtration. Mycelium is homogenized for 5 min with 100 ml of filtrate using a blender. The homogenate is squeezed through cheesecloth and the liquid is centrifuged initially at 2000 *g* for 10 min and then at 38,000 *g* for 30 min. The pellet is discarded and the supernatant fluid is combined with the bulk of the original medium to produce a crude extract which contains the intracellular and extracellular enzymes. A small quantity is retained, dialyzed, and assayed for enzyme activity and the remainder is concentrated by freeze drying.

Step 2: Ammonium Sulfate Fractionation. The freeze-dried extract from step 1 is dissolved in 300 ml of distilled water and the solution is made up to 60% saturation by the slow addition, with stirring, of solid enzyme grade $(NH_4)_2SO_4$ (117 g). After 1 hr, the precipitate is removed by centrifugation at 38,000 *g* for 30 min and the supernatant liquid retained. Solid $(NH_4)_2SO_4$ (94.2 g) is added slowly, with stirring, to the fluid fraction from the first precipitation to bring it to 100% saturation. After stirring for a further hour the protein precipitate is collected by centrifugation and dissolved in the minimum of distilled water before dialysis overnight against 5 liters of 0.05 *M* phosphate buffer, pH 7.0. Further dialysis for

24 hr is done against two changes of 5 liters of 2 m*M* phosphate buffer, pH 7.0. Any precipitate formed during dialysis is removed by centrifugation and the dialysate stored overnight at 4°.

Step 3: Chromatography on DEAE-Cellulose. The dialysate from step 2 (50 ml) containing ca. 30 mg protein is applied to a column (30 × 2.5 cm) of DEAE-cellulose (Whatman DEAE 23) that has been equilibrated with 0.01 *M* phosphate buffer, pH 7.0, and eluted at a flow rate of 120 ml/hr. Pectin lyase had little affinity for the cellulose and is eluted in the void volume (23 ml), which is collected in the first 200 ml obtained by washing the column with 2 liters of 0.01 *M* phosphate buffer, pH 7.0. This fraction is dialyzed overnight against two changes (each of 1 liter) of 5 m*M* phosphate buffer, pH 7.0.

Step 4: Gel Filtration. The dialysate from step 3 is freeze dried and redissolved in 4.0 ml of distilled water and made 17% (w/v) with respect to sucrose before being applied to a Sephadex G-100 column (45 × 2.5 cm) that has been equilibrated with 0.1 *M* phosphate buffer, pH 7.0. Elution is by downward flow (12 ml/hr) and 3.3-ml fractions are collected after discarding the void volume of 60 ml. Pectin lyase activity elutes as a large peak at fraction 10 and a smaller peak at fraction 20, each of which corresponds to a distinct peak of protein. In the experiment summarized in Table I only those fractions with high specific activity are retained (i.e., fractions 6–15 and 18–22). These fractions are combined and used in the next step.

Step 5: Isoelectric Focusing. Eluate from gel filtration (50 ml) is subjected to isoelectric focusing using an LKB (LKB, Bromma, Sweden) 110-ml column with 1% (w/v) LKB ampholines of pH range 3.5–10.0 in a linear sucrose gradient [0–50% (w/v) sucrose]. Enzyme extract is distributed equally between the two gradient component solutions prior to establishment of the gradient using a gradient former. Electrofocusing is done for 72 hr at constant power input at 5 W with the anode at the bottom of the column. The column is then eluted at a rate of 60 ml/hr with 3.3-ml

TABLE I
PURIFICATION OF PECTIN LYASE

Step	Total activity (U)	Total protein (mg)	Specific activity (U/mg)	Purification factor
1. Crude extract	42,840	2,160	19.8	1.0
2. Dialyzed $(NH_4)_2SO_4$ fraction	7,872	61.5	128.0	6.5
3. Eluate from DEAE-cellulose	5,620	13.8	407.3	20.5
4. Eluate from Sephadex G-100	4,692	7.5	669.2	34.3

fractions collected. The pH of the fractions is recorded immediately and they are dialyzed against three changes of 5 m*M* phosphate buffer, pH 7.0, before assaying for enzyme activity and protein. A peak of stable enzyme activity, coinciding with a protein peak, occurs between fractions 22–33, with a p*I* value of 7.9. The combined fractions can be concentrated by freeze drying and stored without loss of activity for about 1 year.

Properties

Purity and Specificity. The enzyme rapidly depolymerized pectin by the elimination of unsaturated residues but much lower activity occurred toward polypectate. Activity against the latter could reflect the general level of pectin contamination in commercial polypectate samples. No polygalacturonase (EC 3.2.1.15) activity was detected in the preparation using either the viscometric or the cup-plate assays with sodium polypectate as the substrate. The rapid capacity for depolymerization by the pectate lyase preparation suggests it may possess endo affinities according to previous terminology.

Since the enzyme obtained after isoelectric focusing yielded a single band of protein with conventional methods of disk gel electrophoresis,[1] and had low activity of other pectolytic enzymes, the preparation was of high purity.

Molecular Weight. When determined by gel filtration,[9] two forms of the enzyme eluted in step 4 with molecular weights of 29,500 and 118,000, which suggested the existence of monomeric and tetrameric components. When these fractions were combined and subjected to isoelectric focusing a single enzyme peaked with a p*I* value of 7.9. Further gel filtration showed the preparation predominantly to contain an enzyme with a molecular weight of 29,500.

pH Optimum. Both forms of the enzyme showed a pH optimum of 7.5 for the reaction.

Activation by Ca^{2+}. The enzyme lacked an absolute requirement for calcium ions at all stages of preparation but maximum enzyme activity required the presence of 1 m*M* Ca^{2+}.

Uses of the Enzyme

The enzyme has uses in studies on cell wall degradation during plant pathogenesis and in the analysis of cell wall polymers. It may be used in the isolation of plant cell protoplasts.

[9] P. Andrews, *Biochem. J.* **91,** 222 (1964).

[38] Pectinesterases from *Phytophthora infestans*

By HELGA FÖRSTER

The fungal plant pathogen *Phytophthora infestans* is known to produce several pectolytic enzymes *in vitro*.[1-4] Pectinesterase (pectin pectylhydrolase, EC 3.1.1.11), which deesterifies the pectin molecule liberating methanol and pectic acid (pectin + nH$_2$O = n methanol + pectate), is synthesized constitutively by the fungus. There is evidence that three different forms of the enzyme are secreted into the culture medium.[5] By the procedure described below, one of these forms can be purified to homogeneity. The other two forms are purified as an enzyme complex.

Assay Method

Various methods to determine pectinesterase activity have been described.[6] Since during enzyme purification numerous samples have to be tested, the following simple and less time-consuming method has been used successfully. Like the titration method it is based on the decrease in pH of the reaction mixture due to increase in free carboxyl groups during enzyme action.

Assay Reagents

Substrate: 2% (w/v) pectin. The pH is adjusted to 7 with 1 and 0.1 *M* NaOH. Since it may take several hours to adjust the pH, the solution is best prepared 1 day before use. Just before use the pH should be checked again

Buffer: Sodium phosphate buffer, 10 m*M*, pH 7

Enzyme solution: Since the decrease in pH is about linear between pH 7 and ca. pH 6.3, the enzyme activity of the sample should not be too high in order to get a good estimate of the enzyme units present. At lower pH enzyme activity decreases because of the pH itself and because end-product inhibition starts

[1] F. Grossmann, *Naturwissenschaften* **50,** 721 (1963).
[2] D. D. Clarke, *Nature (London)* **211,** 649 (1966).
[3] M. Knee and J. Friend, *Phytochemistry* **7,** 1289 (1968).
[4] A. L. J. Cole, *Phytochemistry* **9,** 337 (1970).
[5] H. Förster and I. Rasched, *Plant Physiol.* **77,** 109 (1985).
[6] L. Rexová-Benková and O. Markowič, *Adv. Carbohydr. Chem. Biochem.* **33,** 323 (1976).

Procedure

The assay is done at room temperature in polypropylene tubes which are just large enough to take a pH electrode. The total volume of the reaction mixture is 3.5 ml. To 1.5 ml of the pectin solution 10 m*M* sodium phosphate buffer (pH 7) is added to give a final sodium phosphate concentration of 4.3 m*M*. The amount of buffer depends on the buffer concentration of the enzyme extract. For example, if 0.1 ml enzyme extract in 50 m*M* sodium phosphate buffer is used, 1 ml 10 m*M* sodium phosphate buffer has to be added. The amount of buffer present in the reaction mixture is important and has to be the same in each test, since the decrease in pH is dependent on the buffer concentration. Before adding the enzyme extract, the final volume of the reaction mixture is brought to 3.5 ml with distilled water. The blank consists of pectin solution, buffer and distilled water. Immediately after addition of the enzyme extract to the reaction mixture, the pH is checked. The decrease in pH is measured for 30 min at 10-min intervals. The blank sample is measured at the same intervals, in order to check if the pH of the pectin solution has been completely equilibrated and to make sure that the pH electrode is working correctly in the viscous pectin solution. Changes in pH of the blank, which usually are not more that 0.02 pH units, are taken into consideration when calculating the units of enzyme present in the sample. With this pectinesterase assay 8 to 10 samples can be tested in parallel.

Units and Specific Activity

One unit of enzyme is defined as the amount of enzyme which causes a decrease in pH of the reaction mixture of 0.1 in 30 min. This definition of unit is different from the one proposed by Kertesz[7] (1 mg of CH_3O liberated in 30 min/ml or g of enzyme or mEq of ester hydrolyzed/min/g of enzyme), but it is more convenient when handling many samples during the enzyme extraction procedure, since no titrations have to be done.

Enzyme Purification

Step 1: Fungus Culture and Production of Crude Enzyme Extract. Phytophthora infestans is grown in a chemically defined nutrient broth[8] in Roux bottles at 20° for 2 weeks. Extended growth or addition of pectin to the culture medium does not enhance pectinesterase production by the fungus. In order to obtain 0.1 mg of purified pectinesterase I 5 to 10 liters

[7] Z. I. Kertesz, this series, Vol. 1, p. 158.
[8] H. R. Hohl, *Phytopathol. Z.* **84,** 18 (1975).

of culture filtrate is needed. The culture filtrate is lyophilized. The following steps are carried out at 4°.

Step 2: Ammonium Sulfate Precipitation. Lyophilized culture filtrate (6 liters) is redissolved in 350 ml distilled water and centrifuged for 45 min at 48,000 *g* to remove debris. The supernatant is adjusted to 10^{-4} *M* EDTA. To precipitate proteins, 767 g ammonium sulfate/liter was added to this crude extract to give a saturated solution. The pH is maintained close to pH 6.5 with 2 *M* Tris. After 2 hr of stirring, the solution is centrifuged at 48,000 *g* for 45 min. The protein precipitate is collected, redissolved in distilled water (about 100 ml) and dialyzed against 10 m*M* sodium phosphate buffer (pH 6.5) for 24–36 hr with several buffer changes. Although pectinesterases are precipitated only at high ammonium sulfate concentration, a fractional precipitation does not facilitate further enzyme purification; contaminating proteins which precipitate at lower salt concentrations can be removed in the following purification steps.

Step 3: DEAE-Cellulose Chromatography. The dialyzed solution is passed through a 5 × 12 cm column of DEAE-cellulose (Whatman DE-52) equilibrated and eluted with 10 m*M* sodium phosphate buffer (pH 6.5). The flow rate of the column is 100 ml/hr. Pectinesterases are not adsorbed to the column and are eluted with the starting buffer; most of the pigmented material can be removed by this step. The eluted enzyme extract (ca. 250 ml) is lyophilized to dryness, resuspended in ca. 20 ml distilled water, and dialyzed for 24–36 hr against 10 m*M* sodium phosphate buffer (pH 6.5) with several buffer changes.

Step 4: CM-Cellulose Chromatography. Dialyzed protein is then applied to a 3.5 × 10 cm CM-cellulose column (Whatman CM-52) equilibrated with 10 m*M* sodium phosphate buffer (pH 6.5). The column is washed with 70 ml equilibration buffer and is then eluted with a linear gradient of 0 to 0.2 *M* NaCl in the same buffer at a flow rate of 70 ml/hr. The total volume of the gradient is 460 ml. Fractions (8.7 ml) are collected. Testing the fractions for enzyme activity (50-μl aliquots) and measuring at 280 nm for total protein result in an elution profile as shown in Fig. 1. The active fractions of peaks 1 (pectinesterase I, PE I) and 2 (pectinesterase II, PE II) are combined separately, whereby three to four fractions between the two peaks are discarded.

Step 5: Sephacryl S-200 Gel Filtration of PE II. The combined fractions of PE II from CM-cellulose chromatography are lyophilized, redissolved in a minimum volume of distilled water, and concentrated by collodion bags (Sartorius) to a volume of about 0.7 ml. The concentrated extract is applied to a 1 × 100 cm column of Sephacryl S-200 Superfine (Pharmacia) equilibrated with 50 m*M* sodium phosphate buffer (pH 6.5). The flow rate is adjusted to 2.5 ml/hr and 1.5-ml fractions are collected. Testing for

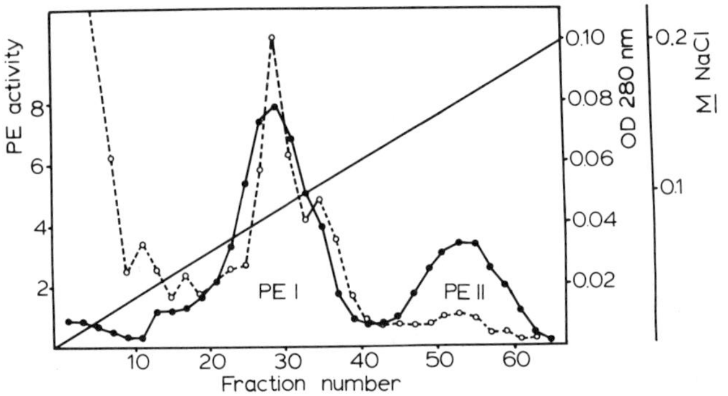

FIG. 1. Chromatography of crude pectinesterase extract on CM-cellulose (●——●) Enzyme activity (U/50 μl) and (O------O) OD 280 nm. Reproduced by permission from Förster and Rasched.[5]

enzyme activity (20-μl aliquots) and measuring at 280 nm for total protein reveal that the enzyme is eluted in front of a minor contaminating protein peak.

Step 6: Chromatofocusing of PE I. The combined active fractions from CM-cellulose chromatography containing PE I are concentrated on a PM 10 membrane (Amicon) or in a dialysis bag using PEG 20,000 (Serva) to about 4 ml and are then dialyzed against 25 m*M* ethanolamine–CH_3COOH buffer (pH 9.5) for 36 hr with several buffer changes. A 1 × 30 cm column of the Polybuffer exchanger PBE 94 (Pharmacia) is packed, equilibrated, and loaded with the dialyzed extract according to the manufacturer's instruction except that before applying the sample to the column no elution buffer is applied to the column. This creates a better pH gradient. The column is eluted at a flow rate of 10 ml/hr with 125 ml Polybuffer 96 (Pharmacia; pH 7.5) diluted 1 : 10 with distilled water. Fractions (5 ml) are collected, and PE activity is tested in 50-μl aliquots (1.45 ml 10 m*M* sodium phosphate buffer is added for the enzyme assay). A typical elution profile is shown in Fig. 2; enzyme activity is recovered as a double peak. The active fractions are combined, concentrated in a dialysis bag using PEG 20,000 (Serva), and afterward in a collodion bag (Sartorius) to a volume of about 0.8 ml.

Step 7: Sephacryl S-200 Gel Filtration of PE I. The extract is then applied to a 1 × 100 cm column of Sephacryl S-200 Superfine (Pharmacia) and run under the same conditions as for PE II.

Step 8: SDS Electrophoresis. SDS electrophoresis[9] of the single active fractions from Sephacryl gel filtration shows which of the fractions contain the pure enzymes. With PE II usually the last two fractions of the enzyme

[9] U. K. Laemmli, *Nature (London)* **227,** 680 (1970).

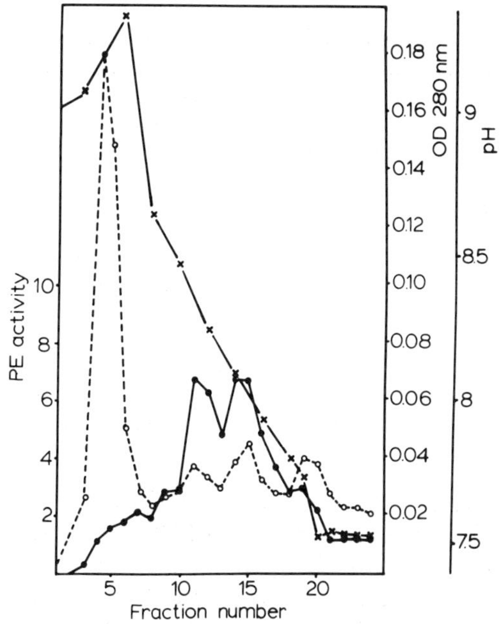

FIG. 2. Chromatofocusing of CM-cellulose-purified pectinesterase I extract on polybuffer exchanger PBE 94. (●——●) Enzyme activity (U/50 μl), (O------O) OD 280 nm, and (X——X) pH. Reproduced by permission from Förster and Rasched.[5]

activity peak are contaminated with other proteins. Pure fractions of PE I are found on the leading part of the activity peak; here a double protein band between 45 and 58 kDa is seen on the gel. Application of the same enzyme activity to the gel (15 U can be detected after silver staining) reveals that in the first fractions of the peak the upper protein band of this double peak dominates, whereas in the last fractions only the lower band is present. This and the fact that there is a double-enzyme activity peak after chromatofocusing of PE I give evidence that both proteins of the purified PE I extract are pectinesterases: thus PE I probably represents an enzyme complex consisting of two pectinesterases.

A summary of the purification procedure is presented in Table I. This enzyme purification was repeated many times by the author. In some cases SDS electrophoresis of PE I showed a faint protein impurity band at 69 kDa.

Enzyme Properties

Molecular Weight. The molecular weight of PE I is 45,000 as estimated by gel filtration and 45,000 to 48,000 as estimated by SDS electrophoresis.

TABLE I
PURIFICATION OF EXTRACELLULAR PECTINESTERASES FROM *P. infestans*

Fraction	Specific activity (U/mg protein)		Recovery (%)
Culture filtrate	13		100
Ammonium sulfate precipitate	32		92
DEAE-cellulose eluate	525		86
	PE I	PE II	Total PE I PE II
CM-cellulose peaks	1,585	1,745	61 = 40 + 21
Chromatofocusing	ND[a]	—	— 68 —
Sephacryl S-200 peak, pure fractions	12,100	15,350	12 = 11 + 1

[a] Not determined (ND), since polybuffer interferes with the Lowry protein assay.

The molecular weight of PE II was determined to be 35,000 to 37,000 by gel filtration, and 40,000 as determined by SDS electrophoresis.

Effects of Salts on Enzyme Activity. The activities of both PE I and PE II are stimulated by the addition of NaCl or $CaCl_2$ to the reaction mixture, 0.1 *M* NaCl or 0.1 *M* $CaCl_2$ being the optimal concentrations. It is assumed that the presence of salts is essential for enzyme activity; the low activity without adding salts might be due to impurities in the pectin preparation.[5]

pH Optimum of Enzyme Activity. The pH optimum can be determined by the titration method.[10] The reaction mixture containing 9 ml of the pectin solution in 0.25 *M* NaCl (adjusted to the pH to be tested) and the enzyme solution is stirred continuously and is kept at the pH to be tested by titration with 0.015 *M* NaOH. Enzyme activity can be expressed by the amount of NaOH consumed within 30 min at a certain temperature. The pH optimum (100% activity) of both PE I and PE II is pH 7. Activity of PE I at pH 5, 6, 8, and 9 is 58, 70, 73, and 64% respectively. Activity of PE II at pH 4, 5, 6, 8, and 9 is 24, 67, 94, 98, and 0% respectively. Since pectin is demethoxylated to some degree in alkaline medium, the enzyme activities given for this pH range might not be accurate.

Temperature Effects. Pectinesterases of *P. infestans* are inactivated after 10 min at 60°.

Further Pectolytic Enzyme Activities. Testing purified pectinesterases PE I and PE II for polygalacturonase (EC 3.2.1.15), pectate lyase (EC 4.2.2.2), and pectin lyase (EC 4.2.2.10) reveals that PE I purified by the

[10] Z. I. Kertesz, *J. Biol. Chem.* **121,** 589 (1937).

described procedure shows a very slight pectin lyase activity (increase in absorbance at 235 nm of 0.018 within 1 hr).

Assay for Glycoproteins. PE I and PE II give positive reactions in a modified silver staining procedure for detecting glycoproteins on SDS-electrophoresis gels.[11] Furthermore, PE I is adsorbed to Con A-Sepharose (for PE II this was not tested). This indicates that pectinesterases from *P. infestans* are glycoproteins.

Further Purification of PE I. The chemical properties of both proteins of PE I seem to be very similar. The double protein band cannot be separated by chromatography on hydroxylapatite, different ion exchangers (DEAE-Sepharose, CM-Sepharose), Con A-Sepharose and cross-linked polypectate,[12] hydrophobic interaction chromatography on octyl sepharose, or adsorption on bentonite and chromatofocusing with Pharmacia Polybuffer exchanger PBE 118.

[11] G. Dubray and G. Bezard, *Anal. Biochem.* **119,** 325 (1982).
[12] L. Marcus and A. Schejter, *Physiol. Plant Pathol.* **22,** 1 (1983).

[39] Polygalacturonase from *Corticium rolfsii*

By Kiyoshi Tagawa and Akira Kaji

Polygalacturonase [poly(1,4-α-D-galacturonide) glycanohydrolase, EC 3.2.1.15) catalyzes the random hydrolytic splitting of internal glycosidic α-1,4 linkages in D-galacturonan chains of pectic substances. In *Corticium rolfsii*, this enzyme is produced inducibly in a medium containing D-galacturonic acid or pectin. The enzyme is unusually stable in acidic conditions and active at pH 2.5.[1,2]

Assay Method

Principle

Two assay methods have generally been employed for the estimation of polygalacturonase activity: release of reducing groups from pectic acid and viscosity reduction of pectic acid. The viscometric assay is more sensitive than the reducing group assay, so it is convenient for estimation of the

[1] A. Kaji and K. Tagawa, *J. Agric. Chem. Soc. Jpn.* **40,** 325 (1966).
[2] A. Kaji and T. Okada, *Arch. Biochem. Biophys.* **131,** 203 (1969).

minor activity. It cannot, however, be used for estimation of the activity in acidic conditions such as at pH below 3.0, where pectic acid forms a gel. In these conditions, it is better to use the reducing group assay using acid-soluble pectic acid[3] as a substrate.

Reducing Group Assay Method

Reagents

1.0–0.8% acid-soluble pectic acid solution, pH 2.5
0.1 *M* citrate–phosphate buffer, pH 2.5
Suitably diluted enzyme solution
1 *M* Na_2CO_3
Somogyi's copper reagent and arsenomolybdate reagent[4,5]

Procedure. The reaction mixture contains 2 ml of the pectic acid solution, 0.5 ml of buffer, and 0.5 ml of enzyme solution. The reaction proceeds at 30° for 30 min. At the end of the time, 0.1 ml of Na_2CO_3 is added to stop further reaction. Reducing groups released are measured colorimetrically with the Somogyi–Nelson reagent[4,5] using D-galacturonic acid as a standard.

Definition of Unit. One unit of polygalacturonase activity is defined as the amount of enzyme which will produce 1 μmol of reducing groups per minute at 30°.

Viscometric Assay Method

The optimum pH for activity of the polygalacturonase from *C. rolfsii* lies at 2.5.[2] Therefore, it is not possible to estimate accurately the enzyme activity by the viscometric assay. It is, however, useful for characterizing the enzyme in order to evaluate the viscosity reducing activity on pectic acid at pH 4.0.[1,2]

Reagents

1.0–0.8% sodium pectate solution, pH 4.0
0.1 *M* citrate–phosphate buffer, pH 4.0
Suitably diluted enzyme solution

Procedure. Sodium pectate solution (3.5 ml) and 1 ml of buffer are pipetted into an Ostwald viscometer (capillary length, 9 cm) kept at 30°,

[3] Acid-soluble pectic acid is prepared from pectin NF by the method of R. M. McCready and C. G. Seegmiller [*Arch. Biochem. Biophys.* **50,** 440 (1954)]. Its average degree of polymerization is about 10.

[4] M. Somogyi, *J. Biol. Chem.* **195,** 19 (1952).

[5] N. Nelson, *J. Biol. Chem.* **153,** 375 (1944).

followed by addition of 0.5 ml of enzyme solution. Initial flow time is obtained by substituting water for enzyme solution. The flow time of the reaction mixture is taken at suitable intervals and the time for 50% viscosity reduction is determined graphically. A linear relationship is obtained between the enzyme concentration and the reciprocal of the time required to reach 50% of the original viscosity.

Definition of Unit. One viscosimetric unit is arbitarily defined as the amount of enzyme that will reduce the viscosity of pectic acid by 50% in 1 min.

Preparation and Purification Procedure

Cultivation of Organism

Corticium rolfsii IFO 6146 (supplied by the Institute for Fermentation, Osaka, Japan) is cultured in a medium composed of 12.5 g of pectin, 10 g of peptone, 0.5 g of K_2HPO_4, 0.5 g of NH_4NO_3, 0.2 g of $MgSO_4 \cdot 7H_2O$, and 0.3 ml of 2% FeCl, in 1 liter of water. The initial pH is adjusted to 5.4 by addition of 1 *N* HCl. Cultivation is carried out at 28° for 72 hr under aerobic conditions in a jar fermenter. The mycelia are filtered off through cloth and the solution is centrifuged at 5°. The crude enzyme solution thus obtained is purified as follows.

Purification Procedure

Unless otherwise specified, purification of the enzyme is carried out at 5°.

Step 1: Salting Out with Ammonium Sulfate. To 10 liters of the crude enzyme solution, 6100 g of $(NH_4)_2SO_4$ (0.85 saturation) is added with constant stirring. After standing for several hours, the resulting precipitate is collected by centrifugation, dissolved in 100 ml of water, and the solution dialyzed for 24 hr against water and then for 24 hr against 0.01 *M* citrate–phosphate buffer, pH 6.0.

Step 2: DEAE-Cellulose Column Chromatography. The dialyzed solution (130 ml) is chromatographed on a column (4 × 50 cm) of DEAE-cellulose equilibrated with 0.01 *M* citrate–phosphate buffer, pH 6.0. After washing with the equilibrating buffer (600 ml), elution is performed with the same buffer containing 0.2 *M* NaCl. The active fraction (200 ml) is collected, dialyzed against water for 48 hr, and concentrated with polyethylene glycol (M_r 60,000). The concentrate is dialyzed against 0.1 *M* acetate buffer, pH 4.0, for 48 hr.

Step 3: Affinity Chromatography on Crosslinked Pectic Acid. The concentrated enzyme solution (30 ml) is purified by affinity chromatography

TABLE I
PURIFICATION OF POLYGALACTURONASE FROM *C. rolfsii*

Step	Volume (ml)	Activity (U/ml)	Protein (mg/ml)	Specific activity (U/mg protein)	Yield (%)
Culture filtrate	10,000	1.8	3.4	0.53	100
Salting out with ammonium sulfate	130	72	17.1	4.2	52
DEAE-cellulose chromatography	30	208	5.2	40.0	34.7
Affinity chromatography	10	332	3.0	110.7	18.4
Sephadex G-100 gel filtration	12	265	2.1	126.2	17.7

on a column (2 × 20 cm) packed with pectic acid crosslinked by epichlorhydrin.[6] The column is equilibrated with 0.1 *M* acetate buffer, pH 4.0, and then the enzyme solution is added. After washing the column with the same buffer (100 ml), the enzyme is eluted with 0.2 *M* acetate buffer, pH 6.0. The active fraction (40 ml) is collected, dialyzed against water for 24 hr, and then against 0.01 *M* acetate buffer, pH 6.0, for 24 hr. On lyophilization, the enzyme gives a white powder and it is dissolved in 10 ml of water.

Step 4: Gel Filtration on Sephadex G-100. The enzyme solution (10 ml) is applied to a Sephadex G-100 column (2 × 40 cm) that was previously equilibrated with 0.05 *M* acetate buffer, pH 6.0, and eluted with the same buffer. The enzyme is eluted as a single peak coincident with the elution of protein. The active fraction (12 ml) is collected and stored at −20°.

The purified enzyme thus obtained appears homogeneous when examined by ultracentrifugation analysis[2] and has a specific activity of 126 U/mg of protein, representing an overall 240-fold purification.

A summary of the above purification procedure is presented in Table I.

Properties

Optimum pH. Maximum enzyme activity is obtained at pH 2.5 with acid-soluble pectic acid. It is also confirmed from measurement of the release of pectin[7] that the enzyme is most active at this pH region when it acts on plant tissues.[1,2,8]

[6] L. Rexová-Benková and V. Tibenský, *Biochim. Biophys. Acta* **268,** 187 (1972).
[7] K. Tagawa and A. Kaji, *Tech. Bull. Fac. Agric., Kagawa Univ.* **17,** 104 (1966).
[8] A. Kaji, N. Mikuni, and K. Tagawa, *Tech. Bull. Fac. Agric., Kagawe Univ.* **16,** 137 (1965).

Stability. The purified enzyme is stable in solutions having a wide range of pH values from 2.0 to 8.0,[2] and can be stored below 5° for a month without any appreciable loss of the activity.[2]

Substrate Specificity. The real substrate for the purified enzyme is pectic acid (nonmethylated polygalacturonan), because, at pH 4.0 where pectic acid is soluble, it hydrolyzes pectic acid more rapidly than acid-soluble pectic acid or pectin. A limited hydrolysis of pectin is attainable,[2] depending on its degree of esterification: pectin NF (68% esterified) 12%, methylated pectin (91% esterified) 1.5%. With prolonged incubation pectic acid is completely hydrolyzed to mono- and digalacturonic acids and the extent of its hydrolysis reaches 70%.[2] The enzyme attacks α-1,4-linked D-galacturonide chains of pectic substances in plant tissues. For this reason fungal maceration is observed in fruits, vegetables, and barks of young trees.[1,2,8]

Comment. The fungal polygalacturonases have been used for production of pulps, juices,[9] and concentrates from fruits and vegetables, and baby and geriatric foods.[10] Recently, this hydrolytic ability has been applied to biomass utilization and biotechnology for extraction of leaf proteins,[11] liquiefaction of agricultural biomass,[12] and isolation of plant protoplasts.[13] For these purposes, the enzyme of *C. rolfsii* is particularly useful owing to the activity in low pH regions where the enzyme reaction proceeds nearly aseptically.

Although pectolytic enzymes other than polygalacturonases from various origins show some activity in plant tissue maceration, singly or in combination, an enzyme preparation from *Clostridium felsineum*[14] reveals a strong macerating activity in spite of a lack of activity against purified pectin or pectic acid. This enzyme presumably acts on native pectin where galacturonan combines with other polysaccharides and should be useful for vegetable biomass utilization.

[9] M. Manabe, A. Kaji, and T. Tarutani, *J. Food Ind. Soc. Jpn.* **13,** 269 (1966).

[10] H. K. Sreenath, M. D. Frey, H. Scherz, and B. J. Radola, *Biotechnol. Bioeng.* **26,** 788 (1984).

[11] M. Oshima and K. Tagawa, unpublished observations.

[12] G. Beldman, M. J. F. Leeuwen, A. G. J. Searle-Van Voragen, F. M. Rombouts, and W. Pilnié, *Comm. Eur. Commun.* **EUR9940,** 41 (1985).

[13] A. Ruesink, this series, Vol. 69, p. 69.

[14] A. Kaji, *Bull. Agric. Chem. Soc. Jpn.* **20,** 8 (1956).

[40] Isozymes of Pectinesterase and Polygalacturonase from *Botrytis cinerea* Pers.

By ABEL SCHEJTER and LIONEL MARCUS

$$\text{Polymethylgalacturonate (pectin)} + H_2O \xrightarrow{\text{pectinesterase}} \text{polygalacturonic acid} + \text{methanol}$$

$$\text{Polygalacturonic acid} + H_2O \xrightarrow{\text{polygalacturonase}} \text{oligogalacturonic acids} + \text{galacturonic acid}$$

$$\text{Polymethylgalacturonate (pectin)} + H_2O \rightarrow \text{oligogalacturonic acids} + \text{galacturonic acid} + \text{methanol}$$

The enzymes pectinesterase (pectyl hydrolase, EC 3.1.1.11) and polygalacturonase [poly(1,4-α-D-galacturonide) glycanohydrolase, EC 3.2.1.15] are secreted by the fungus *Botrytis cinerea* Pers. and are involved in the rotting and maceration of fresh fruit and vegetables.[1]

Pectinesterase

This enzyme has been purified from fruits (banana,[2] orange,[3] tomato[4,5]), fungi (*Corticium rolfsii*,[6] *Fusarium oxysporum*[7]), and bacteria (*Clostridium multifermentans*[7,8]).

Assay Method

Principle. The enzyme activity is measured by continuous automatic titration with NaOH of the carboxyl groups released during the reaction.[3]

Reagents

Pectin, 0.5% in 0.01 *M* acetate buffer, pH 6.5, containing 0.1 *M* NaCl and 0.01% NaN_3

NaOH, 0.1 *M*

[1] D. F. Bateman, *Encycl. Plant Physiol.* **4,** 316 (1976).
[2] C. J. Brady, *Aust. J. Plant Physiol.* **3,** 163 (1976).
[3] C. Versteeg, F. M. Rombouts, and W. Pilnik, *Lebensm.-Wiss. Technol.* **11,** 267 (1976).
[4] M. Lee and J. D. Macmillan, *Biochemistry* **9,** 1930 (1970).
[5] O. Markovič, *Collect. Czech. Chem. Commun.* **39,** 908 (1974).
[6] O. Yoshihara, T. Matsuo, and A. Kaji, *Agric. Biol. Chem.* **41,** 2335 (1977).
[7] L. Miller and J. D. Macmillan, *Biochemistry* **10,** 570 (1971).
[8] M. I. Sheiman, J. D. Macmillan, L. Miller, and T. Case, *Eur. J. Biochem.* **64,** 565 (1976).

METHODS IN ENZYMOLOGY, VOL. 161

Procedure. The pectin solution, 4.9 ml, is brought to 30° in a jacketed vessel. The reaction is initiated by adding 0.1 ml of a solution of enzyme in 0.01 *M* acetate buffer, pH 6.5, and allowed to run for 3 to 5 min, while the volume of NaOH delivered by a TTT 11 radiometer (Copenhagen, Denmark) automatic titrator is recorded. The solution is continuously stirred with a magnetic stirrer and flushed with pure nitrogen.

Definition of Unit and Specific Activity. One unit of enzyme activity is defined as the number of microequivalents of methyl ester groups cleaved by 1 ml of the enzyme solution per minute, under the conditions described above. Specific activity is expressed in units per milligram of protein.

Other Assay Methods. For a rapid test for the presence of enzyme activity, for example in testing fractions eluted from a chromatographic column, it is very convenient to use the following procedure.[3] To 25 ml of the solution of pectin in acetate buffer, five drops of 1% methyl red in ethanol are added. One-milliliter aliquots of this reagent are placed in test tubes, and 20 μl of each fraction is added to them. Those fractions that develop a red color within 1 hr are tested quantitatively with the automatic titrimetric assay.

Purification Procedure

Growth of Fungus. A strain of *Botrytis cinerea* Pers. obtained from the Volcani Institute, Beth Dagan, Israel, is cultured for 7 days in agar slants at 30°, under fluorescent light illumination. An inoculum of 10 ml sterile water washings of three to five slants is added to a glucose-free medium in a 10-liter fermenter (model 19, New Brunswick Scientific, NJ). The medium contains, per liter: 10 g citrus pectin (Sigma Chemical Co., St. Louis, MO), 10 g $NaNO_3$, 5 g KH_2PO_4, 2.5 g $MgSO_4 \cdot 7H_2O$, and 1 g $CaCl_2$, and traces of $FeCl_3$ and vitamin B_1. The culture is aerated and stirred at 30° for 10 days, under constant fluorescent light illumination. The culture fluid is filtered through Micracloth (Chicopee Mills, Inc., Milltown, NJ); the filtrates are centrifuged at 8000 *g* for 20 min and the precipitated mycelium discarded.

First Acetone Precipitation. To the filtrated and centrifuged culture fluid cold acetone is added at 4° to a final concentration of 20% (v : v), and this is left to stand overnight. Most of the clean fluid can be siphoned; the residual is centrifuged at 27,000 *g* for 10 min, and the supernatant added to the collected fluid.

Second Acetone Precipitation. To the fluid, cold acetone is added at 4° to a final concentration of 66% (v : v) and left to stand overnight. The clean supernatant is siphoned, and the remaining suspension is centrifuged at 27,000 *g* for 10 min. The precipitate is collected and suspended in 200 ml of 0.01 *M* sodium acetate buffer, pH 4.2. This is gently stirred overnight, centrifuged (27,000 *g*, 10 min), and the precipitate suspended in another 100 ml of buffer. This procedure is repeated until the extracting buffer solution shows no enzyme activity. The final volume of the enzyme solution is approximately 500 ml. This volume is reduced to about 50 ml by lyophilization, and dialyzed overnight against 0.01 *M* acetate buffer, pH 4.2.

Crosslinked Polypectate Column Chromatography. The support material for the affinity chromatography column, crosslinked polypectate, is prepared in the following way.[9] Sodium polypectate, batch No. 5709, is from Nutritional Biochemicals Corporation, Cleveland, Ohio. It is important that the polypectate by of molecular weight 100,000 or higher, since polypectate with lower molecular weight fails to produce the desired crosslinked material. To a solution of 28 ml epichlorohydrin (1-chloro-2,3-epoxypropane) (from Fluka, A.G., Switzerland) in 150 ml of 95% ethanol, 50 g sodium polypectate is added while stirring. To this, 50 ml of 5 *M* NaOH is added. The suspension is rotated in a 500-ml round-bottom flask kept at 40° for 8 hr, and neutralized to pH 7.0 with 1 *M* acetic acid. The crosslinked polypectate is collected by filtration through a medium sintered glass filter, and washed with 300 ml of a 3:1 mixture (v:v) of absolute ethanol and water, followed by 200 ml of 95% ethanol. The crosslinked polypectate is left to dry in air to constant weight. Once dry, the material is resuspended in 500 ml distilled water in a 2-liter beaker, stirred, and left to settle for 1 min. The suspension is decanted and the material remaining at the bottom of the beaker is collected on a coarse sintered glass filter, washed with 200 ml 95% ethanol, and dried in air to constant weight. This procedure yields a homogeneous powder of crosslinked polypectate.

Twenty grams of crosslinked polypectate is suspended in 200 ml 0.1 *M* acetate buffer pH 4.2, and left to swell for 2 hr. The material is collected on a fine sintered glass filter; it is washed with 500 ml distilled water, in order to reduce salt concentration, and poured on a 1.2-cm-diameter glass column, resulting in a column height of about 27 cm. The column is equilibrated in the cold room at 4°, with 0.01 *M* acetate buffer, pH 4.2, until constant pH.

A 10-ml aliquot of the enzyme solution is applied to the column, followed by buffer. The flow rate is adjusted with a peristaltic pump to 8 ml/hr, and 6-ml fractions are collected and examined for protein content at 280 nm, and for enzymatic activity using the rapid methyl red test. The elution is continued until about 10 protein-free fractions are eluted. At this point, the elution is continued by applying to the buffer a linear NaCl

[9] F. M. Rombouts, A. K. Wissenburg, and W. Pilnik, *J. Chromatogr.* **168,** 151 (1979).

gradient, from 0 to 0.1 *M* salt, until no further protein-containing fractions are obtained. Pectinesterase activity is tested for all protein-containing fractions. Two pectinesterase isoenzymes are obtained: pectinesterase I in the first elution with buffer only, and pectinesterase II in the NaCl gradient elution.

The purification procedure is summarized in Table I. The fractions with high specific activity are pooled, dialyzed against water, and stored at 4°. They can also be preserved as a lyophilized powder in the cold.

Properties of the Purified Isoenzymes

Molecular Weight. Pectinesterases I and II have molecular weights of 28,400 and 27,800, respectively, by SDS-polyacrylamide gel electrophoresis and 28,100 and 27,600, respectively, by gel exclusion chromatography.

Purity. The two isoenzymes migrate as single bands in SDS gel electrophoresis.

Isoelectric Point. The isoelectric point of both isoenzymes is 6.8, as determined by flat bed isoelectric focusing on agarose gel.

Kinetic Constants. K_m values for pectin at 30° are 6×10^{-3} g/100 ml for pectinesterase I and 4×10^{-4} g/100 ml for pectinesterase II. Values of V_{max} are expressed in micromoles of ester groups hydrolyzed per unit of enzyme per minute at 30°: 0.98 and 0.60 for pectinesterase I and II, respectively.

Optimum pH. Maximal activities are observed at pH values 7.0 and 6.5 for pectinesterase I and II, respectively.

Inhibitors. Sodium polypectate is a competitive inhibitor, with K_i values of 2.8×10^{-3} and 1.1×10^{-4} g/ml for pectinesterases I and II, respectively.

Polygalacturonase

This enzyme is an endopolygalacturonase, and acts on polygalacturonic acid (polypectate) but not on pectin. It has been isolated and purified from fruits (tomato[10,11]), yeast *(Saccharomyces fragilis)*,[12] fungi (*Aspergillus japonicus,*[13] *Aspergillis niger,*[14,15] *Botrytis cinerea,*[16] *Fusarium oxysporum,*[17]

[10] H. Takehana, T. Shibuya, H. Nakagawa, and N. Ogura, *Tech. Bull. Fac. Hortic., Chiba Univ.* **25,** 29 (1977).

[11] R. Pressey and J. K. Avants, *Biochim. Biophys. Acta* **309,** 363 (1973).

[12] H. J. Phaff, this series, Vol. 8, p. 109.

[13] S. Ishii and T. Yokotsuka, *Agric. Biol. Chem.* **36,** 1885 (1972).

[14] K. Heinrichova and L. Rexová-Benková, *Collect. Czech. Chem. Commun.* **42,** 2569 (1977).

[15] R. D. Cooke, E. M. Ferber, and L. Kanagasabapathy, *Biochim. Biophys. Acta* **452,** 440 (1976).

TABLE I

PURIFICATION OF PECTOLYTIC ENZYMES FROM *B. cinerea*

Purification step	Volume (ml)	Total protein (mg)	Specific PG activity (U mg^{-1})	Specific PE activity (U mg^{-1})	Purification factor		Yield (%)	
					PG	PE	PG	PE
Culture fluid	8400	5376	258	9.4	0	0	100	100
20% Acetone (supernatant)	9340	4109	322	57	1.25	6.1	76.4	76.4
66% Acetone (powder)	500	960	586	361	2.27	38.4	17.9	17.9
Polygalacturonase I	30	270	2180	—	8.43	—	5.0	—
Polygalacturonase II	30	210	7092	—	27.40	—	3.9	—
Pectylhydrolase I	36	482	—	608	—	64.7	—	9.0
Pectylhydrolase II	36	360	—	710	—	75.5	—	6.7

Rhizoctonia fragariae,[18,19] *Rhizopus arrhizus,*[20] *Trichoderma koningii,*[21] and *Verticillium albo-atrum*[22]), and bacteria (*Erwinia carotovora*[23] and *Pseudomonas cepacia*[24]).

Assay Method

Principle. The enzyme activity is measured by following spectrophotometrically at 530 nm the increase in reducing groups,[25] after these react with 3,5-dinitrosalicylic acid (Bernfeld reagent[26]).

Reagents

0.5% Sodium polypectate, grade II (Sigma Chemical Co., St. Louis, MO), in 0.01 *M* acetate buffer, pH 4.2

Bernfeld reagent[26]: Dissolve at room temperature 1 g of 3,5-dinitrosalicylic acid in 20 ml of 2 *N* NaOH and 50 ml H_2O, add 30 g of Rochelle salt (sodium potassium tartrate monohydrate), and make up to 100 ml with H_2O. Protect this solution from CO_2

D-Galacturonic acid monohydrate (BDH, Ltd., Poole, England)

Procedure. To 1.0 ml of the sodium polypectate solution, 1.0 ml of enzyme solution in the acetate buffer is added, and the mixture is incubated at 30°. After 3 min, 5.0 ml of the Bernfeld reagent solution is added and the solution is kept for 10 min in a boiling water bath. The solution is cooled and its optical absorbance at 530 nm is read in a 1-cm light path cuvette, against the product of a blank test in which 1.0 ml of buffer substitutes for the enzyme solution. If necessary, the brown-colored solution is diluted with water to keep the readings within the 0–1.0 absorbance scale.

Definition and Calculation of Enzyme Unit and Specific Activity. The unit of activity is the amount of enzyme that releases 1 μmol of reducing

---

[16] H. Urbanek and J. Zalewska-Sobczak, *Biochim. Biophys. Acta* **377,** 402 (1975).
[17] L. L. Strand, M. E. Corden, and D. L. MacDonald, *Biochim. Biophys. Acta* **429,** 870 (1976).
[18] F. Cervone, A. Scala, M. Foresti, M. G. Cacace, and C. Noviello, *Biochim. Biophys. Acta* **482,** 379 (1977).
[19] F. Cervone, A. Scala, and F. Scala, *Physiol. Plant Pathol.* **12,** 19 (1978).
[20] Y. K. Lui and B. S. Luh, *J. Food Sci.* **43,** 721 (1978).
[21] C. Fanelli, M. G. Cacace, and F. Cervone, *J. Gen. Microbiol.* **104,** 305 (1978).
[22] M. C. Wang and N. T. Keen, *Arch. Biochem. Biophys.* **141,** 749 (1970).
[23] S. Nasuno and M. P. Starr, *J. Biol. Chem.* **241,** 5298 (1966).
[24] J. M. Ulrich, *Plysiol. Plant Pathol.* **5,** 37 (1975).
[25] L. Rexová-Benková and V. Tibenský, *Biochim. Biophys. Acta* **268,** 167 (1972).
[26] P. Bernfeld, this series, Vol. 1, p. 17.

groups/min at 30°. The concentration of reducing groups is estimated from calibration curves obtained by performing the Bernfeld reaction with solutions of known concentrations of D-galacturonic acid in the acetate buffer. It is recommended to use only the linear portions of these curves. Specific activity is expressed in enzyme units per milligram of protein estimated according to Lowry *et al.*[27]

Other Assay Methods. It is possible to follow the enzyme breakdown of polygalacturonate by measuring the changes in viscosity of a 0.5% solution with an Ostwald viscosimeter.[28] Although this procedure is less convenient for quantitive estimations, it is an important measurement in mechanistic investigations, since it reveals whether the enzyme is an endo- or exopolygalacturonase. Thus, rapid decrease in viscosity relative to reducing groups liberation indicates that internal glycosidic bonds are split by the enzyme, while relatively slow changes in viscosity indicate glycolysis at the ends of the chains. The nature of the products of polypectate degradation can be investigated by thin-layer chromatography or silica gel plates.[29]

Purification Procedure. The procedures for cell growth, processing of the culture fluid, and affinity chromatography are identical to those described above for purification of pectinesterase. During the chromatography, a fraction with polygalacturonase activity, PG-I, is eluted before the first pectinesterase isozyme peak. A second fraction with similar enzyme activity, PG-II, is eluted shortly after the beginning of the NaCl linear gradient. Both isozyme solutions are dialyzed against water and kept at 4°. They can also be preserved as a powder after lyophilization.

Properties of the Purified Isozymes

Molecular Weight. Polygalacturonases I and II have molecular weights 34,600 and 56,800, respectively, by SDS-gel electrophoresis, and 34,100 and 56,200, respectively, by gel-exclusion chromatography.

Purity. The two isozymes migrate as single bands in SDS gel electrophoresis.

Isoelectric Point. The isoelectric points of polygalacturonases I and II determined by agarose gel isoelectric focusing are 7.3 and 7.6, respectively.

Kinetic Constants. Both K_m and V_{max} values are dependent on the salt concentration of the solutions for the two isozymes. The K_m value is minimal for the two isozymes at 5×10^{-2} *M* NaCl, while V_{max} is maximal at 1×10^{-1} *M* NaCl. At 30° and pH 4.2 (0.01 *M* acetate buffer), in 0.1 *M*

[27] O. H. Lowry, N. J. Rosebrough, A. L. Farr, and R. J. Randall, *J. Biol. Chem.* **193,** 265 (1951).

[28] D. F. Bateman, *Physiol. Plant Pathol.* **2,** 175 (1972).

[29] L. Rexová-Benková, *Chem. Zvesti* **24,** 59 (1970).

NaCl, K_m is 1.4×10^{-2} g of sodium polypectate/100 ml for isozyme I, and 2.1×10^{-2}, in the same units, for isozyme II. Under similar conditions, V_{max} is 2500 and 1900 μmol of liberated reducing groups·(mg protein)$^{-1}$·(min)$^{-1}$ for isozymes I and II, respectively.

Optimum pH. Maximal activities are measured at pH values 4.5 and 4.0 for isozymes I and II, respectively.

Inhibitors. Both isozymes are inhibited by the products of their enzymatic activity, and competitively inhibited by pectin (K_i values for pectin are 9×10^{-3} and 5×10^{-3} g/100 ml for isozymes I and II, respectively) and by sodium polyglutamate (K_i values are 6×10^{-3} and 4×10^{-3} g/100 ml for the isozymes I and II, respectively).

[41] Galacturan 1,4-α-Galacturonidase from Carrot *Daucus carota* and Liverwort *Marchantia polymorpha*

By HARUYOSHI KONNO

Poly-1,4-α-D-galacturonate + n H_2O → n D-galacturonic acid

Assay Method

Principle. Exopolygalacturonase [poly(1,4-α-D-galacturonide) galacturonohydrolase; EC 3.2.1.67, galacturan 1,4-α-galacturonidase] activity is determined by measuring the increment in number of reducing groups (galacturonic acids) from polygalacturonate.

Reagents. Prepare a stock solution of the following composition:

Sodium acetate buffer, pH 4.0 and 5.0, 250 m*M*
Acid-insoluble polygalacturonate (sodium salt), 1% (w/v), in distilled water

Acid-insoluble polygalacturonate used as substrate is prepared by partial acid hydrolysis of a commercial citrus pectin as previously described.[1] The acid-insoluble polygalacturonate obtained is further purified by gel filtration on Sephadex G-150. The sample is applied to a column (1.8 × 90 cm) previously equilibrated with 50 m*M* sodium acetate buffer, pH 5.2, containing 25 m*M* EDTA, and eluted with the same buffer. The major peak of carbohydrate is pooled, dialyzed against distilled water, and con-

[1] H. Konno, K. Katoh, and Y. Yamasaki, *Plant Cell Physiol.* **22**, 899 (1981).

centrated with evaporation *in vacuo* below 40°. The polygalacturonate in the concentrated solution is precipitated on the addition of ethanol. The ethanol precipitate, purified acid-insoluble polygalacturonate, is used as substrate in this assay, and has an average degree of polymerization of 52.

Procedure. The reaction mixture contains 0.2 ml of 1% acid-insoluble polygalacturonate solution, 0.2 ml of 250 m*M* sodium acetate buffer, pH 5.0 or 4.0, and enzyme solution in a final volume of 1.0 ml. The enzyme activities for carrot and liverwort are measured at pH 5.0 and at pH 4.0 with sodium acetate buffer, respectively. After incubation for 1 hr at 37°, the reaction is terminated by heating in a boiling water bath for 5 min, and analyzed for reducing groups (galacturonic acids).

The reducing groups formed are measured by the method described by Nelson[2] or Somogyi,[3] using galacturonic acid as a standard. One unit of enzyme activity is defined as the amount that formed 1 μmol of galacturonic acid/min under the above conditions. Specific activitiy is expressed as units per milligram of protein determined by the method of Lowry *et al.*[4]

Purification of Carrot Exopolygalacturonase[5]

Cell Culture. Carrot (*Daucus carota* L. cv. Kintoki) roots are obtained from local gardens, and explanted on Murashige and Skoog's medium.[6] Cell suspension cultures of carrot are then started from the callus formed, and subcultured every 2 weeks using the Murashige and Skoog's medium containing 3% (w/v) sucrose as a carbon source and 4.52 μM 2,4-dichlorophenoxyacetic acid as a growth regulator. No cytokinin is added. Stock cultures are maintained by inoculating 15 ml of cell suspension into 500 ml of fresh medium in 700-ml flat oblong flasks with an inner thickness of 28 mm. For experiments, 9-day-old cells which have been cultured under subcultural conditions are suspended in the medium at the concentration of approximately 0.2 mg cell dry weight/ml. The flask is illuminated perpendicular to the flat surface with a bank of fluorescent lamps (FL 40 W) at a photon flux density of approximately 40 $\mu E/m^2 \cdot sec^{-1}$ (400–700 nm) determined with a Lambda photometer (Lincoln, New England). Aeration and agitation to keep the culture in suspension are done by passing air through the culture (50–100 ml/min) at 25°. After 15

[2] N. Nelson, *J. Biol. Chem.* **153,** 375 (1944).
[3] M. Somogyi, *J. Biol. Chem.* **195,** 19 (1952).
[4] O. H. Lowry, N. J. Rosebrough, A. L. Farr, and R. J. Randall, *J. Biol. Chem.* **193,** 265 (1951).
[5] H. Konno, Y. Yamasaki, and K. Katoh, *Physiol. Plant.* **61,** 20 (1984).
[6] T. Murashige and F. Skoog, *Physiol. Plant.* **15,** 473 (1962).

days of culture, cells are harvested by filtration, washed with distilled water, lyophilized, and stored at $-20°$ prior to utilization for enzyme preparation. All subsequent steps are conducted at 0 to 4° unless otherwise stated, and all buffers used contain 9 m*M* 2-mercaptoethanol.

Enzyme Extraction. Lyophilized cells (69 g) are suspended in 2 liters of 100 m*M* potassium phosphate buffer, pH 7.0, and disrupted by sonication (Ohtake Work, Tokyo, Japan) at 20 kHz for 10 min at 0°. The homogenate is centrifuged at 8000 *g* for 15 min, and the supernatant is collected. The cell wall materials are resuspended in 500 ml of the same buffer, and centrifuged at 8000 *g* for 15 min. The supernatants obtained are combined and dialyzed against 20 m*M* sodium acetate buffer, pH 5.2, containing 10 m*M* NaCl with several changes of the buffer. Any precipitate that formed during dialysis at pH 5.2 is removed by centrifugation. The supernatant is concentrated to 300 ml using a Millipore ultrafiltration device (Pellicon Lab cassette; filter type: PT; pore size: 10,000NMWL; Millipore Corp., Bedford, MA), and dialyzed against 20 m*M* Tris–HCl buffer, pH 7.6, containing 10 m*M* NaCl.

DEAE-Sephadex A-50 Ion-Exchange Chromatography. The dialyzed enzyme solution is applied to a column (3.2 × 40 cm) of DEAE-Sephadex A-50, previously equilibrated with 20 m*M* Tris–HCl buffer, pH 7.6, containing 10 m*M* NaCl. The column is washed with 450 ml of the same buffer, and then eluted with 1.2 liters of a linear NaCl gradient from 10 to 500 m*M* in the same buffer. Fractions of 15 ml are collected at an average flow rate of 0.3 ml/min. The enzyme activity is eluted at approximately 180 m*M* NaCl concentration as a sharp symmetrical peak. Fractions containing enzyme activity are combined, dialyzed against 50 m*M* sodium acetate buffer, pH 6.0, containing 10 m*M* NaCl, and concentrated to around 5 ml using an Amicon ultrafiltration device (UM10 membrane with a molecular weight cut-off of 10,000; Amicon Corp., Lexington, MA).

Sephadex G-150 Gel Filtration. The concentrated enzyme solution is applied to a column (2.8 × 90 cm) of Sephadex G-150, previously equilibrated with 50 m*M* sodium acetate buffer, pH 6.0, containing 10 m*M* NaCl, and then eluted with the same buffer. Fractions of 10 ml are collected at an average flow rate of 0.2 ml/min. Protein is fractionated into at least five peaks, one of which is coincident with the enzyme activity. Fractions containing enzyme activity are combined, and concentrated to around 1 ml using an Amicon ultrafiltration device (UM10 membrane).

Preparative Disk Gel Electrophoresis. The sample from the preceding step is finally purified by the preparative polyacrylamide disk gel electrophoresis. Disk gel electrophoresis is carried out by the method of Davis.[7]

[7] B. J. Davis, *Ann. N.Y. Acad. Sci.* **121,** 404 (1964).

The system used is a 7.5% (w/v) polyacrylamide gel column (3.8 × 10 cm) containing 380 m*M* Tris–HCl buffer, pH 8.9. The enzyme solution is placed on a polyacrylamide gel column, and the electrophoresis is conducted at 60 mA for 5 hr using 20 m*M* Tris–glycine, pH 8.3, as an electrode buffer. After electrophoresis, part of the gel is stained for protein with Amido Black 10 B. The other part is used for enzyme preparation by soaking in ice-cold 50 m*M* sodium acetate buffer, pH 6.0, containing 10 m*M* NaCl for 30 min, and slices corresponding to the protein bands are cut with a gel slicer. The protein contained in each gel slice is eluted overnight with the same buffer. The fraction containing enzyme activity is concentrated to around 5 ml using an Amicon ultrafiltration device (UM10 membrane), and dialyzed against 50 m*M* sodium acetate buffer, pH 6.0, containing 10 m*M* NaCl.

The data in Table I, summarizing a typical purification of exopolygalacturonase, show that a 69-fold increase in specific activity is achieved with a recovery of approximately 24%.

Properties

Chemical and physicochemical properties have already been reported,[1] but in this description some additional information has been included.

Purity and Stability. The enzyme preparation purified in this manner shows a single protein band coincident with the exopolygalacturonase activity on analytical polyacrylamide disk gel electrophoresis. The activity of the enzyme in purified and concentrated solutions is stable for at least 2 months when stored at −20°. Incubation at pH 5.0 for 30 min at temperatures up to 55° barely inactivates the enzyme, but one-half of the activity is destroyed when heated to 65°, and all of it is destroyed when heated at 80°. After incubation at 30° in McIlvaine's buffer at various pH values for 20 hr, the enzyme is found to be stable in the pH range of 3.5 to 7.0.

TABLE I
PURIFICATION OF EXOPOLYGALACTURONASE FROM CARROT *Daucus carota*

Step	Total activity (U)	Total protein (mg)	Specific activity (U/mg protein)	Yiel[illegible] (%)
Extraction	5.90	1286	0.005	10[illegible]
DEAE-Sephadex A-50	5.30	176	0.030	9[illegible]
Sephadex G-150	5.13	30	0.171	8[illegible]
Preparative disk gel electrophoresis	1.42	4.1	0.346	2[illegible]

Molecular Weight and Isoelectric Points. The molecular weight of the enzyme, as estimated by Sephadex G-200 gel filtration, is approximately 48,000. The isoelectric point is found to be pH 4.75 by isoelectric focusing in polyacrylamide gel using ampholine of pH 3 to 10 according to the procedure described by Awdeh *et al.*[8]

Kinetic Characteristics. The apparent K_m and V_{max} values are 4.35×10^{-2} m*M* and 97.1 U/mg protein for polygalacturonate with a degree of polymerization of approximately 52. The pH-dependent enzyme activity is examined in McIlvaine's and sodium acetate buffers, and the pH optimum is found to be 4.8. Half-maximal activities are obtained at pH 4.0 and 6.5. The following cations and EDTA at concentrations of 0.5 or 1.0 m*M* have no effect on the activity: Mg^{2+}, Mn^{2+}, Co^{2+}, Ca^{2+}, Na^+, and K^+. However, 1.0 m*M* Hg^{2+}, Cu^{2+}, and Ba^{2+} reduce the activity by 15–28%.

Substrate Specificity. The polygalacturonate is a far better substrate for the enzyme than are pectic acid and pectin (7.5% methylesterified). The relative rates of degradation are as follows: polygalacturonate, 100; pectic acid, 56.2; and pectin, 6.2, during a 1-hr reaction.

Purification of Liverwort Exopolygalacturonase[9]

Cell Culture. Liverwort (*Marchantia polymorpha* L.) cell line (HYH-2F) in cell suspension culture is routinely subcultured every week using MSK-2 medium described by Katoh *et al.*[10] Stock cultures are maintained by inoculating 15 ml of cell suspension into 500 ml of fresh medium in 700-ml flat oblong flasks with an inner thickness of 28 mm. For experiments, 7-day-old cells from stock culture are suspended in the medium at the concentration of approximately 0.2 mg cell dry w/ml. The flask is illuminated perpendicular to the flat surface with a bank of fluorescent lamps (FL 40 W) at a photon flux density of approximately 90 $\mu E/m^2 \cdot sec^{-1}$ (400–700 nm). Aeration and agitation to keep the culture in suspension are done by passing air containing 1% CO_2 through the culture (50–100 ml/min) at 25°. After 15 days of culture, cells are harvested and stored by the same procedure described above (see Carrot Cell Culture). All subsequent steps are conducted at 0 to 4° unless otherwise stated, and all buffers used contain 9 m*M* 2-mercaptoethanol.

Enzyme Extraction. Lyophilized cells (183 g) are suspended in 3 liters of 100 m*M* potassium phosphate buffer, pH 7.0, and disrupted by sonica-

[8] Z. L. Awdeh, A. R. Williamson, and B. A. Askonas, *Nature (London)* **219,** 66 (1968).

[9] H. Konno, Y. Yamasaki, and K. Katoh, *Plant Physiol.* **73,** 216 (1983).

[10] K. Katoh, M. Ishikawa, K. Miyake, Y. Ohta, Y. Hirose, and T. Iwamura, *Physiol. Plant.* **49,** 241 (1980).

tion at 20 kHz for 10 min at 0°. The homogenate is centrifuged at 8000 *g* for 15 min, and the supernatant is collected. The cell wall materials are resuspended in 700 ml of the same buffer, and centrifuged at 8000 *g* for 15 min. To remove the phenolic matter, the supernatants obtained are combined, and divided into approximately 10 portions; each portion is passed through a Sephadex G-25 column (10 × 30 cm) equilibrated with the same buffer, and the void (protein) fraction is pooled and dialyzed against 20 m*M* sodium acetate buffer, pH 5.2, containing 10 m*M* NaCl with several changes of the buffer. Greenish precipitates that formed during dialysis at pH 5.2 are removed by centrifugation. The supernatant is collected, concentrated to around 100 ml using a Millipore ultrafiltration devise (Pellicon; filter type: PT; pore size: 10,000NMWL), and dialyzed against 50 m*M* sodium acetate buffer, pH 6.0, containing 10 m*M* NaCl.

CM-Sephadex C-50 Batch Chromatography. The dialyzed enzyme solution is passed through a CM-Sephadex C-50 column (3.0 × 40 cm) equilibrated with 50 m*M* sodium acetate buffer, pH 6.0, containing 10 m*M* NaCl. The eluant containing enzyme activity is pooled, and concentrated to around 20 ml using an Amicon ultrafiltration devise (UM10 membrane).

Sephacryl S-200 Gel Filtration. The concentrated enzyme solution is applied to a column (2.8 × 90 cm) of Sephacryl S-200 (Superfine), previously equilibrated with 50 m*M* sodium acetate buffer, pH 6.0, containing 10 m*M* NaCl, and then eluted with the same buffer. Fractions of 10 ml are collected at an average flow rate of 0.3 ml/min. Fractions containing enzyme activity are combined, concentrated to around 20 ml using a Millipore ultrafiltration device (Pellicon; filter type: PT; pore size: 10,000NMWL), and dialyzed against 20 m*M* Tris–HCl buffer, pH 7.6, containing 10 m*M* NaCl with several changes of the buffer.

DEAE-Sephadex A-50 Ion-Exchange Chromatography. The dialyzed enzyme solution is applied to a column (3.2 × 40 cm) of DEAE-Sephadex A-50, previously equilibrated with 20 m*M* Tris–HCl buffer, pH 7.6, containing 10 m*M* NaCl. The column is washed with the same buffer, and then eluted with 1.2 liters of a linear NaCl gradient from 10 to 500 m*M* in the same buffer. Fractions of 15 ml are collected at an average flow rate of 0.3 ml/min. The enzyme activity is eluted at approximately 200 m*M* NaCl concentration as a sharp symmetrical peak. Fractions containing enzyme activity are combined, dialyzed against 50 m*M* sodium acetate buffer, pH 6.0, containing 10 m*M* NaCl, and concentrated to around 3 ml using an Amicon ultrafiltration device (UM10 membrane).

Sephadex G-200 Gel Filtration. The concentrated enzyme solution is applied to a column (2.0 × 90 cm) of Sephadex G-200, previously equilibrated with 50 m*M* sodium acetate buffer, pH 6.0, containing 10 m*M*

NaCl, and then eluted with the same buffer. Fractions of 10 ml are collected at an average flow rate of 0.2 ml/min. Protein is fractionated into at least three peaks, a major peak of which is coincident with the enzyme activity. Fractions containing enzyme activity are combined, concentrated to 5 ml using an Amicon ultrafiltration device (UM10 membrane), and dialyzed against 50 m*M* sodium acetate buffer, pH 6.0, containing 10 m*M* NaCl.

The data in Table II, summarizing a typical purification of exopolygalacturonase, show that an 88-fold increase in specific activity is achieved with a recovery of approximately 9%.

Properties

Chemical and physicochemical properties have already been reported,[9] but in this description some additional information has been included.

Purity and Stability. The purified enzyme shows one major protein band coincident with enzyme activity and several minor protein bands free from enzyme activity on analytical polyacrylamide disk gel electrophoresis. Homogeneous enzyme preparation can be obtained utilizing preparative polyacrylamide disk gel electrophoresis (see Purification of Carrot Exopolygalacturonase). The activity of the enzyme in purified and concentrated solution is stable for at least 1 month when stored at −20°. Incubation at pH 4.0 for 30 min at temperatures up to 55° barely inactivates the enzyme, but one-half of the activity is destroyed when heated to 65°, and all of it is destroyed when heated at 80°. After incubation at 30° in McIlvaine's buffer at various pH values for 20 hr, the enzyme is found to be comparatively stable in the pH range of 3.0 to 6.5.

TABLE II
PURIFICATION OF EXOPOLYGALACTURONASE FROM LIVERWORT *Marchantia polymorpha*

Step	Total activity (U)	Total protein (mg)	Specific activity (U/mg protein)	Yield (%)
xtraction				
Sephadex G-25	39.2	45,047	0.0009	100
Dialysis at pH 5.2	18.3	1,287	0.0142	47
M-Sephadex C-50 batch	11.6	673	0.0172	30
ephacryl S-200	6.68	302	0.0221	17
EAE-Sephadex A-50	4.05	103	0.0393	10
ephadex G-200	3.65	46.2	0.0790	9.3

Molecular Weight and Isoelectric Point. The molecular weight of the enzyme, as estimated by Sephadex G-200 gel filtration, is approximately 76,000, and the isoelectric point is found to be pH 5.20 by isoelectric focusing in polyacrylamide gel using ampholine of pH 3 to 10.

Kinetic Characteristics. The apparent K_m and V_{max} values are 3.23×10^{-2} m*M* and 31.6 U/mg protein for polygalacturonate with a degree of polymerization of 52. The pH optimum is found to be 3.6–3.8, and half-maximal activities are obtained at pH 3.2 and 4.6. The following cations and EDTA at concentration of 0.5 or 1.0 m*M* have no effect on the activity: Mg^{2+}, Co^{2+}, Ca^{2+}, Na^{+}, and K^{+}. However, 0.5 m*M* Hg^{2+}, Mn^{2+}, Ba^{2+}, and Cu^{2+} reduce the activity by 20–53%.

Substrate Specificity. The polygalacturonate is a far better substrate for the enzyme than are pectic acid and pectin (7.5% methylesterified). The relative rates of degradation are as follows: polygalacturonate, 100; pectic acid, 55.4; and pectin, 18.2, during a 1-hr reaction.

Physiological Function

Exopolygalacturonases are produced by higher plants, fungi, and some bacteria. The enzyme in plant tissues is usually bound to cell wall materials via ionic interactions; consequently, solubilization requires a high salt concentration. Exopolygalacturonase from cell cultures of carrot and liverwort can be extracted completely with 100 m*M* potassium phosphate buffer, pH 7.0, free from NaCl or LiCl, and exhibits only one form. The exopolygalacturonase activity is found at all stages of growth, but changes significantly during the cell growth cycle, in relation to the rate of cell cultures growth. The purified exopolygalacturonase from carrot cell cultures, as well as that from liverwort cell cultures, do not degrade the corresponding cell wall preparation during *in vitro* incubation, and only partially degrade native pectic polysaccharides extracted from the walls.[5,9] The degradation results in hydrolysis of 8 and 29% of the glycosyl linkages of the pectic polymers purified from the extracted carrot pectic polysaccharides.[11] Therefore, it appears that exopolygalacturonase serves in cell wall metabolism, for instance in the partial degradation of pectic polysaccharides, during cell growth and/or differentiation.

Acknowledgments

The author is grateful to Dr. K. Katoh (Suntory Institute for Bioorganic Research) for providing cell cultures of carrot and liverwort. Thanks are also expressed to Prof. K. Kurahashi (Osaka University) and Dr. Y. Yamasaki (Okayama University) for valuable suggestions.

[11] H. Konno, Y. Yamasaki, and K. Katoh, *Phytochemistry* **25**, 623 (1986).

[42] Endopectate Lyase from *Erwinia aroideae*

By HARUYOSHI KONNO

Poly-1,4-α-D-galacturonate → *n* (4-deoxy-β-L-*threo*-hex-4-enopyranosyluronic acid)-1,4-D-galacturonic acid (unsaturated digalacturonic acid)

Assay Method

Principle. Endopectate lyase (poly-1,4-α-D-galacturonide lyase, EC 4.2.2.2) activity is determined spectrophotometrically by measuring the increment of unsaturated compounds at 235 nm absorbance as previously reported.[1] The end product produced by the enzyme from polygalacturonate consists mainly of unsaturated digalacturonic acid in addition to smaller amounts of unsaturated trigalacturonic acid and saturated mono- and digalacturonic acids. The method used here is slightly modified in form.

Reagents. Prepare a stock solution of following composition:

Tris–HCl buffer, pH 8.6, 200 m*M*
Sodium acetate buffer, pH 3.6, 100 m*M*
Calcium chloride, dihydrate, 100 m*M*
Acid-insoluble polygalacturonate (sodium salt), 1% (w/v), in distilled water (see chapter [41] in this volume)

Procedure. The reaction mixture contains 0.2% (w/v) acid-insoluble polygalacturonate, 0.5 m*M* $CaCl_2$, 50 m*M* Tris–HCl buffer, pH 8.6, and enzyme solution in a final volume of 2.0 ml. A blank consisting of boiled enzyme is run with each sample. After incubation for 10 min at 37°, the reaction is terminated by adding 2 ml of 100 m*M* sodium acetate buffer, pH 3.6, and analyzed for unsaturated compounds (aldehyde groups) at 232 nm. The increase of 2.3 in the absorbance at 232 nm in a 2-ml reaction mixture under the above conditions is equivalent to the release of 1 μmol of aldehyde groups. The amount of unsaturated compounds formed can also be determined using the thiobarbituric acid assay described by Weissbach and Hurwitz.[2]

One unit of enzyme activity is defined as the amount that formed 1 μmol unsaturated compounds (aldehyde groups) per minute under the

[1] F. Moran, S. Nasuno, and M. P. Starr, *Arch. Biochem. Biophys.* **123,** 298 (1968).
[2] A. Weissbach and J. Hurwitz, *J. Biol. Chem.* **234,** 705 (1959).

above conditions. Specific activity is expressed as units per milligram protein. Protein is determined by the method of Lowry *et al.*[3]

Purification Procedure[4]

Cell Growth. A seed culture of *Erwinia aroideae* is maintained on slants of potato extract agar containing 2% (w/v) sucrose, and grown at 27° for 3 days. A culture medium which contains 0.5% (w/v) pectic acid, 0.5% (w/v) peptone, 0.3% (w/v) meat extract, 15 m*M* KH_2PO_4, and 28 m*M* $Na_2HPO_4 \cdot 12H_2O$ (final pH adjusted to pH 7.2) is used for the enzyme production. The organism grown on the potato–sucrose slant is inoculated into 200 ml of the liquid medium in a 500-ml shake flask, and cultivation is carried at 27° with reciprocal shaking at 110 strokes/min. After 4 days of culture, cells are harvested by centrifugation at 8000 *g* for 20 min, washed with distilled water, and stored at −20°.

Crude Extract. Cell paste (150 g) is suspended in 300 ml of 100 m*M* Tris–HCl buffer, pH 7.0, containing 9 m*M* 2-mercaptoethanol, and disrupted by sonication (Ohtake Work, Tokyo, Japan) at 20 kHz for 10 min at 0°. All subsequent steps are conducted at 2 to 4°. The homogenate is centrifuged at 8000 *g* for 15 min, and the supernatant is collected. The cell debris is resuspended in 150 ml of 10 m*M* Tris–HCl buffer, pH 7.0, and centrifuged at 8000 *g* for 15 min. The supernatants obtained are combined and dialyzed against 20 m*M* sodium phosphate buffer, pH 7.0, with several changes of the buffer.

Ammonium Sulfate Fractionation. Solid ammonium sulfate is added to the dialyzed enzyme solution with constant stirring until 3.4 *M* (80% saturation) is reached, and the turbid solution is left overnight. The resulting precipitate is collected by filtration, dissolved in a minimum volume of 20 m*M* sodium phosphate buffer, pH 7.0, and dialyzed against the same buffer with several changes of the buffer.

CM-Sephadex C-50 Ion-Exchange Batch Chromatography. Any precipitate which formed during dialysis is removed by centrifugation, and the clear solution is applied to a column (3.2 × 20 cm) of CM-Sephadex C-50, previously equilibrated with 20 m*M* sodium phosphate buffer, pH 7.0, and then washed with the same buffer. Adsorbed protein is eluted with the same buffer containing 300 m*M* NaCl, and 10-ml fractions are collected. The fractions containing enzyme activity are combined and dialyzed against 20 m*M* sodium phosphate buffer, pH 7.0, with several changes of the buffer.

[3] O. H. Lowry, N. J. Rosebrough, A. L. Farr, and R. J. Randall, *J. Biol. Chem.* **193,** 265 (1951).

[4] H. Konno and Y. Yamasaki, *Plant Physiol.* **69,** 864 (1982).

CM-Sephadex C-50 Ion-Exchange Column Chromatography. The dialyzed enzyme solution is applied to a second CM-Sephadex C-50 column (3.2 × 20 cm), previously equilibrated with 20 m*M* sodium phosphate buffer, pH 7.0, and then washed with the same buffer. The column is eluted with 900 ml of a linear NaCl gradient from 0 to 300 m*M* in the same buffer. Fractions of 10 ml are collected at an average flow rate of 0.3 ml/min. The enzyme activity is separated into two peaks by this chromatography, and the major activity peak eluted near 100 m*M* NaCl concentration is pooled. The sample is dialyzed against 20 m*M* sodium phosphate buffer, pH 7.0, and concentrated to around 3 ml using an Amicon ultrafiltration device (UM10 membrane with a molecular weight cut-off of 10,000; Amicon Corp., Lexington, MA).

Sephadex G-200 Gel Filtration. The sample from the preceding step is applied to a column (2.0 × 90 cm) of Sephadex G-200, previously equilibrated with 20 m*M* sodium phosphate buffer, pH 7.0, and then eluted with the same buffer. Fractions of 5 ml are collected at an average flow rate of 0.2 ml/min. Protein is fractionated into two sharp peaks, a major peak of which is coincident with the enzyme activity. Fractions containing enzyme activity are combined, concentrated to around 5 ml using an Amicon ultrafiltration device (UM10 membrane), and dialyzed against 20 m*M* sodium phosphate buffer, pH 7.0.

The data in Table I, summarizing a typical purification of endopectate lyase, show that a 200-fold increase in specific activity is achieved with a recovery of approximately 18%.

Properties

Chemical and physicochemical properties have already been reported,[4] but in this description some additional information has been included.

Purity and Stability. The enzyme preparation purified in this manner shows one major protein band coincident with enzyme activity and two

TABLE I
PURIFICATION OF ENDOPECTATE LYASE FROM *Erwinia aroideae*

Step	Total activity (U)	Total protein (mg)	Specific activity (U/mg protein)	Yield (%)
nmonium sulfate fractionation	3,536	13,252	0.27	100
ᴍ-Sephadex C-50 batch	1,508	104	14.5	43
ᴍ-Sephadex C-50	1,219	26.6	45.8	34
phadex G-200	621	11.4	54.5	18

minor protein bands free from enzyme activity on analytical polyacrylamide disk gel electrophoresis. Homogeneous enzyme preparation can be obtained utilizing the preparative polyacrylamide disk gel electrophoresis (see Chapter [41] in this volume). The activity of the enzyme in purified and concentrated solution is stable for at least 1 month when stored at −20°. Incubation at pH 8.6 for 10 min at temperatures up to 37° barely inactivates the enzyme, but one-half of the activity is destroyed when heated to 45°, and all of it is destroyed when heated at 65°. After incubation at 4° at various pH values for 24 hr, the enzyme is found to be comparatively stable in the pH range of 7.0 to 7.4 and least stable below pH 4.0.

Molecular Weight. The molecular weight of the enzyme, as estimated by Sephadex G-200 gel filtration, is approximately 67,000.

Kinetic Characteristics. The apparent K_m and V_{max} values are 11.8×10^{-2} mM and 31.9 U/mg protein for polygalacturonate with a degree of polymerization of approximately 52. The pH optimum is found to be 9.3. The enzyme requires calcium ion for activity. Calcium ion stimulates the enzyme activity with an optimum concentration of 0.5 mM, causing a 21-fold increase over the control (no Ca^{2+} added). While 0.5 mM Mn^{2+}, Co^{2+}, Cu^{2+}, and Na^{+} have no effect on the activity, Hg^{2+}, Mg^{2+}, and Ba^{2+} reduce the activity by approximately 25–44%. The addition of 2.5 mM EDTA inhibits the activity completely.

Substrate Specificity. The polygalacturonate is a far better substrate for the enzyme than are pectic acid and pectin (7.5% methylesterified). The relative rates of degradation are as follows: polygalacturonate, 100; pectic acid, 47.9; and pectin, 21.3, during a 30-min reaction.

Biological Function

The structural features of pectic polysaccharides, which serve as structural major elements of plant cell walls, have been studied extensively by chemical analysis of pectic polymer extracted with reagents such as EDTA, ammonium oxalate, and sodium hexametaphosphate.[5] However, these procedures are time consuming and lack sufficient sensitivity. Albersheim and co-workers[6] have presented evidence that endopolygalacturonase purified from the culture filtrate of *Colletotrichum lindemuthianum* is able to attack directly isolated plant cell walls, with soluble pectic fractions as the reaction products. Therefore, the extraction of pectic polysaccharides from

[5] R. Goldberg, *in* "Cell Components" (H. F. Linskens and J. F. Jackson, eds.), p. 1. Springer-Verlag, Berlin and New York, 1985.

[6] P. D. English, A. Maglothin, K. Keegstra, and P. Albersheim, *Plant Physiol.* **49,** 293 (1972).

cell walls has been made possible by the availability of the purified enzymes, which are capable of degrading isolated plant cell walls. Endopectate lyases are produced by various phytopathogenic microorganisms. In a number of cases, these enzymes have shown to cause extensive cell wall breakdown and maceration, resulting in cell death of infected host tissues. The endopectate lyase purified from the soft rot bacterium *E. aroideae* also has the ability to effectively yield pectic fragments as the reaction products from isolated cell wall preparations.[4,7] Therefore, purified endopectate lyase is useful in elucidating the chemical structure of pectic polysaccharides of plant cell walls.

Acknowledgment

The author wishes to thank Prof. K. Kurahashi of Osaka University and Dr. Y. Yamasaki of Okayama University for valuable suggestions.

[7] H. Konno, Y. Yamasaki, and K. Katoh, *Phytochemistry* **25,** 623 (1986).

[43] High-Performance Liquid Chromatography of Pectic Enzymes

By OTAKAR MIKEŠ and LUBOMÍRA REXOVÁ-BENKOVÁ

Introduction

Plant pathogens, saprophytic fungi and bacteria, grown on media containing pectin, produce a mixture of enzymes which catalyze its degradation, i.e., de-esterification of units of D-galactopyranuronic acid and depolymerization of the D-galacturonan chain by hydrolytic or β-eliminative splitting of the α-1,4-glycosidic linkages. The composition of the enzyme mixture and the content of individual enzymes are determined by the type of microorganism and the cultivation conditions.[1] Individual enzymes occur as multiple molecular forms. Commercial preparations of pectinases used in food industry mostly contain endopolygalacturonase [EC 3.2.1.15, poly(1,4-α-D-galacturonide) glycanohydrolase], exopolygalacturonase [galacturan 1,4-α-galacturonidase, EC 3.2.1.67; poly(1,4-α-D-galacturonide) galacturonohydrolase], pectinesterase [EC 3.1.1.11, pectin pectylhydrolase], and pectin lyase [EC 4.2.2.10, poly(methoxygalacturonide) lyase].

[1] L. Rexová-Benková and O. Markovič, *Adv. Carbohydr. Chem. Biochem.* **33,** 323 (1976).

The first three enzymes also can occur separately or together in several plant organs.[1]

Various methods have been used for the separation of pectic enzymes, such as ion-exchange chromatography on DEAE-cellulose,[2,3] CM-cellulose,[4] cellulose phosphate,[5] gel chromatography,[6] and affinity chromatography on pectic acid cross-linked by epichlorhydrin[7,8] and its amino derivative,[9] or on poly(hydroxyalkyl methacrylate) containing glycosidically bound oligo-D-galactosiduronic acids.[10,11] Only chromatography on DEAE-cellulose leads to the separation of all the enzymes present in the starting material. Because of the variability in the composition of pectic enzyme complexes and the new trends in some industrial processes to apply preparations containing certain components of the pectolytic complex or mixtures of definite composition, rapid and efficient separation procedures on an analytical and preparative scale are required in research and industry.

High-performance liquid chromatography of biopolymers, especially of enzymes (cf., for example, Refs. 12 and 13), has permitted the application of this rapid method to the separation of pectic enzymes of microbial[14–16] and plant[17] origin. Microbial pectic enzymes from three commercial preparations of pectinases were separated in 35–70 min on various types of ion-exchange derivatives of hydrophilic polymer Spheron using a combination of isocratic and linear gradient elution.[14–16] Two forms of endo-

[2] L. Rexová-Benková and A. Slezárik, *Collect. Czech. Chem. Commun.* **31,** 122 (1966).

[3] K. Heinrichová and L. Rexová-Benková, *Biochim. Biophys. Acta* **422,** 349 (1976).

[4] P. J. Mill and R. Tuttobello, *Biochem. J.* **79,** 57 (1961).

[5] A. Koller and H. Neukom, *Eur. J. Biochem.* **7,** 485 (1969).

[6] J. F. Thibault and C. Mercier, *J. Solid-Phase Biochem.* **2,** 295 (1978).

[7] L. Rexová-Benková and V. Tibenský, *Biochim. Biophys. Acta* **268,** 187 (1972).

[8] J. Visser, R. Maeyer, R. Topp, and F. Rombouts, *Colloq. Inst. Natl. Sante Rech. Med.* **86,** 61 (1979).

[9] M. A. Vijayalakshmi, C. Bonaventura, D. Picque, and E. Segard, *in* "Affinity Chromatography" (O. Hoffmann-Ostenhof, ed.), Proc. Int. Symp., p. 115. Pergamon, Oxford, 1978.

[10] L. Rexová-Benková, J. Omelková, K. Filka, and J. Kocourek, *Carbohydr. Res.* **122,** 269 (1983).

[11] L. Rexová-Benková, K. Filka, and J. Kocourek, *J. Chromatogr.* **376,** 413 (1986).

[12] O. Mikeš, *Ernaehr.-Nutr.* **5,** 88 (1981).

[13] O. Mikeš, *in* "Recent Developments in Food Analysis" (W. Baltes, P. B. Czedik-Eysenberg, and W. Pfannhauser, eds.), p. 306. Weinheim, Deerfield Beach, Florida, 1982.

[14] O. Mikeš, P. Štrop, and J. Sedláčková, *J. Chromatogr.* **148,** 237 (1978).

[15] O. Mikeš, J. Sedláčková, L. Rexová-Benková, and J. Omelková, *J. Chromatogr.* **207,** 99 (1981).

[16] L. Rexová-Benková, J. Omelková, O. Mikeš, and J. Sedláčková, *J. Chromatogr.* **238,** 183 (1982).

[17] R. Pressey, *HortScience* **19,** 572 (1984).

polygalacturonase from tomato that differed in molecular weight were separated in 20 min on a strong cation-exchanger Mono S column of the fast protein liquid chromatography (FPLC) system of Pharmacia Fine Chemicals, Uppsala, Sweden, using elution with a salt gradient.[17]

Spheron is a rigid hydrophilic macroporous glycol methacrylate copolymer in the spherical form.[18] Its ion-exchange derivatives, used for rapid separation of pectic enzymes,[14-16] have been described in a series of papers: general chemical description,[19] characterization of the polymeric matrix,[20] diethylaminoethyl derivative,[21] carboxylic cation exchangers,[22] phosphate derivatives,[23] sulfate and sulfo derivatives,[24] and quaternary ammonium derivatives.[25] The application of Spheron ion exchangers for the separation of proteins, including enzymes, has been reviewed[26] and the equipment used for medium-pressure liquid chromatography (MPLC) has been briefly described.[22,27] Spheron ion exchangers are commercially available from Lachema (621 33 Brno, Czechoslovakia).

Chromatographic supports of the same chemical composition, however, in the form of very fine spheres prepacked into stainless steel or pressure-stable glass columns for HPLC can be obtained under the name Separon HEMA[28] (hydroxyethylmethacrylate) from Laboratory Instrument Works (162 03 Prague, Czechoslovakia) and from Tessek (8240 Risskov, Denmark).

Mono S used by Pressey for the separation of tomato polygalacturonases[17] is a strongly acid cation exchanger, developed by Pharmacia (Laboratory Separation Division, S-751 82 Uppsala, Sweden) and delivered in the form of prepacked columns. It is a part of a series of ion exchangers suitable for modern medium-pressure FPLC System, developed and produced by Pharmacia. Both the ion exchangers and the equipment were described by Richey.[29]

[18] J. Čoupek, M. Křiváková, and S. Pokorný, *J. Polym. Sci., Polym. Symp.* **42,** 185 (1973).
[19] O. Mikeš, P. Štrop, J. Zbrožek, and J. Čoupek, *J. Chromatogr.* **119,** 339 (1976).
[20] O. Mikeš, P. Štrop, and J. Čoupek, *J. Chromatogr.* **153,** 23 (1978).
[21] O. Mikeš, P. Štrop, J. Zbrožek, and J. Čoupek, *J. Chromatogr.* **180,** 17 (1979).
[22] O. Mikeš, P. Štrop, M. Smrž, and J. Čoupek, *J. Chromatogr.* **192,** 159 (1980).
[23] O. Mikeš, P. Štrop, Z. Hostomská, M. Smrž, J. Čoupek, A. Frydrychová, and M. Bareš, *J. Chromatogr.* **261,** 363 (1983).
[24] O. Mikeš, P. Štrop, Z. Hostomská, M. Smrž, S. Slováková, and J. Čoupek, *J. Chromatogr.* **301,** 93 (1984).
[25] O. Mikeš, Z. Hostomská, P. Štrop, M. Smrž, S. Slováková, P. Vrátný, L. Rexová-Benková, J. Kolář, and J. Čoupek, *J. Chromatogr.* **440,** 287 (1988).
[26] O. Mikeš, *Int. J. Pept. Protein Res.* **14,** 393 (1979).
[27] O. Mikeš, *Compr. Biochem.* **8,** 205 (1984).
[28] J. Čoupek, *Anal. Chem. Symp. Ser.* **9,** 165 (1982).
[29] J. Richey, *Int. Lab.* **13,** 50 (1983).

Separation of Pectic Enzymes Using Spheron Ion Exchangers

Equipment

The equipment used for the MPLC separation of pectic enzymes[14-16] was constructed in our laboratory using spare parts from an amino acid analyzer and a laboratory micropump. The separation line consisted of this order of components:

1. A plexiglas mixer for linear gradients consists of two similar cylinders. One cylinder, mixed with an electromechanical stirrer, has a side funnel for continuous linking of linear gradients without interrupting the chromatographic operation. A simple linear gradient mixer (Pharmacia) can easily be modified for this purpose.
2. A fritted disk, sealed into a glass balloon, serves as a filter and a bubble collector.
3. A proportional programmable micropump (Instrument Development Workshops, Czechoslovak Academy of Sciences, Prague) has two reverse-operating pistons that can be adjusted to the same output power for reduction of pulses. This can be replaced by a commercial laboratory micropump with two reverse-operating pistons.
4. A manometer (0–6 MPa ≐ 0–60 atm) is connected to the line between the pump and the column and also serves as a pulse damper.
5. The (shorter) jacketed glass column of the amino acid analyzer has dimensions 25 × 0.8 cm (i.d.) and is cooled using tap water (12–15°).
6. The effluent is monitored by means of a tandem system of two through-flow photocells (for the measurement of UV absorbance at 285 and 254 nm), connected to UV-analyzer units (Instrument Development Workshop, Czechoslovak Academy of Sciences, Prague). Absorbance is recorded on E 24 and TZ 21 S line recorders (Laboratory Instrument Works, Prague), which also simultaneously records the collection of fractions.
7. Operating on the time principle, the test-tube fraction collector constructed in the workshop of the Institute of Organic Chemistry and Biochemistry of the Czechoslovak Academy of Sciences, is equipped with a relay yielding recordable pulses for both line recorders after each step.

The equipment has the following technical parameters. When packed with 20- to 40-μm particles of Spheron ion exchangers, the counterpressure is a maximum of 15 atm, through-flow 3–4 ml/min, and separation time for pectic enzymes 35–70 min. Similar equipment can be built using other commercially available liquid chromatograph components and laboratory instruments.

TABLE I

SEPARATION OF PECTIC ENZYMES INCLUDING MULTIPLE FORMS ON SPHERON ION EXCHANGERS[a]

Type of the ion exchanger	Cation exchangers			Anion exchangers	
	Weakly acid	Medium acid	Strongly acid	Medium basic	Strongly basic
Designation	C-1000	Phosphate-1000	S-1000	DEAE-1000	TEAE-1000
Functional group	$-COOH$	$-PO(OH)_2$	$-SO_3H$	$-CH_2N(C_2H_5)_2$	$-CH_2\overset{+}{N}(C_2H_5)_3$
Capacity (mEq/g)	1.85	3.10	1.72	1.50	1.40
Particle diameter (μm)	25–40	40–60	25–40	25–40	25–40
Type of buffers for linear gradient system	Sodium formate, pH 3.5	Sodium formate, pH 3.5	Sodium formate, pH 3.5	Tris–HCl, pH 7.0	Tris–HCl, pH 7.0
Pectolytic[b] preparation					
Pectinex Ultra[c]	$N_1 + \underline{S}; N_2 + L_1 + L_2$	$N_1 + X; N_2 + L_1 + L_2$	$S_1 + L_1 + N_1; \underline{L_2}; \underline{S_2}; \underline{N_2}$	$\underline{N_1}; S; \underline{L}; \underline{N_2}$	$\underline{N_1}; N_1 + S; L + N_2$
Rohament P[c]	$\underline{L}; \underline{S}; \underline{N}$	$N_1; N_1 + X; L + N_2$			
Leozym[d]	$\underline{L}; S; \underline{N}$	$N_1; N_2 + L; \underline{S}$	$\underline{S}; L_1; \underline{N}; L_2$	$\underline{S}; N + L$	$\underline{N_1}; S + L + N_2$

[a] From Refs. 15 and 16. The macroporous Spheron matrix has the exclusion limit 10^6 Da.

[b] Pectinex Ultra was the product of Ferment, Basel, Switzerland; Rohament P of Röhm, GmbH, Darmstadt, F.R.G.; two batches were employed. Leozym was obtained from Slovlik, Leopoldov, Czechoslovakia. Designation of pectic enzymes: L, pectin lyase; N, endopolygalacturonase; S, pectin esterase; X, exopolygalacturonase. The symbols are ordered in the sequence of elution: + means a joint peak, semicolon means separated peaks. The numbers indicate multiple forms of the same type of enzyme and are numbered only in the order of elution. No experiments were made to clarify their mutual identity (cf. discussions in the original papers). The underlined enzymes were obtained in the form of peaks with only one strongly prevailing pectolytic activity. This does not mean, however, that such a fraction was homogeneous from the point of view of claims of protein chemistry.

[c] From Ref. 15.

[d] From Ref. 16.

Selection of the Ion Exchanger

The results obtained in the separation of pectic enzymes are described in Refs. 15 and 16 and are summarized in Table I. With Pectinex Ultra a relatively good separation of individual pectic enzymes is achieved using the moderately basic DEAE derivative. The chromatogram is illustrated in Fig. 1. Three separations of this type are described[15] using Tris–HCl buffers: (1) pH 7 with a course of linear gradients that differ from Fig. 1, (2) pH 5, and (3) pH 7 with adjusted gradients, as shown in Fig. 1. A comparison of the results underlines the importance of optimal adjustments to the gradient system. Using Rohament P and Leozym the three main types of pectic enzymes are fairly well separated on weakly acid carboxylic cation exchangers. Figure 2 illustrates two experiments using Rohament P, differing in load. Leozym was chromatographed using identical conditions[16] and the results are very similar.

Treatment of Spheron Ion Exchangers prior to Use for MPLC, Reequilibration, and Regeneration

The application of MPLC allows easy handling of ion exchangers (particle size $\geqq 20\ \mu$m) in the normally equipped laboratory and packing of the glass columns without any special knowledge or auxiliary instruments. In contrast, for high-pressure liquid chromatography (HPLC), the ion exchangers are too fine (particle size $\leqq 10\ \mu$m) and they do not sediment from the suspension. Packing the columns requires experience in this field as well as packing devices, so that prepacked commercial columns are often required. Some of the procedures described below (with the exception of the application of 2 *M* strong bases and 2 *M* strong acids to chemically extremely stable Spheron derivatives) can also be used for other similar macroporous packings for MPLC, if they are of an aerogel type.

Guidelines for handling gels are summarized in Table II.

1. Because the macrostructure of Spheron is an aerogel type,[19,30] the first step needed when a new material is used is deaeration. The ion exchanger is transferred into a round-bottomed flask, and a thin slurry is prepared in water. The flask is evacuated using a water pump in a warm water bath (~40–50°). The suspension is boiled for several minutes. After boiling, bumping is noticed, which indicates that the deaeration is complete.

2. The slurry is diluted with a 10-fold amount of water and poured into a measuring cylinder, mixed carefully (to prevent bubbles), and sedi-

[30] J. Janák, J. Čoupek, M. Krejčí, O. Mikeš, and J. Turková, *in* "Liquid Column Chromatography" (Z. Deyl, K. Macek, and J. Janák, eds.), p. 187. Elsevier, Amsterdam, 1975.

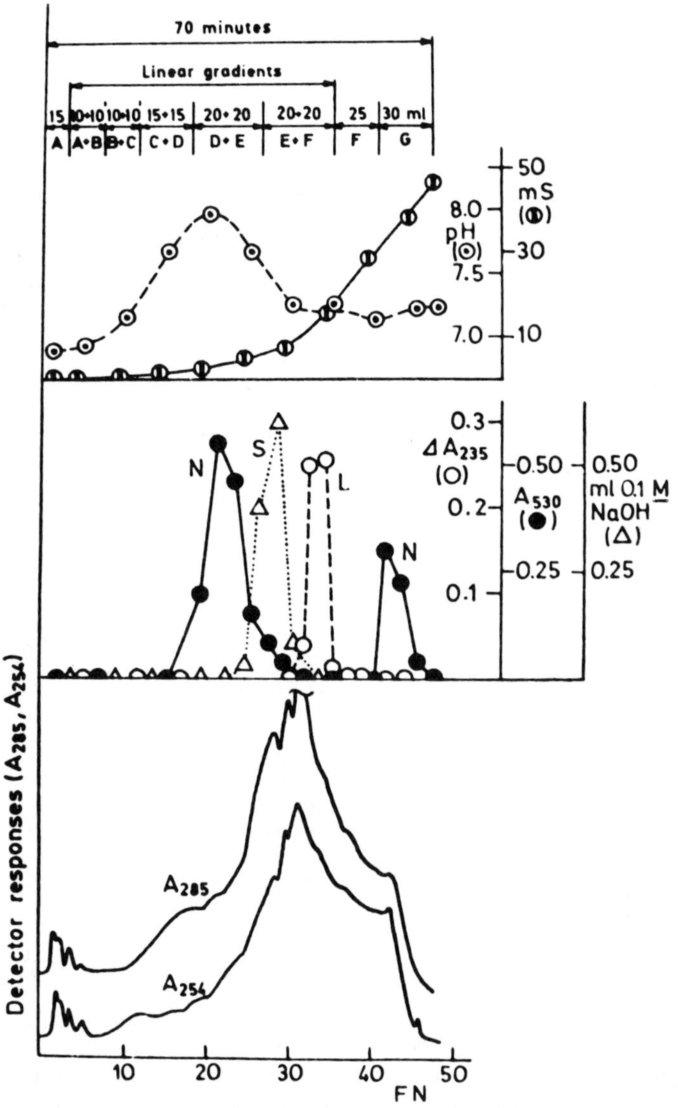

FIG. 1. Chromatography of 25 mg of Pectinex Ultra on Spheron DEAE-1000 (20×0.8 cm) with a linear gradient elution system using buffers of pH 7.0. The buffers were prepared from hydrochloric acid with a given final concentration, which was adjusted with Tris to pH 7.0: A, 0.005 *M*; B, 0.05 *M*; C, 0.1 *M*; D, 0.2 *M*; E, 0.4 *M*; F, buffer E with 1 *M* sodium chloride; G, unbuffered 2 *M* sodium chloride. Labeling of enzymes is described in footnote *b* of Table I. FN, Fraction numbers; mS, electric conductivity in milliSiemens (for the illustration of the ionic strength curve); ΔA_{235}, A_{530}, and milliliters of 0.1 *M* NaOH are values for quantitative enzyme assays for pectin lyase, endopolygalacturonase, and pectin esterase, respectively. Fractions (5.1 ml) were taken in 90-sec intervals, pressure was 0.6 MPa ($\doteq$6 atm). (Reprinted from Ref. 15 with permission.)

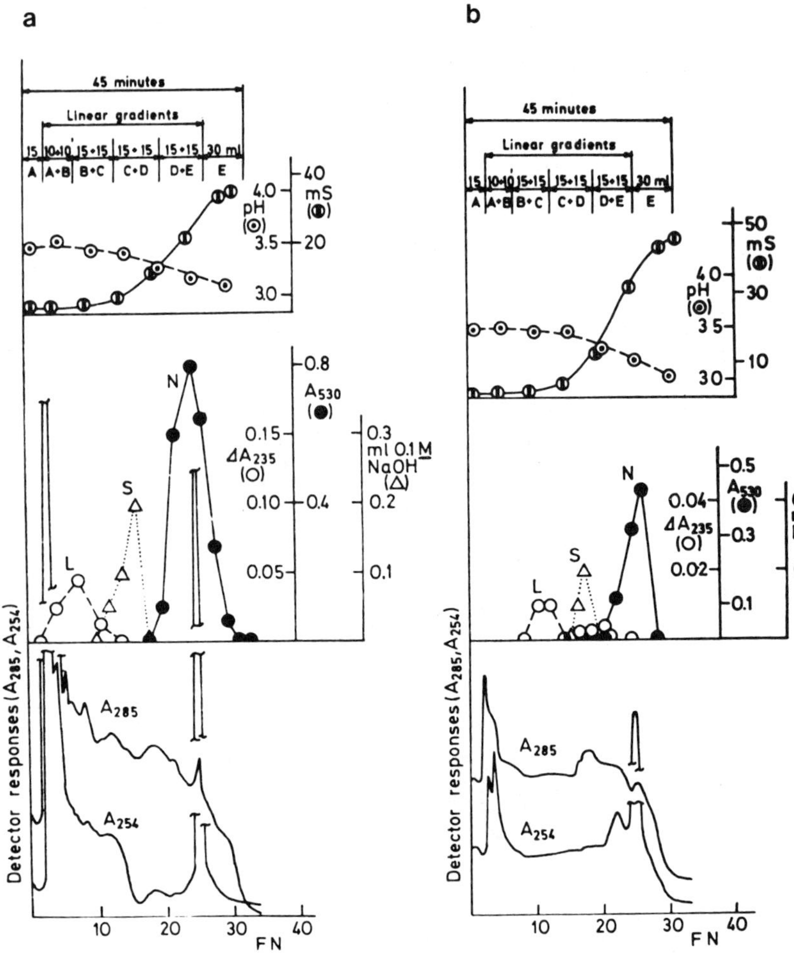

FIG. 2. Chromatography of 25 mg (a) and 5 mg (b) of Rohament P on carboxylated cation exchanger Spheron C-1000 (20 × 0.8 cm) with a linear gradient elution system using buffers prepared from sodium hydroxide solutions with the indicated final concentration, adjusted with formic acid to pH 3.5: A, 0.05 *M*; B, 0.1 *M*; C, 0.2 *M*; D, buffer C (1 *M*) in sodium chloride; E, unbuffered 2 *M* sodium chloride. Fractions (4.8 or 4.7 ml) were taken at 90-sec intervals. Pressure was 0.4 MPa (≐4 atm). All other labeling is the same as that in Fig. 1. (Reprinted from Ref. 15 with permission.)

mented. Small bubbles often attach to particles and prevent their sedimentation. After most of the ion exchanger (roughly 95%) has settled, the turbid supernatant is poured off. Very fine particles in the supernatant block the pores in the column frits. Mixing and sedimentation are repeated several times, until the supernatant is clear.

TABLE II
STEPS FOR HANDLING SPHERON ION EXCHANGERS FOR MPLC[a]

New:	(1) Deaeration (H_2O)	
	(2) Decantation and swelling (H_2O)	
New and used:	(3) Regeneration (2 *M* NaCl) and washing (H_2O)	
	(4) Cycling (laboratory temperature)	
	Cation exchangers	Anion exchangers
	2 *M* NaOH	2 *M* HCl
	H_2O	H_2O^b
	2 *M* HCl	2 *M* NaOH[b]
	H_2O	H_2O^b
	Final form: H^+	Final form: OH^-
	(5) Equilibration (first buffer)	
	(6) Packing into a glass chromatographic column (first buffer)	
	(7) Checking the column equilibration (first buffer)	
	(8) Chromatography (buffers of increasing ionic strength); reequilibration (steep reversed gradient and washing with the first buffer used in chromatography)	
	(9) Emptying the column (H_2O or 2 *M* NaCl)	
	(10) Regeneration (see step 3)	

[a] See text for details.

[b] Boil water for 10 min to remove carbon dioxide and cool prior to use. Sodium carbonate should be washed off the surface of sodium hydroxide pellets before dissolving.

3. The ion exchanger is then poured into a modified sintered-glass funnel[27] and mixed with a regenerant. Usually 2 *M* sodium chloride is used for this purpose. After standing for 5–10 min the ion exchanger is rewashed with the regenerant several times and then washed repeatedly with water. If the ion exchanger becomes colored during chromatography due to sample contaminants, organic solvents can be used for washing. Water-soluble solvents (alcohol, acetone), must be applied first, then solvents immiscible with water, and again water-soluble solvents, followed by water.

4. Cycling of ion exchangers follows the scheme given in Table II. Washing with each solution is repeated several times and contact with the ion exchanger should be about 5–10 min for each washing.

5. The ion exchanger is mixed with the first buffer selected for chromatography (usually having the lowest ionic strength). After 5 min the cation exchanger is titrated (using a glass electrode) with the base component and the anion exchanger with the acidic component of the buffer, until the original pH of the buffer is reached. Ion exchangers are then washed at least five times using a standing interval of 5–10 min or more for each washing.

The pH and conductivity are checked in effluent after each repeated washing: Equilibrium is reached when these values in the effluent are identical with the values of the original buffer used for washing.

6. Stepwise pulse packing of the columns is carried out using a modification of the slurry method.[27] A line consisting of a starting buffer reservoir, a laboratory micropump, a manometer, and a chromatography column is prepared and filled with buffer. A screw stopper for the column effluent is required. Part of the equilibrated ion-exchanger slurry is transferred by means of a pipet into the empty chromatographic column in a vertical position. The space above the ion exchanger slurry is filled with buffer. The head is connected with the column, the pump is turned on, and the column is washed using the speed selected for chromatography. The ion exchanger sediments quickly. After settling is completed, the column effluent is closed with the screw stopper. In a few seconds the pressure elevates rapidly and when it reaches 25–30 atm the screw stopper is quickly opened. The pressure falls suddenly in the form of a pulse and the layer of ion exchanger is somewhat pressed. The column is opened, clear supernatant is removed by suction through a pipet, a new portion of slurry is applied, and the process is repeated until the desired height of the ion-exchange column is reached.

7. Checking the equilibration of the packed column is very simple: Washing with deaerated starting buffer continues until pH, conductivity, and absorbance of the effluent are the same as that of the influent. Buffers should be deaerated prior to use (using round-bottomed flask and a water pump). Strong bumping is an indication of the end of deaeration. Degassing with a stream of helium bubbles is also possible.

8. The principles of chromatography of enzymes on Spheron derivatives are the same as in the case of other ion exchangers. Elution is achieved mainly by the ionic strength gradient. In some cases, however, a higher molarity of elution buffers may be required in comparison with polysaccharide ion exchangers. Deaeration of all buffers prevents formation of bubbles in the effluent at the bottom part of the column, which disturbs the continuous absorbance measurements. After the chromatography is completed, the column can be reequilibrated for the next run by using a short and steep reversed gradient from the last buffer to the buffer of the lowest ionic strength, and then thoroughly washing with this starting buffer. After many runs or if the ion exchanger has been contaminated, a complete regeneration is necessary.

9. To empty the column a supply of the pumped buffer is connected to the effluent part of the column in a reversed position and the ion exchanger is pressed directly from an open column to a modified sintered-glass funnel[27] for regeneration.

Chromatography of Pectic Enzymes[14-16]

The sample (5–25 mg, even more) is dissolved in 300–500 μl of buffer A (lowest ionic strength) and the solution centrifuged. The clear supernatant is cooled to about 1° (to elevate its density) and is carefully transferred using a cooled syringe with a piece of polyethylene tubing into buffer A above the top of the ion exchanger, by the method of underlayering. The glass column is packed with the ion exchanger to a 20-cm height and the space above is filled with buffer A. Then the head is mounted on, pressed down, and the chromatography started. In other cases the fritted disk of the head connector touches the top of the ion exchanger and the sample solution is applied using a six-way valve and sample loop. The underlayering method offers better peak shapes.

The chromatography of pectic enzymes is illustrated in Figs. 1 and 2. The concentration of buffers is referenced to the counter ions. The conductivity of every fifth fraction is measured using a conductivity meter (type OK-102/1, Radelkis, Budapest, Hungary) and the pH of each fifth fraction is evaluated with compensating pH meters. The continuous measurement of UV absorbance is of little value, because it does not provide information on the position of pectic enzymes. The purity of the analyzed technical preparations can be determined, however, by this method.

Enzyme Assay

Endopolygalacturonase (N) and exopolygalacturonase (X) can be assayed spectrophotometrically by the reducing-group method of Somogyi.[31] Sodium pectate (0.5% solution of pectic acid in 0.1 *M* acetate buffer, pH 4.2) is used as the substrate for endopolygalacturonase and di(D-galactosiduronic acid) (1 mmol/liter of 0.1 *M* acetate buffer, pH 4.5) is used for exopolygalacturonase. The reaction mixtures, composed of 0.8 ml of the substrate solution and 0.2 ml of the effluent fraction, is incubated at room temperature for 10 min (endopolygalacturonase), or 30 min (exopolygalacturonase). The reaction is stopped by adding 1 ml of Somogyi reagent.[31] The mixture is then heated for 10 min in a boiling water bath. After addition of 1 ml arsenomolybdate reagent and dilution by 10 ml water, absorbance of the colored products is measured at 530 nm.

The activity of pectinesterase (S) is determined by the titration of the carboxyl groups released from citrus pectin.[32] The reaction mixture, containing 5 ml of 0.5% solution of citrus pectin, 65.1% esterified, in 0.1 *M*

[31] M. Somogyi, *J. Biol. Chem.* **195,** 19 (1952).
[32] V. B. Fish and R. B. Dustman, *J. Am. Chem. Soc.* **67,** 1155 (1945).

acetate buffer, pH 4.4, and 1 ml of the effluent fraction, is incubated at room temperature. After 20 min of incubation pH is measured and if necessary it is readjusted to 4.4. The incubation continues for 40 min more and is followed by titration to pH 7.0 with 0.1 *M* NaOH. The difference between the total volume of NaOH used for the titration of the sample and the volume used for the titration of a control sample containing 1 ml effluent fraction inactivated by 5 min heating in boiling water bath, represents the equivalent of carboxylic groups released by the enzyme.

The activity of pectin lyase (L) is determined in terms of the increase in absorbance at 235 nm[33] using highly esterified pectin as substrate. The reaction mixture, containing 2.5 ml of 0.5% solution of pectin, 93.8% esterified, in 0.1 *M* acetate buffer, pH 5.6, and 0.5 ml of effluent fraction, is incubated at room temperature for 30 min and the absorbance at 230 nm is measured against a control sample containing an effluent fraction inactivated by 5 min heating in a boiling water bath.

Substrates for the Enzyme Assays

Citrus pectin, degree of esterification 65.1%, is purified from Genu Pectin type B, Slow Set (Københavns Pektinfabrik, Copenhagen, Denmark) by washing with 60% ethanol containing 5% hydrochloric acid and then with 60 and 96% ethanol. Pectic acid is prepared from citrus pectin by repeated alkaline deesterification with 0.1 *M* sodium hydroxide and subsequent precipitation at pH 2.5.

The highly esterified pectin (degree of esterification 93.8%) is prepared by esterification of pectic acid with 1 *M* sulfuric acid solution in methanol.[34]

Di(D-galactosiduronic acid) is isolated from the partial acid hydrolysate of pectic acid by gel chromatography on Sephadex G-25 (Fine).[35]

Separation of Tomato Polygalacturonases Using Mono S Cation Exchanger

Introduction

Ripe tomatoes contain two forms of polygalacturonases, PG I and PG II, which have been separated by ion-exchange chromatography on DEAE-Sephadex A-50[36] and by gel-permeation chromatography on Se-

[33] P. Albersheim, H. Neukom, and H. Deuel, *Helv. Chim. Acta* **42,** 1422 (1960).
[34] W. Henri, H. Neukom, and H. Deuel, *Helv. Chim. Acta* **44,** 1939 (1961).
[35] L. Rexová-Benková, *Chem. Zvesti* **24,** 59 (1970).
[36] R. Pressey and J. K. Avants, *Biochim. Biophys. Acta* **309,** 363 (1973).

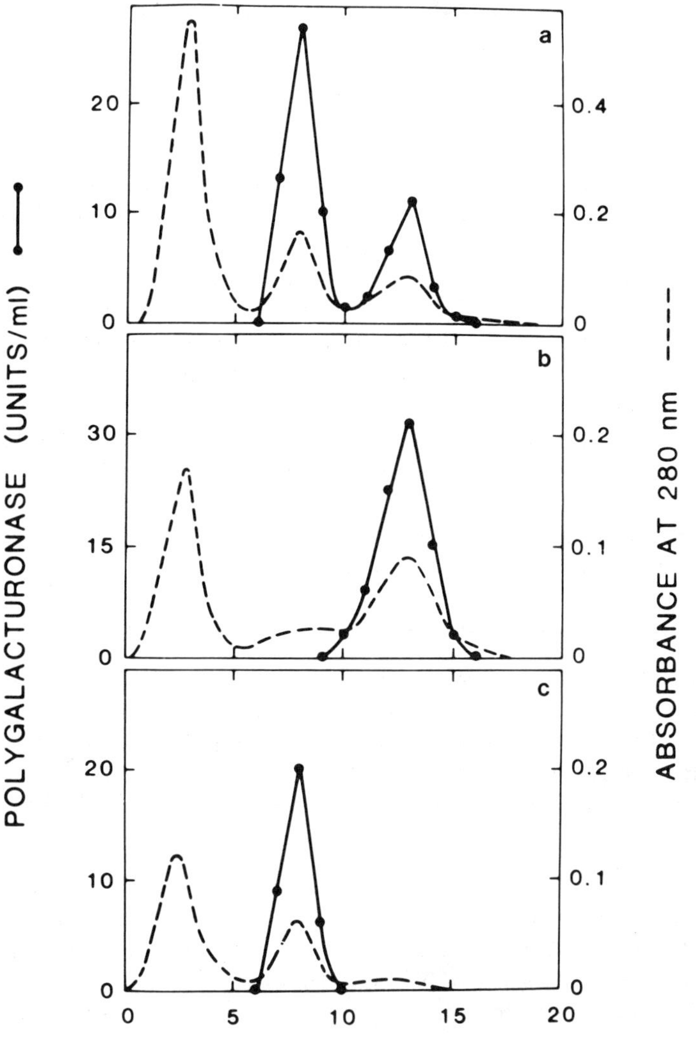

FIG. 3. HPLC of the crude (a) and partially purified tomato extract, containing polygalacturonases PG I (b) and PG II (c); b and c correspond to rechromatographed Sephadex fractions. A strong cation exchanger Mono S HR 5/5 prepacked column was used for all three chromatographic analyses with programmable chromatographic equipment from Pharmacia Fine Chemicals. Buffer A consisted of 0.02 MES [i.e., 2-(*N*-morpholino)ethanesulfonic acid] and 0.15 *M* NaCl, pH 6, and buffer B of 0.02 *M* MES and 0.65 *M* NaCl, pH 6.0. Linear gradient segments were programmed as follows: 0–2 min (0% B), 2–18 min (0–100% B), and 18–20 min (100–0% B) at a flow rate of 1 ml/min. The separation lasted 20 min. (Reprinted from Ref. 17 with permission.)

phadex G-100.[37] Since both methods are time consuming, Pressey described[17] a rapid HPLC method.

Preparation of the Extract

The preparation steps are conducted at about 3°. A quantity (200 g) of tissue from ripe tomato fruit (*Lycopersicon esculentum* Mills cv. "Better Boy") is homogenized with 200 ml of water. The water-insoluble fraction, collected by centrifugation, is suspended in 300 ml of 1.0 *M* NaCl and adjusted to pH 6.0 by 0.1 *M* NaOH. After stirring (1 hr) and centrifugation, proteins in the supernatant are precipitated with ammonium sulfate (75% saturation, i.e., 535.56 g/liter). After centrifugation they are dissolved in 15 ml of 0.15 *M* NaCl and dialyzed overnight against 0.15 *M* NaCl and clarified by centrifugation.

The crude extract is partially purified by gel-permeation chromatography on Sephadex G-100, the fractions are concentrated by ultrafiltration (Amicon PH10 membrane), and preparations PG I and PG II are obtained. Both partially purified preparations and the nonpurified extract are chromatographed on Mono S ion exchanger.

Fast Protein Liquid Chromatography

One milliliter of enzyme solution in 0.15 *M* NaCl is applied to the Mono S HR 5/5 column. The chromatography is illustrated in Fig. 3. One-milliliter fractions are collected and assayed for polygalacturonase. The first enzyme peak in Fig. 3a is identified as PG II and the second as PG I, so that the order of elution is the reverse of that on the Sephadex column.

Enzyme Assay

Polygalacturonase is analyzed in a 1-ml reaction mixture containing 0.20 *M* NaCl, 0.5% pectic acid, pH 4.2.[38] The reducing groups formed after 30 min at 37° are measured by the arsenomolybdate method.[39] One unit of activity is defined as the amount that releases 1 μmol of reducing groups in 30 min.

Discussion

The rapid separation of pectic enzymes is important from an analytical point of view. The determination of the activity of individual pectic en-

[37] G. A. Tucker, N. G. Robertson, and D. Grierson, *Eur. J. Biochem.* **112,** 119 (1980).
[38] R. Pressey and J. K. Avants, *J. Food Sci.* **36,** 486 (1971).
[39] N. Nelson, *J. Biol. Chem.* **153,** 375 (1944).

zymes in pectolytic preparations is considerably complicated by the partial suppression of the activity values due to the simultaneous effect of several enzymes on the same substrate. This is true for all the pectic enzymes with the exception of exopolygalacturonase, the activity of which can be determined by means of a specific substrate, di(D-galactosiduronic acid). The accuracy of the spectrophotometric determination of pectin lyase and polygalacturonase activity is decreased due to the high content of colored and reducing contaminants. MPLC and HPLC methods permit the separation of individual enzymes freed from contaminations and thus eliminate these drawbacks, because the nonionic reducing substances are often eluted within the void volume. The rapid determination of multiple forms of the same type of enzymes, which is of both scientific and technological importance (cf. discussions in Refs. 15 and 16), is also a possibility.

These analytical methods have large-scale preparative and technical applications. The flow rate of chromatography on Spheron ion exchangers is usually about 5 ml/cm^2/min, corresponding to about 30 liters/min on a large, closed filter 80 cm in diameter and 20 cm high (using the middle pressures mentioned above). We believe that the MPLC of enzymes can be carried out even on a semipilot plant scale, with the advantage of having short elution times of less than 1 hr/run.

Section III

Chitin

A. Preparation of Substrates for Chitin-Degrading Enzymes
Articles 44 through 47

B. Assay for Chitin-Degrading Enzymes
Articles 48 through 50

C. Analytical Methods for Chitin
Articles 51 through 55

D. Purification of Chitin-Degrading Enzymes
Articles 56 through 69

[44] Chitin Solutions and Purification of Chitin

By PAUL R. AUSTIN

Chitin in crustaceans is present as a mucopolysaccharide, intimately associated with calcareous shell material, adventitious and covalently bound protein, a small amount of fat, pigments, and trace metals. Hence the first stages of chitin purification involve its separation and isolation from these extraneous materials, usually by successive acid and alkali extractions. Chitins are properly called chitin isolates, as they vary with species, detailed method of preparation, and application requirements, i.e., there is a family of chitins. Brine and Austin have shown that even after rigorous alkali treatments they may still contain minor amounts of tightly bound protein, ash, and denatured chitin.[1,2] Notably, at this initial stage, the chitin has never been in solution or filtered. In this chapter, dissolution and further purification of chitin isolates via the newer anhydrous media are described.

Purification of Chitin

There is no clear line of demarcation between chitin isolation and its purification, but the isolation method affects chitin quality and in part determines any additional dissolution and other purification steps required. Muzzarelli has outlined a variety of systems for the preparation of chitin.[3,4]

Specifications. Because there is no standard chitin, specifications for chitin isolates vary depending on their end use. However, color, ash, acetyl value, solubility, viscosity, specific rotation, and/or molecular weight are usually cited. Nitrogen content is of limited value, as both chitin and its accompanying protein have nearly the same nitrogen content, 6.9 and 6.3% N, respectively.

Pertinent Chemical Factors. Salient properties of chitin that determine its method of isolation and purification include the following:

1. As a polyacetal it is sensitive to and easily hydrolyzed by acids, even more so than cellulose[5]

[1] C. J. Brine and P. R. Austin, *Comp. Biochem. Physiol. B* **69,** 283 (1981).
[2] C. J. Brine and P. R. Austin, *Comp. Biochem. Physiol. B* **70,** 173 (1981).
[3] R. A. A. Muzzarelli, *in* "The Polysaccharides" (G. O. Aspinall, ed.), p. 417. Academic Press, New York, 1985.
[4] R. A. A. Muzzarelli, "Chitin." Pergamon, Oxford, 1977.

2. It is quite stable to dilute alkali
3. It is sensitive to heat, even in the 40–50° range and may denature with time
4. It is a potent metal-sequestering agent; hence water washing should be done with deionized water to the greatest extent possible[3,4]
5. Hydrates, alcoholates, or ketonates of chitin are formed when chitin in solution is precipitated by the appropriate nonsolvent[6]
6. Chlorine bleaches should be avoided, as they form *N*-chloro compounds with chitin; peroxide bleaches are preferred
7. Chitin isolates usually have a few free amino groups, perhaps one for every six glucose units
8. Chitin is a chiral polymer by virtue of both its glucose moieties and its helical structure

Chitin Solvents and Solutions

Many investigators have sought to dissolve and spin chitin to obtain rayonlike fibers for comparison with those from cellulose. Hence a number of solvent systems have been elaborated that permit filtration, extrusion, and coagulation. For many chitin derivatives, such as deacetylation to chitosan or hydrolysis to glucosamine, substantial removal of protein and shell inorganics is usually adequate. However, the current trend toward biomedical applications for chitin, as well as the interest in chitin in enzymology, has focused still further attention on chitin purification methods. Immobilization of enzymes on chitin has progressed dramatically, as summarized by Muzzarelli,[3] making use of a variety of chitin isolates.

Polymer technology has fostered the use of aprotic and other special organic solvents for characterization studies and for fiber and film manufacture. With chitin, these nonaqueous systems provide not only an opportunity for removing extraneous material by a filtration step but, importantly, for regeneration or renaturing the chitin in a conformation approaching its natural crystalline state. Such renaturing usually requires use of a nonaqueous solvent and coagulation system.

LiCl/Amide Solvents for Chitin

A major breakthrough in solvents for chitin came in 1976 when it was found by Austin and Rutherford that a few LiCl/tertiary amides

[5] P. R. Austin, G. A. Reed, and J. R. Deschamps, *Proc. Int. Conf. Chitin/Chitosan, 2nd*, p. 99 (1982).

[6] P. R. Austin, U.S. Patent 4,063,016 (1977).

TABLE I
APROTIC LiCl/AMIDE SOLVENTS FOR CHITIN

Solvent (with 5–7% LiCl)	Chitin solubility[b]	Solvent properties			
		Specific gravity	bp (°C)	Viscosity (cP)	δ^a ($MPa^{1/2}$)
N,N-Dimethylacetamide	†††	0.937	166	1.1	22.1
N,N-Dimethylpropionamide	†	0.927	175	1.6	20.7[c]
N-Methyl-2-pyrrolidinone	†††	1.027	202	1.7	23.1
1,3-Dimethyl-2-imidazolidinone	††	1.044	223	2.0	22.3[c]

[a] δ = Solubility parameter, megaPascal.[11,12]
[b] Qualitative rate and amount of chitin dissolved; ††† is best.
[c] Calculated from bp.[13]

(R—CO—N=R_2) systems would yield at least 5% solutions with most chitins.[7,8] In an art and science in which chitin chemistry owes so much to cellulose, chitin in this instance was able to pay its debt, in part, for subsequently this same LiCl/amide system was found by McCormick to be a solvent for cellulose.[9]

The solubility behavior of chitin was rationalized via the Hildebrand solubility parameters (Table I).[10–13] The effective systems are nondegrading and are useful for purifying chitin, measuring physical properties, preparing filaments and films, and carrying out certain chemical reactions under homogeneous and anhydrous conditions.

Chitin Purified via LiCl/Amide System. In a study of the chitin–protein covalent linkage in mucopolysaccharides, a decalcified alkali-stable fraction was treated successfully with 1.67 *N* H_2SO_3 at 20° for 3 hr, dried, dissolved in 5% LiCl/*N,N*-dimethylacetamide, the chitin precipitated with acetone, dried, and then digested with 1 *N* NaOH at 100° for 48 hr, filtered, and dried. After this rigorous treatment, the residual chitin, isolated from different species, still contained small amounts of protein.[1,2]

[7] P. R. Austin, U.S. Patent 4,059,457 (1977).
[8] F. A. Rutherford and P. R. Austin, *Proc. Int. Conf. Chitin/Chitosan, 1st*, p. 182 (1978).
[9] C. L. McCormick, P. A. Callais, and B. H. Hutchinson, Jr., *Macromolecules* **18,** 2394 (1985).
[10] P. R. Austin, *in* "Chitin, Chitosan, and Related Enzymes" (J. P. Zikakis, ed.), p. 227. Academic Press, New York, 1984.
[11] A. F. M. Barton, "Handbook of Solubility Parameters." CRC Press, Boca Ratan, Florida, 1983.
[12] A. F. Barton, *Chem. Rev.* **75,** 731 (1975).
[13] H. Burrell, *Interchem. Rev.* **14,** 3 (1955).

Other Applications. Examples of the utility of the LiCl/amide solvents are manifold. Usually these studies involved 5–7% LiCl in *N,N*-dimethylacetamide or *N*-methyl-2-pyrrolidinone. Absorbable surgical sutures were developed from chitin by Austin,[14] Kifune *et al.*,[15] and others.[16] Chitin films were cast for study of their permeability to water and salt solutions for reverse osmosis.[17] Similarly, chitin membranes were prepared and their permeability to solutes of various molecular weights (urea, creatinine, and vitamin B_{12}) determined. Membranes cast from LiCl/amide solution were superior to those cast from acid solvents.[18]

Chitin reactions with acid chlorides and isocyanates were found possible in LiCl/amide solvents, and slow-release herbicides were prepared from chitin and a chloroformamide.[19,20]

Physical properties of chitin have been determined in the LiCl/amide solvents, including intrinsic viscosity, Mark–Houwink viscometric constants, and molecular weight,[21–23] specific rotation,[8,21] UV absorption spectra,[23] and NMR for conformational data.[24]

Polyfluorinated Compounds as Solvents for Chitin

Hexafluoroisopropanol (HFIP) and hexafluoroacetone sesquihydrate (HFAS) are effective solvents for chitin and useful for the preparation of films and filaments for surgical elements. Chitin for spinning and film casting and for solubility tests was refined by additional acid and alkali treatments, but the chitin recovered from solution was characterized only by physical means.[21,25] Circular dichroism was determined in films cast

[14] P. R. Austin, U.S. Patent 4,029,727 (1977).

[15] K. Kifune, K. Inoue, and S. Mori, U.S. Patent 4,431,601 (1984).

[16] M. Nakajima, K. Atsumi, and K. Kifune, *in* "Chitin, Chitosan, and Related Enzymes" (J. P. Zikakis, ed.), p. 407. Academic Press, New York, 1984.

[17] F. A. Rutherford and W. A. Dunson, *in* "Chitin, Chitosan, and Related Enzymes" (J. P. Zikakis, ed.), p. 135. Academic Press, New York, 1984.

[18] S. Aiba, M. Izume, N. Minoura, and Y. Fujiwara, *in* "Chitin in Nature and Technology" (R. Muzzarelli, C. Jeuniaux, and G. W. Gooday, eds.), p. 396. Plenum, New York, 1986.

[19] C. L. McCormick and D. K. Lichatowich, *J. Polym. Sci., Polym. Lett. Ed.* **17,** 479 (1979).

[20] C. L. McCormick and M. W. Anderson, *in* "Chitin, Chitosan, and Related Enzymes" (J. P. Zikakis, ed.), p. 41. Academic Press, New York, 1984.

[21] F. A. Rutherford, Thesis. University of Delaware, Newark, Delaware, 1976.

[22] S. H. Sennett, Thesis. University of Delaware, Newark, Delaware, 1985.

[23] J. W. Castle, J. R. Deschamps, and K. Tice, *in* "Chitin, Chitosan, and Related Enzymes" (J. P. Zikakis, ed.), p. 273. Academic Press, New York, 1984.

[24] M. Vincendon, *in* "Chitin in Nature and Technology" (R. Muzzarelli, C. Jeuniaux, and G. W. Gooday, eds.), p. 343. Plenum, New York, 1986.

[25] R. C. Capozza, U.S. Patent 3,989,535 (1976).

from HFIP.[26] The HFIP is an irritant and HFAS is highly toxic, qualities which limit their utility.

Organic Acid Solvents for Chitin

Chitin is soluble in a narrow range of carboxylic and sulfonic acids, namely, formic, di- and trichloroacetic and methanesulfonic acids. All of these acids slowly degrade chitin. Low temperatures around 0–5° may be helpful in dissolution of chitin. The viscous polymer systems may be diluted appreciably with chloroalcohols or chlorohydrocarbons. Developed originally as spinning and film casting solvents, these systems are of continuing interest for physical property studies of chitin and for preparation of chitin derivatives, as well as described.

Formic Acid. Chitin is only sparingly soluble in high-strength formic acid and the system is exceedingly sensitive to water, which reduces chitin solubility dramatically.[10,27] However, dissolution of chitin is improved substantially by addition of a soluble salt such as NaCl, NaBr, $CaCl_2$, LiCl, KCl, or KSCN. Dichloromethane and 2-chloromethanol are useful diluents for chitin solutions in formic acid.

Multifilament chitin fibers have been spun from formic acid systems and the filaments characterized, but tenacities and wet strengths were low. The spinning solutions were prepared by a repeated freeze–thaw technique.[28]

Chloroacetic Acids. Dichloroacetic acid (toxic, corrosive) is the most convenient chitin solvent in this group because it is a liquid, whereas trichloroacetic acid (corrosive, hygroscopic) is a solid and some cosolvent is usually required.[29] A combination of the two is effective and easier to handle. Dichloroacetic acid in admixture with formic acid has been used for fiber spinning.[28]

Methanesulfonic Acid. Chitin dissolved in a mixture of methanesulfonic acid and acetic acid was acetylated readily with acetic anhydride.[28]

Acknowledgment

This work is a result of research sponsored in part by NOAA office of Sea Grant, Department of Commerce, under Grant No. NA85AA-D-SG033 (Project No. R/N-8), and by the University of Delaware. The technical assistance of Susan H. Sennett in developing some of the tabular data is greatly appreciated.

[26] L. A. Buffington and E. S. Stevens, *J. Am. Chem. Soc.* **101,** 5159 (1979).
[27] V. F. P. Lee, Dissertation. University of Washington, Seattle, Washington, 1974.
[28] J. Noguchi, S. Tokura, and N. Nishi, *Proc. Int. Conf. Chitin/Chitosan, 1st,* p. 315 (1978).
[29] P. R. Austin, U.S. Patent 3,892,731 (1975).

[45] Water-Soluble Glycol Chitin and Carboxymethylchitin

By SHIGEHIRO HIRANO

Principle

Chitin is treated with aqueous sodium hydroxide solution to give alkali chitin (sodium alkoxides of chitin). The alkali chitin is allowed to react with ethylene chlorohydrin (2-chloroethanol) to give glycol [*O*-(2-hydroxyethyl)]chitin, and with chloroacetic acid (or its salt) to give carboxymethyl [*O*-carboxymethyl(CM)]chitin.

Glycol Chitin and Carboxymethylchitin

Procedure A[1-4]

Powdered chitin (4 g) is suspended in 42% (w/w) sodium hydroxide solution (80 ml), containing 0.2% sodium dodecyl sulfate, and the suspension is stirred at room temperature for 4 hr under diminished pressure (~20 mmHg), with occasional swirling. When less than 42% (w/w) sodium hydroxide solution is used, the yield of water-soluble product is lessened. The alkali chitin is collected by filtration through a glass fiber filter paper, pore diameter 10 μm, washed with a small volume of 42% (w/w) sodium hydroxide solution, and pressed to a wet weight of <14 g.

The cake of alkali chitin obtained above is transferred to a beaker and vigorously mixed with finely crushed ice (60 g). The mixture is mechanically stirred for 30 min in an ice bath to give a highly viscous alkali chitin. The product is diluted with a solution of sodium hydroxide (28 g) dissolved in water (112 ml). The concentration of sodium hydroxide is 14% (w/w). At this step, a gel is sometimes formed; in this case, the mixture is cooled below 0° and stirred to give a homogeneous solution. The alkali chitin solution is allowed to cool in an ice bath, and ethylene chlorohydrin (39 ml, 24 equivalent mol/*N*-acetyl-D-glucosamine residue) or sodium chloroacetate (28 g, 12 equivalent mol/mol *N*-acetyl-D-glucosamine residue), dissolved in 14% (w/w) sodium hydroxide solution (70 ml) in an ice

[1] R. Senju and S. Okimasu, *Nippon Nogei Kagaku Kaishi* **23,** 432 (1950).
[2] S. Okimasu, *Nippon Nogei Kagaku Kaishi* **32,** 303 (1959).
[3] H. Yamada and T. Imoto, *Carbohydr. Res.* **92,** 160 (1981).
[4] R. Trujillo, *Carbohydr. Res.* **7,** 483 (1968).

bath, is added dropwise, with stirring, over 30 min. The mixture is allowed to stand at room temperature overnight.

The solution is recooled in an ice bath; acetic anhydride (8 ml) is added dropwise, with stirring, over 30 min. If this step is omitted, an appreciable amount of N-deacetylated product is obtained. The mixture is neutralized with acetic acid under cooling, and dialyzed against running water for 2 days, followed by dialysis against distilled water for 1 day. The dialysate is centrifuged at 5000 rpm (2800 g) for 20 min, in order to remove insoluble material, and 3 vol acetone is added to the supernatant. After standing overnight, the precipitate is collected by centrifugation (5000 rpm for 20 min) and washed with acetone. The product is resuspended in ethanol (~30 ml), collected by filtration, and dried to give ~3.5 g of glycol chitin, or CM-chitin sodium salt. The CM-chitin sodium salt is redissolved in water (120 ml) and acidified by addition of hydrochloric acid solution. The resulting solution is dialyzed against distilled water for 1 day and lyophilized to give salt-free CM-chitin.

Procedure B[5,6]

Powdered chitin (4 g) is slurried with 60% (w/w) sodium hydroxide solution (16 ml), containing 0.2% sodium dodecyl sulfate at 4° for 1 hr, and then the slurry is kept in a freezer at −20° overnight.

The slurry of frozen alkali chitin obtained above is added, in portions, to 2-propanol (70 ml), containing ethylene chlorohydrin (16 ml, 10 equivalent mol/*N*-acetyl-D-glucosamine residue) or chloroacetic acid (9.5 g, 5 equivalent mol/*N*-acetyl-D-glucosamine residue) in an ice bath over 30 min. The mixture is mechanically stirred at room temperature for 1 hr, the product collected by filtration through a glass filter paper, and washed well with ethanol to give a powdered material. The material is suspended in water (300 ml) by stirring at room temperature to give a viscous solution.

The product is N-acetylated by addition of acetic anhydride and isolated as described in procedure A.

Characterization

Glycol chitin and CM-chitin are soluble in water and hydrolyzed by chitinase (EC 3.2.1.14) and lysozyme (EC 3.2.1.17).

Glycol chitin: ds 0.8–0.9 for glycol; $[\alpha]_D^{18}$ +8° (*c* 0.7, water); ν_{max}^{KBr} 1660 and 1560 cm^{-1} (C=O and NH of *N*-acetyl); ^{13}C NMR (D_2O): δ 177.5

[5] S. Tokura, N. Nishi, A. Tsutsumi, and O. Somorin, *Polym. J.* **15,** 485 (1983).
[6] S. Tokura, S. Nishimura, and N. Nishi, *Polym. J.* **15,** 597 (1983).

(C=O of *N*-acetyl) and 25.5 (CH_3 of *N*-acetyl). CM-chitin sodium salt: ds 0.7–0.8 for CM; $[\alpha]_D^{18}$ −4–−6° (*c* 1.0, aqueous 0.5% sodium hydroxide solution); ν_{max}^{KBr} 1660 CO_2^- and C=O) of *N*-acetyl; ^{13}C NMR (D_2O): δ 180.0 (C=O of CM), 177.5 (C=O of *N*-acetyl), and 25.5 (CH_3 of *N*-acetyl). CM-chitin: ds 0.7–0.8 for CM; $[\alpha]_D^{18}$ −4° (*c* 0.6 water); ν_{max}^{KBr} 1740 (C=O of CO_2H), and 1660 and 1560 cm^{-1} (C=O and NH of *N*-acetyl).

[46] Isolation of Oligomeric Fragments of Chitin by Preparative High-Performance Liquid Chromatography

By KEVIN B. HICKS

Chitin, a biopolymer composed of β-1,4-linked 2-acetamido-2-deoxy-D-glucose (GlcNAc) residues, is commonly found in the cell walls of fungi and in the exoskeletons of most arthropods. The natural polymer is quite insoluble in water, but partial acid hydrolysis yields a series of soluble β-1,4-linked oligomers that are useful substrates for studies on enzymes (i.e., lysozyme) and that have also been reported[1] to be useful in a variety of medicinal and industrial applications. Large amounts of chitin-containing biomass exist today in the form of shellfish processing wastes. The conversion of these abundant by-products into useful oligomers and derivatives is an important current research goal.

Pure chitin oligomers are usually prepared by the fractionation of acid-hydrolyzed chitin on large columns packed with charcoal,[2] or gel filtration media.[3] Although these techniques produce relatively pure standards, they are tedious and quite time consuming; some separations require more than a week. Recently,[4-6] the technique of high-performance liquid chromatography (HPLC) has been used to separate, on the analytical scale, oligomers from chitin hydrolysates. In this report, we demonstrate the separation of chitin fragments on a variety of HPLC stationary phases, including aminopropyl silica, octadecyl silica, and cation-exchange resins in the Ag^+ and H^+ form, and apply these principles to the prepara-

[1] R. A. A. Muzzarelli, "Chitin." Pergamon, Oxford, 1977.
[2] S. A. Barker, A. B. Foster, M. Stacey, and J. M. Webber, *J. Chem. Soc.*, p. 2218 (1958).
[3] B. Capon and R. L. Foster, *J. Chem. Soc.*, p. 1654 (1970).
[4] P. van Eikeren and H. McLaughlin, *Anal. Biochem.* **77**, 513 (1977).
[5] S. J. Mellis and J. U. Baenziger, *Anal. Biochem.* **114**, 276 (1981).
[6] K. Blumberg, F. Liniere, L. Pustilnik, and C. A. Bush, *Anal. Biochem.* **119**, 407 (1982).

tive HPLC fractionation of milligram to gram quantities of chitin oligomers with degree of polymerization (dp) values of 2–6.

Preparation of Chitin Hydrolysate

Chitin (Pfanstiehl Laboratories, lot 11144) is ground in a mortar grinder (Brinkman Instruments,[7] Model RMO) to a fine powder, then hydrolyzed as follows. Chitin powder (50 g) is added with constant stirring, to 200 ml cold (4°) concentrated HCl in a flask immersed in an ice bath. The flask is stirred at room temperature for 2.5 hr, warmed to 40°, and then stirred for an additional hour. After cooling, 2700 ml of Duolite A-561-free base form (Diamond Shamrock Corporation), slurried in 1 liter of water, is slowly added to neutralize (pH >3) the acid. The solution is filtered through Whatman 1 paper covered with Celite (Johns Manville) and then reduced in volume to 250 ml by evaporation at reduced pressure and low temperature ($<30°$). Free amino groups on the chitin fragments are re-N-acetylated[2] by adding methanol (22.5 ml) and acetic anhydride (6 ml) to the cold (4°) 250 ml filtrate in the presence of 270 ml of Amberlite (Rohm and Haas) IRA-400, carbonate form. After stirring the reaction for 1.5 hr at 4°, the solution is again filtered and deionized by passage through columns containing 200 ml Amberlite IR-120 H^+ and 50 ml Duolite A-561 free base form, respectively. The effluent and wash from the final column is evaporated carefully to 150 ml, then lyophilized. The yield is 9.41 g of white fluffy solid.

Chromatographic Equipment

HPLC is performed on a Gilson gradient HPLC system composed of two model 303 solvent pumps, an Apple IIe controller, a Rheodyne model 7125 fixed loop injector, a Kratos 520 column heater, and a Waters model 403 preparative differential refractometer. For automated injections, a Gilson model 302 pump, controlled by the microprocessor, is used to inject samples. Fractions are collected manually. Chromatograms are recorded on a Houston Instruments recorder. All solvents (HPLC grade) are filtered prior to use and degassed with helium during chromatography.

High-performance cation exchange columns (0.78×30 cm) in three ionic forms, HPX-87-H (H^+ form), HPX-87-C (Ca^{2+} form), and HPX-42-A (Ag^+ form), are purchased from Bio-Rad Laboratories. The former two columns are packed with <10-μm, 8% crosslinked, sulfonated

[7] Reference to brand or firm name does not constitute endorsement by the U.S. Department of Agriculture over others of a similar nature not mentioned.

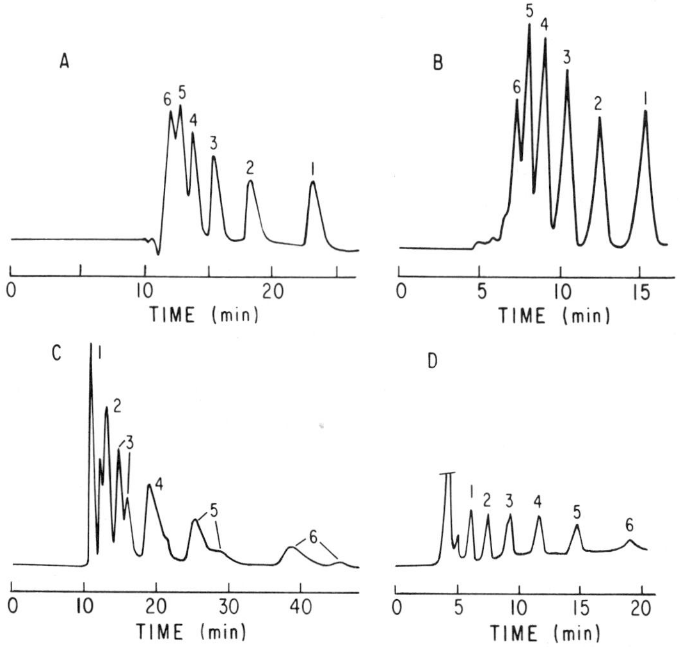

FIG. 1. Comparative separations of chitin oligomers on four HPLC stationary phases. (A) HPX-87H at 75°; mobile phase: 0.01 *N* H_2SO_4 at 0.3 ml/min, 20-μl (1 mg) injection. (B) HPX-42A at 75°; mobile phase: H_2O at 0.6 ml/min, 20-μl (1 mg) injection. (C) Dynamax C-18 at 25°; mobile phase: H_2O at 4 ml/min, 500-μl (35 mg) injection. (D) IBM Amino at 25°; mobile phase: acetonitrile (75)/H_2O (25) at 1 ml/min, 20-μl (0.4 mg) injection. Refractive index detection at various attenuations. Numerals above peaks refer to dp values.

polystyrene. The latter column is packed with 4% crosslinked resin. The analytical (0.46 × 25 cm) aminopropyl silica (5 μm) column is purchased from IBM Instruments. The Dynamax C-18 silica column and the Dynamax custom-packed aminopropyl silica column are obtained from Rainin Instrument Company, and are packed with 7-μm irregular silica into preparative sized (2.0 × 30 cm) axial compression cartridges. Precolumns are used with all columns except the Rainin preparative models.

Comparative Separations on Four HPLC Stationary Phases

Figure 1 shows the analytical scale separation of chitin oligomers on four commercially available stationary phases. On high-performance cation-exchange resins in the H^+ (Fig. 1A) and Ag^+ (Fig. 1B) forms, the oligomers are separated by a combination of size-exclusion, hydrophobic interaction, and ligand-exchange mechanisms,[8] resulting in an elution

[8] K. B. Hicks, P. C. Lim, and M. J. Haas, *J. Chromatogr.* **319**, 159 (1985).

order that follows descending molecular weight. These columns must be operated at ≥75° in order to obtain narrow, resolved peaks. It is noteworthy that the more commonly used Ca^{2+} form of these resins was not a useful stationary phase for this separation; poor resolution resulted under all conditions tested. Complete separation of all six oligomers was also accomplished on a C_{18} bonded silica (reversed-phase) column (Fig. 1C), but as previously noted,[6] peaks are broad, due to the separation of the α- and β-anomeric forms of each oligosaccharide. The aminopropyl silica (normal phase) column (Fig. 1D) provides excellent resolution of the oligomers, completely separating dp 1 through 6 in less than 20 min. On both of the silica-based stationary phases (Figs. 1C and D), chitin oligomers elute in an opposite (and complementary) order to that seen on the cation-exchange resins. In all of the chromatographic systems tested here, each peak appeared to contain only one component, and pure GlcNAc and *N,N'*-diacetylchitobiose coeluted with peaks 1 and 2, respectively, in Fig. 1A–D.

Preparative HPLC of Chitin Oligomers

Each of the stationary phases described above can be used for isolation of quantities of purified chitin fragments. The cation-exchange column in Fig. 1A, is normally used as an analytical column, but its relatively large diameter (0.78 cm) allows it to be used for semipreparative HPLC (Fig. 2A), in which 15 mg of the crude mixture may be fractionated. The

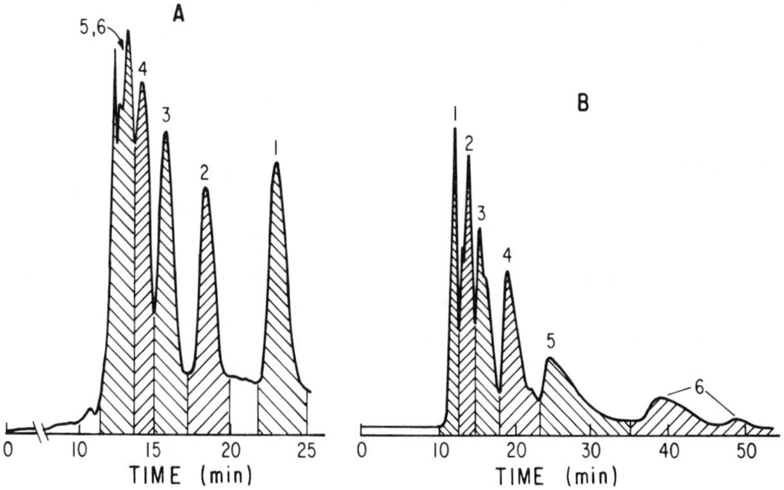

FIG. 2. Preparative HPLC of chitin fragments. (A) HPX-87H column, 200-μl (15 mg) injection; other conditions as in Fig. 1A. (B) Dynamax C-18 column, 2000-μl (140 mg) injection; system pressure: 300 psi, other conditions as in Fig. 1C.

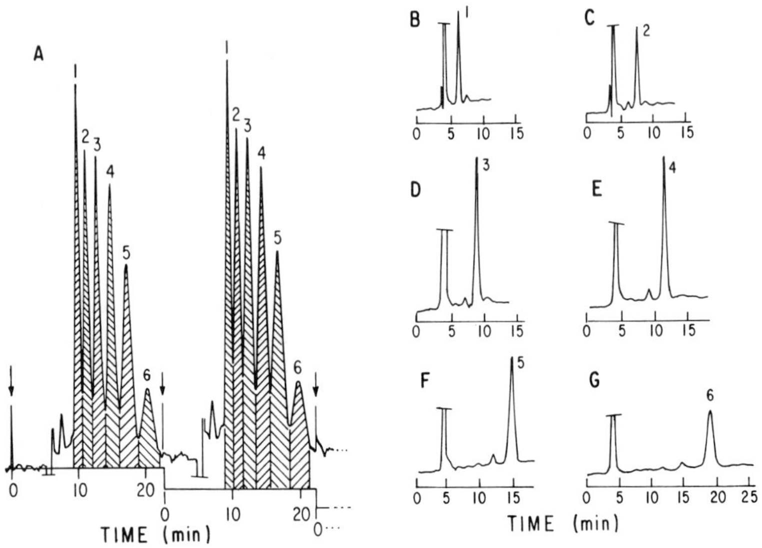

FIG. 3. Preparative HPLC of chitin fragments on aminopropyl silica, using automated, repetitive injections (injections shown by arrows). (A) Sequence of two automated injections (3.5 ml volume, 100 mg) on Dynamax custom-packed NH_2 column, mobile phase: acetonitrile (60)/H_2O (40), at 12 ml/min and 1500 psi. (B–G) Purified fragments collected from A, analyzed by the column described in Fig. 1D.

HPX-42A column gave good analytical separations, but developed high back pressures after several semipreparative-sized injections. Larger sample sizes (100–140 mg) can be injected onto the preparative C_{18} bonded phase column (Fig. 2B) and the preparative aminopropyl silica column (Fig. 3A).

Discussion

Each phase demonstrated in Figs. 2 and 3 has unique characteristics, as shown in Table I. The H^+ form column is quite durable and inexpensive to operate, yet is only useful for isolating milligram amounts of oligomers. The C_{18} column has much greater sample capacity, and can be used to fractionate gram quantities of chitin oligomers. This column must occasionally be washed with 50% acetonitrile to restore its capacity, but otherwise is exceptionally durable in this application. Because this column is operated with inexpensive mobile phase (H_2O), at low flow rates and low back pressures, it can be easily accommodated on standard analytical HPLC systems. However, as with the cation-exchange column, some of the isolated oligomers are only 54–75% pure (Table I), and must be rechro-

TABLE I

PREPARATIVE HPLC OF CHITIN OLIGOMERS ON THREE HPLC STATIONARY PHASES

Column[b]	Sample size (mg)	Analysis time (min)[c]	Milligrams of each oligomer collected per injection (% purity)[a]					
			1	2	3	4	5	6
HPX-87H	15	20	1.8 (95)	1.6 (74)	1.7 (72)	1.6 (75)	2.7	
Dynamax C-18	140	40	15.0 (95)	27.0 (75)	29.0 (57)	29.0 (54)	23.0 (61)	15.0 (95)
Dynamax NH_2	100	23	12.0 (92)	12.0 (92)	15.0 (94)	16.0 (94)	15.0 (94)	8.0 (98)

[a] Purity was determined by analytical HPLC on the basis of peak area.
[b] Chromatographic conditions as in Figs. 2 and 3.
[c] Time needed between repetitive injections.

matographed, individually, to give oligomers with about 95% purity. In contrast, fractions from the aminopropyl silica column are from 92 to 98% pure from a single chromatographic run (Table I); a second purification gives oligomers that are >98% pure. Samples can be processed rapidly (every 23 min) on this column and it is possible to use automated, repetitive injections as shown in Fig. 3A, to fractionate 1 g of chitin hydrolysate in about 270 min. The resulting fractions (Fig. 3B–G) contain between 80 and 160 mg of oligomers at purity levels averaging 94%. A disadvantage of this column is its relatively short lifetime,[9] which is somewhat offset by carefully pretreating samples prior to injection, and following other recommended[9] precautions (see Note Added in Proof). In addition, the requirements for binary mobile phases and higher (12–15 ml/min) flow rates can preclude the use of these columns on some analytical HPLC systems. Surprisingly, the operating back pressure of this column was no greater than that of an analytical column packed with the same phase.

When oligomer fractions from these columns are concentrated, extreme care must be taken to use the most gentle conditions available. The fractions from the cation-exchange column must also be treated with anion-exchange resin, to remove sulfuric acid. Fractions from the aminopropyl column should be evaporated under diminished pressure, and at low temperatures (<30°) to remove acetonitrile. The resulting aqueous fractions from all columns should then be carefully freeze dried, rather than evaporated to dryness. The latter process yields aggregated oligomers which do not redissolve into aqueous solution. Even under the best of conditions, the lyophilized powders will tend to aggregate, and levels of insoluble, aggregated material increase with storage time. Pure fractions should be stored at low temperatures (−20°) prior to use.

In summary, preparative HPLC on commercially available stationary phases can be used to efficiently fractionate gram quantities of oligomers from chitin hydrolysates. The choice of column depends on the amounts of oligomers required, the purity desired, and on the supporting instrumentation that one has available.

Acknowledgment

I acknowledge Scott M. Sondey for excellent technical assistance.

Note Added in Proof

An improved version of the Dynamax NH_2 column is now available. This column is extremely durable and may last approximately 10× as long as previous aminopropyl silica gel columns. This fact makes the use of this column quite practical.

[9] B. Porsch, *J. Chromatogr.* **253,** 49 (1982).

[47] Preparation of Crustacean Chitin

By KENZO SHIMAHARA and YASUYUKI TAKIGUCHI

Isolation of Crustacean Chitin[1]

Principle

In crustacean cuticles (shells), chitin is tightly associated with inorganic salts, such as calcium carbonate, proteins, and lipids including pigments. Therefore, in order to isolate chitin from crustacean shells, the following three steps are required: demineralization with dilute hydrochloric acid (step 1a) or with ethylenediaminetetraacetic acid (EDTA) (step 1b), deproteinization with aqueous sodium hydroxide (step 2a) or by use of the proteolytic activity of a bacterium[2] (step 2b), and elimination of lipids with organic solvent(s) (step 3). These three steps are usually carried out in the above-mentioned order,[1,7] but some investigators proposed a procedure in which deproteinization was followed by demineralization.[8]

For the purpose of eliminating inorganic salts and protein, a procedure which consists of steps 1a, 2a, and 3 is preferable. However, this procedure causes some chemical changes, such as deacetylation and depolymerization, on the associated chitin. On the other hand, a procedure which consists of steps 1b, 2b, and 3 is preferable for isolating less denatured chitin, although elimination of inorganic salts and protein is incomplete (Table I).

Starting Materials

Shells of crabs, prawns, and lobsters serve as the starting materials. The shells are scraped and repeatedly rinsed under running water, crushed into

[1] K. Shimahara, Y. Takiguchi, K. Ohkouchi, K. Kitamura, and O. Okada, *in* "Chitin, Chitosan, and Related Enzymes" (J. P. Zikakis, ed.), p. 239. Academic Press, New York, 1984.

[2] Some investigators studied enzymatic deproteinization by various proteases; however, levels of unremoved protein were found to be considerably higher.[3-6]

[3] M. Takeda and E. Abe, *Norinsho Suisan Koshusho Kenkyu Hokoku* **11,** 399 (1962).

[4] M. Takeda and H. Katsuura, *Suisan Daigakko Kenkyu Hokoku* **13,** 109 (1964).

[5] A. W. Bough, W. L. Salter, A. C. M. Wu, and B. E. Perkins, *Biotechnol. Bioeng.* **20,** 1931 (1978).

[6] C. J. Brine and P. R. Austin, *Comp. Biochem. Physiol. B* **70,** 173 (1981).

[7] R. H. Hackman, *Aust. J. Biol. Sci.* **7,** 168 (1954).

[8] J. N. BeMiller and R. L. Whistler, *J. Org. Chem.* **27,** 1161 (1962).

chips about 0.5 cm^2 in size, and air dried at room temperature. The dried shell chips can be stored without spoilage for several months at room temperature.

Shell of antarctic krill also serves as starting material. As the starting material of processes including step 2b, abdominal shell (pleon cuticle) of prawn is preferable.

Step 1a: Demineralization with Hydrochloric Acid. The dried shell chips (20 g) are immersed in 1 liter of 2*N* hydrochloric acid. The mixture is kept for 2 days at room temperature with occasional stirring using a glass rod. In the initial stage of the reaction, frequent stirring is required to prevent the floating of the shell chips caused by the generation of carbon dioxide gas. In the middle of the treatment, the exhausted hydrochloric acid is exchanged. After 2 days of immersing, demineralized shell chips are collected and washed with deionized water until neutral. The yield of demineralized shell chips from the abdominal shell of prawns is approximately 50%.

By this method, almost complete elimination of the inorganic salts can be attained. However, deacetylation and depolymerization of the associated chitin proceeds to some extent.

Demineralization methods carried out at 0°[7] and with more dilute acid[9] were also proposed.

Step 1b: Demineralization with Ethylenediaminetetraacetic Acid (EDTA). This is a modification of the method of Foster and Hackman.[3,10]

The dried shell chips (20 g) are immersed in 2 liters of 0.1 *M* EDTA solution, previously adjusted to pH 7.5 with aqueous ammonia. The mixture is kept at room temperature for 6 days with occasional stirring using a glass rod. Every 2 days, the exhausted EDTA solution is exchanged for a fresh one. After 6 days of immersion, demineralized shell chips are collected and washed repeatedly with deionized water. The yield of demineralized shell chips from the abdominal shell of prawn is approximately 50%.

This method is preferable for the purpose of isolating less denatured chitin, although elimination of inorganic salts is not complete.

Step 2a: Deproteinization with Aqueous Sodium Hydroxide. This is a modification of the method of Hackman.[7]

The demineralized shell chips (20 g) produced by step 1 are added to 1 liter of 1 *N* aqueous sodium hydroxide.[11] The mixture is boiled for 36 hr with occasional stirring. Appropriate amounts of water should be added as

[9] P. Broussignac, *Chim. Ind., Genie Chim.* **99,** 1241 (1968).

[10] A. B. Foster and R. H. Hackman, *Nature (London)* **180,** 40 (1957).

[11] The use of a stainless-steel vessel is preferable, because hot caustic alkali attacks the surface of glassware or porcelain.

the vaporization proceeds. The exhausted alkaline solution is exchanged for a fresh one 6 hr after the beginning of the reaction. After 36 hr of boiling, the demineralized shell chips or crude chitin chips are collected and washed with deionized water until neutral. The yield of crude chitin from the demineralized prawn-shell chips is approximately 60%.

By this method, almost complete elimination of the protein can be attained. However, deacetylation of the associated chitin proceeds to some extent.

Step 2b: Deproteinization by Use of Bacterial Proteolytic Activity. This method is based on the hydrolysis of chitin-associated protein with protease(s) produced by growing cells of a bacterial strain. The strain employed is *Pseudomonas maltophilia* LC102, which was found to be highly proteolytic but neither chitinolytic nor chitin deacetylating.[12]

The culture medium is prepared by dissolving 2 g of dipotassium hydrogen phosphate into 1 liter of tap water and adding 8.0 g (dry weight) of the demineralized shell chips from abdominal shell of prawn.[13-15] The medium is adjusted to pH 7.0 with hydrochloric acid. Sterilization is performed at 121° for 15 min. A loopful of *P. maltophilia* LC102 is transferred from a nutrient–agar slant into 80 ml of medium contained in a 500-ml Sakaguchi flask. Cultivation is carried out at 30° on a reciprocal shaker. After an appropriate period of incubation (1–4 days, depending on the preparation), the flask is reautoclaved and the demineralized and deproteinized shell chips or crude chitin chips are collected and repeatedly washed with deionized water. The yield from the demineralized prawn-shell chips is approximately 60%.

The time course of deproteinization is shown in Fig. 1. Within 24 hr, protein content of the demineralized shell chips from prawn diminishes to approximately 1%. However, complete elimination of the protein cannot be attained by prolonged incubation. This method does not cause the depolymerization and deacetylation of chitin.

Step 3: Elimination of Lipids and Pigments. The chips (20 g) produced by step 2 are added to 500 ml of 95% ethanol and refluxed for 6 hr. The decolorized chitin chips are then collected, washed with ethanol, and air dried. The dried chips are, if necessary, ground into powder of appropriate size.

[12] M. Ikeda and K. Shimahara, *Nippon Nogei Kagaku Kaishi* **52,** 335 (1978).

[13] The demineralized shell chips should be kept wet, because the use of the dried chips in this step slightly increases the protein content of the product.

[14] The use of the shells of crabs and lobsters should be avoided, because of the high protein content in the products (see Fig. 1).

[15] The abdominal shells of *Metapenaeus affinis* and *Penaeus chinensis* exhibited approximately identical deproteinization curves.

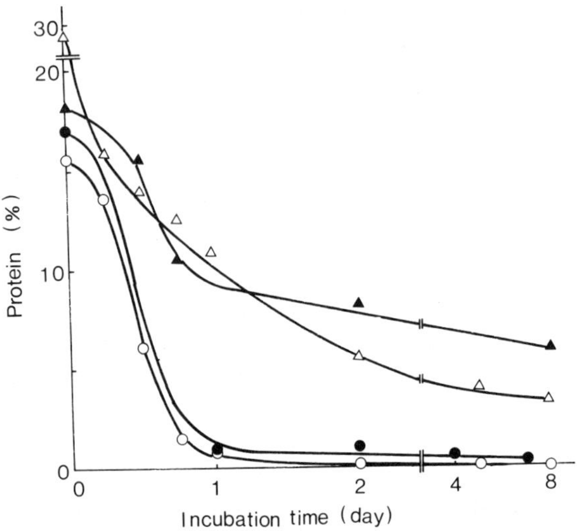

FIG. 1. Time course of deproteinization of demineralized shell chips from various sources by *Pseudomonas maltophilia* LC102.[1] (● and ○) Abdominal shell chips of *Penaeus japonicus* (kuruma prawn) demineralized with HCl *(Step 1a)* and with EDTA *(Step 1b)*, respectively; (△) *Paralithodes camtschatecus* (Alaska king crab) shell chips demineralized with HCl; and (▲) *Panulirus japonicus* (Japanese lobster) shell chips demineralized wtih HCl.

A method which consists of two successive extractions with ethanol and ether at room temperature has been proposed.[7]

Composition and Molecular Weight

Composition and molecular weight of various chitin preparations from abdominal shell of *Penaeus japonicus* (kuruma prawn) are given in Table I. The assay methods employed shall be described in the following item. As

TABLE I
COMPOSITION AND MOLECULAR WEIGHT OF VARIOUS CHITIN PREPARATIONS ISOLATED FROM *Penaeus japonicus* ABDOMINAL SHELL[a]

Isolation procedure (step)	Protein (%)	Ash (%)	Deacetylation (%)	Molecular weight
1a, 2a, and 3	0.0	0.1	17.1	1.8×10^6
1b, 2a, and 3	0.0	0.3	12.4	2.6×10^6
1a, 2b, and 3	0.9	0.9	10.6	3.1×10^6
1b, 2b, and 3	0.3	0.5	9.9	3.5×10^6

[a] From Shimahara *et al.*[1]

shown in Table I, both composition and molecular weight of chitin vary considerably with different isolation procedures.

Assay Methods

Determination of Protein Content[16]

Protein associated with chitin is hydrolyzed with hot aqueous sodium hydroxide and the concentration of ninhydrin-positive substances in the hydrolysate is determined colorimetrically.

Reagents

Sodium hydroxide, 10 *N*
Hydrochloric acid, 12 *N*
Acetate buffer, 0.5 *M*, pH 5.1
Ninhydrin–hydrindantin solution: Ninhydrin (0.5 g) and hydrindantin (0.15 g) are dissolved in and filled up to 100 ml with methyl Cellosolve. The solution should be prepared just before use

A sample (0.3 g) of dried chitin powder (7–14 mesh) is added to 50 ml of 10 *N* sodium hydroxide in a 100-ml Erlenmeyer flask made of polypropylene. The flask is covered with aluminum foil and heated at 121° for 60 min in an autoclave. The reaction mixture is then cooled rapidly, neutralized with hydrochloric acid in an ice bath, and filtered through a 116-3 glass filter (pore size 20–30 μm). The filtrate and the washings are combined and water is added to 150 ml to prepare a sample solution. If the protein content of the chitin powder is too high, the solution is diluted to a suitable concentration. In a test tube, 0.5 ml of the sample solution, 5 ml of buffer, and 5 ml of ninhydrin–hydrindantin solution are added and mixed by shaking. The test tube is covered with aluminum foil and incubated in boiling water for 10 min. After rapid cooling, the absorbance is determined at 564 nm. The reference solution is prepared by the identical procedure except for using 0.5 ml of deionized water instead of the sample solution. The protein content P (in grams) is calculated from Eq. (1):

$$P\,(\%) = 2.37(A_{564}/W) \qquad (1)$$

where A_{564} is the absorbance at 564 nm and W is the weight (in grams) of chitin powder. Eq. (1) is valid when the protein contents are equal to or less than 60%.

[16] Y. Takiguchi, K. Ohkouchi, H. Yamashita, and K. Shimahara, *Nippon Nogei Kagaku Kaishi* **61,** 437 (1987).

Determination of Degree of Deacetylation[17]

This method is based on infrared spectroscopic measurements.

About 3 mg of chitin powder (passed through a 200-mesh sieve) is mechanically blended with 400 mg of potassium bromide powder to prepare a KBr disk. An infrared spectrum is recorded in a range from 4000 to 1200 cm^{-1} and absorbances at 2878 cm^{-1} (the C—H band) and 1550 cm^{-1} (the amide II band) are evaluated by the baseline method. The degree of deacetylation D is calculated from Eq. (2):

$$D\,(\%) = 98.03 - 34.68(A_{1550}/A_{2878}) \tag{2}$$

where A_{1550} and A_{2878} are the absorbances at 1550 cm^{-1} and 2878 cm^{-1}, respectively. An alternative infrared spectroscopic method using absorbances at 3450 and 1655 cm^{-1} was proposed by Domszy and Roberts.[18]

A method based on the oxidation of chitin with sodium periodate was also reported.[19]

Determination of Molecular Weight[20]

This method is based on viscometry. In this method, viscometric constants for chitosan are applied, because those for chitin are not yet available.

Lithium chloride is dissolved in *N,N*-dimethylacetamide to a concentration of 5%. Various amounts of dried chitin powder (100–200 mesh) are weighed and dissolved in the dimethylacetamide solution to prepare chitin solutions having various concentrations. Viscosity of each solution is measured at 30° with a Ubbelohde-type viscometer. Intrinsic viscosity is determined graphically by plotting the viscosity data against concentration. The molecular weight of chitin M is calculated from Eq. (3):

$$[\eta] = 8.93 \times 10^{-4} M^{0.71} \tag{3}$$

where $[\eta]$ is the intrinsic viscosity.

Alternative viscometric constants are reported by Roberts and Domszy.[21]

[17] T. Sannan, K. Kurita, K. Ogura, and Y. Iwakura, *Polymer* **19,** 458 (1978).
[18] J. G. Domszy and G. A. F. Roberts, *Macromol. Chem.* **186,** 1671 (1985).
[19] G. K. Moore and G. A. F. Roberts, *Int. J. Biol. Macromol.* **2,** 115 (1980).
[20] F. A. Rutherford III and P. R. Austin, *Proc. Int. Conf. Chitin/Chitosan, 1st,* p. 182 (1978).
[21] G. A. F. Roberts and J. G. Domszy, *Int. J. Biol. Macromol.* **4,** 347 (1982).

Preparation of Colloidal Chitin

Principle

The procedure consists of the dissolution of chitin into and the reprecipitation from concentrated hydrochloric acid and the dispersion of the reprecipitated chitin into water. Colloidal chitin serves as a substrate of chitinase and a carbon and nitrogen source of chitinolytic microorganisms.[22]

Procedure

A sample (20 g) of chitin powder (passed through a 42-mesh sieve) is added slowly into 800 ml of concentrated hydrochloric acid below 5° with vigorous stirring. After homogeneous dispersion of chitin powder has been reached, the mixture is heated gently up to 37° with moderate stirring. The viscosity of the mixture increases rapidly and then, within a few minutes, begins to decrease. As the viscosity decreases, the appearance of the mixture becomes clearer. At the stage when a small amount of chitin is still undissolved, the mixture is filtered through glass wool and the filtrate is poured into 8 liters of deionized water below 5° with stirring. Within a few minutes, the solution becomes turbid because of the reprecipitation of chitin. After 30 min, the stirring is stopped, and then the suspension is kept overnight below 5°. The supernatant is then decanted out and the remaining mixture is filtered with suction through Toyo Roshi or Whatman No. 2 filter paper. The residue is washed with water until the washings become neutral. The acid-free residue is added into 500 ml of deionized water and resuspended with vigorous stirring to prepare so-called colloidal chitin solution. Chitin content of the solution is determined by drying a sample *in vacuo*. This colloidal chitin solution remains stable when stored for a few weeks in a dark place below 5°.

[22] S. C. Hsu and J. L. Lookwood, *Appl. Microbiol.* **29**, 422 (1975).

[48] Assay for Chitinase Using Tritiated Chitin

By ENRICO CABIB

Chitinase activity has been assayed by a variety of procedures, including the monitoring of changes in the molecular size of substrate by viscosity measurements[1] and the determination of chitooligosaccharides or *N*-acetylglucosamine liberated in the reaction. In the latter case, either reducing power[2] has been measured or some modification of the Morgan and Elson procedure[3,4] has been used. The Morgan and Elson reaction with *p*-dimethylaminobenzaldehyde proceeds only with free *N*-acetylhexosamines; therefore in this case it is necessary to include in the reaction mixture a source of β-*N*-acetylhexosaminidase to convert all oligosaccharides formed into the monosaccharide.[4]

The assay presented here, based on the formation of soluble oligosaccharides from tritium-labeled chitin, is by far the most sensitive, because of the possibility of using substrate of very high specific activity. It is suitable for both endo- and exochitinases, it obviates the need for an auxiliary β-*N*-acetylhexosaminidase, and is extremely simple to carry out.[5]

Assay Method

Principle. [*acetyl*-^{3}H]Chitin is incubated with the chitinase solution. The water-soluble oligosaccharides formed are separated from the insoluble chitin by filtration and their radioactivity is determined.

Reagents

Buffer, 1 *M*. The buffer will depend on the chitinase used and on its pH optimum

[*acetyl*-^{3}H]Chitin, 15 mg/ml, specific activity $\sim 1 \times 10^6$ cpm/mg (see preparation below)

Trichloroacetic acid, 10% (w/v)

Preparation of [acetyl-^{3}H]Chitin. Tritiated chitin is obtained by reacetylation of chitosan with [^{3}H]acetic anhydride.[5,6] Chitosan (1 g) is ground

[1] C. Jeuniaux, this series, Vol. 8, p. 644.

[2] A. C. Chen, R. T. Mayer, and J. R. DeLoach, *Arch. Biochem. Biophys.* **216,** 314 (1982).

[3] J. L. Reissig and L. F. Leloir, this series, Vol. 8, p. 175.

[4] E. Cabib and B. Bowers, *J. Biol. Chem.* **246,** 152 (1971).

[5] J. Molano, A. Duran, and E. Cabib, *Anal. Biochem.* **83,** 648 (1977).

[6] S. Hirano, Y. Ohe, and H. Ono, *Carbohydr. Res.* **47,** 315 (1976).

in a mortar while adding slowly and in small portions 20 ml of 10% acetic acid, until a syrupy solution is obtained. Some samples of commercial chitosan swell with acetic acid but do not go into solution; they are not suitable for acetylation. The chitosan solution is covered with a sheet of Parafilm and allowed to stand overnight at room temperature to complete dissolution of the polysaccharide. The next day, 90 ml of methanol is added slowly with mixing and the cloudy solution is filtered through glass wool on a Büchner funnel. The filtrate is placed in a beaker on a magnetic stirrer and 1.5 ml of acetic anhydride, containing 10 mCi of [^{3}H]acetic anhydride, is added. After stirring for 1–2 min, the mixture gels and the magnet stir bar comes to a standstill. The gel is allowed to stand for about 30 min and then is cut up into small pieces with a spatula. The liquid that oozes out is removed and the gel fragments are transferred to the cup of a motor-driven homogenizer, such as a Waring blender or Sorvall Omni-Mixer. After covering with methanol, the suspension is homogenized for 1 min at maximum speed. The finely divided chitin is filtered with a medium-porosity sintered glass funnel and washed extensively with water, until the filtrate is free from radioactivity. The chitin is suspended in 0.02% sodium azide to a final concentration of about 15 mg/ml (dry weight/volume) and stored at 5°. With time, more radioactivity may leak from the chitin particles into the water; it can be eliminated by filtering and washing the material as described above. Although the chitin is finely divided, the suspension tends to clog pipets. It can, however, be easily measured with an automatic micropipettor fitted with a disposable tip from which a 2- to 3-mm piece has been cut off to enlarge the opening.

For determinations of specific activity, the radioactivity and the *N*-acetylglucosamine content[3,7] are measured after exhaustive incubation with a mixture of chitinase and snail gut extract.[8]

Procedure. The reaction mixture contains 15 μl of [^{3}H]chitin suspension, 5 μl of 1 *M* buffer, variable amounts of chitinase, and water to complete 100 μl. After incubation at 30° with shaking for the desired time, the reaction is stopped by addition of 0.2 ml of 10% trichloroacetic acid. The suspension is filtered through Gelman glass fiber filters (2.5-cm diameter), type A/E, directly into scintillation vials, with a Bio-Rad multiple filtration apparatus. The tube contents and filter are washed into the vial with three 0.3-ml portions of water. After adding 12 ml of a scintillation fluid (e.g., Hydrofluor from National Diagnostics) the samples are counted.

[7] E. Cabib and A. Shurlati, this volume [55].

[8] Glusulase, from DuPont Pharmaceuticals, Wilmington, Delaware.

Comments. Samples may be centrifuged rather than filtered.[5] In this case, 0.3 ml of 10% trichloroacetic acid is added to stop the reaction. After centrifuging for 5 min at 200 *g*, a 200-μl portion of the supernatant fluid is transferred to a scintillation vial. With this modification, however, spurious counts are found occasionally, probably caused by some particles of chitin floating on the surface.

The kinetics of the assay depend on the enzyme used. Some chitinases, such as those from wheat germ[9] and *Streptomyces*,[5] give rise to nonlinear time curves and require very short (1 to 5 min) incubation times for accurate measurements of activity. Others, such as the *Serratia* chitinase,[10] maintain linearity for a longer time.

[9] E. Cabib, this volume [63].
[10] E. Cabib, this volume [56].

[49] Viscosimetric Assay for Chitinase

By Akira Ohtakara

The assay procedures of chitinase activity by the turbidimetric method and colorimetric measurement of *N*-acetylglucosamine have been previously described.[1] Insoluble compounds such as colloidal chitin and chitin are used as substrates in these assay procedures. The assay of chitinase activity by the viscosimetric method was first carried out using the solution of chitosan acetate.[2] Carboxymethylchitin[3] has been reported to be a water-soluble substrate suitable for the assay of chitinase activity. However, both the compounds have the disadvantage that the viscosity is markedly affected by ionic strength and pH. On the contrary, glycol chitin[4] does not have such disadvantage and is a useful substrate for the viscosimetric assay. 6-O-Hydroxypropylchitin[5] is similar to this compound and is also usable as a water-soluble substrate.

[1] C. Jeuniaux, this series, Vol. 8, p. 644.
[2] M. V. Tracey, *Biochem. J.* **61,** 579 (1955).
[3] E. Hultin, *Acta Chem. Scand.* **9,** 192 (1955).
[4] A. Ohtakara, *Agric. Biol. Chem.* **25,** 50 (1961).
[5] Y. Takiguchi, N. Nagahata, and K. Shimahara, *Nippon Nogei Kagaku Kaishi* **50,** 243 (1976).

Procedures for the viscosimetric assay of chitinase activity in which water-soluble derivatives of chitin are used as substrates will be described below.

Assay Methods

Glycol Chitin as Substrate[4,6]

Principle. The assay is based on the measurement of the viscosity of glycol chitin solution, which is reduced by the action of chitinase in an Ostwald viscosimeter.

Substrate. Glycol chitin is prepared from purified chitin according to the procedure originally described by Senju and Okimasu.[7]

Apparatus. Ostwald viscosimeters, with the flow time for water of 42.3 sec at 30°[4] and 22.0 sec at 35.5°,[6] are used in procedures I and II, respectively.

Procedure I.[4] Five milliliters of 0.2% glycol chitin is well mixed with 4 ml of McIlvaine's buffer (0.2 *M* Na_2HPO_4–0.1 *M* citric acid, optimum pH for each enzyme) in an Ostwald viscosimeter and the mixture is preincubated for 10 min at 30°. The reaction is started by adding 1 ml of enzyme solution, properly diluted and preincubated at 30°, into the above substrate buffer mixture in the viscosimeter and the flow time of the reaction mixture is measured at different time intervals. The reaction mixture with 1 ml of distilled water instead of enzyme solution is used as a control.

Typical curves indicating decrease in the viscosity of glycol chitin solution with various amounts of chitinase are shown in Fig. 1, in which specific viscosity (relative viscosity − 1) is plotted against reaction time. Half-life time (in minutes), which is the reaction time required for the specific viscosity to be reduced to one-half its initial value, is given by the point of intersection of these curves with one-half line (F in Fig. 1) for the initial specific viscosity. It is desirable to determine half-life time using three different amounts of enzyme.

If 1 U of chitinase activity is defined as the amount of the enzyme which is required to attain a half-life time in 30 min at 30° and optimum pH, there is a linear relationship between chitinase activity and the amount of enzyme (Fig. 2). Thus, the viscosimetric unit for chitinase can be obtained by dividing 30 min by half-life time.

[6] E. Hultin and G. Lundblad, *Scand. J. Clin. Lab. Invest.* **18,** 201 (1966).
[7] R. Senju and S. Okimasu, *Nippon Nogei Kagaku Kaishi* **23,** 432 (1950).

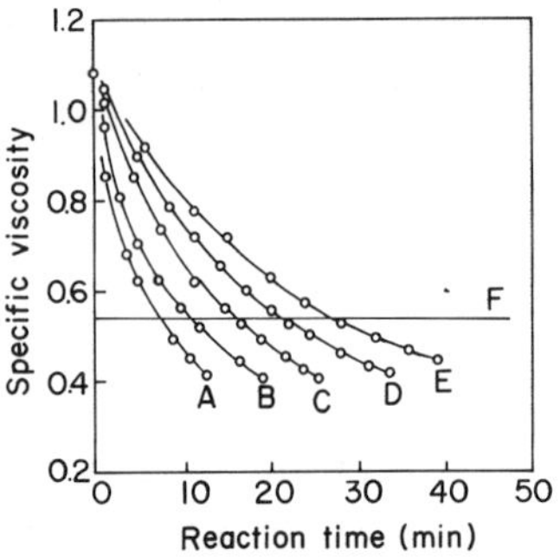

FIG. 1. Changes in the viscosity of glycol chitin solution with various amounts of enzyme. The reaction mixture consists of 5 ml of 0.2% glycol chitin, 4 ml of McIlvaine's buffer, pH 3.6, and 1 ml of chitinase solution purified from *Aspergillus niger*.[4] (A) 1/100 Dilution of the purified chitinase solution, (B) 1/150, (C) 1/200, (D) 1/300, (E) 1/400, (F) one-half line for the initial specific viscosity.

Procedure II.[6,8,9] The reaction mixture consists of 1.5 ml of 0.35% glycol chitin, 1.5 ml of 0.1 *M* McIlvaine's buffer (optimum pH for each enzyme), and 1.0 ml of enzyme solution. The flow time of the reaction mixture is measured in an Ostwald viscosimeter at 35.5° over a suitable time interval. The flow time for water is also measured in the same viscosimeter and specific viscosity is calculated at each reaction time. A plot of the inverse value of the specific viscosity against the reaction time gives a straight line. Chitinase activity can be expressed by the slope of this line.

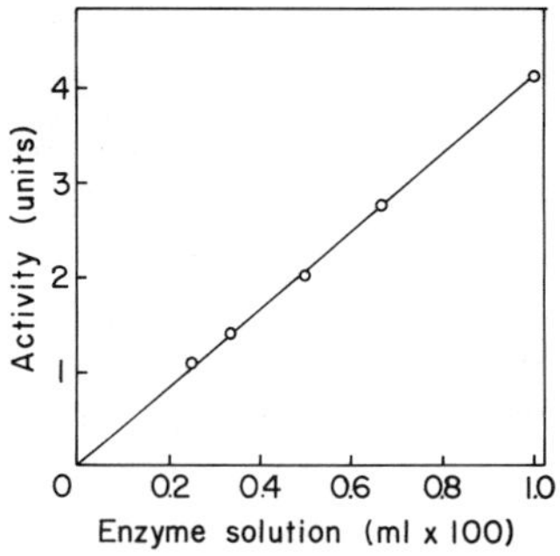

FIG. 2. Relationship between chitinase activity and the amount of enzyme.

[8] G. Lundblad, B. Hedersteadt, J. Lind, and M. Steby, *Eur. J. Biochem.* **46,** 367 (1974).
[9] G. Lundblad, M. Elander, and J. Lind, *Acta Chem. Scand., Ser. B* **30,** 889 (1976).

The viscosimetric unit (Hultin unit[6]) in this procedure is obtained by multiplying the slope of this line ($1/\eta_{sp} \cdot sec^{-1}$) by the square of the concentration (%) of glycol chitin and further multiplying the above value by 10^9 for convenience.

6-O-Hydroxypropylchitin as Substrate[5]

Principle. The principle is fundamentally the same as in the assay method described in procedure I, except that 6-O-hydroxypropylchitin prepared by the method of Sannan and Sobue[10] is used as substrate instead of glycol chitin.

Procedure. A mixture of 5 ml of 1% 6-O-hydroxypropylchitin with 9 ml of 0.6 *M* Na_2HPO_4–0.3 *M* citric acid buffer (optimum pH for each enzyme), and an enzyme solution are incubated separately for 10 min at 30°. The reaction is started by adding 1 ml of enzyme solution into the substrate buffer mixture in a Ubbelohde viscosimeter and the flow time of the reaction mixture is measured at different intervals of time.

The viscosimetric unit can be determined by the same procedure as in the case of glycol chitin. In this assay procedure, however, the reaction time required for the specific viscosity to be reduced to one-third its initial value is used for more accurate determination.

Comments

The viscosimetric assay for chitinase is a more sensitive and effective procedure to detect a slight activity. However, this assay procedure is somewhat troublesome and too time consuming to determine the chitinase activity of numerous samples.

The viscosity of a glycol chitin solution varies depending on the degree of polymerization. Some differences have been observed among the values of chitinase activity determined by procedure I using different batches of substrates.[4] Therefore, the viscosimetric assay of chitinase activity should be carried out using the same batch of glycol chitin in the same experiment. In connection with this, some differences in the degree of polymerization of glycol chitin do not affect the measurement of a Hultin unit in procedure II.[6]

The viscosimetric assay has been applied to the determination of chitinase activity in the purification of chitinases from *Aspergillus niger*[11] and

[10] T. Sannan and H. Sobue, *Rep. Prog. Polym. Phys. Jpn.* **17,** 701 (1974).

[11] A. Ohtakara, *Agric. Biol. Chem.* **25,** 54 (1961).

Vibrio sp.[12] (procedure I), and those from goat serum[8] and bovine serum[13] (procedure II). Endochitinase activity can be determined by these assay procedures,[14,15] but it should be noted that the viscosimetric assay has been also used for the determination of lysozyme activity.[6,16]

[12] A. Ohtakara, M. Mitsutomi, and Y. Uchida, *J. Ferment. Technol.* **57,** 169 (1979).
[13] G. Lundblad, M. Elander, J. Lind, and K. Slettengren, *Eur. J. Biochem.* **100,** 455 (1979).
[14] T. Boller, A. Gehri, F. Mauch, and U. Vögeli, *Planta* **157,** 22 (1983).
[15] Y. Takiguchi, N. Nagahata, and K. Shimahara, *Nippon Nogei Kagaku Kaishi* **59,** 253 (1985).
[16] K. Hamaguchi and M. Funatsu, *J. Biochem. (Tokyo)* **46,** 1659 (1959).

[50] Colorimetric Assay for Chitinase

By Thomas Boller and Felix Mauch

The colorimetric assay described here is applicable to the various types of chitinase present in microorganisms, animals, and plants.[1] We have evaluated it with particular reference to plant chitinases, which are of interest to us because they may function as a defense against chitin-containing pathogens.[2-6] The most widely used colorimetric assay for plant chitinases has been an exochitinase assay, based on the determination of monomeric *N*-acetylglucosamine (GlcNAc) released from colloidal chitin.[7] However, plant chitinases generally are endochitinases and produce chitooligosaccharides as principal products.[2-5,8] Therefore, measurements of plant chitinases with the exochitinase assay should be viewed with caution. For accurate determination, it is essential to measure the chitooligosaccharides produced in the assay. This can be accomplished by the enzymatic hydrolysis of the reaction products to monomeric GlcNAc prior to the colorimetric measurement.

[1] C. Jeuniaux, this series, Vol. 8, p. 644.
[2] T. Boller, A. Gehri, F. Mauch, and U. Vögeli, *Planta* **157,** 22 (1983).
[3] F. Mauch, L. A. Hadwiger, and T. Boller, *Plant Physiol.* **76,** 607 (1984).
[4] T. Boller, *in* "Cellular and Molecular Biology of Plant Stress" (J. L. Key and T. Kosuge, eds.), p. 247. Liss, New York, 1985.
[5] T. Boller, A. Gehri, F. Mauch, and U. Vögeli, this volume [60].
[6] A. Schlumbaum, F. Mauch, U. Vögeli, and T. Boller, *Nature (London)* **324,** 365 (1986).
[7] F. B. Abeles, R. P. Bosshart, L. E. Forrence, and W. H. Habig, *Plant Physiol.* **47,** 129 (1971).
[8] J. Molano, I. Polacheck, A. Duran, and E. Cabib, *J. Biol. Chem.* **254,** 4901 (1979).

Assay Method[2]

Principle. Endochitinase forms soluble chitooligosaccharides from insoluble chitin. After removal of the undigested substrate by centrifugation, the chitooligosaccharides are completely hydrolyzed to monomeric GlcNAc by incubation with snail gut enzyme.[9] The monomeric GlcNAc is then determined with *p*-dimethylaminobenzaldehyde (DMAB).[10]

Reagents

Colloidal chitin, 10 mg/ml, in 3 m*M* NaN_3. The colloidal chitin is prepared according to Berger and Reynolds,[11] as described elsewhere in this volume[12]

Sodium acetate buffer, 1 *M*, pH 4.0

Snail gut enzyme, 30 mg/ml, desalted; 600 mg of the commercial lyophilized snail gut enzyme (Cytohelicase, obtained from Industrie Biologique Française, Clichy, France) is dissolved in 10 ml 20 m*M* KCl and chromatographed on a Sephadex G-25 column (38 × 1.6 cm) using a 10 m*M* KCl solution, containing 1 m*M* EDTA and adjusted to pH 6.8, for equilibration and elution.[9] The first 20 ml eluted after the void volume is collected

Potassium phosphate buffer, 1 *M*, pH 7.1

Sodium borate buffer, 1 *M*, pH 9.8

DMAB reagent[10]: 1 vol of a stock solution of 8 g *p*-dimethylaminobenzaldehyde in 70 ml glacial acetic acid and 10 ml concentrated HCl is mixed with 9 vol of glacial acetic acid immediately before use

GlcNAc standard: *N*-acetylglucosamine, 2 m*M*, in water

Assay Procedure. The following are pipetted into a 1.5-ml Eppendorf tube: 10 μl sodium acetate buffer and 0.4 ml enzyme solution in 10 m*M* sodium acetate buffer, pH 5.0. This buffer mixture yields pH 4.4 in the assay. The reaction is carried out at 37° in a shaking water bath. It is started by the addition of 0.1 ml colloidal chitin and stopped by centrifugation (1000 *g*, 3 min) after 2 hr. An aliquot of the supernatant (0.3 ml) is pipetted into a glass reagent tube containing 30 μl phosphate buffer, and incubated with 20 μl snail gut enzyme for 1 hr. In this reaction, the buffer mixture yields pH 6.8.

After 1 hr, the reaction mixture is brought to pH 8.9 by the addition of 70 μl of the borate buffer. The mixture is incubated in a boiling water bath for exactly 3 min, and then rapidly cooled in an ice-water bath. After

[9] E. Cabib and B. Bowers, *J. Biol. Chem.* **246,** 152 (1971).

[10] J. L. Reissig, J. L. Strominger, and L. F. Leloir, *J. Biol. Chem.* **217,** 959 (1955).

[11] L. R. Berger and D. M. Reynolds, *Biochim. Biophys. Acta* **29,** 522 (1958).

[12] Y. Shimahara and Y. Takiguchi, this volume [47].

addition of 2 ml of the DMAB reagent, the mixture is incubated for 20 min at 37°. Immediately thereafter, A_{585} is measured in the spectrophotometer.

Blanks and Internal Standards. For each enzyme preparation (E) measured, an enzyme blank (EB) and an enzyme blank with internal standard (EI) are carried through the procedure. The enzyme blank contains 0.1 ml water instead of the colloidal chitin. The enzyme blank with internal standard contains 0.1 ml 2 m*M* GlcNAc instead of the colloidal chitin.

In addition, for each series of measurements, a substrate blank (SB), a reagent blank (RB), and a reagent blank with internal standard (RI) are carried through the procedure. The substrate blank contains 0.4 ml 10 m*M* sodium acetate buffer, pH 5.0, instead of the enzyme preparation; the reagent blank contains 0.4 ml of the same buffer and 0.1 ml water instead of chitin; the reagent blank with internal standard also contains 0.4 ml of the same buffer and 0.1 ml of the GlcNAc standard instead of chitin. The amount of GlcNAc equivalents released by the enzyme is calculated as follows:

$$\left[\frac{A_{585}(\mathrm{E}) - A_{585}(\mathrm{EB})}{A_{585}(\mathrm{EI}) - A_{585}(\mathrm{EB})} - \frac{A_{585}(\mathrm{SB}) - A_{585}(\mathrm{RB})}{A_{585}(\mathrm{RI}) - A_{585}(\mathrm{RB})}\right] 200 \text{ nmol}$$

Standard Curve. Formation of reaction products in the chitinase assay is not proportional to the enzyme concentration. Therefore, it is necessary to establish a standard curve relating the amount of enzyme to the rate of product formation. To this end, a freshly made dilution series from an extract containing a high activity of chitinase, prepared in plastic tubes (glassware tends to adsorb chitinase at the high dilutions required), is assayed (Fig. 1). The standard curve has to be renewed for each new preparation of chitin since different preparations of chitin yield different standard curves when assayed with the same enzyme extract.

Standard curves for different plant chitinases are usually congruent.[2,13] However, the standard curves for some microbial chitinases may differ markedly from those for plant chitinases (unpublished observations). Routinely, three different dilutions (e.g., 1:10, 1:20, and 1:40) of each new enzyme preparation are assayed to check the fit of the results with the standard curve.

Definition of Unit. One unit of chitinase is defined as the amount that catalyzes the release of soluble chitooligosaccharides containing 1 μmol of GlcNAc in 1 min at infinite dilution. The initial slope of the standard curve is used to calculate the unit (Fig. 1).

[13] J. P. Métraux and T. Boller, *Physiol. Mol. Plant Pathol.* **28**, 161 (1986).

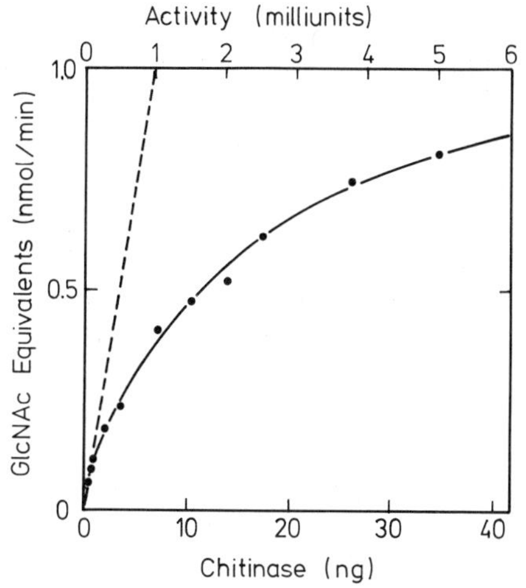

FIG. 1. Standard curve for bean chitinase. The colorimetric chitinase assay is performed with a freshly prepared dilution series of purified chitinase.[2,5] The initial slope of the standard curve (broken line) is used to define the unit. Redrawn from data of Boller *et al.*[2]

Modifications

Variation of Buffers. As long as internal standards are used, the assay conditions can be varied considerably with regard to buffers and pH values. For maximal sensitivity, it is important to adjust the pH of the borate buffer used in the last step to obtain pH 8.9 in the reaction mixture before boiling.

Alternate Substrates. Both colloidal chitin[11,12] and regenerated chitin (i.e., reacetylated chitosan)[14] can be used as substrates. Particularly suitable is colloidal chitin prepared from regenerated chitin (unpublished observations). Chitinase-free purified fungal cell walls containing chitin can be substituted for chitin to study the activity of chitinase on a putative natural substrate.[2]

Variation of the Enzymatic Treatment of the Products Formed by Chitinase. The method described here is based on the complete hydrolysis of the soluble chitooligosaccharides by the snail gut enzyme mixture.

The minimal amount of snail gut enzyme preparation needed to obtain complete hydrolysis is determined in a series of preliminary assays, in which a constant, high amount of chitinase is used, releasing 100–200 nmol GlcNAc equivalents in the form of chitooligosaccharides, and the

[14] J. Molano, A. Duran, and E. Cabib, *Anal. Biochem.* **83,** 648 (1977).

amount of snail gut enzyme is varied (0, 5, 10, 20, 40, 80 μl). We generally find 20 μl of the snail gut enzyme preparation sufficient for complete hydrolysis within 1 hr.

It is important to note that snail gut enzyme is rich not only in hydrolases acting on chitooligosaccharides (chitobiases) but also in chitinase. Therefore, it is important to remove all unreacted substrate by centrifugation before the snail gut enzyme treatment. Inadvertent stirring of the sediment after centrifugation and inclusion of chitin particles in the snail gut enzyme treatment can result in large spurious activities.

Other enzyme preparations may be used to hydrolyze the chitooligosaccharides. Commercial *N*-acetyl-β-D-glucosaminidase (from jack bean)[15] and β-glucosidase (from almond emulsin)[1,16] have been employed. The β-glucosidase preparation is active because it also contains *N*-acetyl-β-glucosaminidase. These plant enzyme preparations may also contain chitinase; therefore, the procedures[1] in which "β-glucosidase" or *N*-acetyl-β-glucosaminidase preparations are directly included in the chitinase assay itself are not recommendable.

Interferences

Some crude plant extracts, e.g., those from *Vicia faba* leaves, contain secondary products that are chromogenic in the assay for GlcNAc. The high A_{585} of the enzyme blank renders an accurate determination of chitinase by the colorimetric assay impossible.

Comparison with Other Chitinase Assays

Colorimetric assays for chitinase based on the determination of reducing sugars[17] are frequently used in microbiology. These assays are unsuitable for crude plant enzyme preparations which usually contain large amounts of reducing sugars. The colorimetric assay described here is specific for GlcNAc and is readily applicable to crude plant extracts as well as to other sources of chitinase.

The viscosimetric assay, described elsewhere in this volume,[18] has the advantages of using a soluble substrate (which facilitates kinetic analyses) and of being more specific for endochitinase (as opposed to exochitinase) than other assays. Disadvantages are the requirement for an artificial

[15] G. Lundblad, M. Elander, J. Lind, and K. Slettengren, *Eur. J. Biochem.* **100,** 455 (1979).

[16] N. Boden, U. Sommer, and K. D. Spindler, *Insect Biochem.* **15,** 19 (1985).

[17] J. Monreal and E. T. Reese, *Can. J. Microbiol.* **15,** 689 (1969).

[18] A. Ohtakara, this volume [49].

substrate and, in the absence of specialized equipment, the rather time-consuming, tedious procedure. Using different crude plant extracts, we obtain qualitatively very similar results with the colorimetric and the viscosimetric assay, indicating that both assays measure the same type of endochitinase activity.[2]

The radiochemical assay using regenerated [^{3}H]chitin as a substrate[14] is described elsewhere in this volume.[19] The colorimetric and the radiochemical assay can be compared directly by using the same preparation of regenerated [^{3}H]chitin as a substrate. Such comparisons[2,3,14] have shown that the two assays are fully equivalent, yielding identical results.

The radiochemical assay is more rapid and simpler than the colorimetric assay. Therefore, the use of the radiochemical assay is recommended for routine work. The colorimetric assay described here is most useful when the radioactive substrate is not available, or when native chitin or chitin-containing materials like fungal cell walls are to be employed as substrates.

Acknowledgment

This work was supported by the Swiss National Science Foundation.

[19] E. Cabib, this volume [48].

[51] Physical Methods for the Determination of Chitin Structure and Conformation

By JOHN BLACKWELL

Chitin is a naturally occurring fiber-forming polymer that functions as the load-bearing component of the skeletal materials of many "lower" animals, most notably the arthropods. Chemically it is a repeating polymer of β-1,4-linked anhydro-2-acetamido-D-glucose, and can be thought of as a naturally occurring cellulose derivative. This is an idealized structure in that as many as one in six of the monomers may be glucosamines,[1] i.e., the acetyl groups are absent. Chitin generally occurs in the presence of protein, and it is possible that these deacetylated residues are involved in some way in the chitin–protein interaction. One source of completely acetylated

[1] C. H. Giles, A. Hassan, M. Laidlow, and R. V. R. Subramanian, *J. Soc. Dyers Colour.* **74**, 647 (1958).

chitin is that from the spines of certain marine diatoms.[2] It is interesting that this highly crystalline plant material occurs free of protein.

The physical structure of chitin at the molecular level has been the subject of continuous study since the 1920s, and the primary methods of analysis have been X-ray crystallography and infrared spectroscopy. X-Ray fiber diagrams of oriented chitin specimens (such as deproteinized lobster tendon) show an obvious similarity to those of cellulose. It is clear from the similarity of the layer line spacings, and the virtual absence of odd order $00l$ reflections, that chitin and cellulose chains have the same basic 2_1 helical conformation, in which two monomer residues repeat in 10.3–10.4 Å. Electron microscopy shows that chitin has a fibrous morphology: the polymer is semicrystalline, and the amorphous regions probably correspond to the less ordered edges of the fibrous crystallites.

Comparison of the X-ray data for chitins from different sources has revealed the existence in nature of more than one polymorphic form.[3] Polymorphism is a common phenomenon in crystalline polymers, because there is frequently relatively little difference in potential energy between several modes of packing the chains. Most chitins, including those from crustaceans, insects, and fungi, are in the so-called α-form. However, a rare second form known as β-chitin has been identified in four sources: the spines of the polychaete *Aphrodite*,[4] the pen of the squid *Loligo*,[5] the tubes of *Pogonophora*,[6] and the spines of certain marine diatoms.[7] A third form, γ chitin, has been reported from the stomach lining of *Loligo*,[3] but this is less well characterized, and it is yet to be established that this is a separate form, rather than a distorted, poorly crystalline α or β structure. All three structures have the same 2_1 helical conformation and differ simply in the mode of packing of adjacent chains.

X-Ray patterns of chitin generally contain 20–50 Bragg reflections. Such patterns are relatively detailed compared to many other fibrous polymers, but there is insufficient information to determine the crystal structure by the methods used for materials that form large crystals, which generally yield 1000 or more X-ray reflections. Instead, fibrous polymer structures have to be determined by "trial and error" methods. These require construction of the model which gives predicted intensities that are in best agreement with those observed for the X-ray reflections. In recent years, trial and error modeling has become automated by use of linked

[2] M. Falk, D. G. Smith, J. McLachlan, and A. G. McInnes, *Can. J. Chem.* **44,** 2269 (1966).
[3] K. M. Rudall, *Adv. Insect Physiol.* **1,** 257 (1963).
[4] N. Lotmar and L. E. R. Picken, *Experientia* **6,** 58 (1950).
[5] K. M. Rudall, *Soc. Symp. Exp. Biol.* **9,** 49 (1955).
[6] J. Blackwell, K. D. Parker, and K. M. Rudall, *J. Mar. Biol. Assoc. U. K.* **45,** 659 (1965).
[7] J. Blackwell, K. D. Parker, and K. M. Rudall, *J. Mol. Biol.* **28,** 383 (1967).

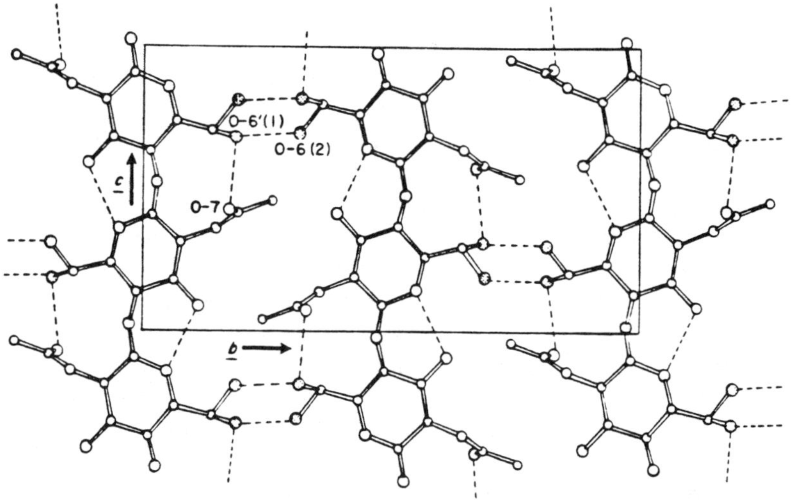

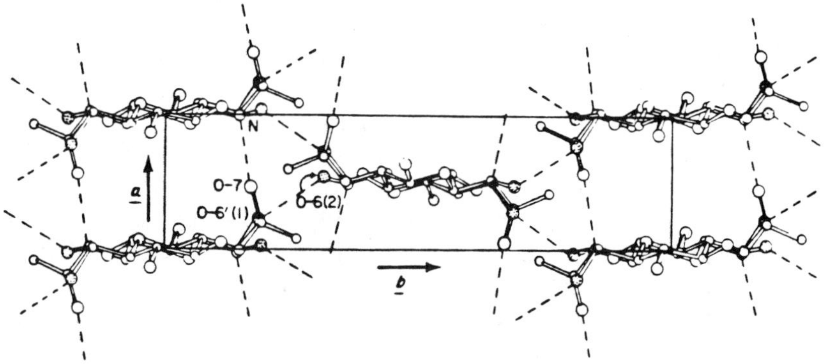

FIG. 1. Structure of α chitin.[9] (a) *bc* Projection and (b) *ab* projection. Note that the CH_2OH groups are present in two torsional orientations, leading to two possible positions for this oxygen. The two positions shown were treated as half-oxygens.

atom least-squares procedures,[8] which adjust the model to obtain the best least-squares agreement between observed and calculated intensities while taking account of the allowed stereochemistry and possible constraints such as the need to form hydrogen bonds.

The structure refined for α chitin by LALS methods is shown in Fig. 1 and was determined based on the intensity data for deproteinized lobster

[8] P. J. C. Smith and S. Arnott, *Acta Crystallogr. A* **34,** 3 (1978).

tendon.[9] The unit cell is orthorhombic with dimensions $a = 4.74$ Å, $b = 18.86$ Å, and $c = 10.32$ Å (fiber axis). The space group is approximately $P2_12_12_1$, which necessitates antiparallel packing of chains. Note that there is a sense to the chemical structure: adjacent chains along the b axis are alternately "up" and "down." This structure is similar to that due to Carlstrom,[10] who refined the structure using optical diffraction masks to model the structure and proposed an orthorhombic packing of antiparallel chains. He arranged the chains in sheets along the a axis linked by N—H···O=C hydrogen bonds between the acetamido groups. The stacking of the chains is further stabilized by the hydrophobic forces between the "surfaces" of the glucose rings. However, optical diffraction methods were insufficient to determine the arrangement of the —CH_2OH side chains, which dictates the type of intermolecular bonding along the b axis. LALS refinement[9] favored a 50/50 random mixture of two rotational conformations for the CH_2OH groups, as shown in Fig. 1. Half of the CH_2OH groups are bonded to carbonyls within the same stack of chains, and half are bonded to the CH_2OH groups on an adjacent stack. The existence of this intersheet bonding is probably responsible for the stability of the α chitin structure, specifically its inability to swell in water.

The β form of chitin was identified by Lotmar and Picken,[4] based on its different X-ray pattern. This form also differs from α chitin in that it is readily swollen in water, and it has been found to form a series of crystalline hydrate structures.[11] The anhydrous form is found in thoroughly dried diatom spines, and has a monoclinic unit cell with dimensions $a = 4.85$ Å, $b = 9.26$ Å, $c = 10.38$ Å (fiber axis), and $\gamma = 97.5°$. The space group is $P2_1$. This unit cell has half the volume of that of α chitin, and contains a disaccharide unit of a single chain. This necessitates a parallel chain structure for β chitin, i.e., all the chains in the crystallite have the same sense. The structure refined[12] for anhydrous β chitin is shown in Fig. 2. The chains have a 2_1 helical conformation and are stacked along the a axis and linked by C=O···H—N hydrogen bonds, as in α chitin. However, the —CH_2OH groups are all hydrogen bonded to carbonyl groups in the same stack of chains, and hence there is no hydrogen bonding between adjacent stacks. Thus the structure is easily swollen by intercalation of water molecules between the stacks of chains. In this regard it is interesting that β chitin is found exclusively in aquatic organisms.

The hydrates are formed by expansion of the b axis, leaving the a and c

[9] R. Minke and J. Blackwell, *J. Mol. Biol.* **120,** 167 (1978).

[10] D. Carlstrom, *J. Biophys. Biochem. Cytol.* **3,** 669 (1957).

[11] J. Blackwell, *Biopolymers* **7,** 281 (1969).

[12] K. H. Gardner and J. Blackwell, *Biopolymers* **14,** 1581 (1975).

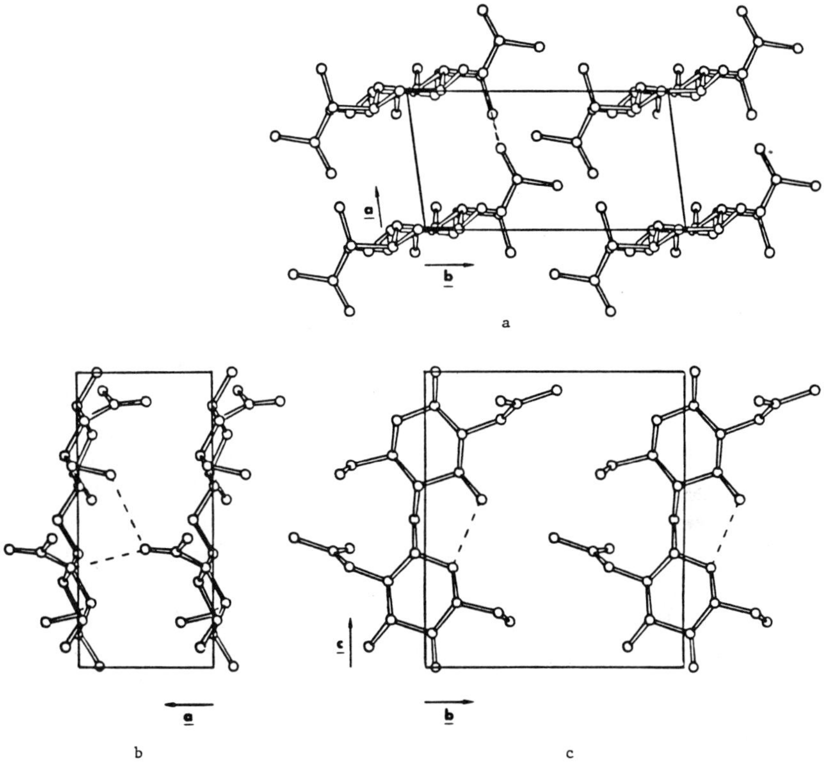

FIG. 2. Structure of anhydrous *β* chitin.[12] (a) *ab* Projection, (b) *ac* projection, and (c) *bc* projection.

axes unchanged. For example, diatom chitin forms two hydrates, in which *b* is increased to 10.3 and 11.0 Å. These hydrates are stable at room conditions, but the anhydrous structure is reformed on vacuum drying. However, the *β* chitin of *Loligo* pen, *Aphrodite* chaetae, and *Pogonophora* tubes exists partially (in some specimens predominantly) as hydrate forms, and these are not dehydrated by drying. This has led to the suggestion[11] that there may be two types of *β* chitin, designated A and B, with the possibility that type B is related to the existence of unacetylated residues. However, at this time the X-ray patterns are insufficiently detailed to resolve these differences.

γ Chitin was identified by Rudall,[3] who showed that a three-chain unit cell was necessary to index the observed reflections. He proposed a structure in which two "up" chains are followed by a "down" chain. However, this form has not been subjected to the detailed analysis given to the *α* and *β* forms, and it has yet to be established that this is a true third structure, and not a distorted version of one of the others, which is not unreasonable given the complexity of the β chitin hydrates.

The parallel-β/antiparallel-α structures are analogous to the chain polarities found in the polymorphic forms of cellulose. Native cellulose also has the cellulose I structure comprised of parallel chains.[13,14] However, swelling native cellulose in alkali or regenerating from solution leads to formation of cellulose II, in which adjacent chains are antiparallel.[15,16] The regenerated form, cellulose II, is the more stable of the two, and thus the large amount of cellulose produced in nature is in a metastable form. Both α and β chitin occur naturally, and of these the α form is the more stable: β chitin is converted to the α form by treatment with acid, but there is no known means by which this transformation can be reversed. The parallel polarity for native cellulose is probably due to the mechanism of biosynthesis, which is thought to involve synthesis and immediate crystallization of the chains from arrays of identical enzyme complexes, i.e., a bundle of chains with the same sense crystallizes to form a fibril. Cellulose I has hydrogen bonds between the stacks of chains and is not swollen in water. This is not the case in the analogous parallel chain β chitin, which is readily swollen, and this feature would be less than desirable for structural materials in nonaquatic animals. Hence it would appear that most chitin is synthesized by a mechanism different from that of native cellulose, so as to produce antiparallel polarity and a nonswellable structure.

From the analytical point of view, α and β chitin are easily distinguished from their X-ray patterns, even for unoriented specimens, in view of the different d spacings for the first observed *hk0* reflection. For α chitin, the first observed reflection in the powder pattern is at $d = 9.4$ Å. For squid and *Aphrodite* β chitin, this reflection occurs at 10.3 Å, due to the greater separation of the stacks of chains in the hydrated structure. For *Pogonophora* β chitin, there are usually two reflections in this region, which are at $d = 10.3$ and 9.2 Å in dry specimens, pointing to a mixture of hydrated and anhydrous structures. In the case of diatom spine β chitin, only the 9.2 reflection for the anhydrous form is seen. *Pogonophora* and diatom chitin are highly crystalline, and hence the experimental accuracy is sufficient to distinguish between 9.4 and 9.2 Å, i.e., between the α and anhydrous β forms.

The α and β chitin can also be distinguished by infrared spectroscopy, due to differences in their hydrogen-bonding networks. The most obvious

[13] K. H. Gardner and J. Blackwell, *Biopolymers* **13,** 1975 (1974).
[14] A. Sarko and R. Muggli, *Macromolecules* **7,** 486 (1974).
[15] F. J. Kolpak and J. Blackwell, *Macromolecules* **9,** 273 (1976).
[16] A. J. Stipanovick and A. Sarko, *Macromolecules* **9,** 851 (1976).

spectral differences are in the frequencies of the amide I modes in the region 1660–1620 cm^{-1}, where we observe a doublet at 1656 and 1621 cm^{-1} for α chitin and a singlet at 1631 cm^{-1} for β chitin.[3] There has been much discussion in the literature of the origin of the splitting for α chitin. One suggestion has been that this is due to the existence of crystalline and amorphous regions,[2] but if this were the case the same effect would be seen for β chitin. More complex possibilities involve coupling of vibrations on adjacent amide groups, analogous to the coupling that occurs in polypeptides,[17] but no satisfactory solution has been obtained consistent with the structures determined by X-ray diffraction. At this point the likely solution is that in α chitin we have two types of amide groups[9] due to the 50/50 random bonding of the CH_2OH groups. Half the groups are involved in only one hydrogen bond (N—H···O=C) and are responsible for the 1656 cm^{-1} mode. The remaining amide groups form this bond plus another to the CH_2OH group, and this additional bonding leads to a lowering of the amide I frequency to 1621 cm^{-1}. In β chitin, all the amide groups form the two bonds, and the frequency of the amide I mode is 10 cm^{-1} higher than that for α chitin because the C=O···H—N distance is slightly longer.

A final point is that the above discussion applies to the structure of pure (or purified) chitin. Most chitins occur in the presence of protein, but this usually has no ordered structure. Hence, the diffraction pattern of an intact chitin–protein complex will show the diffraction characteristics of crystalline chitin, plus an amorphous halo due to the amorphous protein (and chitin). In a few cases, however, the protein is ordered: some of the best characterized examples come from insect cuticles, as well as *Loligo* pen and *Aphrodite* chaetae. Rudall and Kenchington[18] have examined these materials by X-ray methods and have characterized chitin–protein complexes in which the layer line repeat is increased from 10.3 Å in chitin to 31, 41, or 62 Å. Electron microscopy indicates that these systems consist of an ordered (often hexagonal) array of chitin fibrils which are 30–40 Å across and are embedded in a matrix of protein. The diffraction characteristics of α and β chitin are recognizable in the pattern of these complexes and hence the fibrils are crystalline with a structure like that in Figs. 1 or 2. The remaining data in the pattern are due to the protein and hence this component also has an ordered structure. Analysis of the intensity distribution for the protein component of *Megarhyssa* ovipositor[19] points to a helical structure in which six protein subunits are arranged in a helix with a

[17] T. Miyazawa, *J. Chem. Phys.* **32,** 1647 (1960).
[18] K. M. Rudall and W. Kenchington, *Biol. Rev.* **49,** 597 (1973).
[19] J. Blackwell and M. A. Weih, *J. Mol. Biol.* **137,** 49 (1980).

repeat of 31 Å, i.e., every six *N*-acetylglucosamine residues. These chitin–protein systems are unique in that they are fiber matrix components in which both fiber and matrix are ordered. As such they present an opportunity to study the fiber–matrix interaction, and the results could serve as a model for the interactions that occur in cellulose composites in plant cell walls, collagen composites in connective tissue, and also the many fiber-reinforced synthetic polymers.

Acknowledgment

Work on chitin in this laboratory has been supported by the National Science Foundation, most recently through Grant No. DMR84-17525 from the Polymer Program.

[52] Determination of the Degree of Acetylation of Chitin and Chitosan

By DONALD H. DAVIES and ERNEST R. HAYES

Both chitin and chitosan are copolymers of *N*-acetylglucosamine and glucosamine. The copolymer is called chitin when it contains less than 7% nitrogen and chitosan when the nitrogen content exceeds 7%. Both are found in nature, but chitosan is generally obtained by the deacetylation of chitin. In nature, chitin is covalently bonded to proteins which may also serve to link chitin to carbohydrates. It may also be associated with salts, e.g., calcium carbonate in the shells of crustacea. In plants, it serves as an alternative to cellulose and, in animals, as an alternative to collagen.

Chitin is usually prepared from crustacean waste. During the processing, there will be some deacetylation and some chain scission. The properties of the chitin obtained will depend on the process used, the source of the raw chitinous material, and its treatment before processing. Since many studies use chitin/chitosan from a single source or prepared by a single method, the calibration may not apply to other samples. An excellent review of the literature on chitin and chitosan has been published by Muzzarelli.[1]

In any attempt to analyze chitin, it must be noted that the chitin may contain unremoved protein, carbohydrates, and salts. Chitosan will also probably contain salts, but it can be purified by dissolving in aqueous

[1] R. A. A. Muzzarelli, "Chitin." Pergamon, Oxford, 1977.

acetic acid, filtering, precipitating with sodium hydroxide, and washing with methanol. Chitin and chitosan also bond water tenaciously and it has been suggested[2] that for every missing acetyl group there are two water molecules.

Titration

Rutherford and Austin[3] determined the degree of acetylation by hydrolyzing the acetyl groups with strong alkali and converting the salt to acetic acid, which they distilled off as an azeotrope with water and titrated. A sample of chitin/chitosan (0.1 g) and 40 ml of 50% sodium hydroxide is refluxed for 1.5 hr at 100° after which 25 ml of concentrated phosphoric acid is carefully added. The mixture is fractionally distilled using a Vigreux column. As the distilling flask begins to go dry, 15 ml of hot distilled water is added to the flask. This is repeated until 250 ml of distillate is collected. Aliquots (25 ml) are titrated with 0.01 *N* sodium hydroxide using phenolphthalein as an indicator and the volume is extrapolated to the total volume of the distillate. Finely, divided filter paper (cellulose) is used as a blank. The percentage acetyl was determined from the equation:

$$\%\ \text{Acetyl} = V \times 0.04305/w$$

where V is the corrected volume of sodium hydroxide and w is the weight of sample.

Hayes and Davies[4] determined the degree of acetylation by titrating solutions of chitosan hydrochloride prepared by adding an excess of concentrated hydrochloric acid or sodium chloride to a solution of chitosan in acetic acid. The precipitate is filtered, washed with methanol, and dried overnight at 50°. Chitosan hydrochloride is hygroscopic and should be allowed to equilibrate in a constant humidity chamber. Solutions of chitosan hydrochloride have properties characteristic of quaternary ammonium salt solutions. Samples are dissolved in water and titrated potentiometrically with sodium hydroxide. This procedure is tedious because of the long time required to reach equilibrium. The titration may also be performed using phenolphthalein as an indicator. Alternatively, the number of free amine units in the original chitosan can be determined by titrating the chloride ions with silver nitrate using 2,7-dichlorofluorescein as an indicator. Potassium dichromate cannot be used since chromate ions form an insoluble complex with chitosan. Excellent agreement was observed be-

[2] R. A. A. Muzzarelli, "Chitin," p. 88. Pergamon, Oxford, 1977.
[3] F. A. Rutherford and P. R. Austin, *Proc. Int. Conf. Chitin/Chitosan, 1st,* p. 182 (1978).
[4] E. R. Hayes and D. H. Davies, *Proc. Int. Conf. Chitin/Chitosan, 1st,* p. 406 (1978).

tween the three titrations. The acetyl content is calculated by determining the total nitrogen content by the Kjeldahl method and subtracting the nitrogen of the amine. If the sample were free of contaminants, the Kjeldahl determination would not be required.

Moore and Roberts[5] determined the extent of deacetylation by treating chitin/chitosan with a solution of sodium periodate to cleave the α-amino alcohol units and titrating the unconsumed periodate with sodium arsenite.

Broussignac[6] treated chitin/chitosan with hydrochloric acid to form the amine hydrochloride and then titrated the excess hydrochloric acid potentiometrically.

Infrared Spectroscopy

Sannan *et al.*[7] reported that the degree of acetylation was linearly dependent on the ratio of the transmission of the amide band at 1550 cm^{-1} and the transmission of the C—H band at 2878 cm^{-1}. However, further work by Muzzarelli *et al.*[8] showed that the calibration line depends on the deacetylation treatment and that meaningful results can be obtained only if the proper calibration line is used.

First Derivative UV Spectroscopy

Muzzarelli and Rocchetti[9] have described a method to determine the degree of acetylation using first derivative ultraviolet spectrophotometry at 199 nm. First derivative spectra are recorded in the range of 240 to 190 nm for 0.01, 0.02, and 0.03 *M* acetic acid solutions and for four or five standard solutions in the range 0.5 to 3.5 mg of *N*-acetylglucosamine in 100 ml of 0.01 *M* acetic acid. The spectra are superimposed and the heights measured from the zero-crossing point of the acid. The absorbance readings are linearly dependent on concentration and are not influenced by the presence of acetic acid. Typical results for *Euphausia superba* chitosan are a degree of acetylation of 42.6 with a standard deviation of 1.3% and confidence levels of ± 0.07 at the 95% level. For highly deacetylated chitosans, a correction factor can be applied for the concentration of glucosamine.

[5] G. K. Moore and G. A. F. Roberts, *Proc. Int. Conf. Chitin/Chitosan, 1st,* p. 421 (1978).

[6] P. Broussignac, *Chim. Ind., Genie Chim.* **99,** 1241 (1968).

[7] T. Sannan, K. Kurita, and Y. Iwakura, *Polymer* **19,** 458 (1978).

[8] R. A. A. Muzzarelli, F. Tanfani, G. Scarpini, and G. Laterza, *J. Biochem, Biophys, Methods* **2,** 299 (1980).

[9] R. A. A. Muzzarelli and R. Rocchetti, *in* "Chitin in Nature and Technology" (R. A. A. Muzzarelli, C. Jeuniaux, and G. W. Gooday, eds.), p. 385. Plenum, New York, 1986.

Colorimetric Assay

We have attempted to determine the amine content of chitosan using a modification of the colorimetric method proposed by Ride and Drysdale[10] for estimating fungi which contain chitin. The glucosamine units are deaminated with nitrous acid to form 2,5-anhydromannose, whose concentration can be determined colorimetrically. Chitosan (0.20 g) is dissolved in 100 ml of either 0.25 or 0.50% acetic acid and the solution is further diluted 100 times with distilled water. A sample of this solution (1.5 ml) is mixed with 1.5 ml of a 5% (w/v) solution of sodium nitrite and 1.5 ml of a 5% solution of potassium bisulfate. After the solution has been shaken for 15 min, 1.5 ml of a 12.5% solution of ammonium sulfamate is added followed by 1.5 ml of a 0.5% solution of 3-methyl-2-benzothiazolone hydrazone hydrochloride (this solution must be prepared daily). The solution is heated on a boiling water bath for 3 min and allowed to cool. Ferric chloride (1.5 ml of a 0.5% solution) is then added, the solution is allowed to stand for 30 min, and the absorbance is read at 650 nm. Our results gave a linear fit for the percentage amine versus the absorbance with a correlation coefficient of 0.90 and a standard deviation of 0.62.

Pyrolysis Gas Chromatography

Lal and Hayes[11] and Davies *et al.*[12] used a Chemical Data System Pyroprobe 100 connected to a Hewlett-Packard 5880A series gas chromatograph with a 3.658 × 3.175 mm column packed with 10% SP-1000 on 80–100 mesh Chromosorb W AW DMCS to study chitin and chitosan. The interface temperature was 200°; the final temperature was 750°; the interval was 10 sec; the sample size was 0.4–0.45 mg; the column temperature was 75° for 5 min and then rose at 10°/min to 200°, where it was held for 10 min; the flow rate of helium was 30 ml/min. The data were fed into an Apple IIe microcomputer with an Adalab acquisition/control card from Interactive Microware, Inc. They found that the ratios of the areas of a number of peaks were linearly related to the percentage amine. Their best results were for the ratio of the peaks at 13.69 min and at 9.75 min. The former is a major peak in the pyrogram of chitosan and is absent in the pyrograms of chitin and *N*-acetyl-D-glucosamine, while the latter is a major peak in the pyrograms of chitin and *N*-acetyl-D-glucosamine and absent in the pyrogram of chitosan. The correlation coefficient was 1.0 and the

[10] J. P. Ride and R. B. Drysdale, *Physiol. Plant Pathol.* **2,** 7 (1972).

[11] G. S. Lal and E. R. Hayes, *J. Anal. Appl. Pyrol.* **6,** 183 (1984).

[12] D. H. Davies, E. R. Hayes, and G. S. Lal, *in* "Chitin in Nature and Technology" (R. A. A. Muzzarelli, C. Jeuniaux, and G. W. Gooday, eds.), p. 365. Plenum, New York, 1986.

standard deviation of the errors was 0.17 (calculated from the data in the paper).

We have studied chitin and chitosan by capillary column pyrolysis gas chromatography using the system previously described with a 30 m × 0.329 mm capillary column (J & W Durabond DB-5, 1 μm thick, purchased from Chromatographic Specialties, Ltd.). The column was maintained at 50° for 5 min, heated to 250° at 5°/min, and maintained at 250° for 5 min. The helium flow rate was 2 ml/min with a split of 15 ml/min. The peaks were identified using a Finnigan MAT 4500 quadrupole mass spectrometer equipped with an INCOS data system. The percentage amine is a linear function of a number of different peak area ratios, but the best fit was obtained using the ratio of the 1,2-diaminopropane peak (retention time 2.2 min) to the acetamide peak (retention time 9.5 min). The correlation coefficient was 0.99 and the standard deviation of the errors was 0.37.

Thermal Analysis

Alonso *et al.*[13] reported that the principal thermal effect observed on the differential thermal gravimetric analysis of chitin/chitosan is dependent on the molecular weight and the degree of acetylation of the polymer. For samples with a degree of acetylation greater than 80% there is a linear relationship with a correlation coefficient of 0.98 between the percentage of acetyl groups and the percentage weight loss at 320°. For samples with a degree of acetylation less than 80%, the percentage weight loss at 280° is used and the correlation coefficient is 0.998. Their experiments were performed in a derivatograph using 60 mg of sample and a heating rate of 10°/min in an air atmosphere.

Gas Chromatography

Muzzarelli *et al.*[8] have reported that the retention time of methanol on a column composed of chitin/chitosan is proportional to the degree of acetylation, presumably because chitin forms a complex with methanol.[14] The columns (2 m × 4 mm i.d.) were conditioned at 125° with helium; the operating conditions were as follows: thermostat, 125°; detector, 170°; vaporizer, 225°; helium carrier flow, 20 ml/min; volume of injected solvent, 1 μl.

[13] G. Alonso, C. Peniche-Covas, and J. M. Nieto, *J. Therm. Anal.* **28**, 189 (1983).
[14] P. R. Austin, U.S. Patent 4,063,016 (1977).

[53] Determination of Molecular-Weight Distribution of Chitosan by High-Performance Liquid Chromatography

By ARNOLD C. M. WU

Introduction

Harsh chemical processes to convert natural raw materials into chitosan cause the product to consist of polymers with a range of molecular weights (MW). Among the many methods for measuring molecular weight distribution (MWD), the high-performance liquid chromatography (HPLC) described here offers speed, dependability, and convenience, once its conditions and calibration system are determined.

In this chapter, the procedure to derive the optimal conditions of HPLC for measuring chitosan[1] is described. Its application for measuring MWD of different chitosan samples is also presented.

HPLC System

Components purchased from Waters Associates (Milford, MA) include an aqueous sample clarification kit, stainless steel columns ($\frac{1}{8}$-in. o.d. × 2 ft, with 5-μm sintered disks in end caps), series ALC/GPC-244 HPLC system with M440 UV detector, R-401 refractive index (RI) detector, M 6000 A solvent delivery system, and U6K septumless sample injector.

An OmniScribe recorder (Houston Instruments, Austin, TX) and an Autolab Minigrator (Spectra Physics, Santa Clara, CA) are used for recording and integrating the chromatograms. A compact computer, Compucorp Alpha 325 Scientist (Compucorp, Los Angeles, CA), is used for calculating MWD values. The fractionated RI signals could be quantified and overall MWD given automatically if the Minigrator output is interfaced with any suitable PC having a software program to read and calculate the MWD.

Packing material for HPLC columns is Corning's controlled pore glass supports having a covalently bonded glycerol coating (Glycophase-G/CPG) to inactivate free silicic acid groups on the surface of the glass particles. Glycophase-G/CPG supports (pore sizes 40, 100, 250, 550, 1500, and 2500 Å) are purchased from Pierce (Rockford, IL) except for the 2500 Å material which is supplied by Corning Glass (Medfield, MA). The particle size of all the glass packing materials is 37–74 μm (200–400

[1] A. C. M. Wu, W. A. Bough, E. C. Conrad, and K. E. Alden, Jr., *J. Chromatogr.* **128,** 87 (1976).

mesh). Glycophase-G/CPG is pretreated as follows: each bottle of porous glass (about 30 ml) is washed with 300 ml of tetrahydrofuran by shaking in a 500-ml Erlenmeyer flask for 1 min, allowed to settle for 15 min, and the supernatant decanted; 300 ml of distilled water is added, shaken, and the water layer again decanted. The procedure is repeated a third time with another 300 ml of tetrahydrofuran. After decanting the supernatant, the slurry is dried in an oven overnight at 70°. It is then packed into 2-ft stainless steel columns.

The manufacturer's procedures for operating the HPLC should be followed. Solvent flow rate is 1 ml/min unless otherwise specified. Acetic acid (2%) is used as eluent in the solvent delivery system. The recorded speed is set at 1 cm/min. The Autolab Minigrator is in the "simulated distillation mode" in order to obtain peak heights at 10-sec intervals. Data from the RI detector are used to determine elution volumes and molecular weight distributions in this study. The UV detector is used to assist in measuring the separation of different chitosan components. A 254-nm filter is used in the UV light source with the UV sensitivity set at 0.05 AUFS. The RI attenuation is set at 2× to 32× depending on the sample load with the preferred setting at 8× or greater to minimize signal noise. A 100-μl syringe is used for 50- to 100-μl injections; injections of other sizes are made with 25-μl and 1-ml syringes. Sections of 0.009-in. i.d. tubing are used for connecting different columns. The most suitable injection size for dextran standards was determined to be 50 μl at 5 g/liter. The optimal injection load of chitosan sample is studied by comparing the elution volume of the major chitosan component (peak) at various concentrations (0.625–30 g/liters) and injection volumes. Columns are daily preconditioned by injecting 2 mg total load (i.e., two injections of 100 μl at 10 g/liter concentration) of chitosan in order to deactivate any residual active sites present in Glycophase-G/CPG.

Chitosan Samples and Dextran Standards

All samples or standards are first dissolved in 2% reagent grade acetic acid and filtered through an aqueous sample clarification kit with MF-Millipore filter (pore size 0.45 μm). Chitosan samples are prefiltered through a Gelman glass fiber filter (type E) before passing through the Millipore filter kit.

Dextran standards used (T series) are obtained from Pharmacia (Piscataway, NJ) and range in molecular weights from 10,000 (T-10) to 2,000,000 (T-2000) with defined weight average MWs ($\overline{M}_w$) and number average MWs ($\overline{M}_n$)obtained from light scattering and Sephadex gel filtration methods, respectively, for all except T-2000. Chitosan, batch 4-74, was supplied by Food, Chemical and Research Laboratories (Seattle, WA).

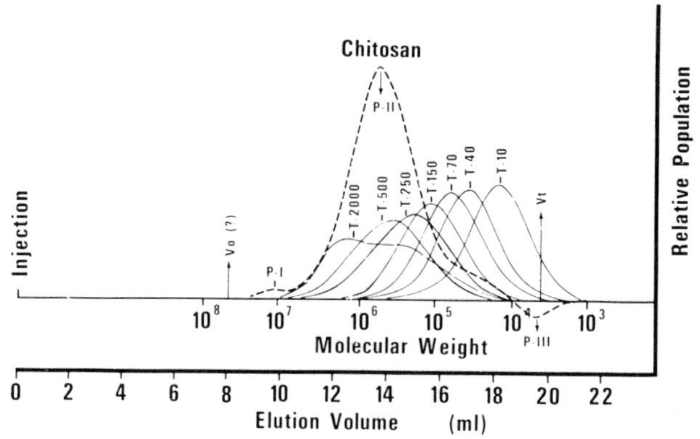

FIG. 1. Refractive index response curves illustrating elution patterns and MW distributions of dextran standards (T-10 through T-2000) and chitosan (4–74) fractionated by HPLC with the optimal column combination described in this chapter.[1]

Calibration Methods Using Dextran Standards

The calibration curves of dextran standards are plotted based on the peak elution volumes of standards vs the logarithms of the weight average, $\ln \overline{M}_w$; arithmetic average of $\overline{M}_w$ and $\overline{M}_n$, $\ln([\overline{M}_w + \overline{M}_n]/2)$; or the geometric average $(\ln \overline{M}_w + \ln \overline{M}_n)/2$ of each standard. An equation describing the linear region of the calibration curves is given as $\ln(M_i) = a + bV_e$ (where M_i = MW of species i, V_e = elution volume). The intercept value, a, and slope value, b, are calculated from the linear regression equation describing the best fit of the elution volumes of standards treated as unknowns. The molecular weights of the chitosan samples used in this study are assumed to be within the linear range of the calibration curve, because the 2500-Å column would extend the void volume to correspond to MW values of approximately 40 million.[2,3] A void volume marker in this range is not available. With the slope and intercept of the calibration curves, and with the heights taken at fixed 10-sec intervals by the Minigrator, the $\overline{M}_w$ and $\overline{M}_n$ of each sample can be calculated from different elution volumes.

The equations for calculating $\overline{M}_w$ and $\overline{M}_n$ are as follows:

$$\overline{M}_w = \Sigma\,(h_i M_i)/\Sigma\,h_i, \qquad \overline{M}_n = \Sigma\,h_i/\Sigma\,(h_i/M_i)$$

where $M_i = \exp(a + bV_e)$, h_i is the height at each 10-sec interval of the chromatogram peak.

[2] G. L. Beyer, A. L. Spartorico, and T. L. Bronson, *in* "Water-Soluble Polymers" (N. M. Bikales, ed.), p. 315. Plenum, New York, 1973.

[3] W. A. Dark and R. J. Limpert, "Evaluation of Available Packings for GPC," Bulletin No. TR 914. Waters Assoc., Milford, Massachusetts, 1973.

At the optimal column combination described in this chapter, $a = 23.9215$, $b = -0.0131$, and V_e is in one-sixtieth of a milliliter (or seconds under 1 ml/min flow rate). These parameters are obtained by plotting the V_e obtained vs geometric average MWs of dextran standards supplied by Pharmacia, which is $\exp[(\ln \overline{M}_w + \ln \overline{M}_n)/2]$.

In using the calibration curve to estimate chitosan MWs based on the dextran standards, MW–MW correlations are used rather than size–size correlations, because the calibration factors (Q, MW U/Å)[4] for the dextran standards as well as chitosan are not available. It is assumed that the calibration forces for dextran standards and chitosan are approximately equal because of their similarity in structure and the difference between them is within the range of precision obtainable by HPLC determination of MW distribution.

Optimal HPLC Conditions

By varying only the column configuration, different calibration equations as well as different elution patterns of chitosan result. By comparing the separation of three RI peaks of chitosan (Fig. 1), the performance of various column combinations can be charted (Fig. 2); the best combination is 2500 Å (1 × 2 ft), 1500 Å (2 × 2 ft), 550 Å (3 × 2 ft), 250 Å (1 × 2 ft), 100 Å (1 × 2 ft), and 40 Å (1 × 2 ft).

When column efficency, $N(=16\ V_e/\text{peak width})$, is used to determine the optimal sample injection size, a total load of 500 μg is optimal, with the consideration that the RI signal is also satisfactory at this injection load. The optimal injection size is either 100 μl at 5 g/liter or 50 μl at 10 g/liter concentration.

Molecular Weight Distribution of Various Chitosan Samples

With the optimal conditions found in this study, the MWD of commercial chitosan sample 4–74 was $\overline{M}_w$ 2,055,000, $\overline{M}_n$ 936,000, dispersity 2.16, and the MW of the highest peak corresponding to the most abundant species in this sample was 1,103,000. The precision (percentage errors) values for estimating the peak MW, $\overline{M}_w$, $\overline{M}_n$, and dispersity of the chitosan solution were 2.01, 1.34, 2.48, and 1.43, respectively. The V_e at peak of the chromatogram obtained for dextran T-250 was repeatable at 15.2 ml, which corresponded to a peak MW of 156,800. Percentage errors for estimating the peak MW, $\overline{M}_w$, $\overline{M}_n$, and dispersity of dextran T-250 were

[4] "Calibration of GPC Systems," Bulletin No. D5074. Waters Assoc., Milford, Massachusetts, 1974.

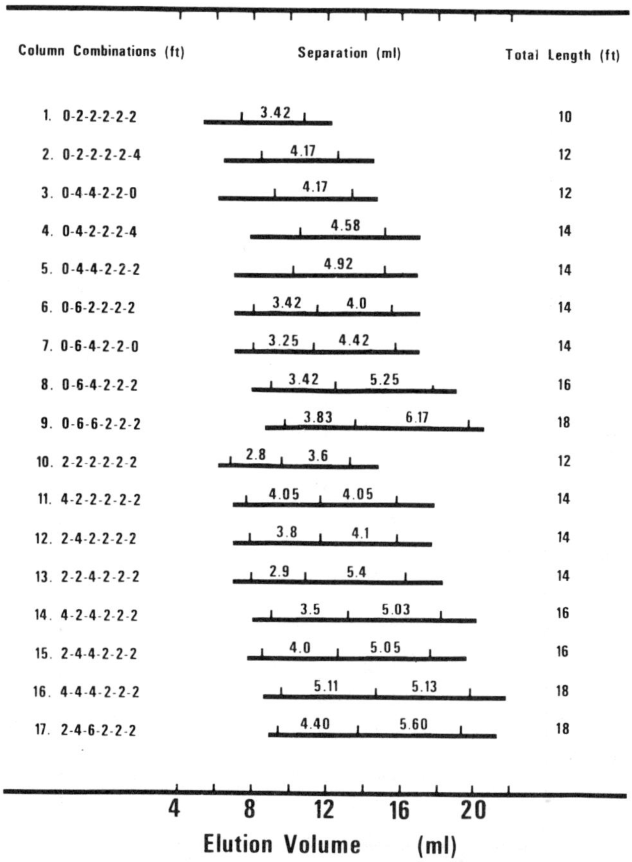

FIG. 2. Comparison of the efficiency of HPLC with different column combinations for separating different MW components in a chitosan solution. Each line represents the location on the HPLC chromatograms of RI peaks (PI, PII, and PIII), and the separation between peaks is indicated (milliliters).[1]

0.0, 0.73, 4.7, and 4.82, respectively. The V_e of dextran T-40 was at 16.8 ml, which corresponded to a peak MW of 36,300. Percentage errors for estimating the $\overline{M}_w$, $\overline{M}_n$, and dispersity of T-40 were 0.82, 1.23, 2.70, and 2.76, respectively. The sources of error included (1) the baseline drift of the RI detector, (2) the performance of the injector, the pump unit, and columns, (3) the precision in determining the calibration curve, (4) the injection size, and (5) the uniformity of the sample solution.

In a separate study,[5] shrimp shells were converted to chitosan with

[5] W. A. Bough, A. C. M. Wu, and W. B. Miller, Tech. Rep. No. 77-7. Georgia Marine Science Center, University System of Georgia, Skidaway Island, Georgia, 1977.

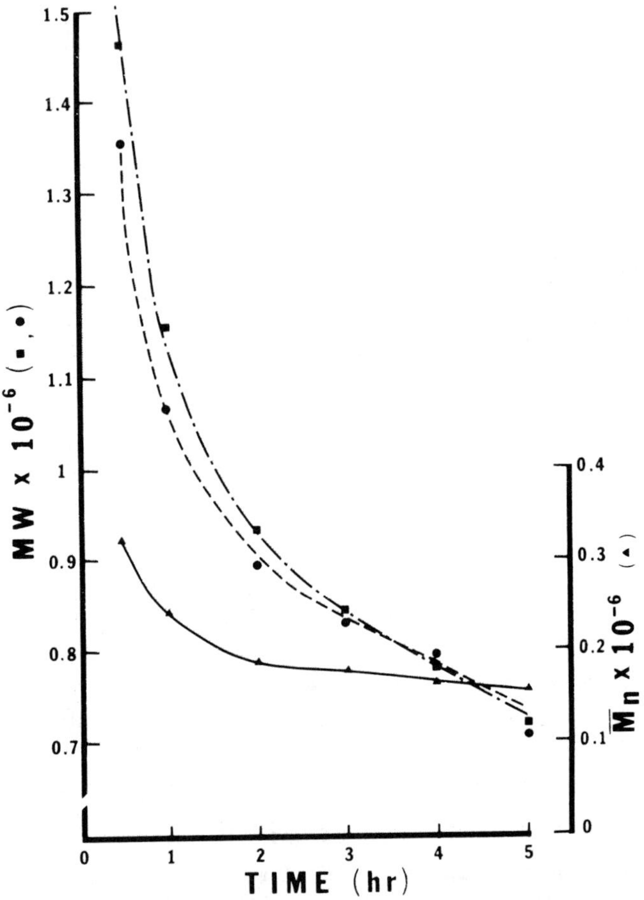

FIG. 3. Molecular weight distributions of chitosan products with different times of deacetylation. The solvent was 2% acetic acid with 0.1 *M* sodium acetate. Values were averages of three replications. $\overline{M}_w$ (squares) and peak molecular weight (circles) refer to the left *y* axis; $\overline{M}_n$ (triangles) refer to the right *y* axis.[5]

various times of deacetylation in 50% NaOH solution, ranging from 0.5 to 5 hr. Their MWD, determined by the same method described in this chapter showed that $\overline{M}_w$ ranged from 1,487,000 (0.5 hr) to 667,000 (5 hr). $\overline{M}_n$ ranged from 322,000 to 139,000. The peak MW ranged from 1,413,000 to 625,000 (Fig. 3).

[54] Analysis of Chitooligosaccharides and Reduced Chitooligosaccharides by High-Performance Liquid Chromatography

By AKIRA OHTAKARA and MASARU MITSUTOMI

Chitooligosaccharides and reduced chitooligosaccharides can be hydrolyzed with chitinase[1] and β-*N*-acetylhexosaminidase[2] from *Pycnoporus cinnabarinus.* These oligosaccharides are rapidly separated by reversed-phase high-performance liquid chromatography (HPLC) on amine-modified silica columns. A technique for separating chitooligosaccharides using HPLC has been recently applied to investigate the lysozyme-catalyzed hydrolysis and transglycosylation,[3,4] and the action pattern of endo- and exochitinase from *Munduca sexta.*[5,6] This chapter describes a rapid and sensitive procedure for simultaneous determination of chitooligosaccharides and reduced chitooligosaccharides by HPLC and its application for studies on the mode of action of chitinase and β-*N*-acetylhexosaminidase.

Simultaneous Determination of Chitooligosaccharides and Reduced Chitooligosaccharides

Apparatus for HPLC. The following apparatus is used in our laboratory, but other comparable apparatus can be also used.

High pressure pump and injector: A Waters Associates model 6000A pump coupled to a U6K universal injector, Waters Associates, Inc.
UV detector: UV spectrophotometer UVIDEC-100-II, Japan Spectroscopic Co., Ltd.
Columns: μBondapak CH (3.9 × 300 mm), Waters Associates, Inc.; μBondapak NH_2 (3.9 × 300 mm) and Radial-PAK μBondapak NH_2 (8.0 × 100 mm) packed within Radial Compression Separation System, Z module, can be also used

[1] A. Ohtakara and M. Mitsutomi, *Proc. Int. Conf. Chitin/Chitosan, 2nd,* p. 117 (1982); see also this volume [57].
[2] A. Ohtakara, M. Mitsutomi, and E. Nakamae, *Agric. Biol. Chem.* **46,** 293 (1982); see also this volume [57].
[3] P. van Eikeren and H. McLaughlin, *Anal. Biochem.* **77,** 513 (1977).
[4] A. Masaki, T. Fukamizo, A. Otakara, T. Torikata, K. Hayashi, and T. Imoto, *J. Biochem. (Tokyo)* **90,** 527 (1981).
[5] D. Koga, M. S. Mai, C. D. Turner, and K. J. Kramer, *Insect Biochem.* **12,** 493 (1982).
[6] D. Koga, J. Jilka, and K. J. Kramer, *Insect Biochem.* **13,** 295 (1983).

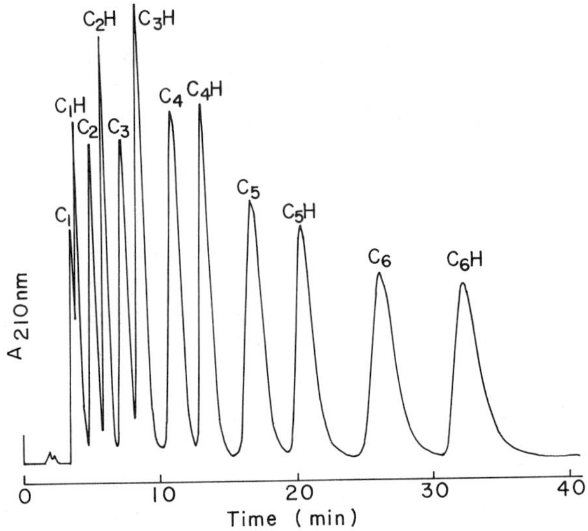

FIG. 1. Separation of chitooligosaccharides and reduced chitooligosaccharides by HPLC. Column, μBondapak CH (3.9 × 300 mm); solvent, acetonitrile–water (75:25, v/v); flow rate, 1.5 ml/min; detector, UV at 210 nm; standard sugars, 25 nmol; C_1, *N*-acetylglucosamine; C_2–C_6, $(GlcNAc)_2$–$(GlcNAc)_6$; C_1H, *N*-acetylglucosaminitol; C_2H–C_6H, reduced $(GlcNAc)_2$–reduced $(GlcNAc)_6$.

Guard column: Bondapak AX/Corasil (37–53 μm), Waters Associates, Inc.

Solvent. A 75:25 (v/v) mixture of acetonitrile and water is used as solvent. Both the reagents are special grade for HPLC. The solvent mixture is degassed by ultrasonic vibration prior to use.

Standard Sugars. Chitooligosaccharides, $(GlcNAc)_n$, are prepared by acid hydrolysis of chitin according to the method of Rupley,[7] followed by gel filtration on BioGel P-4 (Bio-Rad). *N*-Acetylglucosaminitol and reduced chitooligosaccharides, reduced $(GlcNAc)_n$, are prepared by reduction of *N*-acetylglucosamine and corresponding chitooligosaccharides with sodium borohydride and purified by repeated gel filtration on BioGel P-4. The purity of chitooligosaccharides and reduced chitooligosaccharides is ascertained to be above 99% by HPLC using UV detection.

Procedure. Samples (10–25 μl) containing an unknown quantity of chitooligosaccharides and reduced chitooligosaccharides (each 5–70 nmol) are applied to a μBondapak CH column (3.9 × 300 mm) after filtration with a 0.45-μm Millipore filter. The oligosaccharides are eluted with a

[7] J. A. Rupley, *Biochim. Biophys. Acta* **83**, 245 (1964).

75:25 (v/v) mixture of acetonitrile and water, at a flow rate of 1.5 ml/min and the absorbance at 210 nm is monitored using UV spectrophotometer UVIDEC-100-II with 0.16 AUFS. The mixture of standard sugars (20–25 nmol) is also applied to HPLC under the same condition.

A typical elution pattern of chitooligosaccharides and reduced chitooligosaccharides on a μBondapak CH column is shown in Fig. 1. Elution patterns similar to this figure are given by HPLC of these oligosaccharides on μBondapak NH_2 and Radial-PAK μBondapak NH_2 columns. The complete resolution of *N*-acetylglucosamine and *N*-acetylglucosaminitol is unsuccessful in all of the columns. When various amounts of standard sugars are injected, a linear relationship exists between the peak area for each sugar and the injected amount of the sugar. Figure 2 shows standard curves against di-, tri-, and tetrasaccharides. The amount of oligosaccharides in samples can be determined by integrating the peak area of each sugar and converting it to the amount of the sugar with standard curves.

Comments. The column used for separation suffers considerable deterioration throughout continuous use. As a result of this, the capacity factor of the column is gradually reduced and each sugar is more rapidly eluted than before. The resolution among sugars in such a column can be restored by increasing the concentration of acetonitrile in the solvent. Therefore, it is necessary for the determination of oligosaccharides that the mixture of standard sugars should be injected in parallel with sample. The resolution of α and β anomers of chitooligosaccharides has been observed in a Shodex

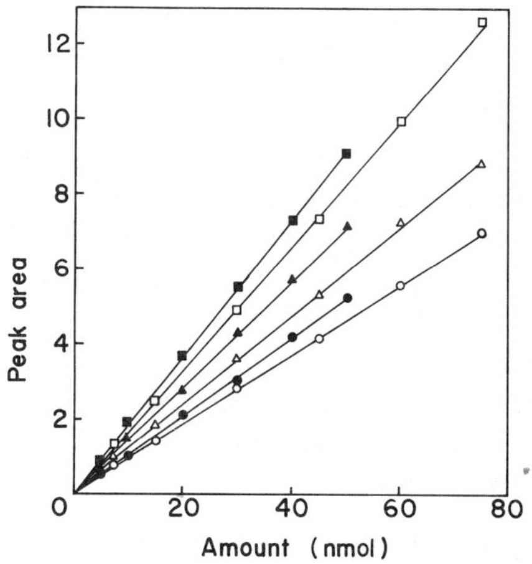

FIG. 2. Standard curves for chitooligosaccharides and reduced chitooligosaccharides. ○, $(GlcNAc)_2$; △, $(GlcNAc)_3$; □, $(GlcNAc)_4$; ●, reduced $(GlcNAc)_2$; ▲, reduced $(GlcNAc)_3$; ■, reduced $(GlcNAc)_4$.

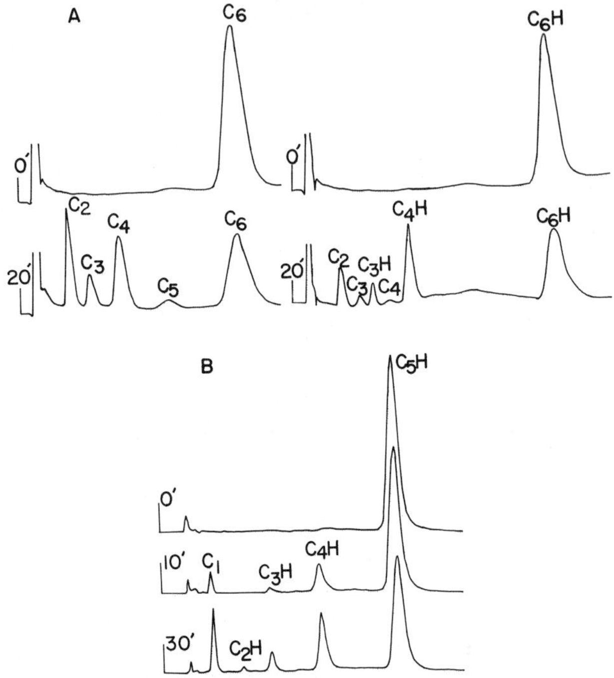

FIG. 3. Separation of the products produced in the hydrolysis of $(GlcNAc)_6$ and reduced $(GlcNAc)_6$ with chitinase (A) and that of reduced $(GlcNAc)_5$ with β-*N*-acetylhexosaminidase (B). The reaction mixture, consisting of 0.2 ml of 5 m*M* substrates, 0.2 ml of 0.05 *M* phosphate buffer (pH 5.5), and 0.1 ml of chitinase (0.046 U) or β-*N*-acetylhexosaminidase (0.06 U) purified from *Pycnoporus cinnabarinus,* is incubated at 37°. A 20-μl sample of the reaction mixture (corresponding to 20 nmol of the substrate) is injected. HPLC is done under the same conditions as in Fig. 1.

Ionpak S-614 column (Showa Denko Co.),[8] and reversed-phase columns,[9] such as model 600-RP (Alltech Associates), Partisil PXS 5/25 ODS (Whatman), and Radial-PAK C_{18} and C_8 (Waters Associates). However, chitooligosaccharides cannot be separated into α and β anomers by HPLC on the amine-modified silica columns mentioned above.

Mode of Hydrolysis of Chitooligosaccharides and Reduced Chitooligosaccharides with Chitinase and β-*N*-Acetylhexosaminidase

Procedure. The reaction mixture, consisting of 0.2 ml of 5 m*M* oligosaccharides, 0.2 ml of 0.05 *M* phosphate buffer of pH 5.5, and 0.1 ml of

[8] T. Fukamizo and K. Hayashi, *J. Biochem. (Tokyo)* **91,** 619 (1982).

[9] K. Blumberg, F. Liniere, L. Pustilnik, and C. A. Bush, *Anal. Biochem.* **119,** 407 (1982).

enzyme solution, is incubated at 37°. After the reaction is stopped by boiling for 3 min, the solution is filtered through a 0.45-μm Millipore filter and 10–25 μl of the solution is subjected to HPLC under the same conditions as described above. The products are identified by comparison with the elution pattern of standard sugars by HPLC under the same condition. The amount of the products can be calculated from the peak area according to the procedure described above.

Application. Application of this procedure for studies on the mode of action of chitinase[1] and β-*N*-acetylhexosaminidase[2] purified from *Pycnoporus cinnabarinus* is described.

Separation of the products produced in the hydrolysis of $(GlcNAc)_6$ and reduced $(GlcNAc)_6$ with chitinase is shown in Fig. 3A. The chitinase hydrolyzes $(GlcNAc)_6$ to $(GlcNAc)_4$ and $(GlcNAc)_2$ faster than to $(GlcNAc)_3$, and also reduced $(GlcNAc)_6$ to reduced $(GlcNAc)_4$ and $(GlcNAc)_2$ faster than to reduced $(GlcNAc)_3$ and $(GlcNAc)_3$. Therefore, the chitinase acts on $(GlcNAc)_6$ and reduced $(GlcNAc)_6$ with the mode of action of an endo-type enzyme, predominantly splitting the second β-*N*-acetylglucosaminide bonds from the nonreducing ends. In addition, a small amount of $(GlcNAc)_5$ and $(GlcNAc)_4$ is detected in the hydrolysis of $(GlcNAc)_6$ and reduced $(GlcNAc)_6$ without the formation of *N*-acetylglucosamine and reduced $(GlcNAc)_2$, respectively. This suggests that these oligosaccharides may be produced by transglycosylation catalyzed by the chitinase.

Figure 3B shows the products produced in the hydrolysis of reduced $(GlcNAc)_5$ with β-*N*-acetylhexosaminidase. This result indicates that the β-*N*-acetylhexosaminidase is an exo-type enzyme acting on reduced $(GlcNAc)_5$ at the nonreducing end to release *N*-acetylglucosamine.

[55] Enzymatic Determination of Chitin

By ENRICO CABIB and ADRIANA SBURLATI

The determination of chitin in biological materials is fraught with problems. There is no direct colorimetric reaction for hexosamines, comparable to the anthrone reaction for hexoses, that can be applied directly to polysaccharides. Complete chemical hydrolysis of chitin requires prolonged treatment with high concentrations of acid that lead to losses in hexosamines. Furthermore, this method is not specific; it liberates hexosamines from glycoproteins as well as chitin. Enzymatic hydrolysis with chitinases has the desired specificity but requires a chitinase that efficiently attacks native chitin. The enzyme from *Serratia marcescens*[1] is adequate for this purpose. An additional problem is caused by the masking of chitin with other materials that prevent access of the chitinase to the substrate (e.g., glucans and mannoproteins in fungal cell walls). Thus, additional steps may be required to uncover the chitin. The procedure presented here has worked well with yeast cell walls but will probably require modifications for other materials.

Assay Method

Principle. Chitin in intact yeast cells or in isolated cell walls is digested with chitinase in the presence of a cell wall lytic preparation (snail gut extract). The enzymes in the snail extract have the double purpose of exposing to chitinase the chitin buried in inner layers of the cell wall and of hydrolyzing the diacetylchitobiose formed by the chitinase to *N*-acetylglucosamine (the snail extract is a good source of *β*-*N*-acetylhexosaminidase). The *N*-acetylglucosamine formed is determined according to Reissig and Leloir.[2]

Reagents

Phosphate buffer, pH 6.3, 1 *M*

Snail gut extract: either Glusulase from DuPont Pharmaceutical, Inc., Wilmington, Delaware, or Cytohelicase from Industrie Biologique Française, Villeneuve-La-Garenne, France. The snail extract is diluted and filtered through Sephadex G-25 as described,[3] which results in an overall 3-fold dilution with respect to the original material

Serratia marcescens chitinase[1]

Reagents for the determinations of *N*-acetylglucosamine, according to Reissig and Leloir[2]

Determination of Chitin in Intact Cells. An aqueous suspension containing 0.6 g of *Saccharomyces cerevisiae* cells (wet weight) per milliliter is prepared. To 60 μl of cell suspension, 60 μl of snail gut extract and 60 μl of *Serratia* chitinase (110 mU) are added. After incubation for 4.5 hr, *N*-acetylglucosamine is measured in a 25-μl aliquot of the digest by the Reissig and Leloir procedure.[2]

Determination of Chitin in Cell Walls. One milliliter of an *S. cerevisiae* suspension containing 0.3 g of cells (wet weight) per milliliter is added to

[1] E. Cabib, this volume [56].

[2] J. L. Reissig and L. F. Leloir, this series, Vol. 8, p. 175.

[3] E. Cabib, this series, Vol. 11, p. 120.

1.5 g of glass beads (0.45–0.5 mm diameter, B. Braun, Melsungen, West Germany) in the chamber of a Mini-Beadbeater, (Biospec Products, Bartlesville, OK). The cell suspension is agitated for 6 min. After removing the broken cell suspension from the glass beads by aspiration with a Pasteur pipet, the beads are washed with several small portions of water. The washings are pooled with the initial extract and centrifuged for 10 min at 4000 *g*. The cell wall pellet is washed three times with distilled water and finally suspended in water to a final volume of 0.5 ml. Aliquots of 30 and 60 μl are digested with chitinase and snail gut extract, and *N*-acetylglucosamine is determined as described in the preceding section.

Because of the possible losses in the preparation of cell walls, it may be desirable to refer the chitin content to the amount of another cell wall component, such as total hexose (glucan + mannan for *S. cerevisiae*). In that case, hexoses can be determined in an aliquot of the cell wall suspension by the anthrone reaction.[4]

Comments. Yeast cell walls may be exposed to the action of chitinase also by heating a suspension of intact cells for 1 hr at 100° in 0.5 *M* NaOH, followed by centrifugation and extensive washing of the treated cells with water. A source of β-*N*-acetylhexosaminidase will still be required to convert all of the chitinase hydrolysis products into free *N*-acetylglucosamine.

With isolated cell walls whose chitin had been specifically labeled with ^{14}C *in vivo* it was possible to solubilize all of the radioactive chitin by prolonged incubation with chitinase only.[5]

Although similar results are obtained for chitin with either intact cells or cell walls, both procedures are given above, because the second one is more appropriate to study composition of cell walls.

The application of the described method to other fungi may require modifications in order to assure accessibility of the cell wall chitin to the hydrolase. In general, an enzymatic preparation (several are commercially available) that converts cells of the organism under study into spheroplasts should be adequate to complement chitinase action. Because of the great variability in the cell wall composition of fungi, no universally adequate preparation or treatment of the cells can be recommended.

[4] R. G. Spiro, this series, Vol. 8, p. 3.

[5] R. L. Roberts and E. Cabib, *Anal. Biochem.* **127**, 402 (1982).

[56] Chitinase from *Serratia marcescens*

By ENRICO CABIB

Serratia marcescens, grown on chitin as a carbon source, secretes into the medium a chitinase of high specific activity that can be purified sufficiently in a single step for use in the enzymatic determination of chitin.[1,2] Because of its stability, wide pH optimum, and linear kinetics over a greater range than that of other chitinases, the enzyme is well suited for analytical use. For the same reasons, it is also a good candidate for the large-scale degradation of chitin for industrial purposes.

Assay Method

The chitinase is assayed by the liberation of tritiated oligosaccharides from [*acetyl*-^{3}H]chitin,[3] with phosphate buffer at pH 6.3, at a final concentration of 0.05 *M* in the reaction mixture (buffer A). A unit of chitinase is that amount of enzyme which catalyzes the release of 1 μmol of soluble product (calculated as *N*-acetylglucosamine) in 1 min at 30°.

Purification Procedure

Principle. Chitinase is purified by the procedure introduced by Jeuniaux,[4] which consists in adsorbing specifically the enzyme on chitin. The chitin–chitinase complex is then incubated at 30° until chitin is digested and the purified enzyme is released into solution.

Growth of S. marcescens. Serratia marcescens QMB 1466[5] is maintained on nutrient agar (Difco) slants, from which it is transferred to YEPD broth medium [1% yeast extract (Difco), 2% peptone (Difco), and 2% glucose] and grown overnight at 30°.

For production of chitinase, flasks containing 800 ml of chitin medium each are inoculated with the broth culture to obtain an initial concentration of 1×10^7 cells/ml. The chitin medium[5] contains, in grams per liter: chitin (practical grade crab shell chitin powder, from Sigma Chemical Co., St. Louis, MO), 15; yeast extract, 0.5; $(NH_4)_2SO_4$, 1.0; $MgSO_4 \cdot 7H_2O$, 0.3;

[1] R. L. Roberts and E. Cabib, *Anal. Biochem.* **127**, 402 (1982).
[2] E. Cabib and A. Sburlati, this volume [55].
[3] E. Cabib, this volume [48].
[4] C. Jeuniaux, this series, Vol. 8, p. 644.
[5] J. Monreal and E. T. Reese, *Can. J. Microbiol.* **15**, 689 (1969).

KH_2PO_4, 1.36. The pH is adjusted to 8.5 with NaOH before autoclaving. The flasks are incubated on a shaker at 30° and the absorbance at 660 nm is monitored. Because the culture contains particles of chitin, prior to the determination of absorbance, a sample is passed through a coarse-grade sintered glass filter which retains the chitin but not the bacteria. Chitinase activity is also assayed in the medium after centrifuging off particulate material and cells at 27,000 *g* for 10 min.

Chitinase activity increases rapidly after the fourth day and peaks after about 10 days, but the cultures are harvested after 8 days to minimize proteolytic degradation of chitinase. Cells and residual chitin are removed from the culture by centrifugation and small particles by filtration through a 1.2 μm pore-size nitrocellulose filter. The resultant particle-free filtrate is concentrated about 3-fold in a Millipore Pellicon unit or similar device, equipped with two membrane packets with an exclusion limit at a molecular weight of 10,000. The volume of the concentrated filtrate is reduced an additional 15 times in an Amicon pressure cell with a YM10 filter.

Affinity Absorption–Desorption on Chitin. Regenerated chitin (reacetylated chitosan[6]) is added to the concentrated filtrate in the amount of 1 mg (dry weight)/770 mU of chitinase. The suspension is allowed to stand for 1 hr on ice and centrifuged at 27,000 *g* for 15 min. The pellet is washed with twice the original volume of buffer A, then suspended up to the original volume in the same buffer and incubated with agitation at 30° for about 1 hr or until the solution is almost clear. The remaining particulate material is removed by centrifugation and the supernatant fluid is dialyzed overnight against buffer A to remove the reaction products. The purification is illustrated in Table I.

Properties

Stability. The enzyme is stable for at least 4 years when stored at −20°.

Physical Characteristics. Upon electrophoresis on nondenaturing polyacrylamide gels the purified preparation yields two closely moving bands, both endowed with chitinase activity. After electrophoresis on sodium dodecyl sulfate-polyacrylamide gels, two major bands are also observed, with apparent molecular weights of 58,000 and 52,500. Minor bands at 40,400 and 21,500 are also detected.

Polysaccharide and Proteolytic Activities of the Preparation. No activity of $\beta(1 \rightarrow 6)$-glucanase, $\beta(1 \rightarrow 3)$-glucanase, mannanase, amylase, or cellulase can be detected. The preparation contains a small amount of β-*N*-acetylglucosaminidase activity (about 2 to 3% of the chitinase activity). The

[6] Obtained as described but using unlabeled acetic anhydride.[3]

TABLE I
PURIFICATION OF *Serratia marcescens* CHITINASE

Step	Volume[a] (ml)	Total protein (mg)	Total activity (mU)	Specific activity (mU/mg protein)	Purification (-fold)	Recovery (%)	Chitinase chitobia ratio
Culture filtrate, after concentration	150	4439	1.01×10^6	2.28×10^2	—	—	7.1
Chitin digest	150	207	2.76×10^5	1.33×10^3	5.9	27.4	41.5

[a] Volume of culture filtrate before concentration was about 7.3 liters.

proportion of β-*N*-acetylglucosaminidase activity is much higher in the crude culture filtrate (see Table I). The purified chitinase contains a small amount of protease activity which, however, is undetectable at pH values below 5.

Kinetics and Reaction Product. The time course of the reaction is linear until about 20% of the substrate is consumed. The K_m for reacetylated chitosan is 4.4 m*M* (expressed as *N*-acetylglucosamine equivalent).

The chitinase has a very broad pH optimum, between pH 4 and 7.

The major reaction product is diacetylchitobiose, with some trisaccharide in the early stages of chitin digestion. Because of the presence of contaminating β-*N*-acetyhexosaminidase activity, the disaccharide is gradually degraded to free *N*-acetylglucosamine upon long incubations.

[57] Chitinase and β-*N*-Acetylhexosaminidase from *Pycnoporus cinnabarinus*

By AKIRA OHTAKARA

The microbial chitinases available commercially have been purified from *Streptomyces griseus,*[1] *Streptomyces antibioticus,*[2] *Serratia marcescens,*[3,4] and *Aeromonas hydrophila* subsp. *anaerogenes* A 52.[5] In addition

[1] L. R. Berger and D. M. Reynolds, *Biochim. Biophys. Acta* **29,** 522 (1958).
[2] C. Jeuniaux, this series, Vol. 8, p. 644.
[3] J. Monreal and E. T. Reese, *Can. J. Microbiol.* **15,** 689 (1969).
[4] R. L. Roberts and E. Cabib, *Anal. Biochem.* **127,** 402 (1982).
[5] M. Yabuki, Y. Yabuki, E. Yanai, A. Ando, and T. Fujii, *Tech. Bull. Fac. Hortic. Chiba Univ.* **32,** 51 (1983).

to chitinase (EC 3.2.1.14), *N*-acetyl-β-glucosaminidase (EC 3.2.1.30) or β-*N*-acetylhexosaminidase (EC 3.2.1.52) is required for the complete hydrolysis of chitin. *Pycnoporus cinnabarinus*, a strain of the chitinase-producing fungi[6] belonging to the basidiomycetes, has produced the chitinolytic enzymes consisting of chitinase[7] and β-*N*-acetylhexosaminidase.[8] Procedures for the purification of both enzymes will be described.

Purification of β-*N*-Acetylhexosaminidase[8]

Assay Methods

p-Nitrophenyl-β-N-acetylhexosaminides as Substrate: Principle. *p*-Nitrophenyl-β-*N*-acetylglucosaminide and β-*N*-acetylgalactosaminide are used as substrates for β-*N*-acetylhexosaminidase. The yellow color of the *p*-nitrophenol released is estimated under alkaline condition in a spectrophotometer at 420 nm.

Reagents

p-Nitrophenyl-β-*N*-acetylglucosaminide (*p*-NPGlcNAc), 4 m*M*
p-Nitrophenyl-β-*N*-acetylgalactosaminide (*p*-NPGalNAc), 2 m*M*
Sodium citrate–HCl buffer, 0.1 *M*, pH 2.2 and 3.7
Sodium carbonate, 0.2 *M*

Procedure. *N*-Acetyl-β-glucosaminidase (β-GlcNAcase) activity is assayed as follows. The reaction mixture, consisting of 0.2 ml of 4 m*M* *p*-NPGlcNAc, 0.2 ml of 0.1 *M* sodium citrate–HCl buffer, pH 2.2, and 0.1 ml of enzyme solution, is incubated for 10 min at 37°. The reaction is stopped by adding 2 ml of 0.2 *M* sodium carbonate, and the yellow color of the *p*-nitrophenol released is estimated by reading the absorbance at 420 nm. β-*N*-Acetylgalactosaminidase (β-GalNAcase) activity is assayed in the same manner except that 2 m*M* *p*-NPGalNAc as substrate and 0.1 *M* sodium citrate–HCl buffer, pH 3.7, are used in the above reaction mixture.

Unit. One unit of the enzyme activities is defined as the amount of the enzyme which releases 1 μmol of *p*-nitrophenol/min under the above conditions.

N,N-Diacetylchitobiose as Substrate: Principle. *N,N*-Diacetylchitobiose is used as substrate for β-*N*-acetylhexosaminidase. *N*-Acetyglucosamine

[6] M. Kawai, *Nippon Nogei Kagaku Kaishi* **47**, 473 (1973).
[7] A. Ohtakara and M. Mitsutomi, *Proc. Int. Conf. Chitin/Chitosan, 2nd,* p. 117 (1982).
[8] A. Ohtakara, M. Yoshida, M. Murakami, and T. Izumi, *Agric. Biol. Chem.* **45**, 239 (1981).

formed in the hydrolysis of *N,N*-diacetylchitobiose is estimated by the Morgan–Elson's reaction.[9]

Reagents

N,N-Diacetylchitobiose, 4 m*M*
Sodium citrate–HCl buffer, 0.1 *M*, pH 2.5
Potassium tetraborate, 0.8 *M*, pH 9.1
p-Dimethylaminobenzaldehyde reagent, 10 g in 100 ml of glacial acetic acid, containing 12.5% (v/v) 10 *N* HCl, diluted one-tenth with glacial acetic acid before use

Procedure. The reaction mixture, consisting of 0.1 ml of 4 m*M* *N,N*-diacetylchitobiose, 0.2 ml of 0.1 *M* sodium citrate–HCl buffer, pH 2.5, and 0.1 ml of enzyme solution is incubated for 10 min at 37°. The reaction is stopped by boiling for 3 min. A 0.1-ml sample of the digests is diluted to 0.5 ml with deionized water, and then 0.1 ml of 0.8 *M* potassium tetraborate is added. The tubes are heated exactly for 3 min in a boiling water bath and cooled in tap water. After the addition of 3 ml of *p*-dimethylaminobenzaldehyde reagent, the tubes are incubated for 20 min at 36–38°, and then cooled. The absorbance at 585 nm is measured, *N*-acetylglucosamine being used as a reference compound.

Unit. One unit of chitobiase activity is defined as the amount of the enzyme which hydrolyzes 1 μmol of *N,N*-diacetylchitobiose/min under the above condition.

Purification Procedure

Step 1: Preparation of Culture Filtrate. Pycnoporus cinnabarinus IFO 6139 is grown and maintained on agar slants in a medium consisting of 25% potato extract and 1% glucose. About 5 mm^2 of mycelia is inoculated into 50 ml of a seed culture medium, pH 5.5, which is composed of 4% sucrose, 2% peptone, 1% yeast extract, 0.5% KH_2PO_4, and 0.05% $MgSO_4 \cdot 7H_2O$, in a 200-ml conical flask and grown for 6 days at 30° on a circular shaker. One milliliter of the seed culture is transferred into a 200-ml conical flask containing 50 ml of a production medium, pH 5.5, consisting of 2% glucose, 3% peptone, 0.5% yeast extract, 1% $(NH_4)_2SO_4$, 0.5% KH_2PO_4, and 0.05% $MgSO_4 \cdot 7H_2O$, and the cultivation is carried out for 6 days at 30° on a circular shaker. The culture filtrate is used as starting enzyme solution.

Step 2: Precipitation with Ammonium Sulfate. To the culture filtrate from step 1, solid ammonium sulfate (516 g/liter) is added under stirring to give 75% saturation. After the mixture has been allowed to stand for 2

[9] J. L. Reissig, J. L. Strominger, and L. F. Leloir, *J. Biol. Chem.* **217**, 959 (1955).

days, the precipitate is collected by centrifugation and dissolved in deionized water. The solution is desalted by passing through a Sephadex G-25 column (5.0 × 50 cm) previously equilibrated with 0.02 *M* phosphate buffer, pH 6.0 (buffer P).

Step 3: Chromatography on DEAE-Sephadex A-50. The enzyme solution containing 1679 mg of protein from step 2 is applied to a DEAE-Sephadex A-50 column (5.0 × 40 cm), which has been equilibrated with buffer P. The column is washed with buffer P until the absorbance at 280 nm decreases below 0.1, and then eluted with a linear salt gradient consisting of 1 liter of buffer P and 1 liter of buffer P containing 1 *M* NaCl, at a flow rate of 65 ml/hr. Eluate is collected in 15-ml fractions, which are assayed for both β-GlcNAcase and β-GalNAcase activities. The main peak of the both enzyme activities appears at the narrow range of NaCl concentrations of 0.1 to 0.15 *M* and the small peak at NaCl concentrations of 0.15 to 0.3 *M*. The latter peak overlaps with the elution profiles of β-glucosidase and α- and β-galactosidase produced by this strain. The main fractions (Nos. 130-137) are pooled and concentrated by ultrafiltration with Amicon membrane UM10.

Step 4: First Gel Filtration on Sephadex G-100. The enzyme solution (6 ml) from step 3 is loaded on a Sephadex G-100 column (2.6 × 96 cm) previously equilibrated with 0.05 *M* citrate buffer, pH 5.0, and eluted with the same buffer, at a flow rate of 20 ml/hr. Fractions of 5 ml are collected. Fractions (Nos. 39–47) containing the enzyme activities are pooled, concentrated by ultrafiltration, and dialyzed against 0.05 *M* citrate buffer, pH 4.2.

Step 5: Chromatography on CM-Sephadex C-50. The enzyme solution (10 ml) from the previous step is applied to a CM-Sephadex C-50 column (1.6 × 23 cm) which has been equilibrated with 0.05 *M* citrate buffer, pH 4.2. The column is completely washed with the same buffer, and then eluted with a linear salt gradient of 100 ml of 0.05 *M* citrate buffer, pH 4.2, and 100 ml of the same buffer containing 0.6 *M* NaCl, at a flow rate of 20 ml/hr. Fractions of 5 ml are collected. The enzyme activities appear at the fractions eluted with 0.3–0.4 *M* NaCl. These fractions (Nos. 77–82) are pooled and concentrated. In this step, the β-*N*-acetylhexosaminidase is completely free from other glycosidase activities.

Step 6: Second Gel Filtration on Sephadex G-100. Half of the enzyme solution (2 ml) from step 5 is again applied to a Sephadex G-100 column (1.6 × 82 cm) previously equilibrated with 0.05 *M* citrate buffer, pH 4.5, and eluted with the same buffer, at a flow rate of 22 ml/hr. Fractions of 5 ml are collected. Fractions (Nos. 13–17) containing the enzyme activities are pooled as the purified enzyme.

The purification procedure is summarized in Table I. The ratio of

TABLE I

PURIFICATION OF β-*N*-ACETYLHEXOSAMINIDASE FROM *Pycnoporus cinnabarinus*

Step	Protein (mg)	β-GlcNAcase				β-GalNAcase		Chitobiase	
		Total units (*A*)	Specific activity (U/mg)	Yield (%)	Purification (-fold)	Total units (*B*)	Ratio (*A/B*)	Total units (*C*)	Ratio (*A/C*)
Culture filtrate	15,500	738	0.048	100	1.0	902	0.82	1,070	0.69
Ammonium sulfate precipitation	2,600	705	0.271	95.5	5.6	858	0.82	948	0.74
DEAE-Sephadex A-50	73.9	214	2.90	29.0	60.4	265	0.81	268	0.80
First Sephadex G-100	12.6	206	16.35	27.9	341	241	0.85	239	0.86
CM-Sephadex C-50	2.7	170	62.96	23.0	1,312	187	0.91	198	0.86
Second Sephadex G-100	2.0	134	67.00	18.2	1,396	149	0.90	176	0.76

β-GlcNAcase activity to β-GalNAcase activity is constant throughout the purification of β-*N*-acetylhexosaminidase and chitobiase activity is also purified in parallel with these activities.

Properties

Homogeneity, Isoelectric Point, and Molecular Weight. The purified enzyme is homogeneous on polyacrylamide disk gel electrophoresis. The isoelectric point of the enzyme is found at pH 5.4, and the molecular weight of the enzyme is estimated to be about 120,000 by gel filtration on Sephadex G-200 and about 65,000 by SDS-polyacrylamide disk gel electrophoresis.

Optimum pH and Stability. The optimum pH of the enzyme is found at 2.2 for *p*-NPGlcNAc, 3.7 for *p*-NPGalNAc, and 2.5 for *N,N*-diacetylchitobiose. The enzyme is stable between pH 2.0 and 4.0 at 5°, but unstable above 50°. The enzyme preparation is relatively stable to freezing and thawing in 0.05 *M* citrate buffer, pH 4.5.

Inhibition. Hg^{2+} shows only slight inhibition to the enzyme activities even in a concentration of 10 m*M*. Both β-GlcNAcase and β-GalNAcase activities are inhibited strongly by *N*-acetylglucosamine but not *N*-acetylgalactosamine. *N,N*-Diacetylchitobiose, *N,N,N*-triacetylchitotriose, and *N,N,N,N*,-tetraacetylchitotetraose inhibit β-GalNAcase activity more than β-GlcNAcase activity.

Michaelis Constants. The K_m values of the enzyme are 0.45 m*M* for *p*-NPGlcNAc and 0.54 m*M* for *p*-NPGalNAc.

Substrate Specificity and Mode of Action. The β-*N*-acetylhexosaminidase does not act on chitin. However, the enzyme hydrolyzes chitooligosaccharides to *N*-acetylglucosamine, at decreasing rate as their molecular weight increases. The enzyme does not attack *N,N*-diacetylchitobiitol, but hydrolyzes reduced chitooligosaccharides more than trisaccharide at the nonreducing end to form *N*-acetylglucosamine.[10] Thus, the mode of action of the β-*N*-acetylhexosaminidase on chitooligosaccharides is exo-type.

Purification of Chitinase[7]

Assay Method

Principle. The assay is based on the estimation of reducing sugars produced in the hydrolysis of colloidal chitin according to a modification[11] of Schales' procedure, with *N*-acetylglucosamine as a reference compound.

[10] A. Ohtakara, M. Mitsutomi, and E. Nakamae, *Agric. Biol. Chem.* **46,** 293 (1982).
[11] T. Imoto and K. Yagishita, *Agric. Biol. Chem.* **35,** 1154 (1971).

Reagents

Colloidal chitin,[12] 0.5%
McIlvaine's buffer, 0.2 *M* Na_2HPO_4, and 0.1 *M* citric acid, pH 4.0
Potassium ferricyanide,[10] 0.05% in 0.5 *M* sodium carbonate

Procedure. The reaction mixture, containing 1 ml of 0.5% colloidal chitin, 2 ml of McIlvaine's buffer, pH 4.0, and 1 ml of enzyme solution, is incubated for 20 min at 37° with shaking. The reaction is stopped by boiling for 3 min. After centrifugation, 1.5 ml of the supernatant fluid is mixed with 2 ml of potassium ferricyanide solution. The tubes are heated for 15 min in a boiling water bath. The absorbance at 420 nm is measured after cooling. Controls without colloidal chitin or enzyme are also tested.

Unit. One unit of chitinase activity is defined as the amount of the enzyme which produces 1 μmol of reducing sugars as *N*-acetylglucosamine per minute under the above condition.

Purification Procedure

Step 1: Preparation of Culture Filtrate and Precipitation with Ammonium Sulfate. The culture filtrate is prepared according to the same procedure as described in the purification of β-*N*-acetylhexosaminidase and brought to 70% saturation (472 g/liter) by adding solid ammonium sulfate. The precipitate is collected by centrifugation, dissolved in deionized water, and desalted by passing through a Sephadex G-25 column (5.0 × 48 cm) previously equilibrated with 0.02 *M* phosphate buffer, pH 6.0. The specific activity of the desalted enzyme solution is about three to four times that of the culture filtrate.

Step 2: Chromatography on DEAE-Sephadex A-50. The enzyme solution, containing 2030 mg of protein from step 1, is applied to a DEAE-Sephadex A-50 column (5.0 × 44 cm) which has been equilibrated with 0.02 *M* phosphate buffer, pH 6.0. The column is completely washed with the same buffer and eluted with a linear salt gradient consisting of 1 liter of 0.02 *M* phosphate buffer, pH 6.0, and 1 liter of the same buffer containing 0.8 *M* NaCl, at a flow rate of 72 ml/hr. Eluate is collected in 15-ml fractions, which are assayed for β-*N*-acetylhexosaminidase and chitinase activities. The β-*N*-acetylhexosaminidase is eluted with 0.1–0.3 *M* NaCl, while chitinase activity appears in the wide range of NaCl concentrations of 0.3 to 0.7 *M*. The chitinase eluted with 0.3–0.5 *M* NaCl separates into two peaks[13] (C-I and C-II in Table II) and overlaps with the tailing of β-*N*-acetylhexosaminidase activities. The chitinase (C-III in Table II)

[12] C. Jeuniaux, *Arch. Int. Physiol. Biochim.* **66,** 408 (1958).

[13] The yield of chitinase activity in the one peak (C-I) is very low in other experiments.

TABLE II
PURIFICATION OF CHITINASE FROM *Pycnoporus cinnabarinus*

Step	Protein (mg)	Total activity (U)	Specific activity (U/mg)	Yield (%)	Purification (-fold)
Ammonium sulfate precipitation	2040	858	0.421	100	1.00
DEAE-Sephadex A-50					
C-I	397	127	0.320	14.8	0.76
C-II	531	115	0.217	13.4	0.51
C-III	463	216	0.467	25.2	1.11
C-III purification					
CM-Sephadex C-50	184	128	0.696	14.9	1.65
Sephadex G-100	54.7	106	1.94	12.4	4.61
$(GlcNAc)_2$-Sepharose 4B	8.2	51	6.22	5.9	14.8

eluted with 0.5–0.7 *M* NaCl is the highest in specific activity, with a recovery of 25.2%. These fractions (C-III, Nos. 229–260) are pooled for further purification, concentrated by ultrafiltration with Amicon membrane PM10, and dialyzed against 0.05 *M* citrate buffer, pH 4.2.

Step 3: Chromatography on CM-Sephadex C-50. The enzyme solution (10 ml) from the previous step is applied to a CM-Sephadex C-50 column (2.6 × 30 cm) which has been equilibrated with 0.05 *M* citrate buffer, pH 4.2. The column is eluted with 400 ml of the same buffer followed by a linear salt gradient elution with 200 ml of 0.05 *M* citrate buffer, pH 4.2, and 200 ml of the same buffer containing 0.5 *M* NaCl, at a flow rate of 20 ml/hr. Fractions of 8 ml are collected. The chitinase passes through the column and is completely separated from the contaminating β-*N*-acetylhexosaminidase eluted with 0.3–0.4 *M* NaCl. Fractions (Nos. 7–13) containing chitinase activity are pooled and concentrated by ultrafiltration.

Step 4: Gel Filtration on Sephadex G-100. The enzyme solution (5.2 ml) from step 3 is loaded on a Sephadex G-100 column (2.6 × 98 cm) previously equilibrated with 0.05 *M* phosphate buffer, pH 6.0, and eluted with the same buffer, at a flow rate of 15 ml/hr. Fractions of 5 ml are collected. Fractions (Nos. 57–74) containing chitinase activity are pooled, concentrated, and dialyzed against 0.05 *M* citrate buffer, pH 5.0.

Step 5: Chromatography on N,N-Diacetylchitobiose-Sepharose 4B. *N,N*-Diacetylchitobiose-Sepharose 4B is prepared according to the same procedure described by Jeffrey *et al.*[14] *N,N,N*-Triacetylchitotriose (0.1 mmol) is reacted with β-(*p*-aminophenyl)ethylamine (3.5 mmol) for

[14] A. M. Jeffrey, D. A. Zopf, and V. Ginsburg, *Biochem. Biophys. Res. Commun.* **62,** 608 (1975).

15 hr under stirring at room temperature, and the product is reduced with sodium borohydride in absolute ethanol to obtain a secondary amine derivative of *N,N*-diacetylchitobiose. The derivative (60 μmol) is coupled to CNBr-activated Sepharose 4B (3 g) in 15 ml of 0.1 *M* $NaHCO_3$, pH 8.2, to give *N,N*-diacetylchitobiose-Sepharose 4B. The enzyme solution, containing 22 mg of protein from step 4, is applied to an *N,N*-diacetylchitobiose-Sepharose 4B column (0.9 × 12.5 cm), previously equilibrated with 0.05 *M* citrate buffer, pH 5.0. The column is washed with 10 times the column volume of the same buffer and eluted with 0.02 *M* phosphate buffer of pH 6.9 containing 1 *M* NaCl, at a flow rate of 6.7 ml/hr. Fractions of 4 ml are collected. Although about 20% of the original activity is found in the protein peak unadsorbed on the column, about 50% of the chitinase activity appears at the fractions eluted with 0.02 *M* phosphate buffer, pH 6.9, containing 1 *M* NaCl. These fractions (Nos. 14–23) are pooled as the purified enzyme.

The purification procedure is summarized in Table II. The enzyme is purified about 15-fold starting from the precipitate with ammonium sulfate, and thus purified about 45- to 60-fold over the culture filtrate.

Properties

Homogeneity, Isoelectric Point, and Molecular Weight. The purified enzyme is nearly homogeneous on polyacrylamide disk gel electrophoresis and completely free from *β-N*-acetylhexosaminidase. The isoelectric point of the enzyme is at pH 3.6 and the molecular weight of the enzyme is estimated to be about 38,000 by gel filtration on Sephadex G-100.

Optimum pH and Stability. The optimum pH of the enzyme is at 4.5 and the enzyme is stable between pH 4.0 and 8.0, but unstable at elevated temperatures.

Substrate Specificity and Mode of Action. The chitinase hydrolyzes colloidal chitin and glycol chitin rapidly and powdered chitin very slowly. Acting upon colloidal chitin, and enzyme produces *N,N*-diacetylchitobiose as the main product accompanied by a slight formation of *N*-acetylglucosamine. The chitinase does not hydrolyze *N,N*-diacetylchitobiose, but cleaves chitooligosaccharides more than trisaccharide and produces *N,N*-diacetylchitobiose and *N*-acetylglucosamine from *N,N,N*-triacetylchitotriose. The enzyme is also capable of hydrolyzing reduced chitooligosaccharides more than tetrasaccharide. From the analysis of the hydrolysis products of these oligosaccharides,[15] the chitinase acts on chitooligosaccharides with the mode of action being endotype, predominantly hydrolyzing the second *β-N*-acetylglucosaminide linkage from the nonreducing end.

[15] A. Ohtakara and M. Mitsutomi, this volume [54].

[58] Chitinase from *Neurospora crassa*

By ANGEL ARROYO-BEGOVICH

Chitin → oligochitosaccharides (chitobiose being predominant)

Assay Method

Chitinase is measured following the release of *N,N′*-diacetyl chitobiose from chitin, according to a modification of the procedure described by Molano *et al.*[1] The enzyme behaves as an endochitinase [poly-1,4-(2-acetamido-2-deoxy)-β-D-glucoside glycanohydrolase, EC 3.2.1.14]; it produces low-molecular-weight, soluble multimers of *N*-acetyl-D-glucosamine, the dimer *N,N′*-diacetylchitobiose being predominant. The substrate employed is nascent chitin which is produced by chitin synthase present in the reaction mixture.

Reagents

Assay buffer: Tris–HCl, 85 m*M*, pH 8.6, containing 33 m*M* *N*-acetyl-D-glucosamine, 0.33 m*M* ATP, 16.5 m*M* $MgCl_2$, 1.6 m*M* UDP-[U-^{14}C]-*N*-acetyl-D-glucosamine (0.29 Ci/mmol), and 640 ng/ml crystalline trypsin

Chitosome particles (source of chitin synthase) are prepared following a modification[2] of the procedure described by Ruíz-Herrera *et al.*[3]: 18 ml of the cytosol fraction described in step 1 of the Purification Procedure is loaded on a BioGel A-5m (Bio-Rad) column (2.5 × 35 cm) equilibrated with Tris–HCl, 50 m*M*, pH 8.6, containing 10 m*M* $MgCl_2$, 1 m*M* EDTA, and 3 m*M* sodium azide (buffer A), and eluted with the same buffer.[4] Chitosome particles elute in the exclusion volume. Active fractions are pooled and centrifuged at 10,000 *g* for 30 min, the chitosome particles remaining in the supernatant. Protein concentration is adjusted at 1.5 mg/ml

DEAE-cellulose chromatographic plates are prepared by spreading a water suspension of DEAE-cellulose (Cellex D, Bio-Rad) on 2.5 × 20 cm glass plates and then drying them at 65° for 5 hr

[1] J. Molano, I. Polacheck, A. Durán, and E. Cabib, *J. Biol. Chem.* **254,** 4901 (1979).

[2] A. Zarain-Herzberg and A. Arroyo-Begovich, *J. Gen. Microbiol.* **129,** 3319 (1983); A. Arroyo-Begovich, unpublished observations.

[3] J. Ruíz-Herrera, E. López-Romero, and S. Bartnicki-García, *J. Biol. Chem.* **252,** 3338 (1977).

[4] Alkaline pH is required to stabilize the activity of *Neurospora* chitosome particles [A. Arroyo-Begovich and J. Ruíz-Herrera, *J. Gen. Microbiol.* **113,** 339 (1979)].

Procedure. The reaction mixture consists of 75 μl of assay buffer, 25 μl of chitosome particles preparation (37 μg protein), and 25 μl of chitinase. The reaction is initiated by the addition of the chitinase, the mixture incubated at 22° for 60 min, and then the reaction is stopped by heating at 75° for 5 min.

The separation of chitinase reaction products is carried out by thin-layer chromatography on DEAE-cellulose plates. The entire reaction mixture is spotted on the plates and subjected to ascending chromatography, employing water as solvent. The undegraded chitin and UDP-*N*-acetyl-D-glucosamine remain at the origin. When the solvent front reaches 1 cm from the end of the plate, plates are removed from the solvent and dried at 100° for 7 min. A segment 3 cm wide corresponding to the solvent front is scraped from the plate, dispersed in 5 ml of a scintillation liquid (2 g PPO, 100 mg dimethyl-POPOP, in 1 liter of toluene), and counted.

Definition of Enzyme Unit and Specific Activity. One unit of chitinase activity is defined as the amount required to release 1 μmol *N,N'*-diacetyl-chitobiose from chitin per hour at 22°. Protein is estimated by the method of Lowry *et al.*[5] Specific activity is defined as units of enzyme activity per milligram of protein.

Materials

KH_2PO_4–NaOH buffer, 50 m*M*, pH 6.2, containing 10 m*M* $MgCl_2$, 1 m*M* EDTA, and 3 m*M* sodium azide (buffer B)

Neurospora crassa, Oak Ridge wild-type 74-OR23-1A,[6] is conveniently grown at 25–28° in 5-gal polycarbonate carboys containing 15 liters of minimal medium[7] plus 2% sucrose. Carboys are inoculated with 2×10^6 conidia (6 days old) per milliliter. The contents of the carboys are agitated for 48 hr by air forced through a sterile cotton filter. The mycelia are collected on a Büchner funnel covered with cheesecloth. After removing as much culture medium as possible, mycelia mats are lyophilized and ground to a fine powder with the aid of a Wiley mill fitted with a 60 mesh sieve and kept at −10° until used. The yield is approximately 35 g of lyophilized powder per carboy

Purification Procedure

The purification procedure given here is essentially that described by Zarain-Herzberg and Arroyo-Begovich.[2]

[5] O. H. Lowry, N. J. Rosebrough, A. L. Farr, and R. J. Randall, *J. Biol. Chem.* **193,** 265 (1951).

[6] Obtained from the Fungal Genetics Stock Center, California State University, Humboldt, Arcata, California.

[7] H. J. Vogel, *Microb. Genet. Bull.* **13,** 42 (1956).

Step 1: Extraction of Mycelia. Mycelium powder (2 g) is suspended in 30 ml of buffer A, mixed with 20 ml of glass beads (0.45 to 0.50-mm diameter), and disrupted for 1 min in a Braun MSK cell homogenizer, cooled with a stream of expanding CO_2. The homogenate is centrifuged at 1000 *g* for 5 min to eliminate the cell wall fraction and then at 68,000 *g* for 60 min. The supernatant (cytosol fraction) is the starting material for the purification procedure. This step, and all subsequent steps, are carried out at 4°.

Step 2: Ammonium Sulfate Precipitation. The cytosol fraction obtained from 10 g (dry mycelium powder) is centrifuged at 160,000 *g* for 1 hr to remove any insoluble material. Ammonium sulfate (242.3 g/liter), is added slowly to the supernatant with stirring. The precipitate, collected by centrifugation at 12,000 *g* for 15 min, is suspended in 7 ml of buffer B and dialyzed against several changes of the same buffer for about 20 hr. The dialysate is centrifuged at 160,000 *g* for 1 hr to remove undissolved material.

Step 3: Chromatography on Sephadex G-100. The supernatant obtained from the previous step (7.5 ml) is applied to a Sephadex G-100 column (99 × 1.9 cm) equilibrated with buffer B. The enzyme is eluted with the same buffer. Fractions of 5 ml are collected, and the fractions containing the enzyme (28–40) are pooled and concentrated to about 10 ml with the aid of an Amicon assembly fitted with a PM10 membrane.

Step 4: Chromatography on Sephadex G-25. The concentrated pool obtained above is loaded on a Sephadex G-25 fine column (87 × 1.9 cm) equilibrated with buffer B, and the enzyme is eluted with the same buffer. Fractions of 5 ml are collected. Under these conditions the enzyme is retained in the column, eluting much later than the exclusion volume (fractions 9–13), probably owing to the interaction of the enzyme with the gel matrix.[2] Active fractions are pooled and concentrated as above with the aid of an Amicon assembly to a volume of 5 ml.

A summary of data averaged from five separate purifications is given in Table I. The specific activity observed for the purified enzyme ranges from 42 to 50 U.

Properties

Stability. The purified enzyme can be stored in buffer B several weeks at 0° or below, and under these conditions it is stable. The optimum pH for the enzyme is 6.7.

Physical Properties. Gel filtration studies employing Ultrogel (LKB) indicated that the enzyme has a molecular weight of approximately 20,600.

TABLE I
PURIFICATION OF *Neurospora crassa* CHITINASE[a]

Fraction	Total protein (mg)	Total activity (U)	Specific activity	Purification (-fold)	Recovery (%)
Cytosol fraction	595	214.2	0.36	—	100
Ammonium sulfate	141	115.6	0.82	2.27	53.9
Sephadex G-100	30.1	90.3	3.00	8.33	42.1
Sephadex G-25	1.1	50.9	46.27	128.52	23.7

[a] Results obtained from the extraction of 10 g of powdered lyophilized mycelia.
[b] Ratio of the specific activity of the enzyme of the given purification step to that of the cytosol fraction.

SDS electrophoresis analysis of the purified enzyme revealed the presence of two main bands and very few minor bands.

Acknowledgments

This work was supported in part by Subvention 001876 from Consejo Nacional de Ciencia y Tecnología, México.

[59] Chitinase from *Verticillium albo-atrum*

By G. F. PEGG

The hyphomycete vascular wilt fungus *Verticillium albo-atrum* Reinke & Berthold produces a constitutive chitinase in culture[1,2] which has been implicated in the *in vitro*[1] and *in vivo*[2-6] lysis of fungal mycelium.

Assay Method

The fungal enzyme poly-1,4-(2-acetamido-2-deoxy)-β-D-glucoside glycanohydrolase (EC 3.2.1.14, chitinase) is effective in hydrolyzing chitin

[1] J. C. Vessey and G. F. Pegg, *Trans. Br. Mycol. Soc.* **60,** 133 (1973).
[2] G. F. Pegg and J. C. Vessey, *Physiol. Plant Pathol.* **3,** 207 (1973).
[3] G. F. Pegg, *in* "Cell Wall Biochemistry Related to Specificity in Host–Plant Interactions" (B. Solheim and J. Raa, eds.), pp. 304–345. Universitetsforlaget, Oslo, Norway, 1977.
[4] G. F. Pegg and D. H. Young, *Physiol. Plant Pathol.* **19,** 371 (1981).
[5] G. F. Pegg and D. H. Young, *Physiol. Plant Pathol.* **21,** 389 (1982).
[6] D. H. Young and G. F. Pegg, *Physiol. Plant Pathol.* **21,** 411 (1982).

and its oligomers chitotetraose, chitotriose, and chitobiose to *N*-acetylglucosamine (NAG) in the absence of *N*-acetylglucosaminidase (NAGase). Similarly chitosan (deacetylated chitin) is hydrolyzed to glucosamine. The reaction product NAG is determined colorimetrically by a modification[7] of Morgan and Elson's method.[8]

Reagents

Colloidal chitin: Derived from ground crab shell[1,8] (Sigma Chemical Company), 1.83 mg ml^{-1}
Buffer: Sodium acetate buffer, pH 5.2, 0.1 *M*
Potassium tetraborate ($K_2B_4O_7$), 0.8 *M*
p-Dimethylaminobenzaldehyde (DMAB): Freshly prepared solution of 1 g in 100 ml glacial acetic acid containing 1.2% (v/v) 10 *N* HCl
Bovine serum albumin (BSA): 3.0 mg ml^{-1} reaction mixture

Procedure

From a purified chitin paste derived from ground crab shell (73 mg dry wt g^{-1} wet wt) a dispersion of 1.83 mg dry wt chitin ml^{-1} is prepared by homogenizing 25 mg chitin in pH 5.2, 0.1 *M* sodium acetate buffer in a Sorvall Omnimixer at 14,000 rpm for 30 sec. The reaction mixture containing 1 ml chitinase solution in acetate buffer, 4 ml chitin suspension, and 3.0 mg ml^{-1} BSA is incubated in a water bath at 37° for 3 hr. One milliliter of reaction mixture and 1 ml of distilled water are boiled in a glass ball-covered centrifuge tube for 10 min and centrifuged. NAG is determined in 0.5-ml aliquots of the supernatant by adding 0.1 ml potassium tetraborate solution and boiling for exactly 3 min in a water bath. DMAB solution (3.0 ml) is added after cooling and read within 10 min at 585 nm. Absorbance is converted to micrograms NAG by reference to an NAG calibration curve, linear over the range 0.1–0.4 *M*.

Definition of a Unit of Enzyme Activity and Specific Activity. One unit of enzyme activity is defined as the amount of enzyme which produces 100 μg NAG ml^{-1} chitin solution in 1 hr at 37° at pH 5.2. Specific activity (SA) is defined as units of enzyme activity per milligram protein.

Purification Procedures

Step 1: Extraction of Fungal Cultures. Stationary cultures of *Verticillium albo-atrum* are grown in 200 ml Czapek–Dox medium in 1-liter Roux bottles for 28 days at 21° in darkness. Cultures are filtered through

[7] J. L. Reissig, J. L. Strominger, and L. F. Leloir, *J. Biol. Chem.* **217,** 959 (1955).
[8] W. T. J. Morgan and L. A. Elson, *Biochem. J.* **28,** 988 (1934).

two layers of washed muslin and through Whatman No. 3 filter paper in a Büchner funnel. The filtrate is centrifuged at 17,000 *g* for 30 min at 4° to pellet spores and the supernatant dialyzed overnight with two changes of 3 liters of pH 9.5, 0.1 *M* glycine–NaOH buffer at 4°.

Step 2: DEAE-Cellulose Chromatography. Dialyzed filtrate (1.5 liters) is stirred with DEAE-cellulose (Cellex D, Bio-Rad Laboratories) and filtered through Whatman No. 1 filter paper and the Cellex pad washed with 200 ml glycine–NaOH buffer. The sample with 37% of the total chitinase has no *exo* 1,3-β-glucanase or 1,3-β-glucosidase activity. *N*-acetylglucosaminidase activity is only 4.3% of the original.

Step 3: Ammonium Sulfate Fractionation. $(NH_4)_2$ SO_4 is added to the filtrate to 1.08 *M* (20% saturation) left for 2 hr at 4°, centrifuged at 17,000 *g* for 30 min at 4°, and the precipitate discarded. The process is repeated after taking the supernatant to 4 *M* (75%) $(NH_4)_2SO_4$ saturation. The precipitate is dialyzed for 24 hr at 4° against 3 × 1 liter changes of 1% glycine buffer.

Step 4: Isoelectric Focusing (IEF). The dialysate from step 3 is run in a sucrose density gradient on an LKB IEF column using carrier ampholytes (LKB) in the pH range 3.5–10. Chitinase appears in a broad band, pH 4–6, with a closely adjacent peak of NAGase at p*I* 5.0. Subsequent purification is based on fractions 44–56, excluding the major NAGase peak. Residual NAGase activity in the pooled chitinase fraction is 0.15% of the original activity.

Step 5: Hydroxyapatite Chromatography. The combined chitinase fractions are dialyzed for 24 hr at 4° against 3 × 1 liter changes of pH 6.8, 0.01 *M* Na_2PO_4 buffer, loaded on a 10.6 × 1.5 cm hydroxyapatite column and eluted in 5-ml fractions in a pH 6.8, 0.01–0.2 *M* Na_2PO_4 linear gradient. Chitinase elutes in approximately 40 ml in 0.08 *M* Na_2PO_4 (80–120 ml total volume).

Step 6: BioGel P-100 Chromatography. The pooled fractions are dialyzed overnight at 4° against 2 × 1 liter changes of pH 8.6, 0.01 *M* glycine–NaOH buffer and freeze dried. The dried sample is dissolved in 1 ml of pH 8.6, 0.1 *M* glycine–NaOH buffer, dialyzed for 12 hr at 4° against 1 liter of the same buffer, and chromatographed on a 40 × 1.5 cm BioGel P-100 column in glycine–NaOH buffer at a flow rate of 5 ml hr^{-1}. Chitinase elutes as a single peak with an estimated M_r of 63,000 by comparison with coeluting standards.

Step 7: IEF Repeat Stage. Fractions 16–26 are pooled and dialyzed for 24 hr at 4° against 3 × 1 liter changes of 1% glycine and the IEF stage of step 4 repeated using carrier ampholytes in the range pH 4–6.0. *Verticillium* chitinase is separated as a single peak with a p*I* 5.0 in 4 × 4-ml fractions of the column volume.

The purification stages are shown in Table I.

TABLE I
PURIFICATION OF *V. albo-atrum* CHITINASE FROM CULTURE FILTRATE

Fraction	Chitinase (U)	Chitinase recovery (%)	Protein recovery (mg)	Specific activity (chitinase units mg^{-1} protein)
Culture filtrate	926		396	2.3
DEAE-Cellex D	342	36.9	138	2.5
20–75% $(NH_4)_2SO_4$	128	37.6	43	3.0
IEF[a] (pH 3.5–10)	98	76.4	15	6.5
Hydroxyapatite chromatograph	41	42.1	1.9	21.6
BioGel P-100 chromatograph	19	45.6	0.49	38.8
IEF (pH 4.0–6.0)	16	85.6	0.37	43.2

[a] IEF, Isoelectric focusing.

Properties

Stability. The enzyme appears stable for several months at 2°. BSA greatly enhances enzyme activity and is essential in the reaction mixture[9] (Fig. 1).

Specificity. Verticillium albo-atrum chitinase is 1.74 times more active on *N,N*-diacetylchitobiose than chitin in a 0.5-ml reaction mixture containing 100 μg substrate and 0.025 U of enzyme in pH 5.2, 0.1 *M* sodium acetate buffer and 3.0 mg ml^{-1} BSA. NAG is measured as the end product after 6 hr at 37°. Similar activity is found on 3,4-dinitrophenyltetra-*N*-acetyl-β-D-chitotetraoside (3,4-DNP-TNAC) with 1 ml containing 50 μg ml^{-1} 3,4-DNP-TNAC and 0.025 enzyme units in pH 6.0, 0.1 *M* sodium citrate–phosphate buffer and 3.0 mg ml^{-1} BSA. The reaction is stopped by boiling after 6 hr at 37° and DNP measured spectrophotometrically at 400 nm.

Chitosan. The enzyme hydrolyzes chitosan. The relationship between reciprocal specific viscosity η_{sp}^{-1} and time is linear from 0.2–0.8 over 120 min.

Chitin. NAG is the only product found in chitin digests. Reducing sugars (NAG equivalents) equal the production of free NAG throughout chitin hydrolysis. The addition of NAGase to the reaction mixture has no effect on the enzyme. No lysozyme (muramidase) activity is shown when assayed on dried *Micrococcus lysodeikticus* cells.[10]

Kinetic Constants. The K_m for chitin hydrolysis is 97 μg dry wt chitin ml^{-1} and the V_{max} is 16.67 μg NAG hr^{-1} ml^{-1} determined from a Lineweaver–Burke plot.

[9] M. V. Tracey, *Biochem. J.* **61,** 579 (1955).
[10] L. R. Berger and R. S. Werser, *Biochim. Biophys. Acta* **26,** 517 (1957).

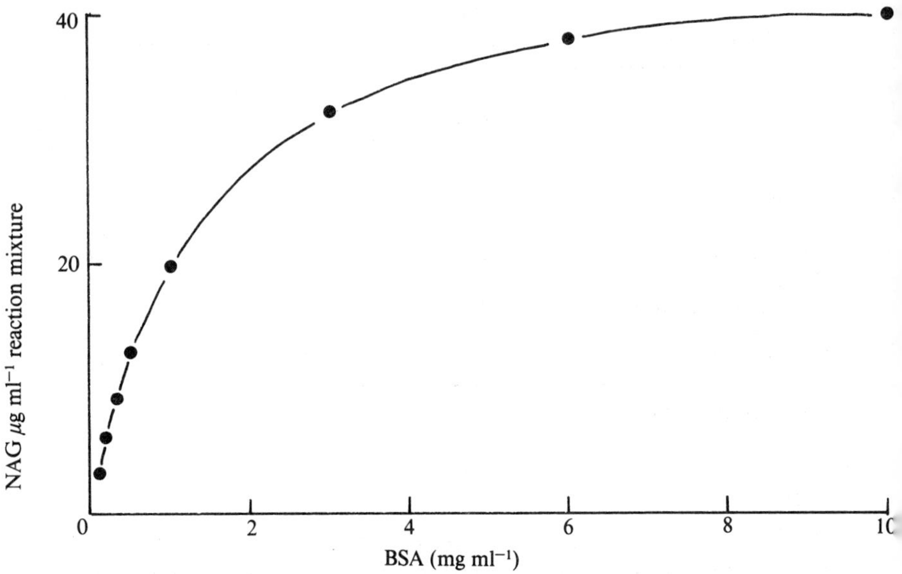

FIG. 1. Effect of bovine serum albumin on *V. albo-atrum* chitinase activity.

TABLE II
EFFECT OF HEAVY METALS, REAGENTS, AND INHIBITORS ON *V. albo-atrum* CHITINASE ACTIVITY AND NAG DETERMINATION

Substance	Concentration	*V. albo-atrum* chitinase		NAG determination	
		Activation (%)	Inhibition (%)	Activation (%)	Inhibit (%)
$CuSO_4$	50 m*M*	—	100	—	92.[illegible]
$HgCl_2$	50 m*M*	—	100	—	82.[illegible]
NaN_3	50 m*M*	—	0.6	3.4	—
KCN	50 m*M*	—	100	—	91.[illegible]
$CaCl_2$	50 m*M*	0.8	—	2.3	—
$MgSO_4$	50 m*M*	7.6	—	0.3	—
KCl	50 m*M*	10.3	—	—	1.4
$MnSO_4$	50 m*M*	—	63.3	—	62.[illegible]
$ZnSO_4$	50 m*M*	—	39.7	—	7.[illegible]
EDTA	50 m*M*	—	12.6	—	2.[illegible]
LKB ampholytes (pH 3.5 – 10 range)	1%	—	6.5	—	4.[illegible]
2-Acetamido-2-deoxy-D-gluconolactone	2 m*M*	—	40.3	2.0	—

pH. Verticillium chitinase has a sharp pH optimum at 3.7 and is inactive above pH 7.7. The response curves are identical for crude and purified preparations.

Temperature. Maximum activity occurs at 51° but at 65° activity is still 40% of maximum. The enzyme is inactivated at 70°.

Molecular Weight. In SDS gel electrophoresis two protein bands are observed. The denser band has a molecular weight of 64,000 and the minor band 58,000 by comparison of their mobilities relative to molecular weight markers.

Effect of Ions and Inhibitors

The effect of heavy metals and biological inhibitors on enzyme activity and substances interfering with NAG determinations is shown in Table II.

Standard assay conditions using chitin as substrate at 1.83 mg ml^{-1} are used but with BSA at 200 μg ml^{-1} reaction mixture. The effect of substances on NAG determination is studied by substituting NAG at 25 μg ml^{-1} for enzyme in the reaction mixture. Effects are shown as a percentage of maximum activity in the absence of the test substance.

[60] Chitinase from *Phaseolus vulgaris* Leaves

By THOMAS BOLLER, ANNETTE GEHRI, FELIX MAUCH, and URS VÖGELI

It is a surprising fact that some plants contain high activities of chitinase, an enzyme that has no known function in plant metabolism. Plant chitinase was first described from almond emulsin[1] and from bean and other seeds.[2] Abeles *et al.*[3] discovered that chitinase is induced in bean leaves by the plant hormone, ethylene, in parallel with β-1,3-glucanase. Since chitin and β-1,3-glucans are important cell wall components of fungi, they have proposed an antifungal function for chitinase and β-1,3-glucanase.[3]

We examined the effect of ethylene on chitinase activity in bean leaves more closely and purified the enzyme in the course of this work.[4] Our

[1] L. Zechmeister and G. Toth, *Enzymologia* **7,** 165 (1939).

[2] R. F. Powning and H. Irzykiewicz, *Comp. Biochem. Physiol.* **14,** 127 (1965).

[3] F. B. Abeles, R. P. Bosshart, L. E. Forrence, and W. H. Habig, *Plant Physiol.* **47,** 129 (1971).

[4] T. Boller, A. Gehri, F. Mauch, and U. Vögeli, *Planta* **157,** 22 (1983).

METHODS IN ENZYMOLOGY, VOL. 161

purification method is derived from the one used for chitinase from wheat germ.[5,6] It is based on affinity chromatography on chitin and is applicable, with slight modifications, to all chitinases that are readily adsorbed by chitin. Some plant chitinases, for example, the major chitinase from cucumber leaves,[7] do not bind well to chitin and cannot be purified by this method. A different method has been used for the purification of chitinase from tomato leaves.[8]

Assay Method

Principle. We routinely employ the radiochemical chitinase assay[9] described elsewhere in this volume,[10] based on the liberation of soluble radioactivity from insoluble [*acetyl*-^{3}H]chitin. By this method, any possible chitin deacetylase activity would be misjudged as chitinase. We have verified that chitin deacetylase is not present in ethylene-treated bean leaves (unpublished observations). We have also used the colorimetric chitinase assay described elsewhere in this volume.[11] We obtain exactly the same results with the colorimetric and the radiochemical assay.[4,11]

Reagents

[^{3}H]Chitin (6 kBq/mg), 10 mg/ml in 3 m*M* NaN_3. The [^{3}H]chitin is prepared by N-acetylation of chitosan with [^{3}H]acetic anhydride[6,9] (regenerated chitin). From this preparation, colloidal [^{3}H]chitin is prepared according to Berger and Reynolds[12] as described elsewhere in this volume[13]

Sodium phosphate buffer, 0.1 *M*, pH 6.5

TCA (trichloroacetic acid), 1 *M*

Procedure. The following are pipetted into an 1.5-ml Eppendorf tube: 50 μl sodium phosphate buffer, 10–100 μl enzyme, and water to a volume of 150 μl. The reaction is carried out in a shaking water bath at 37°. It is started by the addition of 100 μl [^{3}H]chitin and stopped by the addition of 250 μl TCA after 30 min. After centrifugation (1000 *g* for 10 min), the radioactivity of 250 μl of the supernatant is determined.

[5] J. Molano, I. Polacheck, A. Duran, and E. Cabib, *J. Biol. Chem.* **254,** 4901 (1979).

[6] E. Cabib, this volume [63].

[7] J. P. Métraux and T. Boller, *Physiol. Mol. Plant Pathol.* **28,** 161 (1986).

[8] G. F. Pegg, this volume [61].

[9] J. Molano, A. Duran, and E. Cabib, *Anal. Biochem.* **83,** 648 (1977).

[10] E. Cabib, this volume [48].

[11] T. Boller and F. Mauch, this volume [50].

[12] L. R. Berger and D. M. Reynolds, *Biochim. Biophys. Acta* **29,** 522 (1958).

[13] K. Shimahara and Y. Takiguchi, this volume [47].

Definition of Unit and Specific Activity. Since formation of reaction product is nonlinearly related to the enzyme concentration, the activity is calculated from a standard curve prepared with a dilution series of the enzyme, as described elsewhere in this volume.[6,11] One unit of chitinase is defined as the amount that catalyzes the release of soluble chitooligosaccharides containing 1 μmol of *N*-acetylglucosamine residues in 1 min at infinite dilution. Specific activity (units per milligram of protein) is based on protein determinations by the Bradford method.[14]

Plant Material, Ethylene Treatment

Primary leaves of *Phaseolus vulgaris* cv. "Saxa" (bush bean) are employed. The seeds, obtained from Samen Mauser, Zürich, Switzerland, are steeped for 4 hr in running tap water and then grown on trays containing Vermiculite in a greenhouse with a day/night photoperiod of 16/8 hr and temperatures of 25/20°. After 12–14 days, when the primary leaves are fully expanded, the whole trays with about 100 seedlings are put into plastic chambers (100 liters). The chambers are tightly closed. The plants are exposed to an atmosphere containing 10 ppm ethylene by injecting 1 ml pure ethylene gas into the chamber through an inlet port covered with a gas-tight rubber cap. To obtain maximal induction of chitinase, the plants are incubated for 36–48 hr in ethylene. (Longer treatment results in leaf abscission.) About 70 g primary leaves is obtained from one tray.

Purification Procedure[4]

Unless otherwise specified, all operations are carried out at 0–5°.

Step 1: Preparation of the Crude Extract. Leaves from ethylene-treated plants (200 g) are homogenized in 400 ml 0.1 *M* citrate buffer, pH 5.0, in a Sorvall Omnimix homogenizer for 3 × 1 min at top speed. The homogenate is centrifuged (15,000 *g*, 10 min). The supernatant represents the crude extract. It contains most of the chitinase activity but, at the relatively low pH employed, only about 20% of the total leaf protein. Addition of dithiothreitol (2 m*M*) does not interfere but does not increase the yield of chitinase.

Step 2: Heat Treatment. The crude extract is incubated at 50° for 20 min. After cooling, the precipitate is removed by centrifugation (20,000 *g*, 10 min).

Step 3: Ammonium Sulfate Precipitation. To the supernatant, ammonium sulfate is added to give 60% saturation at 0°. After an incubation of

[14] M. M. Bradford, *Anal. Biochem.* **72**, 248 (1976).

1 hr with continuous stirring, the precipitate is collected by centrifugation (20,000 *g*, 20 min) and redissolved in 40 ml 10 m*M* sodium acetate buffer, pH 5. $NaHCO_3$ is added to give a concentration of 20 m*M*, and the pH is raised to pH 8.4 with 0.1 *N* NaOH.

Step 4: Chromatography on Regenerated Chitin. A column (8 × 3 cm) is made of regenerated chitin, prepared according to Molano *et al.*[9] After equilibration with 20 m*M* sodium bicarbonate buffer, pH 8.4, the chitinase preparation is put on the column. The absorption of the eluate is monitored at 280 nm. The column is washed with 150 ml 20 m*M* sodium bicarbonate buffer, pH 8.4, followed by 100 ml 20 m*M* sodium acetate buffer, pH 5.5. Then, chitinase is eluted with 20 m*M* acetic acid (pH 3.2).

Chitinase activity is eluted in a single peak which coincides with the peak of A_{280}. The peak fractions are combined, dialyzed against 10 m*M* sodium acetate buffer, pH 5.0, and lyophilized.

Summary. Table I summarizes the purification protocol. Chitinase is purified about 24-fold, indicating that 4% of the total soluble protein at pH 5 is chitinase. The overall yield is 68%.

Properties

Purity. The enzyme preparation from the chitin column is homogeneous, as judged by SDS-gel electrophoresis. For a further demonstration of purity, the lyophilized chitinase is taken up in 2 ml 0.05 *M* sodium phosphate buffer, pH 6.8, containing 0.1 *M* NaCl, and chromatographed on a 2 × 68 cm column of Sephadex G-75 equilibrated with the same buffer. Chitinase elutes as a single peak with constant specific activity. The position of the peak indicates an apparent molecular weight of 22,000. Obviously, the protein is slightly retarded on the Sephadex column since its molecular weight, estimated with SDS-gel electrophoresis and analytical ultracentrifugation, is about 30,500–32,500 (see below).

TABLE I
PURIFICATION OF ETHYLENE-INDUCED CHITINASE FROM BEAN LEAVES

Step	Volume (ml)	Protein (mg)	Chitinase (U)	Specific activity (U/mg)
Crude extract	500	340	2130	6.3
Heat treatment	500	200	2000	10.0
$(NH_4)_2SO_4$ Precipitation	50	154	2000	13.0
Chitin column	20	10	1440	144

Specificity and Mode of Action. Like other plant chitinases,[4,9,15] bean leaf chitinase initially forms chitooligosaccharides of various chain lengths from colloidal or regenerated chitin.[4] After exhaustive digestive of chitin, the main end products are chitobiose and chitotriose; only about 7% of the product is in the form of the monomer, *N*-acetylglucosamine. Obviously, the exochitinase assay used by Abeles *et al.*,[3] which monitors the production of the monomer only, is unsuitable for bean chitinase.[11]

Bean leaf chitinase readily depolymerizes glycol chitin; it can be measured viscosimetrically on this basis.[4] The enzyme does not hydrolyze chitosan, glycol chitosan, or *p*-nitrophenyl-*β*-*N*-acetylglucosaminide; it has no cellulase or *β*-1,3-glucanase activity.[4]

Bean leaf chitinase releases chitooligosaccharides from purified chitin-containing fungal cell walls. It also attacks bacterial cell walls, acting as a lysozyme.[4,15]

Antifungal Activity. Purified bean chitinase is a potent growth inhibitor of *Trichoderma viride.*[16] The smallest concentration that inhibits growth of this fungus is 2 $\mu g\ ml^{-1}$.

Physical Properties. The molecular weight of bean leaf chitinase, estimated by SDS-gel electrophoresis, is 32,500. Analytical ultracentrifugation yields a similar molecular weight (30,500), indicating that the enzyme occurs as a monomer in solution. It is a basic protein with an isoelectric point of about 9.4, as determined by chromatofocusing. Incidentally, most plant chitinases are basic proteins,[15] although there are exceptions.[7]

Interestingly, the N-terminal primary amino acid sequence of bean leaf chitinase shows a high degree of similarity with WGA, a chitin-binding lectin.[17] The entire amino acid sequence of bean chitinase has been deduced from the nucleotide sequence of a chitinase cDNA clone.[18]

Effect of pH. Bean leaf chitinase has a broad pH optimum around pH 6.5 for the hydrolysis of colloidal or regenerated chitin; its activity is more than 50% of the maximum between pH 3 and 9.[4,15] In contrast, its pH optimum for the hydrolysis of the bacterial cell wall peptidoglycan is quite narrow, centered at pH 4.8, with less than 50% of the maximal activity above pH 5.25 and below pH 4.[15]

Effect of Temperature. The enzyme is stable up to 40°. It loses about 10% of its activity per hour at 50° and about 60%/hr at 60°. Between 0 and

[15] T. Boller, *in* "Cellular and Molecular Biology of Plant Stress" (J. L. Key and T. Kosuge, eds.), p. 247. Liss, New York, 1985.

[16] A. Schlumbaum, F. Mauch, U. Vögeli, and T. Boller, *Nature (London)* **324,** 365 (1986).

[17] J. Lucas, A. Henschen, F. Lottspeich, U. Vögeli, and T. Boller, *FEBS Lett.* **193,** 208 (1985).

[18] K. E. Broglie, J. J. Gaynor, and R. M. Broglie, *Proc. Natl. Acad. Sci. U.S.A.* **83,** 6820 (1986).

40°, its apparent energy of activation, determined from an Arrhenius plot, is 39 kJ mol^{-1}.

Stability. Purified bean chitinase is stable for at least 2 years when kept lyophilized or frozen at −20°.

Activators and Inhibitors. Bean leaf chitinase does not require any cofactors. It is not affected by EDTA. The chitin-hydrolyzing activity is inhibited about 50% by 100 m*M* *N*-acetylglucosamine and less than 10% by 10 m*M* histamine. The lysozyme activity is inhibited less than 10% by 100 m*M* *N*-acetylglucosamine but more than 90% by 10 m*M* histamine.

Subcellular Localization. In bean leaves, chitinase is located in the central vacuole of the cells.[19]

Acknowledgment

This work was supported by the Swiss National Science Foundation.

[19] T. Boller and U. Vögeli, *Plant Physiol.* **74,** 442 (1984).

[61] Chitinase from Tomato *Lycopersicon esculentum*

By G. F. Pegg

Chitinases are found in a number of higher plants,[1–5] including tomato (*Lycopersicon esculentum* Mill.).[6,7] Chitin per se is not synthesized by vascular plants, but *N*-acetylglucosamine as a constituent of other plant polymers, e.g., as a glycoprotein,[8,9] is widespread in many spermatophytes, especially in seeds, including tomato.[9–12]

Sideris *et al.*[13] have reported the presence of free hexosamine in pineapple but this uncombined form appears to be exceptional.

[1] F. B. Abeles, R. P. Bosshart, L. E. Forrence, and W. H. Habig, *Plant Physiol.* **47,** 129 (1970).

[2] W. Grassmann, L. Zechmeister, R. Bender, and G. Toth, *Ber. Dtsch. Chem.* **67,** 1 (1934).

[3] J. Molano, I. Polacheck, A. Duran, and E. Cabib, *J. Biol. Chem.* **254,** 4901 (1979).

[4] T. Boller, A. Gehri, F. Mauch, and U. Vögeli, *Planta* **157,** 22 (1983).

[5] T. Tsukamoto, D. Koga, A. Ide, T. Ishibashi, M. Horino-Matsushige, K. Yagishita, and T. Imoto, *Agric. Biol. Chem.* **48,** 931 (1984).

[6] G. F. Pegg and J. C. Vessey, *Physiol. Plant Pathol.* **3,** 207 (1982).

[7] G. F. Pegg and D. H. Young, *Physiol. Plant Pathol.* **21,** 389 (1982).

Assay Method

Chitinase (EC 3.2.1.14, poly-1,4-(2-acetamido-2-deoxy)-β-D-glucoside glycanohydrolase)[14] hydrolyzes polymers of *N*-acetylglucosamine (NAG), including tetramers and, less effectively, trimers.[15] Complete hydrolysis of the resulting dimers (chitobiose) and trimers (chitotriose) to NAG requires a second enzyme, *N*-acetyl-β-glucosaminidase (EC 3.2.1.30)[14] [2-acetamido-2-deoxy-β-D-glucoside acetamidodeoxyglucohydrolase (NAGase)]. NAGase hydrolyzes terminal nonreducing 2-acetamido-2-deoxy-β-glucose residues in chitobiose and nonspecifically higher analogs of NAG and NAG-containing glycoproteins.

NAG is measured by the colorimetric method of Morgan and Elson[16] modified by Reissig *et al.*[17]

Chitobiose is not detected by dimethylaminobenzaldehyde (DMAB) and since purified tomato chitinase does not contain NAGase it is necessary to add this as an authentic enzyme in quantity sufficient to release 240 μg NAG ml^{-1} hr^{-1} to estimate chitinase activity equivalent to the hydrolysis of 32 μg chitin ml^{-1} hr^{-1} using 1.25 mg native chitin.[18]

Reagent grade β-1,4-glucosidase (Sigma Chemical Co.) from almond, containing chitobiase as an impurity, is added to the reaction mixture at 7.5 μg ml^{-1}.

Preparation of Colloidal Chitin

Colloidal chitin is prepared by a modification of Tracey's method.[19] Crab shell (200 g), finely ground, is washed in 1 liter distilled water for 1 hr and centrifuged. The residue is extracted for 15 min in 505 ml acidified ethanol–ether solution (250 ml each ethanol and diethyl ether plus 5 ml

[8] D. P. Delmer and D. T. A. Lanport, *in* "Cell Wall Biochemistry Related to Specificity in Host–Plant Pathogen Interactions" (B. Solheim and J. Raa, eds.), pp. 85–104. Universitetsforlaget, Oslo, Norway, 1977.

[9] A. Pusztai, *Nature (London)* **201,** 1328 (1964).

[10] R. F. Powning and H. Irzykiewicz, *Comp. Biochem. Physiol.* **14,** 127 (1965).

[11] D. Racusen and M. Foot, *Can. J. Bot.* **52,** 2111 (1974).

[12] J. P. Ride and R. B. Drysdale, *Physiol. Plant Pathol.* **1,** 409 (1971).

[13] C. P. Sideris, H. Y. Young, and B. H. Kraus, *J. Biol. Chem.* **126,** 233 (1938).

[14] Int. Union Pure Appl. Chem. Int. Union Biochem., "Enzyme Nomenclature," p. 443. Elsevier, Amsterdam, 1972.

[15] P. H. Clark and M. V. Tracey, *J. Gen. Microbiol.* **14,** 188 (1956).

[16] W. T. J. Morgan and L. A. Elson, *Biochem. J.* **28,** 988 (1934).

[17] J. L. Reissig, J. L. Strominger, and L. F. Leloir, *J. Biol. Chem.* **217,** 959 (1955).

[18] C. Jeuniaux, "Chitine et Chitinolyse: Un Chapitre de la Biologie Moléculaire." Masson, Paris, 1963.

[19] M. V. Tracey, *Biochem. J.* **61,** 579 (1955).

1 M HCl, all analytical grade). Following centrifugation at 14,000 g for 15 min the residue is bleached in 500 ml 0.2 M $NaClO_2$ at 75° for 1 hr with continuous stirring. After centrifuging at 14,000 g for 15 min at 5°, 10 ml acetone is added to the residue followed by concentrated HCl at 0° until the chitin is dissolved. The solution is centrifuged as before and chitin in the supernatant is precipitated over 2 hr by the addition of 1.5 liters distilled water and ice. The precipitate is washed with 3 × 1 liter distilled water centrifuged at 14,000 g for 15 min at 5° after each wash and stored as a thick paste in a sealed glass jar at 2°.

Preparation of Chitosan. Chitosan (1.0 g, Sigma Chemical Company) is dissolved in 90 ml 10% acetic acid and dialyzed against 3 × 1 liter changes of distilled water for 48 hr at 4°. Following dialysis the preparation contains 4.76 mg chitosan ml^{-1} in a volume of 210 ml.

Reagents. Native chitin from ground crab shell (Sigma Chemical Company) is decalcified by grinding with 1 N HCl and treated with 0.8 N NaOH, 0.5% $KMnO_4$, or 0.2 M sodium chlorite ($NaClO_2$) and washed to give a paste of 73 mg dry wt chitin g^{-1} wet wt. Alternatively crude crab or lobster carapace may be used as described above.

Buffer: Sodium acetate buffer, 0.1 M, pH 5.2
N-Acetylglucosaminidase (NAGase): In β-1,4-glucosidase from almond (Sigma Chemical Company), 7.5 $\mu g\ ml^{-1}$ reaction mixture
Potassium tetraborate ($K_2B_4O_7$): 0.8 M
p-Dimethylaminobenzaldehyde (DMAB): 1 g in 100 ml glacial acetic acid containing 1.25% (v/v) 10 N HCl (freshly prepared)
Bovine serum albumin (BSA): 0.2 mg ml^{-1}, pH 5.2, 0.1 M sodium acetate buffer
Chitosan (partially deacetylated chitin): 3.0 mg ml^{-1}
Inhibitor: 2-Acetamido-2-deoxy-D-gluconolactone (Koch-Light Laboratories, Ltd.), 2 mM

Procedure

A chitin suspension of 1.83 mg ml^{-1} is prepared by dispersing 25 mg chitin in pH 5.2, 0.1 M sodium acetate buffer in a Sorvall Omnimixer at 1400 rpm for 30 sec. The reaction mixture, containing 1.0 ml chitinase in sodium acetate buffer and 4.0 ml chitin suspension, 7.5 $\mu g\ ml^{-1}$ almond β-1,4-glucosidase (3.75 μg), and 0.2 mg ml^{-1} BSA (0.1 mg), is incubated at 37° for 3 hr in a water bath. The NAGase (almond β-1,4-glucosidase) is inactive on chitin.

One milliliter of sample and 1 ml distilled water are boiled in a glass ball-covered centrifuge tube for 10 min and centrifuged. NAG is determined in 0.5-ml aliquots of the supernatant by adding 0.1 ml potassium tetraborate solution and boiling for exactly 3 min in a water bath. After cooling, 3.0 ml DMAB solution is added, mixed, and kept for 20 min at 37°. Samples are cooled and read within 10 min at 585 nm and converted to NAG by reference to a calibration curve of authentic NAG in the range 0.1–0.4 *M*.

Definition of Unit of Enzyme Activity and Specific Activity (SA)

One unit of chitinase activity is defined as the amount of enzyme which produces 100 μg of NAG ml^{-1} chitin reaction mixture in 1 hr at 37° at pH 5.2. Specific activity (SA) is defined as units of enzyme activity per milligram of protein.

Purification Procedures

Step 1: Tissue Extraction. Freshly harvested stems (800 g) from 6-week-old tomato plants (*Lycopersicon esculentum* cv. "Craigella") are finely homogenized in 800 ml of pH 9.5, 0.1 *M* glycine–sodium hydroxide buffer in a Waring blender for 5 min. The homogenate is filtered through two layers of washed nylon muslin (25 μm), the filtrate centrifuged at 17,000 *g* for 30 min at 4°, and the supernatant dialyzed overnight against 2 × 3 liter changes of pH 9.5, 0.1 *M* glycine–NaOH buffer at 4°.

Step 2: DEAE Extraction. The dialyzed extract is mixed with DEAE-cellulose (Cellex D, Bio-Rad Laboratories) in the ratio of 1 g Cellex D:33 mg proteins, stirred, and filtered through Whatman No. 1 filter paper in a Büchner funnel and the Cellex D pad washed with 200 ml buffer. Recovery of chitinase at this step is 78% of total activity.

Step 3: Ammonium Sulfate Fractionation. Ammonium sulfate is added to the filtrate to 1.34 *M* (20% saturation), left for 2 hr at 4°, centrifuged at 17,000 *g* for 30 min at 4°, and the precipitate discarded. The supernatant is brought to 4 *M* (75% saturation) with ammonium sulfate and left overnight at 4°. The ammonium sulfate precipitation is collected by centrifuging at 17,000 *g* for 30 min at 4° and redissolving in a few milliliters of glycine–NaOH buffer. This solution is dialyzed for 24 hr against 3 × 1 liter changes of 1% glycine at 4°.

Step 4: Isoelectric Focusing (IEF). The enzyme sample is loaded onto a 110 LKB IEF column in a sucrose gradient together with carrier ampholytes with a pH range 3.5–10 at a final concentration 1% (w/v). After running at 500 V for 72 hr at 4° the ampholyte sucrose solutions are pumped out and collected in 4-ml fractions in a fraction collector. Protein

TABLE I
PURIFICATION OF TOMATO STEM CHITINASE

Fraction	Chitinase (U)	Chitinase recovery (%)	Protein recovered (mg)	Specific activity (U mg^{-1} protein)
Crude stem extract	117	—	1474	0.08
Cellex D	92	78.5	541	0.17
20–75% $(NH_4)_2SO_4$	51	55.8	146	0.35
IEF[a] (pH 3.5–10)	28	55.0	27	1.04
Hydroxyapatite	25	88.0	3.9	6.41
IEF (pH 3.5–10)	23	91.8	3.1	7.42
BioGel P-100	22	95.0	1.8	12.22

[a] IEF, Isoelectric focusing.

is measured by relative absorbance at 260 and 280 nm, since the ampholytes interfere with the Lowry[20] method—even after prolonged dialysis. Chitinase has a p*I* of 8.5 but the peak overlaps with β-1,3-glucanase, β-1,4-glucosidase, and β-*N*-acetylglucosaminidase activities.

Step 5: Hydroxyapatite Chromatography. Fractions 92–116, representing the chitinase peak from IEF, are pooled, dialyzed for 24 hr at 4° against 3 × 1 liter changes of pH 6.8, 100 m*M* sodium phosphate buffer. The dialyzed fraction is loaded onto a 10.6 × 1.5 hydroxyapatite column and eluted in 80 ml of a linear gradient, pH 6.8, 200 m*M* Na_2PO_4 buffer, at a flow rate of 15 ml hr^{-1}. Most chitinase elutes early in the gradient (21 m*M* Na_2PO_4 in 3-ml fractions free of glucanase, glucosidase, and *N*-glucosaminidase activity).

Step 6: IEF Repeat Stage. Fractions from the hydroxyapatite column are pooled, dialyzed for 24 hr against 3 × 1 liter changes of 1% glycine and run on IEF as in step 3. Fractions representing the chitinase peak with a p*I* of 8.5 are combined, dialyzed overnight against 2 liters pH 8.6, 100 m*M* glycine–NaOH buffer and freeze dried.

Step 7: BioGel P-100 Gel Permeation Chromatography. The freeze-dried enzyme is redissolved in 1 ml of pH 8.6, 10 m*M* glycine–NaOH buffer at 4° and chromatographed on a 40 × 1.5 cm column of BioGel P-100 in glycine–NaOH buffer at 5 ml hr^{-1}. A single peak of chitinase elutes between 35 and 45 ml of the eluate.

The efficiency of the purification procedures is shown in Table I.

[20] O. H. Lowry, N. J. Rosebrough, A. L. Farr, and R. J. Randall, *J. Biol. Chem.* **193,** 265 (1951).

Properties

The tomato chitinase is inactive on *N,N*-diacetylchitobiose, but the tetramer is readily cleaved. Deacetylated chitin (chitosan a smaller, soluble molecule) shows initial rapid hydrolysis, followed by a leveling off. The reciprocal specific viscosity (η_{sp}^{-1}) increases from 0.15 to 0.60 after 6 min, but does not reach unity until 120 min. Microbial chitinases by comparison show a linear relationship between η_{sp}^{-1} and time.

During the initial phase of chitin hydrolysis, reducing sugars[21] exceed NAG, but after a 10-hr incubation, free NAG accounts for all reducing groups detected. The addition of NAGase (present as a constitutive enzyme in tomato) to the purified enzyme doubles the yield of NAG.

Purity. During electrophoresis in pH 9.0, 10 m*M* Tris borate buffer at 220 V for 5 hr with pH 8.4 Tris borate as the electrode buffer, chitinase does not migrate to the anode consistent with the p*I* of 8.5. On native and SDS gels only a single band of enzymatic protein is found.

Stability. Both crude and purified tomato chitinases are stable for many months in the frozen state. The presence of substrate prolongs activity at higher temperatures. Addition of BSA helps to stabilize the enzyme reaction.[18,19]

Physical Properties. By coelution on BioGel P-100, with the molecular markers BSA, ovalbumin, and chymotrypsinogen, the molecular weight of the tomato enzyme is 27,000. By SDS-gel electrophoresis, the mobility of the enzyme relative to markers gave a molecular weight of 31,000.

The kinetic constants for colloidal chitin hydrolysis, calculated from a Lineweaver–Burke plot, are as follows: $K_m = 10.46$ mg dry wt chitin ml^{-1} in the presence and absence of NAGase. The V_{max} values are 97.8 and 55.6 μg NAG hr^{-1} ml^{-1} in the presence and absence of NAGase, respectively.

Temperature. Maximum activity in the absence of NAGase is 44°; with added NAGase, peak activity occurs at 37° with a shoulder at 50°.

pH. The purified enzyme has a peak of maximum activity at pH 5.1, with a small peak at pH 3.4. A similar curve is found in the presence of NAGase but with enhanced activity between pH 4 and 6.

Inhibitors. Unlike microbial chitinases, the purified tomato enzyme was unaffected by 2 m*M* 2-acetamido-2-deoxy-D-gluconolactone.

[21] N. Nelson, *J. Biol. Chem.* **153,** 375 (1944).

[62] Chitinase–Chitobiase from Soybean Seeds and Puffballs

By JOHN P. ZIKAKIS and JOHN E. CASTLE

Chitin, a cellulose-like biopolymer, was discovered in 1811 but most of our knowledge in this field originated in the last 35 years. Chitin and its chitinolytic enzymes are widely distributed in the marine and terrestrial environments. For the complete hydrolysis of chitin, most chitinolytic microorganisms as well as plants and animals[1–3] have an enzyme system consisting of two hydrolases: Chitinase (chitin glycanohydrolase, EC 3.2.1.14) and chitobiase (chitobiose acetylaminodeoxyglucohydrolase, EC 3.2.1.30). Chitin is first attacked by chitinase, releasing mostly *N,N'*-diacetylchitobiose (chitobiose), oligomers of *N*-acetylglucosamine (GlcNAc), and a small quantity of GlcNAc.[3] Chitobiase then hydrolyzes chitobiose to GlcNAc and can hydrolyze chitotriose and chitotetraose at decreasing rates as the number of GlcNAc residues increases.[3]

The unique properties of chitin (including toughness, bioactivity, biodegradability, and nonallergenicity) and the recent commercial interest in the utilization of chitin and its derivatives,[4–7] have demonstrated the need for inexpensive reliable sources of active and stable chitinase preparations.[8,9] Although chitinase is widely distributed in the plant and animal kingdoms, commercially purified chitinases are being obtained from microorganisms, especially *streptomycetes*.[8,10–14] In vertebrates, chitinases are

[1] C. Jeuniaux, this series, Vol. 8, p. 644.

[2] C. Jeuniaux and C. Cornelius, *Proc. Int. Conf. Chitin/Chitosan, 1st,* p. 542 (1978).

[3] R. A. Muzzarelli, "Chitin." Pergamon, Oxford, 1977.

[4] J. P. Zikakis (ed.), "Chitin, Chitosan, and Related Enzymes." Academic Press, New York, 1984.

[5] P. R. Austin, C. J. Brine, J. E. Castle, and J. P. Zikakis, *Science* **212,** 749 (1981).

[6] R. A. A. Muzzarelli, *Carbohydr. Polym.* **3,** 53 (1983).

[7] J. P. Zikakis, W. W. Saylor, and P. R. Austin, *in* "Chitin and Chitosan," (S. Hirano and S. Takura, eds.), p. 233. Jpn. Soc. Chitin Chitosan, Sapporo, Japan, 1982.

[8] P. A. Carroad and R. A. Tom, *J. Food Sci.* **43,** 1158 (1978).

[9] S. A. Wadsworth and J. P. Zikakis, *J. Agric. Food Chem.* **32,** 1284 (1984).

[10] A. Ohtakara, Y. Uchida, and M. Mitsutomi, *Proc. Int. Conf. Chitin/Chitosan, 1st,* p. 587 (1978).

[11] M. V. Tracey, *Biochem. J.* **61,** 579 (1955).

[12] L. R. Berger and D. M. Reynolds, *Biochim. Biophys. Acta* **29,** 522 (1958).

[13] J. Monreal and E. T. Reese, *Can. J. Microbiol.* **15,** 689 (1951).

[14] R. F. Morrissey, E. P. Dugan, and J. S. Koth, *Soil Biol. Biochem.* **8,** 23 (1976).

synthesized in and secreted by the pancreas and the gastric mucosa of insectivorous fishes, amphibians, and reptiles.[15] In some insectivorous birds and mammals, the enzyme is secreted by the gastric mucosa.[15] Recently, it was reported using bacterial chitinase for the bioconversion of shrimp processing waste into yeast single-cell protein.[16] However, the cost of chitinase production played a significant role in the overall economic feasibility of the bioconversion process.

Commercially purified chitinases isolated from microorganisms are expensive, selling for over $250/g. In addition, the level of specific activity of these chitinases varies between suppliers (and batches from the same supplier) from 0 to over 280 ImU/mg.[9] Thus, to stimulate industrial applications, it is necessary to increase the supply of active chitinase while reducing the cost of production. To achieve this goal, investigators have followed two approaches: (1) Use of recombinant DNA technology to develop chitinase-overproducing microbial strains and (2) a search for nonmicrobial, low-cost, readily available sources for extracting the enzyme. Progress has been reported using the first approach as illustrated by the following two studies. The chitinase genes from *Serratia marcescens* have been cloned in *Escherichia coli* and work is underway to develop chitinase-overproducing strains.[17] Furthermore, Soto-Gil and Zyskind[18] reported the construction and isolation of recombinant plasmids that contained the *Vibrio harveyi* (a marine bacterium) structural genes for chitinase, chitobiase, and possibly a permease for chitobiose. In induction studies with *V. harveyi*, the same investigators found that synthesis of these enzymes begins within minutes after the addition of chitobiose.

With respect to nonmicrobial sources of chitonolytic enzyme, Smirnoff[19,20] described a semiindustrial scale method for the isolation of chitinase from the gastric juice and intestinal chyme of chickens. The enzyme has been isolated from bean seeds[21] and, more recently, chitinase and chitobiase were isolated and purified from soybean seeds.[9] This chapter will include descriptions of the methods for the isolation, purification, and assay of chitinase and chitobiase from soybean seeds and puffballs.

[15] C. Jeuniaux, *Nature (London)* **192,** 135 (1961).

[16] I. G. Cosio, R. A. Fisher, and P. A. Carroad, *J. Food Sci.* **47,** 901 (1982).

[17] M. Horwitz, J. Reid, and D. Ogrydziak, *in* "Chitin, Chitosan, and Related Enzymes" (J. P. Zikakis, ed.), p. 191. Academic Press, New York, 1984.

[18] R. W. Soto-Gil and J. W. Zyskind, *in* "Chitin, Chitosan, and Related Enzymes" (J. P. Zikakis, ed.), p. 209. Academic Press, New York, 1984.

[19] W. A. Smirnoff, U.S. Patent 3,862,007 (1975).

[20] W. A. Smirnoff, *Proc. Int. Conf. Chitin/Chitosan, 1st* p. 550 (1978).

[21] R. F. Powning and H. Irzykiewicz, *Comp. Biochem. Physiol.* **14,** 127 (1965).

Enzyme Preparation from Soybean Seeds

The isolation and purification of the chitinase enzyme system from soybean seeds were first reported by Wadsworth and Zikakis[9] and the methodology is described here. Soybean seeds [*Glycine max* (L) Merrill cv. "Ware"] are ground in a Wiley mill to pass a 20-mesh sieve. The ground beans (40–60 g) are then extracted with 0.05 *M* citrate buffer, pH 4.5 (10 g of beans/100 ml of buffer), for 2 hr at 4° on a magnetic stir plate. The mixture is then centrifuged for 20 min at 12,500 *g* and the residue discarded. The supernatant is passed through Schleicher and Schuell No. 595 filter paper. Solid ammonium sulfate is added to 50% saturation and the solution placed on a magnetic stir plate at 4° for 1 hr. The suspension is then centrifuged for about 20 min at 12,500 *g* and the supernatant discarded. The precipitate is redissolved in 0.05 *M* citrate buffer (pH 4.5) at 4° overnight on a magnetic stir plate. The suspension is centrifuged and filtered as above, and the filtrate (400–600 ml) chromatographed.

Affinity Chromatography

The type and size of column to be used in the affinity chromatography may differ from the one described below in order to meet the needs and objectives of a given laboratory. Ground chitin (40 mesh) is the ligand which was extracted from tanner crab shells (Bioshell, Inc.) using the method of Block and Burger.[22] The chitin is washed successively with tap water, 0.05 *N* HCl, and 1% sodium carbonate. The clean chitin is then packed into a glass column (20 × 50 cm) to give a bed volume of about 70 ml. The column is then equilibrated with 0.05 *M* citrate buffer (pH 4.5) and the final flow rate of 2 ml/min is regulated with a Manostat peristaltic cassette pump. Following the application of the soybean extract obtained in the previous section, the column is successively eluted with 1 liter 0.01 *M* Tris-HCl–1 *M* NaCl (pH 7.4), 1 liter of 0.01 *M* Tris–HCl (pH 7.4), 2 liters of 0.05 *N* HCL adjusted to pH 3.3, and 1 liter of 0.05 *M* citrate buffer (pH 4.5) as is modified from Block and Burger.[22] Chromatography is carried out at room temperature (25°) and 8-ml fractions are collected using a Büchler Fractomette Alpha 2000 or an ISCO Foxy fraction collector. Optical density of the eluents is monitored with a Büchler Fractoscan set at 280 nm. All fractions containing protein are stored at 4° for further analyses. The chitin column yields a reproducible elution profile which has five protein peaks containing chitinase, with three peaks containing chitobiase activity as well.[9] The first three peaks contain most

[22] R. Block and M. M. Burger, *Biochem. Biophys. Res. Commun.* **58,** 13 (1974).

of the chitinase and all the chitobiase. The purification achieved with this column is almost 260-fold with an average chitinase specific activity of 89 ImU/mg and an average yield of about 12%. Although chitinase as purified is of high purity (has higher specific activity than two out of three commercial chitinases tested), discontinuous polyacrylamide gel electrophoresis and sodium dodecyl sulfate electrophoresis indicated the presence of residual protein impurities.[9] If homogeneous chitinase is desired, we recommend using high-purity chitin as column packing material rather than technical grade chitin. In fact, high-purity chitin may even increase the enzyme yield.

Substrates

Regenerated chitin is the substrate used in most studies to measure chitinase activity. The synthetic substrate is prepared by the reacetylation of chitosan (Sigma Chemical Co.) according to the method described by Molano *et al.*,[23] using acetic anhydride. However, because the resulting dried chitin suspension tends to form intractable particles that could not be resuspended, we devised the following procedure to obtain the desired concentration.[9] The suspension is placed on a magnetic stir plate to ensure a uniform distribution of material and 1 ml removed with an automatic pipettor (the suspension tends to clog glass pipets). This aliquot is vacuum filtered through a preweighed piece of Schleicher and Schuell No. 595 filter paper and the paper dried overnight at room temperature (25°). The paper is then weighed, the initial weight subtracted, and the remainder is the weight of chitin present in 1 ml of the original suspension. The suspension of regenerated chitin is then adjusted to bring the concentration to 15 mg/ml.[23] Other substrates such as technical grade chitin, purified chitin, microcrystalline chitin, and chitosan may be used without modification.

Chitinase Assay

The assay described below is essentially identical to a previously published procedure.[9] Chitinase (250 μl) is incubated in 15-ml test tubes with 1 ml of 0.05 *M* citrate buffer (pH 3.5), 50 ml of chitobiase (Sigma Chemical Co.), and 0.5 ml of regenerated chitin suspension (15 mg/ml). The volume is adjusted to 1 ml with glass-distilled, deionized water and the tubes are placed in a magnetic stirring plate (Tri-R Instruments, Inc., Rockville Centre, NY). This apparatus is capable of stirring up to eight test tubes simultaneously. The mixtures are incubated for 3 min at room

[23] J. Molano, A. Duran, and E. Cabib, *Anal. Biochem.* **83,** 648 (1977).

temperature (25°) with constant stirring and then centrifuged for 3 min at 900 *g*. The supernatants are then assayed for GlcNAc by a modification of the method of Reissig *et al.*[24]

Samples (0.5 ml) of each supernatant are mixed with 0.25 ml of 0.27 *M* potassium tetraborate solution, followed by heating in a boiling water bath for exactly 8 min. The solutions are cooled with tap water, 3 ml of a 1% dimethylaminobenzaldehyde solution is added, and the solutions are vortexed. After 20 min at 37°, the absorbance of the samples is measured at 585 nm on a Gilford 250 spectrophotometer. A reagent blank containing 0.5 ml of water and appropriate enzyme and substrate blanks were treated in the same fashion. The concentration of GlcNAc in solution is determined from a standard curve prepared with four GlcNAc reference solutions (0.1, 0.4, 0.7, and 1.0 μmol/ml) treated as above. All enzyme activities are expressed in International Units (IU) per milliliter, with 1 IU defined as the production of 1 μmol of GlcNAc/min at 25°. Determination of chitinase activity on native chitins is carried out as described, with the substitution of 2.5 mg of chitin and 0.5 ml of glass-distilled, deionized water for the 0.5 ml of regenerated chitin suspension.

Chitobiase Assay

Enzyme solution (250 μl) is incubated with 125 μl of chitobiose solution (2.5 mg/ml) and 125 μl of 0.05 *M* citrate buffer (pH 3.5) for 30 min at room temperature (25°),[9] as modified from Ohtakara.[25] Chitobiase activity is present in the first three protein peaks of the elution profile of the chromatographed soybean seeds extract and its ranges from 1.7 to 2.2 ImU/ml.[9]

In summary, the described method for purifying chitinase from soybean seeds is simple, inexpensive, and yields chitinase of higher specific activity than two out of three commercially available chitinases tested.[9] Further tests indicate the enzyme is an endochitinase and that the preparation contains isoenzymes. The chitinases have a relatively low pH optimum in the range of 3.3 to 4.0. Activity on regenerated chitin is considerably greater than on more crystalline natural substrates.[9]

Enzyme Preparation from Puffballs

Chitin is a structural constituent of the cell walls of nearly all fungi. Thus, it is not surprising that certain fungi would generate chitinase when

[24] J. L. Reissig, J. L. Strominger, and L. F. Leloir, *J. Biol. Chem.* **217,** 959 (1955).
[25] A. Ohtakara, *Agric. Biol. Chem.* **27,** 454 (1963).

self-destruction is a normal and necessary event in their life cycles. Such is the situation with puffballs, especially of the family Lycoperdaceae, since these fungi must release their complement of spores at the end of the growing season through rupture of their rinds (peridia).[26] Anyone who has walked through a meadow in the fall has probably experienced the clouds of brown or purple spore "smoke" emitted by puffball sporangia crushed underfoot.

The presence of significant chitinase in maturing puffballs was first reported by Tracey in *Lycoperdon* species.[27] The fungi were collected when nearly ripe but before the contents were dry, and the liquid contents separated by squeezing through cloth or by centrifuging. Up to 70% of the fresh weight was obtained as liquid, and a single specimen of *Lycoperdon giganteum* was said to have produced several liters of enzyme. *Lycoperdon pyriforme* was also recommended as a source. Tracey[27] described the enzyme extract from the puffballs as "powerful" and said it lost only half its activity when stored at 5° under toluene. His procedure involved incubating chitin with 0.1–0.5 ml of dialyzed extract in 0.08 *M* acetate buffer at pH 5.0 and 35°. Samples containing up to 5 mg of chitin were completely hydrolyzed to GlcNAc by these amounts of enzyme preparation in 10 days. Thus, it is obvious that the extracts also contained chitobiase.[5]

Enzyme Preparation

The work described in this and subsequent sections with chitinase from puffballs has not been previously published.[28] Two local species of puffballs *Calvatia cyathiformis* and *Lycoperdon candidum,* were collected from a meadow in southern Delaware. *Calvatia* was employed exclusively, since preparations from *Lycoperdon* had very low, nonreproducible activity. As *Calvatia* matured in the fall, the glebas (spore mass chambers) began to turn yellow. At this stage, the puffballs were harvested and the glebas and their contents were removed and suspended in a minimum amount of 0.5 *M* Tris buffer at pH 7.4 for 1 hr at 4°. The liquid was then collected by pressing the suspension on a 400-mesh Nytex screen (Tetko, Inc., Elmsford, NY), followed by passage through a 1.2-μm glass fiber filter. This liquid, containing the chitinolytic enzymes, could be preserved in the frozen state without apparent diminution in activity.

[26] N. L. Marshall, "The Mushroom Book," p. 123. Doubleday-Page, New York, 1923.

[27] M. V. Tracey, *in* "Modern Methods of Plant Analysis" (K. Paech and M. V. Tracey, eds.), p. 271. Springer-Verlag, Berlin and New York, 1955.

[28] J. R. Deschamps, Ph.D. dissertation. University of Delaware, College of Marine Studies, Lewes, Delaware, 1984.

Chitinase Assay

The enzyme preparations obtained in this manner were assayed in comparison with soybean seed chitinase and with commercially available chitinases. The procedure was that described above for soybean seed chitinase. The *Calvatia* preparation had an activity of 870 ImU/mg, comparing favorably with highly purified soybean chitinase at 1100 ImU/mg. The purchased chitinases averaged 90 ImU/mg. Since the puffball chitinase extract had an activity of 175 ImU/ml, it could be employed directly in chitin hydrolysis studies and did not need to be purified further. Protein impurities present did not interfere with the chitinolytic activity in any detectable manner.

Chitin Preparation

Hydrolysis studies were conducted with chitin isolated from horseshoe "crab" shells, *Limulus polyphemus,* by a modification of Rutherford's procedure.[29] Air-dried shells were ground in a Waring Blender. To remove protein, 200 g of the coarse material was suspended in 3 liters of 0.5% sodium hydroxide at room temperature for 1 hr. The solid was separated, washed with tap water, and resuspended in 3 liters of 5% sodium hydroxide for 5 hr. After separation and washing, the solid was again suspended in 5% sodium hydroxide for 18 hr. The product was collected on a filter, washed with deionized water until neutral, air dried, and then ground in a Wiley mill to pass a 40-mesh screen. These procedures remove nearly all of the protein, leaving only the protein covalently bonded to the polysaccharide chains.[30] Since *L. polyphemus* shells do not contain calcium carbonate, a demineralization step with acid is not required.

Chitin Hydrolysis

The enzymatic digestion of the deproteinized chitin was conducted at pH 6.0 for 24 hr. About 10 mg of the 40-mesh chitin and 1 ml of the enzyme preparation (175 ImU) were placed in the water-jacketed reaction vessel of a Fisher model 90 recirculating water bath held at 35°. Pooled products (35 ml) from these digestions were adjusted to pH 2 with acetic acid and partially purified by chromatography through a column (1 × 25 cm) packed with Amberlite XAD-7 resin (Rohm and Haas, Inc., Bris-

[29] F. A. Rutherford, M.S. thesis. University of Delaware, College of Marine Studies, Newark, Delaware, 1975.

[30] C. J. Brine, *in* "Chitin and Chitosan" (S. Hirano and S. Tokura, eds.), p. 105. Jpn. Soc. Chitin Chitosan, Sapporo, Japan, 1982.

tol, PA). Elution was with 150 ml of dilute acetic acid (pH 2) followed by 150 ml of methanol, and UV absorptions at 280 and 260-nm on successive 21-ml fractions were employed for product detection. Samples of the original chitinase were similarly chromatographed to identify chitinase-containing fractions. The chitinase-degraded fractions were not characterized at this stage, but were subsequently converted through action of trypsin to glycopeptides containing one or two molecules of combined GlcNAc. This was accomplished by first lyophilizing combined fractions from the same peak. These solids were dissolved in 0.5 *M* phosphate buffer (pH 7.0) and trypsin (Research Plus Chemical Co.) was added to a concentration of 0.075 mg/ml (about 800 U/ml). Proteolytic digestion was carried out at 25° for 6 hr in the water-jacketed vessel described above. The minimal proportions of amino sugar remaining in these peptides, as revealed by HPLC separation, hydrolysis, and colorimetric analysis, testify to the efficacy of the puffball chitinase in degrading the polysaccharide chains.

It is suggested that chitinase preparations from selected puffball species represent a most convenient source of chitinolytic activity. Extensive purification is not required for purposes such as described here, and the preparations maintain their activity when frozen, at least for periods of several months. We made no attempt to concentrate the enzymes present or to remove any possible inhibitors.

Note. Although "true" puffballs are reported to be nonpoisonous, there is confusion regarding the edibility of some genera.[31] Thus, it would be prudent to avoid ingesting any parts of these fungi, including the extracts.

Acknowledgment

The authors are indebted to Dr. J. R. Deschamps for the field collection and laboratory analysis of chitinase from puffballs. This work was partially supported by the U.S. Department of Commerce, National Oceanic and Atmospheric Administration, Office of Sea Grant. Published as Miscellaneous Paper No. 1141 of the Delaware Agricultural Experiment Station.

[31] O. K. Miller, Jr., "Mushrooms of North America," p. 298. Dutton, New York, 1981.

[63] Endochitinase from Wheat Germ

By ENRICO CABIB

An electrophoretically homogeneous preparation of chitinase can be easily obtained in milligram amounts from defatted wheat germ.[1] The enzyme is a true endochitinase, i.e., it liberates only oligosaccharides of various chain length from chitin. Its fluorescein isothiocyanate derivative is suitable for the localization of chitin by fluorescence microscopy.[2]

Assay Method

The chitinase can be assayed by the liberation of tritiated oligosaccharides from [*acetyl*-^{3}H]chitin,[3] with phosphate buffer at pH 6, final concentration in the reaction mixture, 0.05 *M*.

Purification Procedure

Defatting of Wheat Germ. One kilogram of wheat germ (Sigma Chemical Co., St. Louis, MO) is suspended into 2500 ml of petroleum ether and the suspension is shaken for 15 min at 2°. The solvent is decanted and another extraction is carried out in the same way. The defatted yeast germ is spread out on aluminum foil and dried at room temperature under a hood.

Step 1: Extraction. All operations are carried out at about 2°, except where indicated otherwise. Defatted wheat germ, obtained from 1 kg of untreated material, is suspended in 4 liters of 20 m*M* sodium bicarbonate and the mixture is shaken for 1 hr at 2°. After eliminating large particles by passage through a Büchner funnel, the residue is extracted again in the same fashion with 2.5 liters of solution. The combined filtrates are centrifuged for 20 min at 11,000 *g*, and the pellet is discarded.

Step 2: Precipitation at pH 4.5. The crude extract (see Table I) is brought to pH 4.5 with 2 *M* acetic acid. After standing at room temperature for 15 min, the suspension is centrifuged in the cold (5°) for 20 min at 11,000 *g*. The pellet is discarded and the supernatant liquid is adjusted to pH 8.5 with 4 *N* NaOH. Sodium azide is added to a final concentration of

[1] J. Molano, I. Polacheck, A. Duran, and E. Cabib, *J. Biol. Chem.* **254,** 4901 (1979).
[2] J. Molano, B. Bowers, and E. Cabib, *J. Cell Biol.* **85,** 199 (1980).
[3] E. Cabib, this volume [48].

TABLE I
PURIFICATION OF WHEAT GERM CHITINASE

Purification step	Volume (ml)	Total units[a]	Total protein	Yield (%)	Specific activity
Crude extract	4,500	2,610	54,000	100	0.048
Supernatant (pH 4.5)	4,450	2,540	38,300	97	0.066
Chitin eluate	300	660	180	25	3.7
Sephadex G-50 eluate	65	474	32	18	14.60

[a] One unit of enzyme is defined as the amount that catalyzes the liberation of 1 μmol of oligosaccharide (calculated as *N*-acetylglucosamine)/min at 30°.

0.02%. All solutions used in subsequent steps contain sodium azide in the same concentration to prevent growth.

Step 3: Chromatography on Chitin Column. Chromatography on a chitin column is performed at room temperature (~25°). The acid precipitation supernatant (4450 ml) is added to a chitin[4] column, 6.2 cm in diameter and 20 cm long, previously equilibrated with 20 m*M* sodium bicarbonate. Passage of the extract through the column takes 9 to 10 hr, with retention of about 70% of chitinase activity. After washing successively with 1.3 liters of 20 m*M* sodium bicarbonate and with 1.8 liters of 20 m*M* sodium acetate at pH 6.3, a linear gradient is applied with 1 liter of 20 m*M* acetic acid, pH 3.3, in the reservoir and 1 liter of 20 m*M* sodium acetate, pH 5.3, in the mixing flask. Upon completion of the gradient, an additional 1.5 liters of 20 m*M* acetic acid is applied. During the elution 17-ml fractions are collected. Fractions corresponding to the main protein peak are adjusted immediately to pH 8.3 to 8.5 with a mixture containing 2 *M* Tris and 3.3 *M* NaOH. Chitinase and red blood cell-agglutinating activity are eluted together in this column. Fractions containing chitinase activity are pooled and concentrated at 2° in an Amicon filtration assembly fitted with a PM10 filter, to a volume of 13.5 ml. If a precipitate forms, it is removed by centrifugation in the cold for 15 min at 24,000 *g*.

Step 4: Chromatography on Sephadex G-50 Column. This chromatography is carried out at 2°. The concentrated chitin column eluate, 13 ml, is applied to a Sephadex G-50 (fine) column (2.5 × 95 cm) previously equilibrated with 5 m*M* hydrochloric acid. Elution is carried out with the same solvent and 7-ml fractions are collected. In this column, the chitinase emerges first, followed by the agglutinating activity (Fig. 1). Immediately after measuring the absorbance at 280 nm, fractions 30 to 60 are brought

[4] Prepared by reacetylation of chitosan, as described in this volume [48], but using unlabeled acetic anhydride.

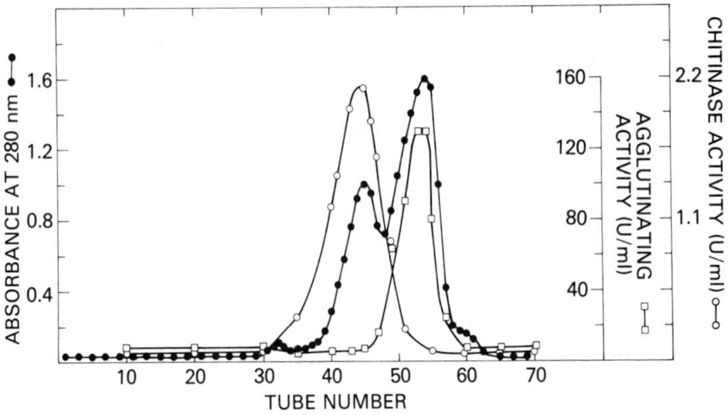

FIG. 1. Chromatography of wheat germ chitinase and agglutinin on Sephadex G-50 column. The agglutinin activity was measured as described.[1]

to pH 8.5 with 1.5 *M* Tris, containing 0.5 *M* NaOH. After measuring chitinase and agglutinating activity, fractions containing chitinase are pooled. The enzyme is stored in this solution, in the presence of 0.02% azide.

The results of a typical purification are shown in Table I.

Properties

Stability. The purified enzyme is stable for at least several weeks when stored at 2 or −20° and at least for 7 years when stored at −80°.

Physical Characteristics. Upon sodium dodecyl sulfate-polyacrylamide gel electrophoresis, the purified enzyme gives rise to a single band with an estimated molecular weight of 30,000. The value obtained by sedimentation equilibrium was 33,000 and that calculated from gel filtration 29,000. Thus, the enzyme is a monomer in solution.

Kinetics. Plots of product formation vs time are curved, therefore a short time assay (5 to 10 min) is recommended for accurate measurements of activity. The pH optimum is 6, with 75% of the activity at pH 3 and 53% at pH 9. The K_m for chitin (reacetylated chitosan: see Assay procedure) is 2 m*M* (chitin concentration calculated as *N*-acetylglucosamine).

The products of the reaction, with reacetylated chitosan[3] as substrate, are oligosaccharides, from diacetylchitobiose to tetraacetylchitotetraose. With nascent chitin (see below) higher saccharides are also formed. Thus, the enzyme behaves as an endochitinase.

The wheat germ enzyme was the first chitinase in which differential activity on nascent or preformed chitin was detected. The chitinase hydrolyzes chitin produced by chitin synthetase in the same reaction mixture much faster (about two orders of magnitude) than preformed chitin.[1] This result has been attributed to the great susceptibility to chitinase hydrolysis of the single chains of nascent chitin, compared to the bundles of chains that form later by hydrogen bonding.[1] This differential activity was also found in chitinases from other sources[1,5] but is most marked in the wheat germ enzyme. Therefore, this chitinase is rather inefficient for the total hydrolysis of native chitin and is not suitable for the enzymatic determination of this polysaccharide.[6]

[5] J. U. Correa, N. Elango, I. Polacheck, and E. Cabib, *J. Biol. Chem.* **257,** 1392 (1982).
[6] E. Cabib and A. Sburlati, this volume [55].

[64] Chitosanase from *Bacillus* Species

By YASUSHI UCHIDA and AKIRA OHTAKARA

Several chitosanases have been isolated from some microorganisms,[1-6] including fungi, bacteria, and actinomycetes. *Bacillus* sp. No. 7-M (FERM-P-8139), which is a mutant isolated from soil, produces potent extracellular chitosanase when grown on colloidal chitosan as an inducible substance. The purified chitosanase can be used for the preparation of glucosamine oligomers[7] having antibacterial and antifungal activities.[8] Procedure for the purification of chitosanase from *Bacillus* sp. No. 7-M and the properties of the enzyme are described below.

[1] A. Hedges and R. S. Wolfe, *J. Bacteriol.* **120,** 844 (1974).
[2] Y. Tominaga and Y. Tsujisaka, *Biochim. Biophys. Acta* **410,** 145 (1975).
[3] J. S. Price and R. Storck, *J. Bacteriol.* **124,** 1574 (1975).
[4] D. M. Fenton, B. Davis, C. Rotgers, and D. E. Eveleigh, *Proc. Int. Conf. Chitin/Chitosan, 1st,* p. 525 (1978).
[5] D. M. Fenton and D. E. Eveleigh, *J. Gen. Microbiol.* **126,** 151 (1981).
[6] A. Ohtakara, H. Ogata, Y. Taketomi, and M. Mitsutomi, *in* "Chitin, Chitosan, and Related Enzymes" (J. P. Zikakis, ed.), p. 147. Academic Press, New York, 1984.
[7] M. Izume and A. Ohtakara, *Agric. Biol. Chem.* **51,** 1189 (1987).
[8] D. F. Kendra and L. A. Hadwiger, *Exp. Mycol.* **8,** 276 (1984).

Assay Methods

Chitosanase Activity

Principle. The assay is based on the estimation of reducing sugars produced in the hydrolysis of soluble chitosan by a modification of Schales' method[9] with glucosamine as a reference compound.

Reagents

Soluble chitosan, 1% (w/v), 1.0 g of chitosan (Katakura Chikkarin Co., Ltd., Tokyo, 99% deacetylated) suspended in 40 ml of deionized water, 9 ml of 1.0 *M* acetic acid added, dissolved under stirring for 2 hr, adjusted to pH 6.0 with 1.0 *M* sodium acetate, and made up to 100 ml with 0.05 *M* acetate buffer, pH 6.0

Schales' reagent, 0.5 g of potassium ferricyanide in 1 liter of 0.5 *M* sodium carbonate, stored in a brown bottle

Procedure. The reaction mixture consisting of 1 ml of 1% soluble chitosan and 1 ml of enzyme solution is incubated for 10 min at 30°. The reaction is stopped by boiling for 3 min. A 1.5-ml sample of the above solution is mixed with 2 ml of Schales' reagent in test tubes. The tubes are covered with aluminum foil and heated in a boiling water bath for 15 min. After cooling, the solution is filtered and the absorbance at 420 nm measured versus water.

Unit. One unit of chitosanase activity is defined as the amount of the enzyme which produces 1 μmol of reducing sugar as glucosamine per minute under the above conditions.

Purification Procedure

Step 1: Preparation of Culture Filtrate. Bacillus sp. No. 7-M is grown and maintained on agar slants in a medium, pH 6.8, consisting of 0.5% yeast extract, 0.5% polypeptone, and 0.3% meat extract. The strain is inoculated into a test tube (2.6 × 20 cm) containing 10 ml of a seed culture medium of the same composition as described above and grown for 1 day at 30° on a reciprocal shaker. One milliliter of the seed culture is transferred into a 200-ml conical flask containing 50 ml of an enzyme production medium, pH 7.2, consisting of 0.5% colloidal chitosan, 0.8% yeast extract, 0.4% polypeptone, and 0.2% meat extract and cultivated for 1 day at 30° on a circular shaker. Forty milliliters of the culture is further transferred into a 5-liter conical flask containing 1 liter of the enzyme production medium and cultivated for 4 days at 30° on a circular shaker

[9] T. Imoto and K. Yagishita, *Agric. Biol. Chem.* **35,** 1154 (1971).

for the production of chitosanase. The culture filtrate is obtained by centrifugation from the above culture broth. All operations of purification are performed at 0–5° unless otherwise specified.

Step 2: Precipitation with Ammonium Sulfate. The culture filtrate (1810 ml) is brought to about 80% saturation (561 g/l) by adding solid ammonium sulfate. After standing overnight, the precipitate produced is collected by centrifugation, dissolved in deionized water, and dialyzed against 0.02 *M* phosphate buffer, pH 6.0 (buffer P) for 2 days.

Step 3: Chromatography on CM-Sephadex C-50. The dialyzed enzyme solution (210 ml) is applied to a column (2.6 × 45 cm) of CM-Sephadex C-50 previously equilibrated with buffer P. The column is completely washed with buffer P (350 ml) and then eluted with a continuous linear gradient of 300 ml of buffer P and 300 ml of the same buffer containing 0.5 *M* sodium chloride, at a flow rate of 25 ml/hr. The eluate is collected in 5-ml fractions which are assayed for the chitosanase activity. The fractions (tubes 220–240) containing chitosanase are pooled and concentrated by ultrafiltration with Amicon membrane PM10.

Step 4: Gel Filtration on Sephadex G-100. The enzyme solution (5.8 ml) from the preceding step is loaded on a Sephadex G-100 column (1.4 × 80 cm) previously equilibrated with buffer P. The column is eluted with buffer P containing 0.25 *M* sodium chloride, at a flow rate of 20 ml/hr. Fractions of 5 ml are collected. The enzyme is eluted at a position of about 1.6 times of the void volume. The active fractions (tubes 49–62) are pooled and dialyzed against buffer P.

Step 5: Rechromatography on CM-Sephadex C-50. The enzyme solution (70 ml) obtained from step 4 is again applied to a column (0.9 × 15 cm) of CM-Sephadex C-50 previously equilibrated with buffer P. The column is eluted with a linear salt gradient of 100 ml of buffer P and the same buffer containing 0.5 *M* sodium chloride, at a flow rate of 30 ml/hr. Fractions of 4 ml are collected. The enzyme activity appears at fractions eluted with about 0.26 *M* sodium chloride as a single peak. Fractions (tubes 38–49) are pooled as purified enzyme. The results of purification are summarized in Table I. Chitosanase thus obtained is purified about 320-fold.

Properties of the Enzyme

Homogeneity and Molecular Weight. The purified chitosanase is homogeneous on polyacrylamide and SDS–polyacrylamide disk gel electrophoresis. The molecular weight of the enzyme is estimated to be about 41,000 by SDS–polyacrylamide disk gel electrophoresis, and about 30,000 by gel filtration on Sephadex G-100.

TABLE I
PURIFICATION OF CHITOSANASE FROM *Bacillus* sp. NUMBER 7-M

Step	Protein (mg)	Activity (U)	Specific activity (U/mg)	Yield (%)	Purificatio (-fold)
Culture filtrate	4742	4733	1.0	100	1.0
Ammonium sulfate precipitation	591	7238	12.2	153	12
Dialysis solution	271	4738	17.5	100	18
CM-Sephadex C-50 (I)	8.5	3473	409	73.4	409
Sephadex G-50	5.3	1726	326	36.5	326
CM-Sephadex C-50 (II)	2.7	862	319	18.2	319

Optimum pH and Stability. The optimum pH of the enzyme activity is found at pH 6.0. The enzyme is relatively stable between pH 5.0 and 11.0 when kept at 37° for 3 hr, but completely inactivated below pH 2.0 and above pH 12.0.

TABLE II
SUBSTRATE SPECIFICITY OF CHITOSANASE

Substrates (0.25%)	Decomposition products[a] (mg/mg protein/60 min)	
	Total reducing sugar[b]	Amino sugar[c]
Powdered chitosan	0	0
Colloidal chitosan	1059	190
Soluble chitosan	779	189
Glycol chitosan	630	92
Powdered chitin	0	0[d]
Colloidal chitin	0	0[d]
Glycol chitin	0	0[d]
Cellulose	0	ND[e]
CMC	147	ND[e]
Methylcellulose	0	ND[e]

[a] The reaction mixture consisting of 1 ml of 1% substrate, 2 ml of 0.05 *M* acetate buffer, pH 6.0, and 1 ml of enzyme solution is incubated at 37° for 60 min.

[b] Total reducing sugar is determined by a modification of Schales' method [T. Imoto and K. Yagishita, *Agric. Biol. Chem.* **35,** 1154 (1971)].

[c] Glucosamine is determined by the method of Rondle and Morgan [C. J. M. Rondle and W. T. J. Morgan, *Biochem. J.* **61,** 586 (1955)].

[d] *N*-Acetylglucosamine is determined by the method of Reissig *et al.* [J. L. Reissig, J. L. Strominger, and L. F. Leloir, *J. Biol. Chem.* **217,** 959 (1955)].

[e] ND, Not determined.

Optimum Temperature and Stability. The optimum temperature of the enzyme activity is at 50°. The enzyme is stable below 50° when kept at pH 6.0 for 15 min, but loses about 40% of original activity at 60° and complete activity at 70°.

Effect of Substrate Concentration. The apparent K_m values for soluble chitosan and glycol chitosan are approximately 3.3 and 2.0%, respectively.

Substrate Specificity. The substrate specificity of the enzyme is presented in Table II. The purified enzyme hydrolyzes colloidal chitosan, soluble chitosan, and glycol chitosan, but does not attack powdered chitosan, powdered chitin, colloidal chitin, and glycol chitin. Amino sugar is produced in the hydrolysis of chitosans excluding powdered chitosan, but no *N*-acetylglucosamine is detected in the hydrolysis of any compounds of chitin. The enzyme hydrolyzes carboxymethylcellulose (CMC) at a considerably slower rate, but does not act on cellulose and methylcellulose.

Mode of Action. The products formed in the hydrolysis of soluble chitosan with the enzyme are analyzed by HPLC using a TSK-GEL NH_2-60 column. Glucosamine oligomers with the degrees of polymerization from 2 to 8 are produced during the hydrolysis and a slight amount of glucosamine is detected after prolonged reaction. In the decomposition of soluble chitosan with chitosanase, a rapid decrease in the viscosity is observed before the formation of reducing sugar and amino sugar. Therefore, the mode of action of chitosanase from *Bacillus* sp. No. 7-M on chitosan is endo-type and essentially the same as that of chitosanase from *Streptomyces griseus.*[10]

[10] A. Ohtakara, this volume [65].

[65] Chitosanase from *Streptomyces griseus*

By AKIRA OHTAKARA

Chitosanases are useful enzymes for the lysis of the cell walls of the fungi belonging to Mucorales and have been purified from some microorganisms.[1-4] The purified chitosanases are classified into two

[1] A. Hedges and R. S. Wolfe, *J. Bacteriol.* **120,** 844 (1974).
[2] J. S. Price and R. Storck, *J. Bacteriol.* **124,** 1574 (1975).
[3] Y. Tominaga and Y. Tsujisaka, *Biochim. Biophys. Acta* **410,** 145 (1975).
[4] D. M. Fenton and D. E. Eveleigh, *J. Gen. Microbiol.* **126,** 151 (1981).

groups[5]: the enzymes hydrolyzing only chitosan and the enzymes hydrolyzing both chitosan and carboxymethylcellulose. A strain of *Streptomyces griseus*[6] has produced chitosanase in culture broth with chitosan as the single carbon and nitrogen source. The purified enzyme is able to hydrolyze chitosan and carboxymethylcellulose, and produces glucosamine oligomers in the hydrolysis of chitosan. A procedure for the purification of chitosanase from *Streptomyces griseus* is described below.

Assay Methods

Chitosanase Activity[7]

Principle. The assay is based on the estimation of amino sugars produced in the hydrolysis of glycol chitosan, a water-soluble derivative of chitosan, by the method of Rondle and Morgan,[8] using glucosamine as a reference compound.

Reagents

Glycol chitosan (Wako Pure Chemical Co., Ltd., Osaka, Japan), 2% in 0.1 *M* sodium phosphate buffer, pH 7.5

Acetylacetone, 1 ml in 50 ml of 0.5 *N* Na_2CO_3, use within 1 hr

Ethanol, absolute ethanol

Ehrlich reagent, 0.8 g of *p*-dimethylaminobenzaldehyde in 30 ml of ethanol, added to 30 ml of concentrated HCl, stored in a cold room

Procedure. The reaction mixture, consisting of 1 ml of 2% glycol chitosan in 0.1 *M* phosphate buffer, pH 7.5, and 1 ml of enzyme solution, is incubated for 10 min at 37°. The reaction is stopped by boiling for 4 min. A 1-ml sample of the above solution is mixed with 1 ml of deionized water and 1 ml of acetylacetone reagent in 10-ml graduated tubes. The tubes are covered with glass caps and heated for 20 min in a boiling water bath. After cooling, 5 ml of ethanol and then 1 ml of Ehrlich reagent are added. The tubes are filled up to the 10-ml gradation with ethanol and mixed gently. After incubation for 10 min at 65–70°, the solution is centrifuged to remove undecomposed substrate and the absorbance at 530 nm is measured. Controls on substrate and enzyme only are also tested.

[5] D. M. Fenton, B. Davis, C. Rotgers, and D. E. Eveleigh, *Proc. Int. Conf. Chitin/Chitosan, 1st,* p. 525 (1978).

[6] A. Ohtakara, H. Ogata, Y. Taketomi, and M. Mitsutomi, *in* "Chitin, Chitosan, and Related Enzymes" (J. P. Zikakis, ed.), p. 147. Academic Press, New York, 1984.

[7] Chitosanase activity can be also assayed using soluble chitosan as substrate at the reaction pH below 6.0 (see this volume [64]).

[8] C. J. M. Rondle and W. T. J. Morgan, *Biochem. J.* **61,** 586 (1955).

Unit. One unit of chitosanase activity is defined as the amount of the enzyme which produces 1 μmol of amino sugars as glucosamine per minute under the above condition.

Carboxymethyl Cellulase (CMCase) Activity

Principle. The assay is based on the estimation of reducing sugars produced in the hydrolysis of carboxymethylcellulose by the Somogyi–Nelson method,[9] with glucose as a reference compound.

Reagents

Carboxymethylcellulose, DE 0.6, 1%
Sodium phosphate buffer, 0.1 *M*, pH 6.5
Reagents for Somogyi–Nelson method:
  Alkaline copper reagent
  Arsenomolybdate reagent

Procedure. The reaction mixture containing 1 ml of 1% carboxymethylcellulose, 2 ml of 0.1 *M* phosphate buffer, pH 6.5, and 1 ml of enzyme solution is incubated for 20 min at 37°. The reaction is stopped by boiling for 4 min. A one-milliliter aliquot of the digest is added to 1 ml of alkaline copper reagent in 25-ml graduated tubes. The tubes are covered with glass-caps and heated for 10 min in a boiling water bath. After cooling, 1 ml of arsenomolybdate reagent is added. The solution is diluted to 25 ml and the absorbance at 660 nm is measured against substrate and enzyme blanks.

Unit. One unit of CMCase activity is defined as the amount of the enzyme which produces 1 μmol of reducing sugars as glucose per minute under the above condition.

Purification Procedure[6]

Preparation of Culture Filtrate. Streptomyces griseus HUT 6037 is grown and maintained on agar slants in a medium, pH 7.0, consisting of 1.0% mannitol, 0.2% polypeptone, 0.1% meat extract, 0.1% yeast extract, and 0.05% $MgSO_4 \cdot 7H_2O$. The strain is inoculated into a test tube (2.6 × 20 cm) containing 10 ml of a seed culture medium with the same composition as in the above medium, except that 0.05% $MgSO_4 \cdot 7H_2O$ is omitted, and grown for 2 days at 30° on a reciprocal shaker. One milliliter of the seed culture is transferred into a 200-ml conical flask containing 50 ml of a production medium, pH 7.0, consisting of 0.5% chitosan, 0.05% KCl, 0.1% K_2HPO_4, 0.05% $MgSO_4 \cdot 7H_2O$ and 0.001% $FeSO_4$, and the cultiva-

[9] M. Somogyi, *J. Biol. Chem.* **195,** 19 (1952).

tion is carried out for 3 days at 30° on a circular shaker. The culture filtrate is obtained by centrifugation followed by filtration from the production medium.

Precipitation with Ammonium Sulfate. The culture filtrate[10] (3800 ml) is concentrated to about one-third of the original volume by immersing cellophane bags containing saturated ammonium sulfate solution (767 g/l) with several changes and then adding solid ammonium sulfate to complete saturation. After standing overnight, the precipitate produced is collected by centrifugation, dissolved in deionized water, and desalted by passing through a Sephadex G-25 column (5.0 × 50 cm) previously equilibrated with 0.02 *M* phosphate buffer, pH 6.0 (buffer P).

Chromatography on CM-Sephadex C-25. The desalted enzyme solution (300 ml) is applied to a CM-Sephadex C-25 column (2.6 × 24.5 cm), which has been previously equilibrated with buffer P. The column is completely washed with buffer P and eluted with a linear gradient of each 250 ml of buffer P and the same buffer containing 0.6 *M* NaCl, at a flow rate of 20 ml/hr. Eluate is collected in 5-ml fractions, which are assayed for chitosanase activity and CMCase activity. Both the activities appear in parallel at the fractions eluted with 0.1–0.2 *M* NaCl. Fractions (Nos. 138–158) are pooled and concentrated by ultrafiltration with Amicon membrane UM2.

Gel Filtration on Sephadex G-75. The enzyme solution (3.8 ml) from the previous step is loaded on a Sephadex G-75 column (2.6 × 87 cm) previously equilibrated with buffer P. The column is eluted with buffer P, at a flow rate of 23 ml/hr. The enzyme appears at a position of about 2.5 times the void volume, separate from a protein peak with a small amount of the enzyme activities. The main fractions (Nos. 71–95) are pooled and concentrated by ultrafiltration. The concentrated enzyme solution (3.5 ml) is again subjected to gel filtration on Sephadex G-75 according to the same procedure as described above. The enzyme is eluted with buffer P as a single peak. Fractions (Nos. 73–92) are pooled as the purified enzyme.

The purification procedure is summarized in Table I. The chitosanase is purified 7.7-fold with a yield of 22% starting from ammonium sulfate precipitation. The ratio of chitosanase activity to CMCase activity remained constant throughout purification.

Properties of the Enzyme[6]

Homogeneity, Isoelectric Point, and Molecular Weight. The purified enzyme is homogeneous on polyacrylamide and SDS–polyacrylamide

[10] The culture filtrate can be directly applied to a CM-Sephadex column (5 × 50 cm), after dilution with deionized water to three times the original volume.

TABLE I

PURIFICATION OF CHITOSANASE FROM *Streptomyces griseus*[a]

Step	Protein (mg)	Chitosanase				CMCase		Activity ratio (*A*/*B*)
		Total units (*A*)	Specific activity (U/mg)	Yield (%)	Purification (-fold)	Total units (*B*)	Yield (%)	
Ammonium sulfate precipitation	301	108	0.36	100	1.0	425	100	0.25
Sephadex G-25	161	76.7	0.48	71	1.3	337	79	0.23
CM-Sephadex C-25	28.9	54.5	1.89	50	5.3	222	52	0.25
First Sephadex G-75	15.3	39.4	2.58	36	7.2	160	38	0.25
Second Sephadex G-75	8.7	24.0	2.76	22	7.7	95	22	0.25

[a] Reproduced from Ohtakara *et al.*[6]

disk gel electrophoresis. The isoelectric point of the enzyme is approximately at pH 9.7. The molecular weight of the enzyme is estimated to be about 35,000 by SDS–polyacrylamide disk gel electrophoresis. However, the values of the molecular weight estimated by gel filtration on Sephadex G-75 and G-100, and TSK Gel Toyopearl HW 55 (Toyosoda Manufacturing Co., Ltd.) are about 10,000.

Optimum pH and Stability. The optimum pH of the enzyme is found to be about 8.0 for chitosanase activity and about 6.5 for CMCase activity. The enzyme is stable between pH 6.0 and 8.0 at 37°, but unstable at the elevated temperature.

Inhibitors. Both chitosanase and CMCase activities are inhibited by metal ions, strongly by Cu^{2+} and Fe^{2+}, and almost completely by Ag^{+} and Hg^{2+} at a concentration of 1 m*M*.

Michaelis Constants. The apparent K_m values of the enzyme are calculated to be 0.21% for glycol chitosan and 0.18% for carboxymethylcellulose.

Substrate Specificity and Mode of Action. The enzyme hydrolyzes the following various chitosan compounds: soluble chitosan, colloidal chitosan, glycol chitosan, and carboxymethylchitosan. The chitosanase is also capable of hydrolyzing carboxymethylcellulose, but does not act on acetylated polymers such as glycol chitin and colloidal chitin. In the decomposition of glycol chitosan and soluble chitosan with the enzyme, a rapid decrease in the viscosity occurs without increased formation of reducing sugars. During the hydrolysis of soluble chitosan the enzyme produces glucosamine oligomers with degrees of polymerization from 2 to 6,[11] but not glucosamine. The mode of action of this chitosanase is endo-type.

[11] M. Izume and A. Ohtakara, *Agric. Biol. Chem.* **51,** 1189 (1987).

[66] Chitin Deacetylase

By YOSHIO ARAKI and EIJI ITO

$$\underset{\text{(chitin)}}{\beta\text{-1,4-Linked (GlcNAc)}_n} \rightarrow \underset{\text{(chitosan)}}{\beta\text{-1,4-linked (GlcN}_m\text{, GlcNAc}_{n-m})} + m\,CH_3CO_2H$$

Chitosan is a unique polysaccharide in that it possesses free amino groups,[1] whereas most of the amino sugar residing in naturally occurring polysaccharides is believed to be N-acylated. An enzyme catalyzing the

[1] S. Bartnicki-Garcia and W. J. Nickerson, *Biochim. Biophys. Acta* **58,** 102 (1962).

conversion of chitin to chitosan was first demonstrated in an extract of *Mucor rouxii*.[2] Later, a similar enzyme was also found in the culture filtrate of a plant pathogen, *Colletotrichum lindemuthianum*.[3]

Assay Method

Principle. The chitin deacetylase activity is assayed by measuring the radioactivity of [^{3}H]acetic acid liberated from a water-soluble chitin derivative, glycol [*acetyl*-^{3}H]chitin.

Labeled Substrate. Forty milligrams of commercially available glycol chitosan (partially O-hydroxyethylated chitosan, obtained from Wako Junyaku Co., Osaka, Japan) is treated at 4° in a mixture containing 400 mg $NaHCO_3$ and [^{3}H]acetic anhydride (6.7 mCi, 200 μmol) in a total volume of 4.5 ml. After 24 hr, 200 μl of unlabeled acetic anhydride is added, and the mixture is allowed to stand for an additional 24 hr at 4°. After thorough dialysis, the product, glycol [*acetyl*-^{3}H]chitin (1.4×10^7 cpm/mg), is used together with unlabeled glycol chitin as substrate for the assay of chitin deacetylase. Unlabeled glycol chitin is prepared from glycol chitosan as described above using unlabeled acetic anhydride. The glycol chitin preparation contains 3,6-di-*O*-hydroxyethylglucosamine, 3-*O*-hydroxyethylglucosamine, 6-*O*-hydroxyethylglucosamine, and O-unsubstituted glucosamine in a molar ratio of 0.22:0.35:0.76:1.00. A colloidal suspension of chitin is prepared according to the method of Jeuniaux.[4]

Reagents

TES–NaOH buffer, 0.5 *M*, pH 5.5
Glycol [*acetyl*-^{3}H]chitin (specific activity, 24,000 cpm/μg)
HCl, 0.1 *M*
Acetic acid, 1 *M*
Ethyl acetate

Procedure. The incubation mixture contains 5 μl of TES–NaOH, 48 μg of glycol [*acetyl*-^{3}H]chitin (142 nmol *N*-acetyl groups, 114,000 cpm), and less than 1 mU of deacetylase in a final volume of 50 μl. After incubation at 30° for 10 min, the reaction is terminated by the addition of 20 μl of HCl, 5 μl of acetic acid, and 100 μl of water. Ethyl acetate (0.5 ml) is added to the mixture, and the solution is vigorously mixed with a vortex mixer for 5 min and centrifuged at 2000 rpm. The organic phase is gently transferred to another tube with a Pasteur pipet. The residual water phase is further

[2] Y. Araki and E. Ito, *Biochem. Biophys. Res. Commun.* **56,** 669 (1973).
[3] H. Kauss, W. Jeblick, and D. H. Young, *Plant Sci. Lett.* **28,** 231 (1983).
[4] C. Jeuniaux, this series, Vol. 8, p. 644.

extracted twice by the same method. Eight milliliters of toluene-based liquid scintillation cocktail is added to the combined organic phase solution and swirled. The cloudy solution is centrifuged at 3000 rpm for 10 min to remove a small volume of water. The clear solution is transferred to a vial and measured for radioactivity in a liquid scintillation counter.

Definition of Unit. One unit of enzyme releases 1.0 μmol of acetic acid from glycol chitin per minute under the conditions described above. Specific activity is defined as the units of enzyme per milligram of protein.

Purification Procedure

The purification procedure given here is essentially the same as that described by Araki and Ito.[5] All operations are performed at 0–4°.

Materials

Buffer A: 5 m*M* sodium acetate, pH 5.2
Buffer B: 20 m*M* Tris–HCl, pH 7.2

Step 1: Crude Enzyme. Mucor rouxii AHU 6019, furnished by Dr. S. Takao, Hokkaido University, is grown with shaking for 2 days at 30° in a medium (15 liters) containing peptone (1%), yeast extract (0.2%), and glucose (2%) at pH 4.5. The mycelia are harvested by filtration through a sintered glass funnel. The frozen mycelia are disrupted by grinding with glass powder (50 g/25 g of damp mycelia) in a mortar for 1 hr. After dilution with buffer A (final volume of 400 ml/25 g of damp mycelia), the homogenate is centrifuged at 2500 *g* for 5 min, and the supernatant is further centrifuged at 20,000 *g* for 20 min. The resulting supernatant (total volume of 1 liter) is used as the crude enzyme.

Step 2: Ammonium Sulfate Precipitation. Solid ammonium sulfate is slowly added with continuous stirring to the crude enzyme solution (200 ml) to 2.5 *M* (63% saturation). The pH is maintained at 7.0 with ammonia. After 15 min the precipitate is removed by centrifugation at 20,000 *g* for 15 min, and the concentration of ammonium sulfate is increased to 3.3 *M* (85% saturation). The precipitate collected by centrifugation is dissolved in 10 ml of buffer A and dialyzed against four changes of 500 ml of buffer A.

Step 3: CM-Cellulose. The dialyzed enzyme solution is centrifuged at 20,000 *g* for 20 min to remove the precipitate formed during dialysis. The solution is applied on a 2 × 6 cm column of CM-cellulose (Serva) equilibrated with buffer A. Unbound proteins are washed out with 30 ml of buffer A at a flow rate of 60 ml/hr. The adsorbed proteins are eluted with a linear gradient of sodium acetate (pH 5.5, 200 ml) from 5 to 100 m*M* at

[5] Y. Araki and E. Ito, *Eur. J. Biochem.* **55,** 71 (1975).

the same flow rate. The enzyme is eluted at about 50 m*M* of sodium acetate.

Step 4: Concentration with Collodion Bag. The active fractions in step 3 are pooled and concentrated to 1 ml by filtration with a collodion bag (SM 13,200, Sartorius).

Step 5: DEAE-Cellulose. The concentrated enzyme solution is dialyzed against buffer B and applied on a 1 × 3.5 cm column of DEAE-cellulose (Serva) equilibrated with buffer B. Unbound proteins are washed out with 10 ml of buffer B. The adsorbed proteins are eluted with a linear gradient (40 ml) of NaCl from 0 to 0.3 *M*, 15 ml of 0.3 *M* NaCl, 30 ml of 0.45 *M* NaCl, and 20 ml of 0.8 *M* NaCl, each in buffer B. The enzyme is eluted at the elution step with 0.45 *M* NaCl. The active fractions are pooled, dialyzed against buffer B, and used as the purified enzyme. A summary of the purification procedure is given in Table I.

Properties

Substrate Specificity. The chitin deacetylase can liberate the *N*-acetyl groups from *N*-acetylglucosamine residues in glycol chitin, a water-soluble chitin derivative. The limit of deacetylation is as low as about 30%, probably owing to the presence of *O*-hydroxyethyl groups at C-3 and C-6 of the *N*-acetylglucosamine residues in the substrate. A colloidal or a powdered preparation of chitin, the likely natural substrate of this enzyme, is less active because of its insolubility. The deacetylase activity depends on the number of monosaccharide units in the substrate molecule. The trimer of *N*-acetylglucosamine reacts much more slowly than glycol chitin, and the dimer is virtually inactive as substrate. The pentamer is as active as glycol chitin, and the limit of deacetylation is about 70% or more. This enzyme is inactive toward the *N*-acetylgalactosamine residues in poly(*N*-acetylgalactosamine) and the *N*-acetylglucosamine residues in a heparin derivative,

TABLE I
PURIFICATION OF CHITIN DEACETYLASE

Step and fraction	Protein (mg)	Total activity (U)	Specific activity (U/mg)
1. Crude enzyme (200 ml)	852	1127	0.13
2. Ammonium sulfate precipitate	186	842	0.45
3. CM-cellulose	19.6	386	1.97
4. Collodion bag	8.24	204	2.48
5. DEAE-cellulose	0.52	92.6	17.8

peptidoglycan, *N*-acetylglucosamine 6-phosphate, *N*-acetylglucosamine 1-phosphate, and UDP-*N*-acetylglucosamine.

pH Optimum. The enzyme is inactive at pH values below 4 and above 9. It is most active at pH 5.5, and the activity at this pH varies markedly with the buffer employed. This effect of the buffer on deacetylase activity appears to be ascribable to inhibition by the anionic component of the buffer.

K_m Value and Stability. The K_m value for glycol chitin is 0.87 g/liter or 2.6 m*M* with respect to monosaccharide residues. The enzyme is stable for a year when stored at $-18°$. However, freezing in the presence of ammonium sulfate causes inactivation of the enzyme.

Metal Ion Requirement and Inhibitors. The enzyme does not require any metal ions: Co^{2+}, Mn^{2+}, Na^+, or EDTA inhibits the chitin deacetylase reaction. Acetic acid is a competitive inhibitor of the enzyme with a K_i value of about 13.3 m*M*. The reaction is inhibited by glycol chitosan, a reaction product.

Distribution. Mucor rouxii possesses an extremely great activity as compared with other microorganisms. This is consistent with the role of the deacetylase in the formation of chitosan, which occurs in this fungus as the major cell wall component.

[67] Poly(*N*-acetylgalactosamine) Deacetylase

By Yoshio Araki and Eiji Ito

$$\underset{\text{[poly(GalNAc)]}}{\alpha\text{-1,4-Linked (GalNAc)}_n} \rightarrow \underset{\text{(polygalactosamine)}}{\alpha\text{-1,4-linked (GalN}_m\text{, GalNAc}_{n-m}\text{)}} + m\ CH_3CO_2H$$

In 1960 Distler and Roseman found that about two-thirds of the galactosamine residues in polygalactosamine (galactosaminoglycan) excreted from *Aspergillus parasiticus* are unsubstituted at their amino groups.[1] Later, a similar but probably different galactosaminoglycan was found in a conidial mutant of *Neurospora crassa*[2] and in *Aspergillus sojae*.[3] The former galactosaminoglycan appears to be linked to a protein and is believed to function as a cell growth regulator or a bioflocculant. The poly(*N*-acetylgalactosamine) [poly(GalNAc)] deacetylase present in *A.*

[1] J. J. Distler and S. Roseman, *J. Biol. Chem.* **235,** 2538 (1960).
[2] J. L. Reissig and J. E. Glasgow, *J. Bacteriol.* **106,** 882 (1971).
[3] J. Nakamura, S. Miyashiro, and Y. Hirose, *Agric. Biol. Chem.* **40,** 619 (1976).

parasiticus catalyzes the hydrolysis of acetamido groups of poly(GalNAc).[4] A similar enzyme was demonstrated in the extract of *N. crassa.*[5]

Assay Method

Principle. The poly(GalNAc) deacetylase activity is assayed by measuring the radioactivity of [^{3}H]acetic acid liberated from the labeled substrate, *N*-[^{3}H]acetylgalactosamine polymer.

Labeled Substrate. Aspergillus parasiticus AHU 7165, furnished by Dr. S. Takao, Hokkaido University, is grown with shaking for 3–6 days at 30° in a medium (8 liters) described by Beadle and Tatum.[6] The mycelia are separated from the culture fluid by filtration through a sintered glass funnel. The crude enzyme and polygalactosamine are isolated from the above mycelia and culture fluid, respectively. The crude polygalactosamine is obtained from the culture fluid as fibrous clots by the addition of 1 vol of acetone or ethanol as described by Distler and Roseman.[1] The yield is 1.3–2.1 g. About 1.2 g of the crude preparation is dissolved in 500 ml of 1% Na_2CO_3, and the solution is heated for 1 hr in a boiling water bath. After cooling to room temperature, the solution is centrifuged at 20,000 *g* for 30 min, and the resulting supernatant is thoroughly dialyzed against deionized water. This preparation contains 5.4 μmol galactosamine residues/mg.[7] The molar ratios of *N*-acetyl groups to galactosamine vary among the preparations obtained at different cultivation periods: 0.64, 0.46, 0.40, and 0.35 for the preparations obtained after cultivation for 3, 4, 5, and 6 days, respectively.

To prepare N-acetylated polygalactosamine labeled in *N*-acetyl groups (*N*-[^{3}H]acetylgalactosamine polymer), the polygalactosamine preparation (18.5 mg, 100 μmol galactosamine residues) is treated at 4° for 24 hr in 7 ml of a mixture containing 700 mg $NaHCO_3$ and [^{3}H]acetic anhydride (6.3 mCi, 30 μmol). The treatment is continued for a further 24 hr with an additional 100 μl of unlabeled acetic anhydride. After thorough dialysis, the product, *N*-[^{3}H]acetylgalactosamine polymer (5.0 μmol galactosamine residues/mg, 2×10^7 cpm/mg), is used as substrate for the assay of poly(GalNAc) deacetylase.

Reagents

MES–NaOH buffer, 0.5 *M*, pH 5.3

N-[^{3}H]Acetylgalactosamine polymer (specific activity, 18,000 cpm/μg) (0.5 mg/ml)

[4] Y. Araki, H. Takada, N. Fujii, and E. Ito, *Eur. J. Biochem.* **102,** 35 (1979).

[5] J. A. Jorge, S. G. Kinney, and J. L. Reissig, *Branz. J. Med. Res.* **15,** 29 (1982).

[6] G. W. Beadle and E. L. Tatum, *Am. J. Bot.* **32,** 678 (1945).

[7] H. Takada, Y. Araki, and E. Ito, *J. Biochem. (Tokyo)* **89,** 1265 (1981).

Ammonium molybdate, 10 m*M*
HCl, 0.1 *M*
Acetic acid, 1 *M*
Ethyl acetate

Procedure. The incubation mixture contains 5 μl of MES–NaOH, 10 μl of *N*-[^{3}H]acetylgalactosamine polymer (25 nmol *N*-acetyl groups, 90,000 cpm), 5 μl of ammonium molybdate, and less than 0.2 mU of the deacetylase in a final volume of 50 μl. The mixture is incubated at 30° for 10 min, and the reaction is terminated by the addition of 20 μl of HCl, 5 μl of acetic acid, and 100 μl of water. [^{3}H]Acetic acid liberated is extracted with ethyl acetate and measured for the radioactivity in a liquid scintillation counter, as described in chapter [66] of this volume.

Definition of Unit. One unit of enzyme releases 1.0 μmol of acetic acid from poly(*N*-acetylgalactosamine) per minute under the conditions described above. Specific activity is defined as the units of enzyme per milligram of protein.

Purification Procedure

The purification procedure given here is essentially the same as that described by Araki *et al.*[4] All operations are performed at 0–4°.

Materials

Buffer A: 5 m*M* sodium acetate, pH 4.8
Buffer B: 20 m*M* Tris–HCl, pH 7.2

Step 1: Crude Enzyme. The frozen mycelia of *A. parasiticus* AHU 7165 (damp weight, 100 g) obtained from an 8-liter culture as described above, are disrupted by grinding with glass powder (10 g/5 g of mycelia) in a mortar at 4° for 1 hr. After addition of 160 ml of buffer B, the homogenate is centrifuged at 2500 *g* for 10 min to remove glass powder and unbroken mycelia, and the supernatant is further centrifuged at 20,000 *g* for 30 min. The resulting supernatant is used as the crude enzyme.

Step 2: Ammonium Sulfate Precipitation. Solid ammonium sulfate is slowly added with continuous stirring to the crude enzyme solution (150 ml) to 1.6 *M* (40% saturation). The pH is maintained at 7.0 with ammonia. After 15 min the precipitate is removed by centrifugation at 20,000 *g* for 15 min, and the concentration of ammonium sulfate is increased to 75% of saturation. The precipitate collected by centrifugation is dissolved in and dialyzed against buffer A. The precipitate formed is removed by centrifugation.

Step 3: CM-Cellulose. The enzyme solution in step 2 is applied on a 3 × 12.8 cm column of CM-cellulose (Serva) equilibrated with buffer A.

The enzymatic activity is unadsorbed on the column. The effluents are pooled.

Step 4: Concentration with Collodion Bag. The pooled enzyme solution is concentrated to 1 ml by filtration through a Diaflow membrane (Amicon) followed by filtration with a collodion bag SM 13,200 (Sartorius).

Step 5: DEAE-Cellulose. The concentrated enzyme solution is neutralized with 1 *M* Tris base and applied on a 0.8 × 4 cm column of DEAE-cellulose (Serva) equilibrated with buffer B. The column is eluted with buffer B and then with a linear gradient (40 ml) of NaCl from 0 to 0.3 *M* in buffer B. The enzyme is eluted with about 0.15 *M* NaCl. The active fractions are collected, dialyzed, and used as the purified enzyme. A summary of the purification procedure is given in Table I.

Properties

Substrate Specificity. The poly(GalNAc) deacetylase can deacetylate the acetamido groups of *N*-acetylgalactosamine residues in poly(*N*-acetylgalactosamine), and the limit of deacetylation is about 70%. The oligosaccharides with 14 or more monosaccharide units are deacetylated as fast as the polymer. On the other hand, the oligosaccharides with three to eight monosaccharide units react much slower than the polymer. The dimer and monomer of *N*-acetylgalactosamine as well as *N*-acetylgalactosamine 1-phosphate and UDP-*N*-acetylgalactosamine are inactive. The enzyme is also inactive toward *N*-acetylglucosamine and the *N*-acetylglucosamine residues in peptidoglycan, chitin, a heparin derivative, UDP-*N*-acetylglucosamine, *N*-acetylglucosamine 1-phosphate, and *N*-acetylglucosamine 6-phosphate.

pH Optimum. The enzyme shows double pH optima of 5.3 and 9.3. The reason for this apparently complicated dependence of the enzyme activity on pH remains to be resolved. The recovery of total enzyme

TABLE I
PURIFICATION OF Poly(GalNAc) DEACETYLASE

Step and fraction	Protein (mg)	Total activity (U)	Specific activity (U/mg)
1. Crude enzyme (150 ml)	716	2650	3.7
2. Ammonium sulfate precipitate	343	2230	6.5
3. CM-cellulose	20.7	1190	59.2
4. Collodion bag	7.5	903	120
5. DEAE-cellulose	2.3	1200	520

activity measured at each purification step under the alkaline assay conditions is coincident with the value obtained from the assay under the acidic conditions. Under the acidic assay conditions, the deacetylase is stimulated about 2-fold by the addition of ammonium molybdate at about 1 m*M*. Some metal ions, including Co^{2+}, stimulate the deacetylase with rather high optimal concentrations of about 80 m*M*. Under the alkaline assay conditions, ammonium molybdate has a small effect and metal ions are rather inhibitory.

K_m Value and Stability. The K_m value for poly(*N*-acetylgalactosamine) determined under either the acidic or the alkaline assay conditions is about 0.15 g/liter or 0.54 m*M* with respect to monosaccharide residues. The enzyme is stable for a year when stored at $-18°$.

Distribution. The enzyme is localized in the cytoplasmic fraction of several *Aspergillus* species which secrete polygalactosamine into culture fluids. In strains unable to produce polygalactosamine, no significant activity of this enzyme is detected. This fact is consistent with the probable role of the deacetylase in the formation of polygalactosamine.

[68] Chitin Deacetylase from *Colletotrichum lindemuthianum*

By HEINRICH KAUSS and BÄRBEL BAUCH

Chitin, a β-1,4-linked homopolymer of *N*-acetyl-D-glucosamine, is accompanied in the walls of various fungi by chitosan, a polymer similar to chitin by lacking the *N*-acetyl groups. It has been shown[1] with cell-free extracts from *Mucor rouxii* (Zygomycetes) that chitosan is most likely synthesized by partial deacetylation of the nascent chain of chitin once the latter is formed by chitin synthase from UDP-GlcNAc. After crystallization the microfibrillar chitin appears to be a relatively poor substrate for the respective enzyme chitin deacetylase. This mode of biosynthesis suggests that the chitin/chitosan complex occurring in fungal cell walls is not a mixture of the two pure polymers but is represented by chitin chains which are deacetylated to a variable degree.

In *M. rouxii*[2,3] and *Mucor miehei*[4] the chitin deacetylase is predomi-

[1] L. L. Davis and S. Bartnicki-Garcia, *Biochemistry* **23,** 1065 (1984).
[2] Y. Araki and E. Ito, *Biochem. Biophys. Res. Commun.* **56,** 669 (1974).
[3] Y. Araki and E. Ito, *Eur. J. Biochem.* **55,** 71 (1975).
[4] H. Kauss, W. Jeblick, and D. H. Young, *Plant Sci. Lett.* **28,** 231 (1982/1983).

nantly found in cell extracts. An enzyme with similar specificity was demonstrated[4] in the plant pathogen *Colletotrichum lindemuthianum,* which is the causal agent of anthracnose in beans and is most likely a member of the Ascomycetes. Since with this fungus the activity in the culture filtrate was between 6- and 25-fold higher than in the corresponding cell extracts and the protein content of the filtrate was also rather low, the latter represents a convenient source for the rapid isolation of chitin deacetylase.[4] This enzyme can be easily assayed using as a substrate the water-soluble glycol chitin, either unlabeled (assay I) or ^{3}H-labeled in the acetyl group (assay II).

Culture Conditions

The kappa strain of *C. lindemuthianum* is used for most of the work and is available from ATCC (No. 56676). Chitin deacetylase activity is also demonstrated in the beta strain of the same fungus and in *Colletotrichum lagenarium,* a pathogen of cucumbers and melons. The enzyme from the latter two sources is, however, not investigated further. The cultures may also be available from Departments of Plant Pathology or from Agricultural Research Stations. They are kept on neopeptone agar (2.8 g glucose, 1.23 g $MgSO_4 \cdot 7H_2O$, 2.72 g KH_2PO_4, 2 g neopeptone, 20 g agar/liter deionized water). If low sporulation is a problem, this might be enhanced by illumination (12-hr light/12-hr dark cycles using "black light" fluorescent lamps, NUV, 320–420 nm).

A spore suspension (at least 5×10^7 spores) is used for heavy inoculation of 200 ml of liquid medium (10 g malt extract, 4 g yeast extract and 4 g glucose/liter of deionized water) in 500-ml Erlenmeyer flasks exhibiting an indentation. The cultures are rotated vigorously (116 rpm, 5 m radius) at 22 to 25°. The enzyme activity in the filtrate is maximal after about 5 days of incubation and might decline later on.

Substrates

Partially O-hydroxyethylated chitosan (glycol chitosan) is available from Sigma and Wako. If this substance is directly added to water, it tends to form gumlike aggregates which are difficult to completely solubilize. Glycol chitosan, however, can be easily solubilized by stirring in 1% (v/v) acetic acid which is then dialyzed against several batches of water. The resulting solution of glycol chitosan (10 mg/1.5 ml water) is supplied with 100 mg of sodium bicarbonate followed by 25 mCi [^{3}H]acetic anhydride (500 mCi/mmol).[3] After standing for 24 hr at 4° 50 μl of unlabeled acetic anhydride is added and the solution is allowed to stand for an additional

24 hr. The pH is then adjusted to 3.0 with acetic acid, followed by dialysis against six batches of water. The resulting substrate is finally diluted to contain 5 μg [^{3}H]acetylglycol chitin/5 μl solution.

Unlabeled glycol chitin can be prepared as above, but omit the tritiated compound. Chitin oligomers up to the trimer are available from Sigma, whereas oligomers of a higher degree of polymerization *(DP)* can be prepared by partial acid hydrolysis of chitin.[5] After neutralization the oligomer mixture is reacetylated as described above for the synthesis of glycol chitin and desalted by passage through a 2.5 × 60 cm Sephadex G-15 column. If the fractions containing the first traces of salt are discarded this step also removes most of the GlcNAc and the respective dimer. The resulting mixture of chitin oligomers with a $DP \geq 3$ might be freeze dried and used as such or further fractioned by gel permeation chromatography.[5] Colloidal chitin[6] must also be reacetylated at least once to avoid unfavorably high zero-time blanks.

Assay I

Comparatively high enzyme activities can be assayed with a modification of the colorometric determination described for chitosan.[7] The assay is performed in glass tubes with 100 μl of 50 m*M* sodium tetraborate/HCl buffer, pH 8.5, 100 μg of substrate in 100 μl of water, and 50 μl of enzyme (e.g., a 1 : 10 diluted 30–80% ammonium sulfate precipitate or 1 : 5 diluted culture filtrate). After incubation at 30° the reaction is terminated by the addition of 250 μl of 5% (w/v) $KHSO_4$.

For color formation 250 μl of 5% (w/v) $NaNO_2$ is added under the hood, allowed to stand with occasional shaking for 15 min, and 250 μl of 12.5% (w/v) ammonium sulfamate is further added. After another 5 min 250 μl of 0.5% (w/v) 3-methyl-2-benzothiazoline hydrazone (freshly prepared each day) is added and the mixture is heated in a boiling water bath for 3 min. The tubes are cooled in tap water and 250 μl of 0.5% (w/v) $FeCl_3$ is added. The latter is stable for about 3 days at 4°. The developing color is read after 30 min at 650 nm.

The time course of enzyme action is followed from 0 to about 10 min with chitin oligomers ($DP \geq 4$), to about 1 hr with glycol chitin, and up to 3 hr with colloidal chitin as a substrate. Standard curves are prepared with 0 to 6 μg of glucosamine · HCl. On a weight base glycol chitosan gives about 65% of the color developed with chitosan; for a comparison of

[5] B. Capon and R. L. Foster, *J. Chem. Soc., Sect. C,* p. 1654 (1970).
[6] C. Jeuniaux, this series, Vol. 8, p. 644.
[7] J. P. Ride and R. B. Drysdale, *Physiol. Plant Pathol.* **2,** 7 (1972).

substrates a correction has to be made. With colloidal chitin the color developed is partially absorbed to the substrate flocculating out; the enzyme activity might be underestimated with this substrate.

Assay II

A more sensitive assay using labeled glycol chitin[2,3] is required for the fractions arising during ion-exchange chromatography. This assay also has the advantage that it is not affected by the yellow color of the culture filtrate. It is performed in 2-ml glass tubes with 50 μl of 50 mM sodium tetraborate/HCl buffer, pH 8.5, and 10 to 50 μl of the enzyme in a total volume of 100 μl. The reaction is started by addition of 5 μl of [^{3}H]acetylglycol chitin (about 200,000 cpm), incubated for 2 to 30 min at 30° and terminated by addition of 60 μl of 0.2 N HCl and 5 μl of 1 N acetic acid. Controls in which the acids were added at zero time before the addition of substrate are run in parallel and the values are subsequently subtracted.

The [^{3}H]acetic acid liberated by the chitin deacetylase is extracted three times in 0.5 ml of ethyl acetate by shaking and the combined extracts are counted after addition of 3 ml of water-miscible liquid scintillation cocktail (for instance, 10 g 2,5-diphenyloxazole + 500 ml of Triton X-100/2 liters of xylene).

The specific radioactivity of the labeled glycol chitin cannot be readily determined. The activity of the enzyme might, therefore, be calculated in relative units, for instance as cpm ^{3}H liberated/30 min.

Enzyme Preparation

Ammonium Sulfate Precipitation. The mycelium from two of the above 200-ml cultures is harvested by suction over a glass fiber filter (e.g., Whatman GF). The filtrate is collected and brought to 30% saturation by addition of the salt in small portions and the precipitate is centrifuged out and discarded. The salt concentration is increased to 80% saturation and the resulting pellet solubilized in 40 ml of 20 mM sodium borate/HCl, pH 8.5, followed by dialysis against the same buffer and centrifugation to remove some precipitate. About 60% of the enzyme activity contained in the filtrate is recovered. The slightly colored preparation is stable at 30°.

Ion-Exchange Chromatography. Chitin deacetylase of a higher purity can be prepared by ion-exchange chromatography. The culture filtrate is adjusted to pH 7.2 and is directly loaded on a column of DEAE-Sephacel (Pharmacia, 2.2 × 16 cm, 20 ml/hr) which was previously equilibrated with 20 mM HEPES/NaOH, pH 7.25. After washing with 200 ml of the same buffer a linear gradient from 0 to 1 M NaCl (total volume 120 ml) in the above buffer is applied (60 ml/hr, 2.5-ml fractions). The enzyme is normally eluted under a rather broad peak between 0.05 to 0.6 *M* NaCl, sometimes with a tendency to show subfractions. Colored substances are partly retained on the column material which, therefore, can be used only once. The fractions with the highest specific activity of chitin deacetylase are pooled, dialyzed against 5 m*M* sodium acetate buffer, pH 4.0, and applied on a SP-Sephadex C-25 column (1.0×20 cm, 50 ml/hr). After washing with 30 ml of the acetate buffer the enzyme is eluted with a linear gradient from 0 to 0.5 *M* NaCl (total volume 50 ml) in the same buffer (40 ml/hr, 1-ml fractions). The results from the latter purification step may vary within the experiments. Sometimes part of the chitin deacetylase may not bind and the remainder may be eluted between 0.1 and 0.2 *M* NaCl. In other experiments almost all of the enzyme may bind to the column and then be eluted in two peaks. In the latter case protein in the first peak may be below the limits of detection with the method recommended, indicating a higher specific activity than in the second peak. Calculating with the results for the second peak eluting from the SP-Sephadex column the overall purification in the two column steps might be 8- to 12-fold over the crude filtrate, corresponding to a several thousand-fold purification over the respective cell extract.[4] The two subsequent ion-exchange steps result in a colorless enzyme preparation but have the disadvantage that due to an often rather small recovery from the columns and to the production of various subfractions the overall yield may be very low. This preparation is less stable at 30°; the time course of assays should be controlled.

Protein Determination

The culture filtrate contains yellow–brown substances which render an exact determination of its protein content difficult. Part of the protein precipitates with 50% (w/v) trichloroacetic acid and can then be estimated according to Lowry *et al.*[8] In the eluate from the SP-Sephadex column protein can be directly determined using a dye-binding method.[9]

Some Properties of the Enzyme

When the pooled fractions eluting between 0.1 to 0.2 *M* NaCl from the SP-Sephadex column are dialyzed against 20 m*M* sodium tetraborate/HCl, pH 8.5, they may be kept frozen at −20° for several months without a great loss in activity. On prolonged incubation (15 hr) the enzyme in this

[8] O. H. Lowry, N. J. Rosebrough, A. L. Farr, and R. J. Randall, *J. Biol. Chem.* **193,** 265 (1951).

[9] M. M. Bradford, *Anal. Biochem.* **72,** 248 (1976).

fraction is able to release a maximum of about 30% of the total [^{3}H]acetyl groups present in assay II. This is similar to the chitin deacetylase from *M. rouxii*.[2,3] The chitin deacetylase from *C. lindemuthianum*[4] shows optimal activity at pH 8.5, in contrast to the enzyme from *M. rouxii*,[2,3] which operates optimally at pH 5.5. The latter enzyme was also found to be inhibited[2,3] by sodium acetate, whereas the chitin deacetylase from *C. lindemuthianum* does not exhibit such a property.[4]

Using the 30–80% ammonium sulfate preparation and assay I the activity of the chitin deacetylase toward substrates decreased in the order $(GlcNAc)_5$/$(GlcNAc)_4$/glycol chitin/$(GlcNAc)_3$/colloidal chitin/$(GlcNAc)_2$ and was 100/98/13/6/2/1%, respectively.

Possible Biological Implications

The question arises, why in plant pathogens of the genus *Colletotrichum* can high activities of chitin deacetylase be found in the culture filtrate? It appears possible that in addition to the formation of chitosan as a structural element of the cell wall, the secreted enzyme may also play a role in the production of shorter chain chitosan fragments. The chitin deacetylases from *M. rouxii*[2,3] and from *C. lindemuthianum* (see above) are highly active on oligomers derived from chitin. Similar fragments arise by the action of endochitinases of plant or fungal origin and might represent a readily accessible substrate for a secreted chitin deacetylase. The resulting chitosan fragments are diffusible to some extent and exhibit a variety of unexpected biological properties. They are, for instance, fungistatic[10] and have been implicated in host pathogen interaction.[11] They also can alter the permeability of plant plasma membranes.[12] Such an action appears to be, on the one hand, of nutritional value for a plant pathogen. On the other hand it elicits, in the plant cells, a syndrome of defense reactions which comprises a rapid sealing of injured cell walls by the 1,3-β-glucan callose, as well as lignification of walls and the synthesis of phytoalexins, substances which are potentially toxic to plant pathogens.[11,12] For the localized induction of callose synthesis the chain of events appears to involve an increase in Ca^{2+} influx into the cytoplasm, causing a direct stimulation of the Ca^{2+}-dependent 1,3-β-glucan synthase[12] located in the plasma membrane.

[10] D. F. Kendra and L. A. Hadwiger, *Exp. Mycol.* **8,** 276 (1984).
[11] L. A. Hadwiger, J. M. Beckman, and M. J. Adams, *Plant Physiol.* **70,** 249 (1981).
[12] H. Kauss, *Annu. Rev. Plant Physiol.* **38,** 47 (1987).

[69] *N,N′*-Diacetylchitobiase of *Vibrio harveyi*

By RAFAEL W. SOTO-GIL, LISA C. CHILDERS, WILLIAM H. HUISMAN, A. STEPHEN DAHMS, MEHRDAD JANNATIPOUR, FARAH HEDJRAN, and JUDITH W. ZYSKIND

Chitin is an insoluble polysaccharide that is the second most commonly occurring organic molecule in the biosphere. The hydrolysis of chitin to its monomeric unit, *N*-acetylglucosamine (GlcNac), requires two enzymes, chitinase and *N,N′*-diacetylchitobiase (EC 3.2.1.30, chitobiase, *N*-acetyl-*β*-glucosaminidase), with chitobiase catalyzing the hydrolysis of *N,N′*-diacetylchitobiose (chitobiose) to GlcNac. We have isolated recombinant plasmids that contain the gene coding for chitobiase from the gram-negative marine bacterium *Vibrio harveyi.*[1] The chitobiase gene appears to be unlinked to the gene(s) encoding chitinase activity in that there is no chitinase activity associated with plasmids containing the *V. harveyi* chitobiase gene. We have found that chitobiase is located in the membrane of *V. harveyi* cells and in the outer membrane of *Escherichia coli* cells containing plasmids with the cloned *V. harveyi* chitobiase gene *(chb).*[1] As a consequence of the outer membrane location of chitobiase, the detergent *N*-lauroylsarcosine has proved effective in removing chitobiase activity from *E. coli* cells containing the cloned *chb* gene and is employed in the isolation procedure described here.

Chitobiase Assays

Activity can be measured by monitoring either the rate of formation of *p*-nitrophenol from *p*-nitrophenyl-2-acetamido-2-deoxy-*β*-D-glucopyranoside[2] or by the increase in reducing power during the hydrolysis of chitobiose with *p*-dimethylaminobenzaldehyde.[3] The former method, although the least sensitive and nonspecific for any *N*-acetylglucosaminidase, is easier and less expensive to perform.

[1] M. Jannatipour, R. W. Soto-Gil, L. C. Childers, and J. W. Zyskind, *J. Bacteriol.* **169,** 3785 (1987).
[2] A. Ohtakara, M. Mitsutomi, and Y. Uchida, *J. Ferment. Technol.* **57,** 169 (1979).
[3] J. L. Reissig, J. L. Strominger, and L. F. Leloir, *J. Biol. Chem.* **217,** 959 (1955).

p-Nitrophenyl-2-acetamido-2-deoxy-β-D-glucopyranoside (PNAG) as Substrate

Reagents

Tris–HCl buffer (10 m*M*, pH 7.3)
PNAG (Sigma), 12 m*M* in Tris–HCl (10 m*M*, pH 7.3)
Tris base, 1 *M*

Procedure. Assays are conducted in 10 m*M* Tris–HCl, pH 7.3, and 666 μ*M* PNAG in a volume of 1 ml at 25°. Enzyme (10–30 μl) is added at zero time to each of the reaction mixtures, which are subsequently incubated for 3 to 15 min before termination of the reaction by the addition of 3 ml of Tris base, resulting in a pH > 10.5. *p*-Nitrophenol release is measured immediately at 400 nm employing a molar absorptivity of 18,100 liter/mol·cm. Controls lacking enzyme are substracted from assay values.

Definition of Unit and Specific Activity. One unit of chitobiase activity is defined as the amount of enzyme that catalyzes the formation of 1 μmol of *p*-nitrophenol in 1 min in the assay described above. Specific activity is expressed as units per milligram of protein. Protein is determined by the method of Lowry[4] with bovine serum albumin as a standard.

Chitobiose as Substrate[3]

Reagents

Tris–HCl buffer (10 m*M*, pH 7.3)
Chitobiose (Pfanstiehl Laboratories, Inc.), 2 m*M* in Tris–HCl (10 m*M*, pH 7.3)
Potassium tetraborate, 0.8 *M*, adjusted to pH 9.1 with KOH
p-Dimethylaminobenzaldehyde (DMAB) reagent: A stock solution of 10% (w/v) DMAB (Aldrich) is prepared in glacial acetic acid that contains 1.25 *M* HCl. The DMAB reagent is prepared by diluting 1 vol of this stock solution with 9 vol of glacial acetic acid shortly before use.

Procedure. The enzyme is appropriately diluted in buffer in a total volume of 1 ml. After preincubation at 37° for 5 min, 0.5 ml of substrate is added to initiate the reaction, and incubation at 37° is continued for 3 to 15 min. The sample is then boiled for 10 min to stop the reaction, and the amount of GlcNac released is measured using the DMAB reagent.[3] To

[4] O. H. Lowry, N. J. Rosebrough, A. L. Farr, and R. J. Randall, *J. Biol. Chem.* **193,** 264 (1951).

0.5 ml of sample, blank, and standards, 0.1 ml of 0.8 *M* potassium tetraborate, pH 9.1, is added. The samples are heated in a vigorously boiling water bath for exactly 3 min, cooled in tap water, and 3 ml DMAB reagent is added. The reaction product is measured immediately at 585 nm wavelength, and the amount of GlcNac is estimated from a standard curve constructed using GlcNac standards in the range of 40–160 nmol. Controls lacking enzyme are subtracted from assay values.

Definition of Unit and Specific Activity. One unit of activity is defined as the amount of enzyme that catalyzes the formation of 1 μmol of GlcNac in 1 min in this assay, and the specific activity is expressed as units per milligram of protein. Protein was determined as described above.

Detergent Removal of Chitobiase from *E. coli* Cells

Escherichia coli containing the plasmid pRSG192[1] is grown in 500 ml L broth [1.0% (w/v) tryptone (Difco), 1.0% (w/v) yeast extract (Difco), and 0.5% (w/v) NaCl] containing 1 m*M* isopropylthio-β-D-galactoside (IPTG) at 37° until an OD_{450} of 1.0 is reached. Plasmid pRSG192 (Fig. 1) contains the *V. harveyi* chitobiase gene with its promoter on a 3.6-kb *Eco*RI fragment cloned into the *Eco*RI site in pUC19.[5] Chitobiose acts as an inducer of chitobiase activity in *V. harveyi.* With pRSG192, however, we have found that there is no significant change in the activity of chitobiase after the addition of chitobiose to the growth medium,[1] indicating that regulatory elements are not included in the cloned DNA, so induction with chitobiose is unnecessary. The amount of chitobiase activity associated with plasmid pRSG192 is 25-fold more than the amount found with *V. harveyi* cells fully induced with chitobiose.[1] The chitobiase gene is positioned in front of, and in the same transcription direction, as the *lac* promoter in pUC19, leading to an increase in chitobiase activity with the addition of IPTG, which is why IPTG is included in the growth medium. The cells are centrifuged, washed twice with 50 ml of 10 m*M* Tris–HCl (pH 7.3), and resuspended in 100 ml of 10 m*M* Tris–HCl (pH 7.3) containing 1% (w/v) *N*-lauroylsarcosine (sarcosyl, Sigma). After incubating for 1 hr at 37° with gentle mixing, the cells are removed by centrifugation at 27,000 *g* for 15 min. The supernatant fluid, called the sarcosyl cell extract, is stored at −20°. There is no loss of chitobiase activity over a 2-week period under these conditions. This procedure, which does not require cell lysis, removes approximately 75% of the chitobiase activity from the cells. Sarcosyl at a concentration of 1% has no effect on the activity of chitobiase.

[5] C. Yanisch-Perron, J. Vieira, and J. Messing, *Gene* **33,** 103 (1985).

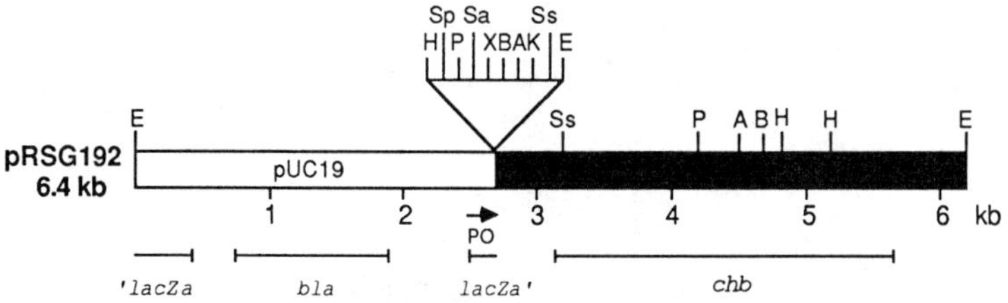

FIG. 1. Restriction and genetic map of pRSG192. The solid bar represents cloned *V. harveyi* DNA. The chitobiase gene-coding region *(chb)* is indicated. Restriction sites are designated by the following: A, *Ava*I; B, *Bam*HI; E, *Eco*RI; H, *Hin*dIII; K, *Kpn*I; P, *Pst*I; Sa, *Sal*I; Sp, *Sph*I; Ss, *Sst*I; and X, *Xba*I.

HPLC Purification of Chitobiase

Preparative isolation of chitobiase was carried out on a 4.6 × 250 mm column of Waters QMA ACCELL ion-exchange medium (37–55 μm, 500-nm pore size). The packed column was washed successively with methanol (50 ml), acetonitrile, 0.1% trifluoroacetic acid in Nanopure (Barnstead) water, and Nanopure water before equilibration with 200 ml of 20 m*M* sodium acetate (pH 6.5, solvent A). An absorbance baseline for the column was established by running a linear gradient from 0.0 to 1.0 *M* NaCl in 20 m*M* sodium acetate (pH 6.5, solvent B) over 45 min at a flow rate of 2 ml/min.

Aliquots of the sarcosyl cell extract (2 × 5 ml, 0.4 mg/ml) were loaded onto the column at 1 ml/min and washed through with solvent A until a baseline 260-nm absorbance (0.5 AUFS) was obtained. The flow rate was increased to 2 ml/min before initiation of the following gradient program: 0–5 min, 100% A; 5–12 min, 0–25% B; 12–40 min, 25–65% B; 40–45 min, 65–100% B; 45–50 min, 100% B; 50–55 min, 100 to 0% B. Three-milliliter fractions were collected. The enzyme eluted in the 8- to 14-min region of the gradient program. Data from a typical purification with chitobiose as a substrate at 37° are presented in Table I. The purified enzyme has a specific activity of 107 U/mg and can be obtained in up to 30% yield. Coomassie blue-stained SDS-polyacrylamide gels exhibited a polypeptide with an apparent M_r 92,000 (confirmed by nucleotide sequencing of the chitobiase gene) comprising about 95% of the applied protein.

Properties of Chitobiase

Escherichia coli cells without plasmid pRSG192 do not contain chitobiase activity; therefore, the sarcosyl cell extract contains only the chitobiase activity encoded by the *V. harveyi* DNA cloned in pRSG192. Mini-

TABLE I

PURIFICATION OF CHITOBIASE FROM *V. harveyi*

Treatment	Volume (ml)	Protein (mg/ml)	Units/ml[a]	Specific activity	Purification (-fold)
Untreated whole cells	40	4.600	9.90	2.15	—
Sarcosyl cell extract	180	0.390	5.39	13.90	6.2
HPLC[b]	6	0.013	1.42[c]	106.70	≈50

[a] Assay method was one using chitobiose as substrate.

[b] Only 10 ml of the sarcosyl extract was applied to the HPLC column.

[c] Only one fraction from the peak is included.

cell analysis of proteins encoded by pRSG192 demonstrated the presence of two new proteins when compared to vector-encoded proteins.[1] The apparent M_r values were 95,300 and 92,100, and because the insert DNA in this plasmid contains only enough coding space for one protein of this size, the larger protein probably corresponds to the unprocessed form of the enzyme.

During transport of chitobiase to the outer membrane of *E. coli,* a signal peptide is removed from the amino terminus of the enzyme. Based on DNA sequence and open reading frame analyses, the probable amino terminus of the unprocessed enzyme is composed of the following amino acid sequence:

MetLeuLysHisSerLeuIleAlaAlaSerValIleThrThrLeuAlaGlyCysSerSerLeuGlnSerSerGluGln

The underlined amino acids are identical to the amino acids adjacent to the processing site in the outer membrane lipoprotein of *E. coli*[6,7] and may represent the region where processing of chitobiase occurs.

The enzyme cleaves both *N,N′*-diacetylchitobiose and the artificial substrate, PNAG, but with differing apparent V_{max} values; the native substrate is hydrolyzed with a turnover number 7.5-fold greater than that for the artificial substrate. The pH optimum at 25° in Tris–HCl is broad over the pH 6–9 region. The enzyme exhibits classical Michaelis–Menten kinetics and an apparent K_m at 25° for the artificial substrate of 250 μM. The enzyme is unaffected by a variety of cations including Na^+, NH_4^+, K^+, and Mg^{2+} and can be assayed in citrate or phosphate buffers. The enzyme is unaffected by the presence of 10 m*M* EDTA and exhibits 80% of its original activity in 1 *M* NaCl. The enzyme can be stored at 4° or at −20° for up to 2 weeks without loss of activity, and lyophilization results in 40 to 50% loss of activity.

Acknowledgments

This work is a result of research sponsored in part by NOAA, National Sea Grant College Program, Department of Commerce, under Grant NA80AA-D-00120, through the California Sea Grant College Program, and in part by the California State Resource Agency, project number R/F-93. Funds for this research were also provided by the California Metabolic Research Foundation.

[6] V. Braun and V. Bosch, *Proc. Natl. Acad. Sci. U.S.A.* **69,** 970 (1972).
[7] K. Nakamura and M. Inouye, *Cell* **18,** 1109 (1979).

Author Index

Numbers in parentheses are footnote reference numbers and indicate that an author's work is referred to although the name is not cited in the text.

A

B

G

H

I

L

M

T

U

V

W

Y

Z

Subject Index

A

B

F

I

J

K

L

M

T